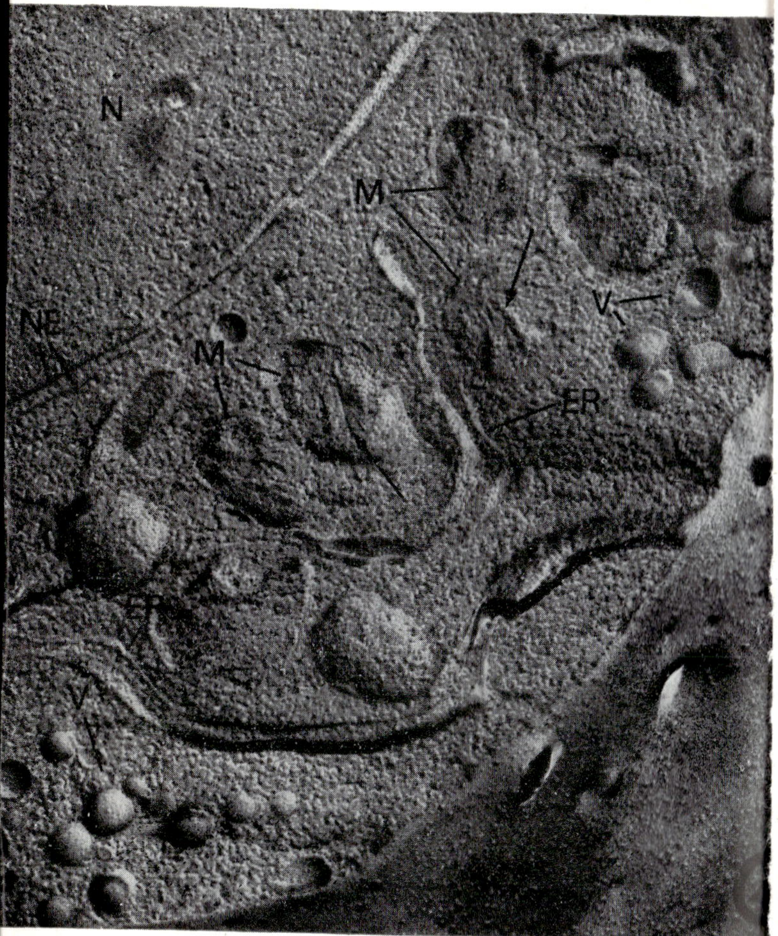

eeze-etched preparation of a portion of a mesenchymal cell. The cytoplasm conta
itochondria (M), tracts of endoplasmic reticulum (ER), and numerous vesicles (
me mitochondria show internal cristae (arrows). A pinocytic invagination (Pi) of
asma membrane is also seen. A portion of the nucleus (N) is seen surrounded by the t
embranes (NE) of the nuclear envelope. × 39 000. Courtesy of Dr. E. Katchburian, T
ndon Hospital Medical College.

The Cell in Medical Science

Volume 2:

Cellular Genetics, Development and Cellular Specialization

The Cell in Medical Science

Volume 2: Cellular Genetics, Development and Cellular Specialization

edited by
F. BECK, *Department of Anatomy, University of Leicester*
and
J. B. LLOYD, *Research Unit in Biochemistry, Keele University*

1974
Academic Press
London New York
A Subsidiary of Harcourt Brace Jovanovich, Publishers

ACADEMIC PRESS INC. (LONDON) LTD.
24/28 Oval Road,
London NW1

United States Edition published by
ACADEMIC PRESS INC.
111 Fifth Avenue
New York, New York 10003

Library of Congress Catalog Card Number: 78-172366
ISBN: 0 12 084202 5

PRINTED IN GREAT BRITAIN BY
WILLMER BROTHERS LIMITED, BIRKENHEAD

Contributors to Volume 2

R. BELLAIRS, *Department of Anatomy, University College London.*
D. BELLAMY, *Department of Zoology, University College, Cardiff.*
C. E. BLANK, *Sub-Department of Medical Genetics, University of Sheffield.*
D. BRYNMOR THOMAS, *Department of Anatomy, University of St. Andrews.*
C. E. FORD, *Sir William Dunn School of Pathology, South Parks Road, Oxford.*
H. J. GAMBLE, *St. Thomas' Hospital Medical School, London.*
J. M. GRAHAM, *London Hospital Medical College, London.*
E. G. GRAY, *Department of Anatomy, University College, London.*
W. R. KEATINGE, *London Hospital Medical College, London.*
D. N. LANDON, *Institute of Neurology, Queen Square, London.*
U. MITTWOCH, *Department of Human Genetics, University College, London.*
A. R. MUIR,* *Royal (Dick) School of Veterinary Studies, Edinburgh.*
C. W. PARR, *Department of Biochemistry, London Hospital Medical College, London.*
J. D. W. TOMLINSON, *Faculty of Medicine, Memorial University of Newfoundland, St. Johns, Newfoundland.*

*Deceased

Preface

The increasing importance of cell biology in Medical Science is becoming clear to clinicians and laboratory scientists alike. It is the meeting ground of many traditional disciplines and forms a central theme for many others. Its impact on subjects as diverse as immunology and neurobiology is already very great and one cannot but appreciate the potential that the application of its techniques and discipline must have for the future.

Many excellent introductory books of cell biology are available but beyond these one has in general to pass either to reviews or to original articles in order to probe more deeply. The present volumes are designed for readers who already have an elementary knowledge of cell biology; they present various aspects of the subject in depth and try to indicate some of the directions in which contemporary cell biology is moving and the methods it uses. No attempt is made to provide a comprehensive cover of cell biology, but the topics are chosen so as to produce a coherent work rather than a series of unconnected essays. We are greatly indepted to our contributors for their willingness to work within a number of constraints and for their patience with a multitude of editorial requests made to help us achieve our aims.

Volume 1 of this book is concerned with the generalized cell unit, and its chapters deal with the biophysical and biochemical basis of the structure and function of the chief subcellular organelles. Volume 2 contains a series of chapters on the relationship of the cell to developmental processes both within the cell itself and in the organism as a whole. There follow chapters on specific cellular specializations, particularly within the neuromuscular system. Here morphological adaptation for specific functional purposes is described in detail. This theme is again followed in Volume 3 where connective tissues, various endocrines, absorptive and secretory cells are dealt with. Volume 4 begins with three chapters on metabolic control mechanisms, and these are followed by chapters on the relationship between the cell and its environment in various pathological states; immunological processes, inflammation, wound healing and carcinogenesis are treated within this framework.

We believe this book will be of value to senior undergraduate students and to research workers looking for summaries on a variety of related topics concerned with cell structure and function. The contributors have been asked to provide only brief bibliographies which enable the reader to develop his own interest; the chapters do not attempt to include an extensive review of the literature.

Like many before us, we are indebted to Academic Press for the patience, forbearance and unfailing courtesy of numerous members of their staff.

F. BECK
University of Leicester
J. B. LLOYD
University of Keele

February 1974

Contents of Volume 2

Cellular Genetics and Development

11. An Introduction to Human Cytogenetics

C. E. BLANK

12. Sex Chromatin

URSULA MITTWOCH

13. The Kinetics of Cell Production

D. BRYNMOR THOMAS

Cellular Specialization

17. The Neurone

J. D. W. TOMLINSON

18. The Synapse

E. G. GRAY

Contents of Volume 1

The Cell and its Organelles

Contents of Volume 3

Cellular Specialization (continued)

Contents of Volume 4

Cellular Control Mechanisms

Cellular Responses to Environment

Cellular Genetics and Development

9. Biochemical Genetics

C. W. PARR
Department of Biochemistry, The London Hospital Medical College, London, England

I. Introduction

Like classical genetics, the study of genetics at the biochemical level is based on the concept that all living organisms contain within themselves the discrete hereditary determinants or "elements" that were first proposed by Mendel. These factors, now called genes, have permanence in that they are inherited unchanged. Mendel was fortunate in that he chose to study characters that could be identified and counted in subsequent generations in an exact manner, and he was able to show that such characters reflected the activities of single genes. Much of our present understanding of the biochemical mechanisms of inheritance has also come as a consequence of the study of clearly segregating characters, determined by one or at most by a few genes. Characters such as height and intelligence, which do not segregate in a clear-cut manner and are generally considered to be controlled by a number of genes, are too complex, as yet, for analysis in biochemical terms.

Some of the earliest information pointing to a relationship between biochemistry and genetics was acquired by studying certain chemical abnormalitites in humans. Garrod (1909) in a classical review proposed that albinism (in which there is a lack of melanin pigments), alkaptonuria, cystinuria and pentosuria were all conditions in which there is either a lack of production, or an excessive accumulation, of a normal metabolite. He proposed the term "inborn errors of metabolism" for these conditions, and suggested that the errors were, basically, inherited defects in the appropriate types of catalytic activity, thus implying for the first time that genes are responsible for the production of enzymes. Subsequent investigations have fully justified Garrod's predictions and many other conditions (e.g. phenylketonuria and galactosaemia) in which there is excessive accumulation of a metabolite have now been shown to fall into the same category.

The concept "one gene, one enzyme", was more specifically proposed by Beadle and Tatum (1941) to explain their findings concerning the bread mould *Neurospora crassa.* This mould was found to grow on a synthetic medium containing inorganic salts, a source of carbon and energy (for example, sucrose) and the vitamin biotin. Beadle and Tatum induced mutations with ultraviolet light and X-rays, and then picked out individual spores and grew them on a medium enriched with all the amino acids and vitamins then available. The cultures originating from the single spores were then transferred back to the basal medium. Now and again (there were two examples in the first two thousand transfers) a mutant strain was found that would no longer grow in the basal medium. It had acquired an obligatory requirement for one or another of the additional components present in the enriched medium. In fact it had acquired, by mutation, an enzyme defect that prevented it from synthesizing that particular component.

A more sophisticated dictum, "one gene, one polypetide chain", followed when it was realized that proteins are frequently agglomerations of polypeptides, not always identical and sometimes, when arising from the activities of different gene loci, substantially different. For instance adult haemoglobin (Hb A), the most common of the haemoglobins occurring in man, comprises two each of two different polypeptide chains (α and β), the different chains being specified by two different loci.

Much of our knowledge of the nature of the actual genetic material has come from the study of micro-organisms such as *Escherichia coli.* It is less easy to study the genetic material itself in multicellular diploid organisms such as man. However, such evidence as is available suggests that the genetic material in all forms of life is the same. In all species it is the particular arrangement of the nucleotide units of the deoxyribonucleic acid (DNA) of the chromosomes that constitutes the genes. Structural genes, via the agency of molecules of messenger ribonucleic acid (*m*RNA) which are cast in their own image, function as templates for the synthesis of polypeptide chains. These chains, sometimes in their monomeric forms but more usually after association with other (similar or dissimilar) polypeptides, and sometimes after incorporation of prosthetic groups, form the completed proteins and enzymes of the cell, which collectively dictate the observed characteristics of an individual.

Regulatory genes exert their influence by controlling the activities of the structural genes. Jacob and Monod have proposed a mechanism for the regulation of protein biosynthesis by regulatory genes in microorganisms, which has now gained general acceptance. It seems inescapable, however, that the greater complexity of higher organisms requires that a substantially modified regulatory mechanism should operate. There have been some recent speculations about the nature of such mechanisms in higher organisms, but it must be said that at present no clear model, reasonably supported by experimental evidence, has yet emerged (see p. 12).

II. Principles

A. CHROMOSOMES AND GENES

The weight of DNA in a single human somatic cell is about $6{\cdot}4 \times 10^{-12}$g, and is constant for all human diploid cells irrespective of type. This finding is a pleasing confirmation of one of the basic generalizations of genetics—that the different somatic cell types within an individual multicellular organism all contain within their nuclei an identical set of genes (the genome)—that set which was created at the time of conception. It can also be demonstrated experimentally that there is no turnover of genetic material throughout the lifetime of a cell, and this also provides evidence that the genes have permanence. Knowing the amount of DNA per cell, one may calculate the number of nucleotide base pairs: it comes to something in excess of 5 000 million. Thus an "average" human chromosome (there being 46 chromosomes per diploid genome) contains about 120 million base pairs, and is thought to comprise one very long linear macromolecule.

Since each DNA anticodon, specifying one amino acid, comprises three base pairs, an average chromosome could conceivably be held responsible for the synthesis of as many as 100 thousand different polypeptide chains each of about 400 amino acids in length. This supposes, however, that the chromosome contains nothing but structural genes: in fact it is now generally believed that structural genes comprise only a few per cent of the chromosome. A proportion of the remaining chromosomal material will exercise a regulatory function.

Forty-four of the 46 chromosomes of the human are autosomal chromosomes (or autosomes) and they comprise 22 homologous pairs (one of each pair inherited from each parent). Homologous pairs are so called because each member associates with the other during the process of meiosis, and participates in the "crossing over" process. This association is presumably associated with the high degree of similarity (or homology) that exists between them, which somehow allows each to recognize the other. It is important to realize, however, that the degree of homology is not absolute. Perhaps ten per cent of the structural genes of one member of a homologous pair will specify polypetides that are very slightly different from the polypetides specified by the companion genes of the other homologous chromosome. Nevertheless the remaining ninety per cent of structural genes are present as pairs of identical genes, one of each pair to a chromosome. Furthermore the positions of the gene pairs (the gene loci) along the lengths of the two autosomes are identical. One presumes that the positions of the regulatory genes are likewise identical.

Thus each autosomally sited gene locus in diploid organisms is represented by two genes, one on each homologous autosome. Where the two genes are absolutely identical then that locus is said to be homozygous. Where the two genes are different the locus is said to be heterozygous and the genes are known as alleles. There can be multiple alleles (A, A′, A″ A‴ etc.) at any given locus but of course only two can be present per locus in any given individual.

The twenty-two homologous pairs of autosomal chromosomes of the human are present in males and females alike, but the sexes differ with respect to a pair of so called sex chromosomes—the female having two X-chromsomes whilst the male has one X and one (dissimilar) Y-chromosome. The X and Y chromosomes are not homologous with each other and, as far as is known, do not share any common loci.

B. MUTATION

Permanent modifications to chromosomes and genes are known as mutations. Mutational events are rare, but their occasional occurrence must be regarded as of vital significance because they are an essential pre-requisite to the establishment of both variation between individuals and variation between species.

Chromosomal rearrangements

Mutations may involve rearrangements, breakages and junctions of chromosomes and the duplication of gene loci. Such gross changes are frequently not inherited, either because they do not occur in the germ cells, or because, when present in germ cells, the latter do not survive to take part in fertilization. When inherited, however, the morphological effects of chromosomal mutations are usually considerable and it is believed that such mutations have played a large and important rôle in evolution.

Point mutations

Point mutations are those that occur at a single gene locus, and they could involve the deletion, addition, transposition or substitution of one or more of the bases.

The addition or deletion of a single base in a structural gene will result in a "frameshift"; i.e. all the subsequent bases will become out of phase, thus constituting a new series of triplets specifying the production of a totally different polypeptide. If one of the newly constituted triplets codes for chain termination, there will be premature termination of polypeptide synthesis. Such greatly modified and possibly abbreviated chains might well be undetectable, in which case an investigator might conclude that there was a total lack of polypeptide synthesis.

The consequences of a point mutation involving the substitution of one base for another will depend solely on the nature of the new triplet, because the rest of the triplets comprising the gene will remain unchanged. If the new triplet codes for chain termination then, as described above, the abbreviated chain is likely to be undetectable. Another possibility is that the new triplet may code for the same amino acid as the old triplet. This possibility arises because the genetic code is "degenerate" in that most of the amino acids are represented by several alternative base triplets, usually differing one from another by a single base. Thus (see Table 1) the third DNA base in the anticodons for lysine, glutamate and valine may be either cytidine (C) or thymine (T), so that a base substitution of C→T or of T→C at this third position would in neither of these instances result in any change in the amino acid incorporated (although it is possible that the rate of synthesis of the polypeptide might be altered).

Nevertheless the great majority of point mutations of which we have any knowledge appear as single amino acid substitutions in the polypeptides, probably the result of single base substitutions in the genes. For instance the inherited variant form of haemoglobin known as sickle cell haemoglobin (Hb S) contains a pair of α-chains identical with the α-chains of normal adult haemoglobin (Hb A) and a pair of slightly modified β-chains known as β^S chains which differ from the β chains of Hb A solely by the presence of the amino acid residue valine in the place of the glutamate residue that is normally found in position six (counting from the amino terminal end) of the β-chain (see p. 45).

There are two alternative DNA triplets (anticodons) for glutamate: CTT and CTC; and two also for valine: CAT and CAC: it will be seen that this particular mutation can be explained by the single DNA base change T⟶A at the middle position of the anticodon, irrespective of which of the alternative triplets for glutamate was originally present. Table 1 summarizes this sequence of events (and shows also the complementary role played by *m*RNA in polypeptide production). It also depicts a mutation at the first position of the anticodon which provides an explanation for the presence of a lysine residue, also in position six along the β-chain, which is found in another inherited haemoglobin variant, Hb C.

TABLE 1. Representation of mutational events whereby a β chain of Hb A is converted into a β^S or a β^C chain (of Hb S or Hb C respectively) by single base substitutions in the gene. (These events occur at the sixth anti-codon, the sixth codon and the sixth amino acid residue of the β chain, respectively.)

Anti-codon sited on the sense strand of the DNA	Corresponding codon of the *m*RNA	Corresponding amino acid residue
TTT ↑	- - - → AAA	
or	or	===→Lysine (in β^C)
TTC ↑	- - - → AAG	
CTT	- - - → GAA	
or	or	===→Glutamate (in β)
CTC	- - - → GAG	
↓ CAT	- - - → GUA	
or	or	===→Valine (in β^S)
↓ CAC	- - - → GUG	

——— Mutation producing a single base change
– – – – – Transcription
===== Translation

Since Hb S (of structure $\alpha_2\beta_2{}^s$) and Hb C ($\alpha_2\beta_2{}^c$) both differ from Hb A ($\alpha_2\beta_2$) by a single amino acid substitution, fortuitiously at the same position along the β-polypeptide chain, it might be asked whether both Hb S and Hb C arose by independent mutations from Hb A, or whether (for instance) Hb C was formed by direct mutation from Hb S. As Table 1 shows, the circumstantial evidence favours the former because it can be accomplished by means of a single base change only. It would have required two base changes to go from Hb S to Hb C: a less likely alternative.

The possibility must always be born in mind that a point mutation might occur at a regulatory locus, and cause polypeptide synthesis to be increased, decreased or stopped.

Point mutation and evolution

In general, deleterious genes will tend to be selectively eliminated from a population, whereas beneficial ones will be retained, and will increase. Indeed, the theory of evolution rests heavily on this assumption. It has generally been assumed that fresh mutations will usually be deleterious in some way or another and that only the exceptional one will be advantageous, a view probably conditioned by the presumption that organisms have presently evolved to some degree of perfection. It has therefore come as something of a surprise to find that many of the point mutations uncovered recently by the use of biochemical techniques appear to be "neutral" ones, or at any rate to possess only very slight advantages or disadvantages. For instance there is an unexpectedly high incidence of variation between individuals in respect of their red cell enzymes, but most of the variations appear to be innocuous.

Some mutational events appear to be unique, i.e. to have only ever occurred once. The fact that Hb C (see above) occurs only in one well-defined area of West Africa suggests that the particular amino-acid substitution implicated (Glu→Lys) may have arisen as a consequence of a single and unique mutation in just one West African. The variant has now become established in that area of the world, by the operation of natural selection, by virtue of some advantage it presumably confers. [It should be noted that, to persist in a population, a mutation must first gain a foothold, and this is bound to be a matter of chance. Even the most highly desirable mutation, if it occurs but once in a single individual who does not procreate, will be lost for ever.]

It is also conceivable that a particular amino acid substitution at a particular site along a polypeptide chain may recur quite independently on a number of different occasions in different individuals in different parts of the world. This could be caused by some special susceptibility to mutation at a special position in the gene. If that mutation were to confer some slight disadvantage, then the frequency of the variant gene in a population would be the result of a balance set up between its rate of production by fresh mutation and its rate of elimination by natural selection.

Examples of *de novo* mutations in man (i.e. mutations not possessed by the parents) are reported from time to time. Such events, however, are rare. (The mutation rate for vertebrates has been calculated to be about 1×10^{-9} per amino acid site per year.) Variations in characteristics such as blood groups, haemoglobins and enzymes in individuals can usually be shown to have been inherited from the parents. Indeed one would hesitate to accept as genuine an apparently *de novo* mutation (i.e. one not possessed by the parents) until a careful study of other inherited characters had been undertaken to establish proof of parentage.

C. CONTROL OF GENE EXPRESSION

Since every somatic cell of an organism possesses the same genome, every cell type must have the latent capability of producing all the polypeptides specified by all the genes making up that genome. However, in multicellular organisms this potential is never exercised to the full; instead the activities of a number of the loci are partially or wholly blocked in a manner that varies with the cell type. The consequences are that certain polypeptides are produced in all cell types, others in some types only, and yet others in but a single cell type. This, basically, is the biochemical means whereby cell differentiation is achieved.

It is sometimes found in higher organisms that two (or more) structurally different polypeptides are produced, specified by different gene loci, that nevertheless form proteins that have the same or very similar functions. Sometimes only one polypeptide type is synthesized within a cell, the other being repressed, and at other times both types are produced within the same cell. Although such polypeptide types usually differ one from another in a number of respects, they frequently possess a degree of similarity in amino acid sequence, molecular size and conformation as to suggest that the loci concerned arose, during the course of evolution, from a single primordial locus.

Proteins determined by a single gene locus

In cases where there is a single gene locus coding for the polypeptides of a particular protein or enzyme, then every cell of that organism requiring that particular variety of protein or enzyme must perforce manufacture the same structural variety of it. Many proteins fall into this category, including several of the enzymes discussed in Sections III and IV (galactose-1-phosphate uridyl transferase, phosphogluconate dehydrogenase and glucose-6-phosphate dehydrogenase). Thus it follows that the galactose-1-phosphate uridyl transferase (for instance) isolated from human erythrocytes will be structurally identical with the transferase produced in the liver of that same individual.

Proteins determined by two or more loci

Where there are two or more loci within a genome producing polypeptides that are structurally different but that form proteins with similar functions, it is possible for one locus to function in one or more tissues and for the alternative locus or loci to function in other tissues. Mammalian phosphorylase serves as a good example. There is a condition in the human (McArdle's syndrome) in which there is an inherited deficiency of muscle phosphorylase, which results in an excessive accumulation of glycogen in skeletal muscle. Yet no abnormal concentration of glycogen is found in the liver, and the activity of the liver phosphorylase is not reduced. These findings, and the fact that the enzyme present in the liver differs somewhat in properties from the enzyme produced in muscle, suggest that there are two separate loci coding for phosphorylase subunits, the locus active in liver being permanently repressed in muscle cells. Nevertheless, although different, the two enzymes have a degree of structural similarity that makes it highly likely that they are both derived from the same evolutionary precursor. It has been reported that glycogen synthetase (responsible for the building up of the glycogen molecule) also bears a structural similarity to the phosphorylases, suggesting that it, also, is derived from the same precursor polypeptide.

The enzyme pyruvate kinase is another protein represented by two loci. Different forms of this enzyme have been found in erythrocytes and white cells.

In other cases two (and sometimes more) loci are active at the same time within the same cell. A well known example of this is provided by the enzyme lactate dehydrogenase (LDH). Two loci, producing two different types of LDH subunits, are active in all cells of the mamalian organism; however their relative activities differ from tissue to tissue. Both types of subunit associate randomly, with their own types as well as with the alternative subunits, to produce tetrameric LDH isoenzymes of which there may be five varieties. The situation is further complicated by the existence of a third locus, but this latter is only active in sperm cells. (The LDH isoenzymes are dealt with in more detail in Section IV.)

By contrast, the polypeptides from the three different loci that have been reported for phosphoglucomutase do not apparently associate with each other. Nevertheless, in some tissues all three loci function at one and the same time. There is probably some degree of differential expression of these loci in different cells as for LDH.

The haemoglobin loci

It is not necessary for cells other than erythrocytes to contain haemoglobin, and in fact they do not do so. The genes that code for haemoglobin subunits are prevented in some manner from functioning in all cells other than erythrocyte precursor cells. Nevertheless there is a great richness of provision, within the genome, of gene loci for the fabrication of haemoglobin subunits. Two such loci, coding for the α and β chains of human Hb A, have already been mentioned. A third locus specifying δ chains is also operative, but at a lower level of activity, in adults, and loci for γ and ε chains provide subunits that form the haemoglobins present at certain stages of fetal life. The various human haemoglobins are discussed in more detail in Section IV.

Physiological activation and repression of gene loci

In bacteria there is good evidence for the existence of a control system whereby the transcription of small groups of adjacently placed genes is regulated by a so-called operator gene sited to one side of the linearly placed genes, the genes together with the operator gene comprising what is known as an "operon". This model, proposed by Jacob and Monod in 1961, is dependent on the existence of polypeptide repressor molecules which, when bound to the operator gene, repress the transcription of all the genes of the operon. Small metabolites modulate the ability of the polypeptide repressor to bind to the operator gene and hence modulate transcription.

The patterns of gene action in multicellular organisms are more complex and cannot be explained by the Jacob and Monod model. We have seen that there are many gene loci within any given cell type that are permanently inactive (e.g. the loci for the various haemoglobin subunits, which are non-functional in all cell types except erythroblasts). Some loci are active only during a particular phase of the development of an organism, e.g. the loci for the ε and γ haemoglobin subunits become permanently "switched off" during early and late fetal life respectively (p. 43). There are other loci that appear to be subject to a fluctuating level of activity, in phase with day to day variations in the pattern of metabolism.

Both protein hormones (e.g. insulin and glucagon) and non-protein hormones (e.g. thyroid hormones and steroids including cortisone) are known to affect protein synthesis and could conceivably fit into a repressor (or a promoter) mechanism. Histones (proteins with a high content of the basic amino acids lysine and arginine) are known to be attached, presumably by electrostatic bonds, to considerable amounts of nuclear DNA and it seems likely that only the uncovered lengths of chromosomal DNA will be active in transcription. The product of transcription—*m*RNA, does not immediately pass to the cytoplasm for at any one time more than half is to be found in the nucleus. This has led to the suggestion that *m*RNA molecules may only be transported into the cytoplasm after they have become bound with some protein moiety to form ribonucleoprotein complexes known as "informosomes". The selective way in which polypeptides are produced within a cell may be related to the selective way in which informosomes are formed. Also, different *m*RNA molecules have different degrees of resistance to the degradative attack of the enzyme ribonuclease. Furthermore, small molecules (e.g. the haem present in reticulocytes) have been shown to be capable of regulating the activity of the cellular ribonuclease.

It is now generally accepted that structural genes account for only a few per cent of the total nuclear DNA, so that more than ninety per cent of the chromosomal material could have a regulatory function. Crick has recently speculated that chromosomes have many alternate fibrous and globular sections, which exercise structural and regulatory functions respectively. The fibrous sections are lengths of double helix able to code for polypeptides of molecular weight 30 000–40 000. They are bound at each end to compact globular sections in which the double helical structured DNA is further intricately folded and twisted in convolutions. The central idea of Crick's proposals is that sections of the duplex DNA in the globular regions are unwound to expose single stranded stretches of perhaps one hundred bases, with specific sequences, and it is these lengths of unpaired DNA that constitute "recognition sites" for either protein or RNA molecules, whichever it be that bears the regulatory message. One fibrous and one globular section adjacent to each other comprises a "genetic complimentation group". One may reasonably expect that significant advances will be made during the next decade in our understanding of the mechanisms of gene regulation in higher organisms.

D. ALLELIC VARIATION, INDIVIDUALITY AND GENE DOSAGE

Historical

It is manifest that individuals differ from one another. The view that human individuality is a consequence of inherited differences in chemical constitution was first proposed by Garrod, who wrote:

> The thought naturally presents itself that these (inborn errors of metabolism) are merely extreme examples of variations of chemical behaviour which are probably everywhere present in minor degrees, and just as no two individuals of a species are absolutely identical in bodily structure, neither are their chemical processes carried out on exactly the same lines. There can be no question that between the families, genera and species both of the animal and vegetable kingdoms, differences exist both of chemical composition and of metabolic processes. If, then, the several genera and species thus differ in their chemistry we can hardly imagine that within the species, when once it is established, a rigid uniformity exists. Such a conception is at variance with all that is known of the origin of the species.

Differences between species

The non-identity of the chemical processes carried out by different individuals and by different species can now be interpreted as a manifestation of the non-identity of the respective genomes. Very small differences between genomes suffice to give individuals their individuality. Differences between species result from a distinctly greater degree of non-identity of the genomes. Nevertheless a substantial degree of similarity remains, particularly between closely related species. For instance the structures of the α-chains of the haemoglobins of chimpanzee and man have been found to be identical, whilst the α-chains of man and gorilla differ by a single amino acid residue. Once two species have separated, the lack of homology increases as evolution works its course. Thus the α-chains of horse haemoglobin and human haemoglobin differ from each other in respect of 17 amino acids and represent, so it it thought, changes that have occurred over something like 80 million years of separation.

Differences between individuals

The two genes present at every autosomal locus are thought to be identical at something like nine out of ten of the loci, but there may well be a small difference (usually a single base change brought about by a so-called point mutation) at the remainder of the loci. Also, there will be a few loci where both alleles differ slightly from the most usually occurring variety. Since there are likely to be at least 10 000 separate autosomal loci in the human (and there could conceivably be 10 times this number) the possibilities for individual variation are immense, and quite compatible with the thesis that every human individual (with the exception of identical twins) possesses a unique and individual set of genes.

This view, which assumes the incidence of heterozygous variation to be greater than had previously been thought, has developed largely as a consequence of numerous reports from many quarters over the past decade of the surprisingly frequent occurrence of genetically determined electrophoretic variation in red cell enzymes. (The variant enzymes are catalytically active and are not, as a rule, particularly injurious to the individual carrying them.) Harris (1969) reported that polymorphism (regarded as the presence of a variant allele with a frequency in excess of one per cent) was present in something like one in three of the red cell enzymes that had been investigated. There seems to be no reason why this degree of polymorphism should not extend over the whole field of proteins and enzymes in all tissues. Of course not all the loci known to show polymorphism will vary from the usual in any particular individual, but a not insignificant proportion may do so.

Thus Garrod's contention, that there are chemical differences between all individuals, is now generally accepted. The differences are caused for the most part by the presence of different alleles at a proportion of the gene loci, thus producing different polypeptides that have subtly different physiological effects. One presumes, as Garrod did, that the outward observable differences between individuals are in some way a consequence of these chemical differences.

Mendelian inheritance: dominance and recessiveness

It will be recalled that one of Mendel's findings was that a stock of tall pea plants that bred true, when crossed with a stock of dwarf plants that also bred true, produced hybrid plants that were all tall (and not of intermediate height, as might have been expected). When these first generation hybrids were self-fertilized, three quarters of the second generation turned out to be tall, and one quarter dwarf. Now some at least of the first generation hybrid plants, that all grew tall, must have contained within themselves a hereditary factor (or gene) that specified dwarf stature, because some of the second generation plants derived from them were, in fact, dwarfs. The character of dwarfness was described as being recessive because it could be present in a plant without being given outward expression; similarly the character of tallness was described as being dominant because it was shown by all the first generation plants, although in fact they were all hybrids.

In fact the characters of tallness and dwarfism are allelic, and all Mendel's first generation and half his second generation plants were heterozygous for these two allelic characters. The terms dominant and recessive refer to the outward appearance and behaviour of a heterozygote. Dominance and recessiveness can only arise, of course, when the cells are diploid. At the biochemical level one is concerned more particularly with the genotype and with the actual gene products. Since we now know that the structural genes produce polypeptides, we may speculate that Mendel's hybrid plants carrying alleles for tallness and dwarfism produced two types of polypeptides, one being active (perhaps catalytically) in promoting growth, and the other inactive. It may be that the character of tallness is dominant because the yield of polypeptide from one allele for tallness is all that is required for the production of tall plants, so that the higher yields of polypeptide obtainable in homozygous plants with two alleles for tallness would not further increase the rate of growth.

Recessive conditions occurring in the human may be regarded in a similar way. Thus galactosaemia (see Section III) is caused by the existence of two defective alleles for the enzyme galactose-1-phosphate uridyl transferase. The condition is recessive because heterozygous individuals who possess only one functioning allele possess nevertheless sufficient transferase for a healthy existence and are clinically indistinguishable from the large majority of individuals who possess two active genes. However, heterozygotes can be readily identified when biochemical methods are used, for they can then be demonstrated to possess half normal levels of the enzyme. Examples are given later in this chapter of other recessive conditions in the human where heterozygous individuals appear to be perfectly normal or usual, but can be identified as heterozygotes by biochemical measurements.

An instructive example of the nature of dominant and recessive states at the biochemical level is provided by the ABO blood group system, described below. Certain other aspects of Mendelian inheritance are also illustrated well by a consideration of this system.

The ABO blood group system

The ABO system provided the first clear demonstration of multiple allelism, for there are three common alleles, A, B and O, at the same locus. (Additional alleles are now known at this locus: thus A has been subdivided into A_1 and A_2 subgroups.) Alleles A and B are considered to be dominant over O because the heterozygous states A/O and B/O appear as A and B respectively during normal blood group typing, and hence are indistinguishable from the homozygous states A/A and B/B (see Table 2). The recessive condition O/O is only manifest when both O alleles are present. The heterozygous state A/B shows both A and B activity, and hence A and B are said to be co-dominant.

TABLE 2. ABO blood group types in English and Sikh populations

Blood group phenotype	Genotype(s)	Frequency in English population	Frequency in Sikh population
A (Red cells agglutinated by anti-A serum)	A/A and A/O	0·437	0·244
B (Red cells agglutinated by anti-B serum)	B/B and B/O	0·085	0·324
AB (Red cells agglutinated by anti-A and by anti-B sera)	A/B	0·033	0·083
O (Red cells *not* agglutinated by anti-A or by anti-B sera)	O/O	0·444	0·348
Totals:		0·999	0·999

Nature of the ABO blood group antigens. Although the ABO blood group substances have been the subject of intensive study over a number of years, they remain intractable substances from a chemical point of view. The main reason for this is that they are tightly bound to the erythrocyte membrane and cannot be removed by extraction with aqueous media. However, a soluble form is often present in body fluids (e.g. gastric juice, saliva and particularly in ovarian cyst fluid) and most of the investigations carried out have been on soluble materials from these sources.

The water-soluble antigens have molecular weights of about 500 000 and are glycopeptide in nature, some 85% of their substance being carbohydrate. It seems likely that the insoluble antigens attached to the erythrocyte membrane contain lipid moieties rather than (or in addition to) protein. Nevertheless there is now ample evidence that the immunological activities of both types of antigens are attributable to their carbohydrate moieties, which are thought to be similar in soluble and in erythrocyte-bound antigens. The soluble blood group antigens appear to consist of large numbers of short oligosaccharide side chains attached at intervals to a polypeptide backbone. The chains contain seven or eight monosaccharide units and the immunological specificity (i.e. the blood group A or B character) depends on whether units of N-acetylgalactosamine or of galactose are attached to penultimate galactose units of the oligosaccharide chains of a precursor glycopeptide called "H-substance". With very rare exceptions, H substance (which has only weak immunological activity) is formed in all individuals.

Watkins and co-workers have shown that the N-acetylgalactosamine and the galactose that respectively confer the A- and B-type of immunological specificity on H-substance are attached by means of specific enzymes (glycosyltransferases) one of which is specific for N-acetylgalactosamine and the other for galactose. These two types of glycosyltransferase are the products of different alleles at the ABO locus: allele A produces the acetylgalactosamine transferase and allele B the galactose transferase. Allele O does not produce an active transferase. Thus the (genetically determined) blood group antigens, although themselves largely carbohydrate in nature, are formed by the activities of enzymes which are themselves the direct products of the genes.

Variations between populations. It is not only individuals who vary, one from another, in their blood groups: there is also considerable variation in the frequencies of the different groups between populations. Table 2 gives figures for the frequencies of the commoner ABO phenotypes in the English and a Sikh population. It will be seen that Sikhs are four times as likely as the English to have blood group B phenotype.

If there is no special selection for an allele from one generation to the next, then the frequency of a given allele should remain unchanged from generation to generation. This dictum was laid down in 1908, independently, by Hardy and Weinberg. One may therefore compute the frequencies of homozygotes and heterozygotes within a population, provided it is assumed that mating is random. Let us assume, for the ABO blood group system, that p represents the frequency of the A allele, q the frequency of the B allele, and r the frequency of the O allele. [A frequency of 100% is taken as 1.] Thus $p+q+r=1$. The frequencies of the homozygous states A/A, B/B, and O/O will be p^2, q^2 and r^2, and the frequencies of the heterozygotes A/B, A/O and B/O will be 2pq, 2pr and 2qr respectively.

It is an interesting exercise, knowing the frequencies of the various blood group phenotypes, to compute the frequencies of the alleles. If such computations are made using the data given in Table 4 it will be found that the frequency of the B allele in the Sikh population ($q=0{\cdot}230$) is nearly four times that for the English population ($q=0{\cdot}061$).

A knowledge of the gene frequency allows one to calculate the proportion of individuals of blood groups A and B who are homozygous for the A and B alleles respectively. Thus with regard to the B allele, only 5% of the English subjects of group B are of the homozygous B/B genotype (the rest having the heterozygous B/O genotype) whereas 20% of Sikhs of blood group B are homozygotes.

Criteria of individuality

As well as being of value in clinical situations where transfusion is involved, the assignment of an individual's blood group is a step towards defining his individuality. If an extensive blood grouping analysis is undertaken employing a number of different systems (e.g. MNS, Kelly, Lutherian, Duffy, Kidd, Lewis and the Rhesus system), then the chances become high that any two individuals, chosen at random, will be different. These chances become higher still when other parameters that exhibit inherited variation, such as haemoglobins, haptoglobins, transferrins, immunoglobulins and ther erythrocyte enzymes, are also characterized.

Some practical use can be made of this ability to identify an individual by inherited chemical characteristics. For instance, a child of blood group AB could not have had a father of blood group O (unless a *de novo* mutation had occurred) and such information is of value in resolving disputes about paternity.

Autosomal gene-dosage effects

Thus far it has been tacitly assumed that each of the two alleles at an autosomal locus directs the synthesis of polypeptide chains independently of the activity, or lack of activity, of the other. This implies that at a homozygous locus each allele is responsible for half the total product, and that where a locus is heterozygous with only one of the alleles producing active product, the total effective gene product will be half normal. For mammalian organisms this simple "gene-dosage" effect appears to apply more often than not. For instance, mention has already been made of the fact that the enzyme lacking in galactosaemia (see also p. 26) has about half normal activity in the erythrocytes of a proportion of the relatives of galactosaemics (including the parents) and this allows heterozygotes to be identified irrespective of the fact that, clinically, they are normal.

The existence of this simple sort of gene-dosage relationship suggests that the synthesis of the gene product in the mammalian organism is frequently not influenced by any sort of feed-back from the cell. The exactitude with which cells often show half normal activity when the loci concerned are heterozygous for an enzyme deficiency is paralleled by a similar exactitude in the amount of product formed at homozygous loci and suggests that frequently (although not invariably) a mammalian cell will synthesize the polypeptides characteristic of it at a pre-determined rate that is not much influenced by changes in the cell's physiological state.

It might be argued that enzymes are usually produced in quantities that far exceed the cell's requirements, so there is no need for a mechanism to compensate, by increased enzyme synthesis, for the reduced levels of activity in cells of individuals heterozygous for an enzyme deficiency. However, there are some enzymes in the mammalian organism that are well known to be subject to adaptive changes and such enzymes tend not to be present in excess: it would be of interest to have information on the levels of an enzyme of this type in individuals known to be heterozygous for a deficiency of it. Liver glucose-6-phosphatase is such an enzyme: it is synthesized in mammals in greater amounts than normal in diabetes and starvation, by an adaptive process, and in lesser amounts than normal when dietary carbohydrate is plentiful. Furthermore there is a type of glycogen storage disease associated with an almost total deficiency of this enzyme, and, on the assumption that the condition is a simple autosomally linked one, both the parents of an affected individual would be expected to be obligatory heterozygotes for the enzyme deficiency. Such a parent was recently investigated and found to have a level of liver glucose-6-phosphatase that was not reduced: a most interesting finding. More information is needed about enzyme levels in this and in analogous situations.

There are, of course, many well substantiated examples in microorganisms where a metabolic signal can induce polypeptide synthesis with great facility (see p. 12).

X-linked inheritance and gene-dosage effects

If the gene-dosage effects observed for autosomally linked loci were to apply also for X-linked genes, it would be expected that X-linked gene products would be produced in normal females, who have two X-chromosomes, in double the amounts found in normal males, who possess only one X-chromosome. But this is not the case. Human glucose-6-phosphate dehydrogenase (G6PD) for instance, which is X-linked, is present in much the same amounts in the erythrocytes of both males and females. This constitutes important evidence in favour of a hypothesis, advanced on other grounds by Mary Lyon, which postulates that at an early stage in the embryonic existence of a female one or other of the two X-chromosomes in each cell, at random, becomes inactivated and forms the so-called Barr body (Chapter 12). Each female somatic cell, therefore, has only one surviving functional

X-chromosome, and hence protein synthesis takes place at much the same rate as in the male cell.

In fact, females heterozygous for G6PD deficiency will have two types of cells, containing either fully active or fully deficient G6PD, depending upon which of the cell's two original chromosomes was inactivated. There is a histochemical procedure which stains cells with active G6PD but not cells with deficient G6PD, and when applied to erythrocytes from females heterozygous for G6PD deficiency two populations, one active and the other inactive for G6PD can be demonstrated. This would not have been the case had the condition been autosomally linked; rather, every cell would have had an intermediate level of activity.

When the red cells of females heterozygous for G6PD deficiency are assayed, the spread of activities for G6PD is so wide, ranging over the whole span from fully active to fully deficient, that identification as heterozygotes is often uncertain and sometimes impossible. The wide spread of activities is at first sight puzzling, but can be explained by the Lyon hypothesis, which postulates that the X-inactivation occurs when only a small number of embryonic cells have been formed. Let us consider the consequences of X-inactivation occurring in a group of eight cells (which we will presume will ultimately provide all the erythropoetic tissues). There will be a small chance (actually 1 in 256) that all the X-chromosomes with the defective G6PD will be condensed to Barr bodies, and another small chance (again 1 in 256) that all of the X-chromosomes with the active G6PD will be condensed. Thus 1 in 128 of female heterozygotes will be indistinguishable, on any basis that involves the measurement of enzymic activity, from normal or from fully deficient individuals and these two types of female, in fact, form the two extremes of a continuous spectrum of G6PD levels. Of course, irrespective of their levels of G6PD activity, female heterozygotes may sometimes be identified positively by family data. Thus, a woman with apparently normal G6PD activity who gives birth to a G6PD deficient son must be heterozygous.

There is some evidence that certain abnormal X-chromosomes are "Lyonized" preferentially, but it does not appear that possession of a gene for G6PD deficiency constitutes sufficient of an abnormality to bring about preferential condensation.

Many other characters are known to be controlled by a locus on the X-chromosome. They include the enzyme phosphoglycerate kinase, red-green colour blindness, haemophilia and one of the blood group systems (the X_g system). Since males have only one X-chromosome whereas females have two, it follows that a recessive X-linked condition will affect males much more frequently than females, for posession of a single defective gene in males inevitably leads to fully expressed defectiveness, whereas a single defective gene in females constitutes a heterozygous state which may show no physical symptoms. Thus it is that red-green colour blindness is much more common in men than in women, and that the much rarer recessive condition of haemophilia is extremely rare in the fully expressed form in women, almost all recorded cases being in men. Since fully expressed G6PD deficiency occurs in about one in ten of male American negroes, one may calculate that it will occur with a frequency of about $1/10 \times 1/10 = 1/100$ in Negro females, which is about what is observed. It should be noted that, since males inherit their X-chromosomes from their mothers, females are obligate carriers for the various X-linked defects expressed in males.

Nothing is known about the biochemical consequences of the possession of a Y-chromosome.

III. Some Typical Inborn Errors of Metabolism

A. ALKAPTONURIA

Alkaptonuria is a rare, although easily recognized, condition in which the urine turns black on standing. As long ago as 1858 a substance called alkapton, subsequently identified as homogentisic acid, was isolated from the urine of alkaptonurics and shown to be the cause of the colour change. Alkaptonuria is a benign condition, of little bother to the subject except that collagenous tissues also darken and arthritis tends to occur in later life.

Alkaptonuria was first described in genetical terms by Garrod in 1902, who observed that it often occurred in families, particularly in brothers and sisters. About half the then known alkaptonurics had parents who were first cousins. The parents, however, did not usually suffer from the complaint themselves. Garrod therefore concluded that the condition was inherited as a recessive trait and that the unaffected parents were carriers.

Alkaptonuria is extremely rare, affecting only about two per million of the population. When a condition is very rare, like this, a recessive (heterozygous) carrier stands a much greater chance of mating with another heterozygote by contracting a consanguinous marriage (e.g. with a first cousin) than by selecting a partner at random. This follows from the fact that first cousins share two common grandparents, and if one of these grandparents were a heterozygous carrier then a first cousin will bear a one in four chance of also being a carrier, whereas the frequency of heterozygous carriers in the population at large may well be one hundred times less than this. Garrod's finding that consanguinity was present in parents in about 50% of cases of alkaptonuria reflects this. One in four of the offspring of a marriage between heterozygotes is likely to be a fully expressed alkaptonuric, so the occurrence of the condition in sibs, particularly in large families, is to be expected.

Garrod also investigated the mechanism of production of homogentisic acid. He showed that its production by alkaptonurics was increased when protein was fed and especially when the aromatic amino acids phenylalanine and tyrosine were given. This suggested that homogentisic acid was a metabolite of these aromatic amino acids. It would be further metabolized in normal individuals, but would accumulate in alkaptonurics by reason of some metabolic block. Confirmation for these ideas was obtained by feeding homogentisic acid to normal individuals and alkaptonurics: it was metabolized by the former but excreted unchanged by the latter.

With considerable foresight Garrod, who described the condition as a "chemical sport", attributed the block to the lack of an enzyme,

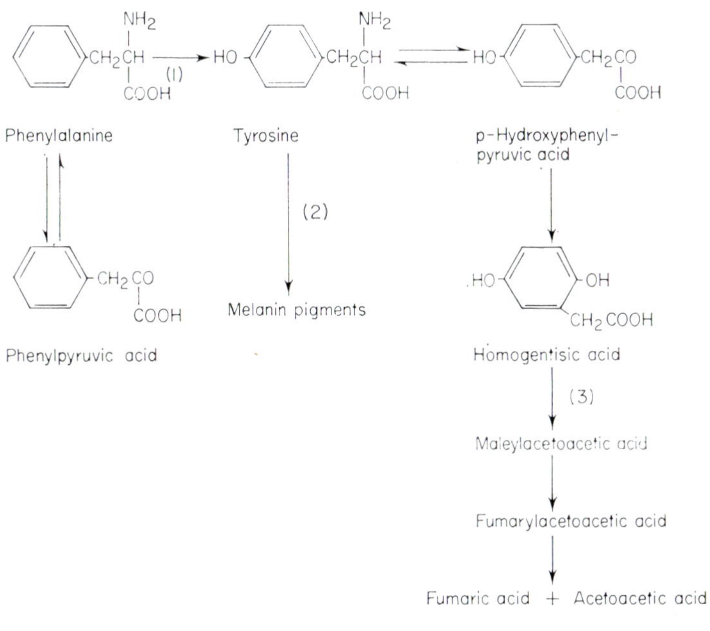

Fig. 1. Some steps in the metabolism of phenylalanine and tyrosine. Enzyme routes that may be blocked: (1) phenylalanine hydroxylase. (2) tyrosinase. (3) homogentisic acid oxidase.

caused in turn by some modification of the Mendelian factor or gene responsible for the production of that enzyme. The metabolic route by which homogentisic acid is produced and metabolized in the human is depicted in Fig. 1. Garrod was not able to identify the missing enzyme, and indeed this was not achieved until many years later. Eventually La Du in 1958 demonstrated that homogentisic acid oxidase, present in normal liver, was almost undetectable in alkaptonuric liver (there was less than 0·02% of the activity found in a control liver).

We still do not know whether the lack of catalytic activity in alkaptonuria is attributable to the production of a defective enzyme (as a consequence of a defective allele at a structural locus) or to a lack of production of enzyme of normal structure and activity (by reason of a mutation at a regulatory gene locus). In other conditions, however, some evidence on this point is available.

B. PHENYLKETONURIA

The presence of phenylketone (phenylpyruvate) in the urine of some mentally deficient patients was reported by Folling in 1934. Recognition of phenylketonuria thus occurred much later than alkaputonuria, although phenylketonuria is a much more serious condition and occurs more frequently. This was probably because phenylketonuria is not accompanied by such definitive symptoms as those observed in alkaptonurics. The urine of phenylketonuric patients, however, has a characteristic musty odour, and it is reported that chemical studies were initiated on the urines of two of the earliest discovered sufferers (who were sibs) as a consequence of the mother drawing attention to the malodour. Several of the earliest described patients were brothers and sisters, and the parents were unaffected.

Symptoms

Phenylketonurics exhibit neurological symptoms such as restlessness and poor coordination. Untreated patients are usually severely mentally deficient and often helpless: some cannot walk, or talk, or control their excretory sphincters. Epileptic seizures and severe eczema are other symptoms. However there is considerable variability in the extent to which the disease may be expressed, for other patients (a minority) have normal intelligence, although they may have a severely affected relative such as a brother or a sister, which would suggest that their milder symptoms cannot be attributed to the possession of an alternative allele.

Biochemistry of phenylketonuria

Phenylalanine accumulates in the plasma of phenylketonurics at a level some thirty times higher than normal (normal being about 1 mg per 100 ml) and is converted in quantity to phenylpyruvate. This conversion occurs also in normal individuals, but only to a small extent. Various other metabolites related to phenylalanine and phenylpyruvate are also produced, in amounts much larger than normal.

It was demonstrated by Jervis that the condition was caused by failure of the liver to convert phenylalanine to tyrosine by hydroxylation (reaction (1) of Fig. 1) with the result that phenylalanine accumulates. The livers of phenylketonurics were found to contain no detectable activity of the enzyme phenylalanine hydroxylase. The condition is therefore an example of an "inborn error" of metabolism. It is not known whether the lack of hydroxylase activity is caused by lack of enzyme production, or whether the enzyme is produced, but in an inactive form.

A minor feature of the condition is that sufferers tend to have blue eyes and fair skins. This is because the phenylalanine that accumulates inhibits the conversion of tyrosine into melanin pigments via route (2) of Fig. 1. Adequate amounts of tyrosine are usually present in the diet; the phenylketonuric does not suffer by his inability to form tyrosine from phenylalanine.

It seems that the major symptoms, the mental ones, are caused by defective myelination. Because myelination occurs during the first years of life, it is during the early years that the phenylketonuric child needs the greatest care and attention. The defective myelination appears to be associated with a deficiency of brain cerebrosides. How the accumulation of phenylalanine or phenylpyruvate or any of the associated metabolites in the plasma is connected with the production of brain cerebrosides is not clear. However, the great variation observed in the severity of the mental deficiency, apparently associated with the same degree of enzymic defect, indicates that the mental symptoms are probably a fairly remote consequence of the enzymic defect.

Diagnosis

The simplest test involves treating a sample of the suspected urine with a solution of ferric chloride. A transient greenish colour indicates the presence of phenylpyruvate. Simpler still, the test may be made using commercially available strips of paper impregnated with ferric chloride. It is important for the diagnosis to be made soon after birth because irreversible brain damage will have occurred by six months of age if the condition is untreated. Nowadays new-born children are routinely tested in many countries. Where the facilities are available, more sophisticated tests are used which detect raised levels of phenylalanine in the plasma. Such tests are preferred because a raised plasma phenylalanine is the more basic defect.

Treatment

Infants diagnosed as phenylketonurics are placed, soon after birth, on special diets low in phenylalanine. The diets are palatable but unexciting and expensive. Plasma phenylalanine levels are checked from time to time to ensure that they remain low. However, it is important that the diet contains some phenylalanine in order that normal growth may take place. On this regime phenylketonuric children do not develop the irreversible mental retardation. Special diets were first instituted in 1954 so that it is as yet rather early to appraise the long term prospects. It would appear, however, that it may be possible to abandon the special diets after the first few years of life, without untoward effects.

Genetics of phenylketonuria

The condition, like alkaptonuria, is an autosomally inherited recessive one that only becomes manifest when the subject is homozygous for the defective allele.

In Europe the incidence is 1 in 20 000 to 1 in 25 000. Until recently (when dietary treatment became possible) the great majority of sufferers were to be found in institutions, where they made up something like 1% of the total inmates. In something like 8% of cases the parents are found to be consanguinous, usually first cousins. It may be calculated that something like 1 in 80 of the population will be heterozygous carriers for this condition: heterozygotes show none of the symptoms of phenylketonuria but they may be identified by their abnormal responses to phenylalanine tolerance tests.

C. GALACTOSAEMIA

Galactosaemia is another serious although rare disease affecting young infants (see also pp. 16 and 19). The child has a profound intolerance to galactose, and the condition is accompanied by a raised level of plasma galactose, an abnormal galactose tolerance curve, and the presence of galactosuria. Early symptoms are vomiting and diarrhoea followed by dehydration, poor appetite and retardation of growth. When untreated the condition often results in early death. Children surviving are severely mentally retarded and often partially blind through cataract formation. Since galactose is a component of lactose, the sugar present in milk, an infant normally takes in substantial amounts of it. If recognized early and a special galactose-free milk is fed, the symptoms of galactosaemia are prevented.

The liver is the organ where most of the galactose is metabolized, and is the tissue that suffers most in galactosaemia. Jaundice is common. Abdominal enlargement caused by hepatomegaly is a clinical feature. Cirrhosis of the liver, preceded by fatty infiltration, is a common cause of death.

Galactose is normally metabolized by conversion to galactose-1-phosphate (Gal-1-P) which then reacts with uridine diphosphate glucose (UDPG) to give UDP-galactose, according to the equation:

$$\text{Gal-1-P} + \text{UDPG} \rightarrow \text{UDPGal} + \text{G-1-P}$$

The enzyme concerned (galactose-1-phosphate uridyl transferase) is inactive in liver specimens from galactosaemic individuals. It seems that it is the intracellular accumulation of Gal-1-P that produces the toxic effects in galactosaemia, rather than the build up of free galactose in the extracellular fluids, which also occurs.

It was mentioned earlier that all cell types contain the same genome, and hence if there is a gene locus coding for the transferase that is active in liver cells, that same locus will also be present in other tissues such for instance as erythrocyte precursors. Of course, the locus may not be active in other tissues but it happens, in the case of the transferase, that it is "switched on" in erythroblasts, and the enzyme is normally present in erythrocytes, and can be measured. Galactosaemic individuals, who lack transferase activity in the liver, are found also to be lacking in transferase activity in the erythrocytes. Thus a definitive diagnosis of galactosaemia may be established by carrying out an enzymic assay of a blood sample, and it is not necessary to perform a liver biopsy. This has meant that the families of galactosaemics may be extensively surveyed for transferase activity. The usual finding is that both the parents and a proportion of the other relatives of a galactosaemic child, although symptomless themselves, possess something like half the normal levels of red cell transferase activity, in keeping with the concept that galactosaemia is an autosomally inherited recessive condition, like alkaptonuria and phenylketonuria.

It has recently been demonstrated by immunological means that galactosaemics do produce the transferase protein, although of course in an enzymically inactive variant form.

IV. Some Inborn Variations of Proteins and Enzymes

A. TECHNIQUES

Now that it is generally recognized that the gene products are polypeptides, a new significance attaches to investigations into the nature and extent of variation shown by polypeptides and their oligomers (i.e. by proteins and enzymes). A number of human proteins and enzymes have now been investigated for the existence of variation, and some examples are discussed later. For such investigations the most useful single technique has undoubtedly been electrophoresis, and this technique has the advantage that it can often be applied to crude extracts. The tissue used has frequently been blood, since this is so readily available. Measurements of enzyme activities and of other kinetic parameters have also proved to be of value in differentiating between variant proteins and isoenzymes. The determination of amino acid sequences and the working out of three dimensional structures are ultimate goals, but pure specimens of the polypeptides concerned are required before such studies can be undertaken, and they have so far been achieved for only a few proteins, notably the haemoglobins.

Electrophoresis

Molecules will migrate at different rates in an electric field if they possess different charges. Pauling and co-workers in 1949 subjected haemolysates from normal and from sickle cell anaemia subjects to free electrophoresis, in a classical Tiselius type apparatus and found that the haemoglobins possessed different mobilities. Furthermore, haemolysates from sickle cell trait subjects were found to possess two haemoglobin components, with mobilities that corrresponded with those of normal haemoglobin (Hb A) and sickle cell haemoglobin (HbS) respectively. The haem units from the two types of haemoglobin were shown to be identical, and it was concluded that the globin moieties differed in some way in respect of their ionizable groups, the Hb S carrying a greater number of positive charges over the whole of the pH range (about 5·8 to 8·0) at which the experiments were conducted. These important findings demonstrated for the first time that there was a molecular difference between the two types of haemoglobin, and prompted Pauling and colleagues to describe sickle cell anaemia as a "molecular disease".

Nowadays the electrophoretic separation of proteins and enzymes is more conveniently carried out in a medium such as filter paper, strips of cellulose acetate, polyacrylamide gel or starch gel, rather than in free solution. Starch gel is a particularly useful medium, and with it proteins may be separated by virtue of differences in molecular size (by a molecular sieving effect) as well as by virtue of differences in charge.

The migration of coloured proteins, such as the haemoglobins, can be followed visually. Other proteins can be seen after staining with certain protein binding dyestuffs (e.g. amido black). In addition there are a number of special procedures which give colours with specific proteins. For instance the plasma haptoglobins can be made visible as faint pink bands by adding a small amount of haemoglobin, with which the haptoglobins combine. The haemoglobin-haptoglobin complexes (and also the haemoglobins themselves, when present in the low concentrations that investigators like to use to get sharp separations) give strong blue colours when treated with hydrogen peroxide and a diaminobiphenyl such as dianisidine or benzidine.

Highly specific staining procedures are often possible for enzymes. One procedure, of general applicability to enzyme systems that generate reduced NAD or reduced NADP, involves the use of the redox active compound phenazine methosulphate (PMS) and the tetrazolium salt "MTT": these react very rapidly with $NADH_2$ (or $NADPH_2$), transferring the hydrogen via the PMS to the MTT and converting the latter to an insoluble deep purple "formazan". The staining mixture for the enzyme lactate dehydrogenase (an enzyme that produces $NADH_2$) contains lactate, NAD, PMS and MTT (together with an appropriate buffer); and the mixture for phosphogluconate dehydrogenase (which produces $NADPH_2$) contains 6-phosphogluconate, NADP, PMS and MTT (see Fig. 2).

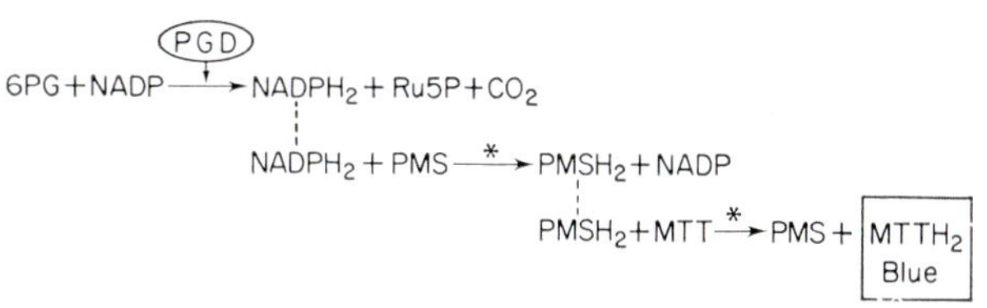

Fig. 2. Procedure for specific staining for the enzyme phosphogluconate dehydrogenase. 6PG = 6-phosphogluconate. Ru5P = ribulose-5-phosphate. PGD = phosphogluconate dehydrogenase. PMS = phenazine methosulphate, an active redox agent. MTT = a tetrazolium salt, which turns blue and becomes insoluble when reduced. * = non-enzymic reactions.

Electrophoretic variation indicates molecular diversity, but gives no indication of its extent. For instance Hb A and fetal haemoglobin (Hb F) differ with respect to some dozens of amino acids and yet they separate less well, electrophoretically, than do Hb A and Hb S, which differ with respect to one single amino acid on each of the two β chains. Neither is electrophoretic homogeneity any guarantee of molecular homogeneity, for two out of every three amino acid substitutions are not accompanied by a charge difference and hence are not likely to be detected by electrophoresis. Nevertheless, in spite of its limitations, electrophoresis has played a major rôle in recent work.

Kinetic studies of enzymes

Determinations of enzymic activity constitute the simplest type of kinetic characterization of an enzyme. Ideally, purified enzyme would be used and the catalytic activity expressed in enzyme units per mg of protein (i.e. specific activity). However, for practical reasons enzyme assays often have to be made on haemolysates, plasma samples or crude tissue extracts: in such cases the amount of enzyme protein is not known and the activity must be expressed in other terms, e.g. as units per ml of plasma or as units per g of Hb.

Let us suppose that the activity of an enzyme normally present in the serum (and measured in units per ml of serum) is present in reduced amount in a particular individual, (and the possibility that the reduced activity is caused by an inhibitor had been ruled out). The depressed level of enzymic activity could be attributed either to reduced production of the usual form of the enzyme, or to production of a structurally modified form with a lower specific activity, and it is often difficult to decide between these two in any given situation. A decision between the two alternatives can sometimes be arrived at by a different type of experimental approach, which will be illustrated by a discussion of the classical case of suxamethonium (succinyl choline) sensitivity.

Suxamethonium is used as a muscle relaxant during surgery and electro-shock treatment. Its effects usually last for only a few minutes because the drug is removed from the plasma quite rapidly by the action of plasma cholinesterase. Certain rare individuals constitute a problem, however, because they take an inordinate length of time to recover from the effects of this drug. The serum cholinesterase levels of these sensitive individuals are found to be significantly reduced, yet there is appreciable residual activity. Unaffected relatives are also found (as a class) to have somewhat reduced levels of cholinesterase, falling in the area between sensitive and normal individuals, and it seemed that some of these relatives might be heterozygous for a condition that was present in homozygous form in the rare sensitive individuals. The difficulty was that within affected families where normal, affected, and heterozygous individuals would all be expected to occur, there was a continuous spread of cholinesterase values so that affected individuals could not be differentiated from heterozygotes, nor heterozygotes from normal individuals. This difficulty was overcome by Kallow and co-workers, who measured serum cholinesterase levels under standard conditions both in the presence and absence of a fixed concentration of dibucaine (a cholinesterase inhibitor), with benzoylcholine as substrate, and calculated the percentage inhibitions. Each individual tested now segregated clearly into one of three categories: normals, where the percentage inhibition ranged around 79; intermediates, where the percentage was about 62, and sensitive individuals, with a percentage inhibition of about 16. The percentage inhibitions (called dibucaine numbers) represent a qualitative character of the enzyme (the affinity for dibucaine relative to the affinity for the substrate). The existence of the three different dibucaine number categories confirms the presence of three different phenotypes, as had been suspected. It follows that suxamethonium-sensitive individuals are homozygous for a structural variant of plasma cholinesterase which has different physical properties, one of which, clearly, is a much reduced affinity for suxamethonium.

Many other kinetic approaches have been used to differentiate between structural variants of enzymes. They include comparisons of Michaelis constants, heat stabilities, pH optima and relative rates with alternative substrates. Examples are given later.

Amino acid composition and sequence

Really practical procedures for the determination of amino acid compositions and sequences have only fairly recently become available. The first great advance was the publication by Moore and Stein, in 1950, of an ion exchange chromatographic procedure whereby all the amino acids of a protein hydrolysate could be separated and estimated quantitatively.

The protein under examination is first hydrolysed to its component amino acids, usually by prolonged heating in 6N HCl. (Tryptophan is completely destroyed by these conditions but may be determined separately.) Meanwhile a column is packed with an ion exchange resin, usually a cation exchanger of the sulphonated polystyrene type. The protein hydrolysate (adjusted to a weakly acidic pH so that all the amino acids will be positively charged and hence will bind to the exchanger) is placed on the column. Each amino acid will have its own special affinity for the ion exchanger, determined not only by its acid strength but also by its non ionic absorption characteristics. The column is eluted with buffers of increasing pH and ionic strength. The amino acids come off one by one as the elution proceeds: the dicarboxylic amino acids are eluted early, followed by the neutral ones, with the dibasic amino acids coming off towards the end. The eluate is collected in a large number of separate portions by means of an automatic fraction collector. Each fraction is analysed by heating with ninhydrin, which produces colours with amino acids. (Nowadays the eluate is analysed continuously, using a flow-through colorimeter, and the absorbances are plotted automatically on a moving chart.) This procedure gives quantitative estimates of the amounts of most of the amino acids. However, no one set of hydrolysis conditions will release intact all the amino acids, and for this reason recourse must usually be made to alternative hydrolysis conditions followed by additional chromatographic runs to obtain complete data for any one protein.

This procedure has now been automated to a high degree of perfection, and today the amino acid compositions of a great number of proteins have been made known.

The proper characterization of a protein, however, requires also a knowledge of its amino acid sequence. The first practical procedure for determining sequences was reported by Sanger in 1955, who elucidated the sequence of amino acids in bovine insulin. Insulin is a small protein of minimum molecular weight about 5 700, containing two short polypeptide chains, bound to each other by two cystine sulphur-sulphur bridges. The bridges were split by oxidation with performic acid, and the two chains separated by chromatography. Each chain was treated with the reagent fluorodinitrobenzene, which combines with N-terminal amino acids. Acid hydrolysis gave yellow dinitrophenyl (DNP) derivatives of the N-terminal amino acid and also the DNP-derivatives of the di-, tri- and tetra-peptides, allowing the several amino acids at the N-terminal end of each chain to be specified in sequence. Partial hydrolysis (both acidic and enzymic) of the untreated polypeptides gave small peptides that were separated by chromatography and electrophoresis and themselves subjected to the DNP procedure outlined above. Digestion of the intact insulin molecule with chymotrypsin followed by partial acid hyrolysis gave peptides with intact sulphur-sulphur bridges, which in turn were subjected to the DNP procedure. By these means, the amino acid sequences of a number of small peptides became known. Overlapping amino acid sequences were fitted together like a jig-saw puzzle, and resulted in a definitive detailed structure for the insulin molecule in which each single amino acid had its special place.

It had not been proven, until this time, that a polypeptide had a unique and specific amino acid sequence; the amino acids could conceivably have been assembled together with some degree of randomness.

The structures and sequences of other insulin molecules, from man, pig, sheep, horse and whale, were worked out shortly afterwards. They were all found to contain an identical number of amino acids, arranged in the same two short polypeptide chains, but the amino acid sequences were not quite identical. Amino acid substitutions were present at one or more of just four positions (out of fifty-one positions). One presumes that these positions are not important for the proper physiological functioning of insulin, and are thus sites where evolutionary changes could take place without harm. It is of interest to note that insulin is synthesized as a single precursor called proinsulin, of molecular weight 9 100. This is converted to insulin by the splitting of two peptide bonds in the interior of the molecule, releasing a peptide chain of 33 amino acids, and leaving the B and A insulin subunits, which had previously formed the beginning and the end respectively of the preinsulin chain.

Another procedure for the end group analysis of proteins and peptides employs phenyl isothiocyanate—the so called Edman reagent. Like DNP, this reagent reacts with the N-terminal amino acid of a polypeptide chain, but mildly acidic conditions then serve to hydrolyse off this terminal amino acid (as its phenylthiohydantoin derivative, which can be identified) without breaking down the remainder of the polypeptide chain. This allows the Edman degradation to be repeated so that the penultimate N-terminal amino acid can be identified, and then the third residue from the end, and so on.

The sequencing process was successfully applied to other proteins, starting with the simpler monomers such as ribonuclease and cytochrome *c*. As the size of the polypeptides increased it became necessary to rely on enzymic hydrolysis for the production of the small peptides. This was because the non-specific nature of acid hydrolysis yields too many products for convenient handling, whereas the proteolytic enzymes are specific for certain types of peptide bond and produce a strictly limited number of peptides. For instance, crystalline trypsin only hydrolyses peptide bonds that involve the carboxyl groups of lysine and arginine.

An important landmark was the complete sequence determination of the polypeptide chains of haemoglobin. This has now been achieved for the haemoglobins of many animal species including man. By themselves the sequence studies of the haemoglobins are not particularly informative, but in conjunction with X-ray crystallographic studies, which have been carried out so brilliantly by Perutz and co-workers, it has been possible to construct a three-dimensional model of haemoglobin in which the positions of every one of the 574 component amino acids and the four haem groups are specified exactly. Helical and non-helical regions of the molecule are defined, and it is known that the haem groups lie within crevices which are hydrophobic in nature, with the iron atoms coordinated to the imidazole side chains of histidine residues. The exterior of the molecule is hydrophilic, giving the molecule its high solubility. These characteristics are common to the haemoglobins from a variety of species, although the haemoglobins may differ from one another in respect of a number of their amino acids. Presumably, as with insulin, evolutionary changes involving replacements of individual amino acids have occurred, although the shape and general properties of the molecule have been preserved.

Methods for the elucidation of amino acid sequences have now achieved a high degree of sophistication, and indeed are carried out by automated pieces of equipment known as "sequenators". By such means the sequences of large numbers of polypeptides and proteins are rapidly becoming known.

It should be added that information about the nature of single amino acid substitutions can often be obtained by a simpler procedure known as "finger-printing". A polypeptide is enzymically hydrolysed with trypsin to give a series of peptides, which are separated by a combined process of paper chromatography (in one direction) and electrophoresis (in a direction at right angles) and stained with ninhydrin. The resulting peptide map may be regarded as a "finger-print" characteristic of the original polypeptide, and is examined alongside a similar peptide map derived from a variant polypeptide. It is likely that the maps will be identical except for one pair of peptides, and hence only these two peptides need to be analysed. Frequently, these two peptides differ in respect of a single pair of amino acids. It was by this procedure that it was first demonstrated that a glutamate residue in the β-chain of Hb A had been replaced by a valine residue in the β-chain of Hb S.

B. PHOSPHOGLUCONATE DEHYDROGENASE

Phosphogluconate dehydrogenase (PGD) is a fairly uncomplicated example of an enzyme that occurs in the human in a number of genetically determined variant forms and is controlled by a single autosomal locus. It shows several types of variation of general interest: electrophoretic variations (from which deductions about the subunit structure can be drawn), variations in stability, and decreased catalytic activity. A procedure for specific staining for PGD, after electrophoresis, is outlined in Fig. 3.

The two common human PGD alleles

Most humans possess, in all tissues, a single electrophoretic type of

PGD designated "a". Such individuals may be said to be of A phenotype, the "a" enzyme being the product of an allele PGD^A. People of A phenotype are usually (not invariably) homozygous for the PGD^A allele. Some individuals (about 4% in London) possess three PGD isoenzymes: "a", "b" and "c" (see Fig. 3). Rare subjects (about 1 in 2 500 in London) show a single "c" isoenzymic band after electrophoresis, attributable to the activities of an allele PGD^C, and such subjects are described as of C phenotype.* Usually they are homozygous for the PGD^C allele, although other genotypes are possible (see Table 3). Returning to the triplet phenotype: it occurs with the same frequency in males and females and it segregates in families in such a manner as to indicate that it is an autosomal co-dominant heterozygous condition of genotype $PGD^A PGD^C$.

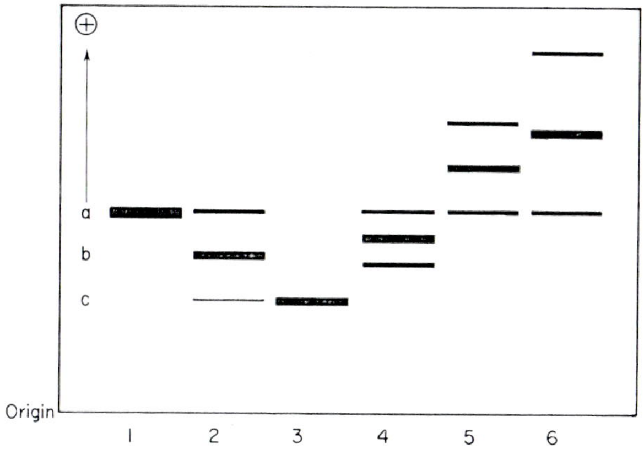

Fig. 3. Starch gel electrophoresis of some variant human PGD types. The electrophoresis is carried out in a phosphate buffer system at pH 6·8. In the interests of clarity satellite bands (see footnote to text) are not shown.

1 A electrophoretic type
2 CA electrophoretic type
3 C electrophoretic type
4 Neath heterozygote
5 Richmond heterozygote
6 Elcho heterozygote

The three human isoenzymes "a", "b" and "c" of the heterozygous triplet (the CA phenotype) have similar molecular weights, of about 130 000. The simplest way to explain the presence of the three components is to suppose that the products of the two alleles (PGD^A

* When the electrophoresis is performed on haemolysates, as it frequently is, certain additional weaker isoenzymic components appear in all three phenotypes, migrating more anodically than the major bands but always taking up positions in a fixed relationship to the major bands, and for this reason they have been called "satellite bands". Satellite isoenzymes are strongest in old haemolysates and are probably chemically modified versions of the major isoenzymic components. Satellite bands are absent, or present at much lower intensities, in extracts from white cells and many other tissues.

and PGD^C) are subunits, say α^A and α^C, which combine together in pairs to form dimers, which constitute the active isoenzymes (Table 3). Two homodimers $\alpha^A\alpha^A$ and $\alpha^C\alpha^C$ are possible, and one heterodimer, $\alpha^A\alpha^C$, constituting the isoenzymes "a", "b" and "c" respectively. If the subunits combine together at random there should be twice as much heterodimer as of each individual homodimer, and indeed it is observed that the middle electrophoretic band, isoenzyme "b", stains with greater intensity than either "a" or "c".

The "c" isoenzyme in heterozygotes stains rather less strongly than the "a" band, indicating that it has less activity. Enzymic assays on

TABLE 3. Mechanism of formation of the three common human PGD isoenzymes

Electrophoretic phenotype	A	CA	C
Sub-unit(s) produced	α^A	α^A (50%) α^C (50%)	α^C
Dimeric isoenzyme(s) produced	$\alpha^A\alpha^A$ = "a"	$\alpha^A\alpha^A$ = "a" (25%) $\alpha^A\alpha^C$ = "b" (50%) $\alpha^C\alpha^C$ = "c" (25%)	$\alpha^C\alpha^C$ = "c"
Alleles producing sub-units	PGD^A	PGD^A and PGD^C	PGD^C
Possible genotypes	PGD^{A}PGD^A PGD^{A}PGD^I*	PGD^{A}PGD^C	PGD^{C}PGD^C PGD^{C}PGD^I*

* The Ilford allele PGD^I, which is rare, makes no contribution to the enzymic activity (see text). It appears either not to produce a subunit, or to produce one that is not only inactive but is incapable of participating in dimer formation.

TABLE 4. Percentage residual activities shown by AA, CA and CC types of human erythrocyte PGD after various treatments. All procedures were carried out at pH 8·0

	Percentage residual activities ± standard deviation			
PGD type	Incubation for 15 min at 37° in 0·2 M tris	Incubation for 15 min at 25° in 1·5 M urea	Incubation for 10 min at 25° in 0·01 M iodoacetate	Assay carried out in the presence of 0·1 per cent "Teepol"
AA	73 ± 6	63 ± 10	80 ± 4	93 ± 4
CA	51 ± 9	30 ± 7	50 ± 3	48 ± 8
CC	38	5 ± 6	20 ± 3	17 ± 5

haemolysates from individuals of genotypes $PGD^A PGD^A$, $PGD^A PGD^C$ and $PGD^C PGD^C$ give relative activities of about 100:90:80. The explanation for the slightly reduced activities may well be that isoenzymes containing α^C subunits are less stable and lose activity during their sojourn in the cell, for there is *in vitro* evidence for reduced stability of the isoenzymes containing α^C subunits (Table 4). Indeed individuals of CA and of C electrophoretic types can be identified by tests for instability with just as much certainty as electrophoretic screening.

Rare electrophoretic variants of human PGD

In addition to the CA type (the common triplet phenotype) there are a number of rare triplet electrophoretic patterns for PGD. Several of these (the Richmond, Elcho and Neath types) are shown in Fig. 3, and several others are known. They all possess the "a" component, but the other two components differ from the "b" and "c" isoenzymes of the CA type. All the triplets have a more intensely staining middle band, and outer bands that are about equidistant from the middle one, and they can be attributed to the existence of a series of rare alleles acting in concert with a PGD^A allele.

PGD variation in other species

Three banded electrophoretic patterns for PGD have also been found in a number of animals: in rats, mice, pigeons, pigs, monkeys, rabbits and cats. Other individuals of the same species have single bands with electrophoretic mobilities that correspond with either the fast or the slow bands of the triplet types. Breeding experiments have shown that the triplet phenotypes, as in the human, are autosomally inherited heterozygous conditions, and the electrophoretic component with the intermediate mobility is presumably a heterodimer containing one subunit from each of the two alleles. Thus it seems that PGD from many species is dimeric in structure. This sort of deduction illustrates a contribution that genetic studies can make to knowledge of enzyme structure.

If an *in vitro* procedure could be found to dissociate two homodimers, and if the conditions were then reversed, the subunits might associate to give a proportion of heterodimer. This has been achieved with two homozygous electrophoretic variants of PGD from the fruit fly *Drosophila melanogaster*, although not, as yet, with PGD from higher organisms.*

* Note that, in the fruit fly, the PGD is X-linked. This does not affect the argument put forward. Since there are only four pairs of chromosomes in the fruit fly, it is hardly surprising that a number of characters that are autosomally linked in higher organisms are X-linked in the fly.

Human PGD alleles conferring reduced activity

Two alleles will be described: one causes reduced activity because of instability and the other, apparently, because of a lack of production of an adequate polypeptide product.

The Whitechapel allele (PGD^W) is a rare allele conferring much greater instability than the PGD^C allele discussed above. Homozygous subjects for this allele have an almost zero erythrocyte PGD level and white cell extracts show reduced activities (about one quarter to one half normal). Heterozygous relatives (of genotype $PGD^A PGD^W$) have about three quarters of normal PGD levels in erythrocytes and white cell extracts show normal levels: both haemolysates and white cell extracts show instability when incubated. These facts are explicable on the hypothesis that the enzyme in the homozygote is the very unstable homodimer $\alpha^W\alpha^W$: this very quickly loses all activity in red cells; in white cells however, by virtue of continuing resynthesis, this enzyme manages to maintain a reduced level of activity. It is assumed that heterozygotes have the usual three isoenzymes; their structures will be $\alpha^A\alpha^A$, $\alpha^A\alpha^W$ and $\alpha^W\alpha^W$ and they will be formed in the proportions of 1 : 2 : 1. It is further asumed that the heterodimer $\alpha^A\alpha^W$ is able to maintain its activity in the red cell by virtue of the stabilizing influence of the α^A subunit, so that only the $\alpha^W\alpha^W$ component becomes inactivated: this explains the three quarters of normal level of activity found in erythrocytes of heterozygotes. The $\alpha^A\alpha^W$ heterodimer, nevertheless, has some degree of instability, which can be demonstrated by *in vitro* incubation.

The Whitechapel homozygote has reduced red cell survival and, in consequence, a relatively young red cell population. There is also some degree of haemolytic response to primaquine (primaquine sensitivity is discussed in Section IV). It is interesting to note that these quasi-pathological symptoms (which are not very serious) were discovered as a consequence of a survey for enzyme variations and not (as is usually the case) as a result of investigations into a clinical abnormality.

The second allele giving rise to reduced activity is the Ilford allele (PGD^I), but this has only been found in the heterozygous state: it confers about half normal activity on the PGD of both haemolysates and white cell extracts. Heterozygotes of genotype $PGD^A PGD^I$ show the A electrophoretic phenotype, indicative of the production of α^A subunits only, and heterozygotes of genotype $PGD^C PGD^I$ show the C electrophoretic type, suggesting that only α^C subunits are formed. This provides circumstantial evidence that a subunit α^I is not produced, for if it were one would expect at least one of the heterodimers $\alpha^A\alpha^I$ or

$\alpha^C\alpha^I$ to migrate independently and to be visible after specific staining. This is the sort of situation that one would expect to result from a "frameshift" mutation at the PGD locus. If the postulated α^I subunit is indeed formed, it must be regarded as non-viable in that it is itself enzymically inactive and is incapable of associating with other subunits.

No individual with the genotype PGD^IPGD^I has ever been found: however one might expect that such an individual, with no PGD activity at all in any tissue, would not be able to survive beyond an early embryonic stage.

C. LACTATE DEHYDROGENASE

Whereas the variant PGD types result from the presence of variant alleles at one locus, and occur as differences between individuals, lactate dehydrogenase (LDH) in most tissues is controlled by two loci, and it is the wide variation in the relative expression of these two loci from one tissue to another that is of particular interest. Variant alleles have been reported at each of the main LDH loci, but they are rare.

Formation of tetramers

In general, it may be said that all cells produce both H subunits (from the LDH_H locus) and M subunits (from the LDH_M locus). The H and M subunits differ from one another in a number of ways, and yet they resemble each other in so many other respects that it is generally accepted that they both evolved from a common precursor. The subunits combine together within the cells in which they are formed to give five possible varieties of tetrameric molecule of structures HHHH, HHHM HHMM, HMMM and MMMM (see Table 5). Where the amounts of the two types of subunit produced are the same, the relative amounts of the five LDH isoenzymes, designated 1,2,3,4 and 5 respectively, may be calculated on the assumption of random association to be present in the ratio of 1 : 4 : 6 : 4 : 1.* In certain tissues (e.g. reticulocytes and leucocytes) the LDH isoenzymes do in fact occur in about these proportions. If equal amounts of LDH-1 (of structure HHHH) and LDH-5 (of structure MMMM) are mixed and frozen overnight in 1M NaCl, the subunits dissociate and reassociate (i.e. hybridize) to form the five isoenzymes in approximately the relative proportions given above. In most tissues, however, one or the other of the LDH loci is the more active and produces more subunits than the other, so that either

* These ratios are the numbers that precede the five terms resulting from the expansion of the expression $(H + M)^4$.

TABLE 5. Mechanism of formation of the five LDH isoenzymes from the H and M loci for human LDH. The third column illustrates the consequences of an allelic variation at one of these loci

Genotype and description	$LDH_H LDH_H/LDH_M LDH_M$ Found in the great majority of humans		$LDH_H LDH_H^*/LDH_M LDH_M$ A rare variant heterozygous at the LDH_H locus	
Derivation of sub-units	$LDH_H \longrightarrow H$ $LDH_M \longrightarrow M$		$LDH_H \longrightarrow H$ $LDH_H^* \longrightarrow \overset{*}{H}$ $LDH_M \longrightarrow M$	
Tetrameric isoenzymes formed, with structures	*LDH-1*	HHHH	*LDH-1* *components*	HHHH $HHH\overset{*}{H}$ $HH\overset{*}{H}\overset{*}{H}$ $H\overset{*}{H}\overset{*}{H}\overset{*}{H}$ $\overset{*}{H}\overset{*}{H}\overset{*}{H}\overset{*}{H}$
These components are given in order of their electrophoretic mobilities. LDH-1 migrates the furthest towards the anode.	*LDH-2*	HHHM	*LDH-2* *components*	HHHM $HH\overset{*}{H}M$ $H\overset{*}{H}\overset{*}{H}M$ $\overset{*}{H}\overset{*}{H}\overset{*}{H}M$
	LDH-3	HHMM	*LDH-3* *components*	HHMM $H\overset{*}{H}MM$ $\overset{*}{H}\overset{*}{H}MM$
	LDH-4	HMMM	*LDH-4* *components*	HMMM $\overset{*}{H}MMM$
	LDH-5	MMMM	*LDH-5*	MMMM

LDH-1 or LDH-5 predominates and there is a progressive diminution in the amounts of the other isoenzymes.

Heart cells produce mostly H subunits, and thus extracts of heart tissue show a very strong LDH-1 component and progressively decreasing amounts of the others, with the LDH-5 component not in evidence. Similarly skeletal muscle produces more than ten times as many M subunits as H, so that muscle extracts show a very strong LDH-5 component after electrophoresis. A clear LDH-4 band and a just visible LDH-3 band are also present, but the LDH-2 and LDH-1 components are usually not to be seen. The nature of the control mechanisms that regulate the activities of the LDH_H and LDH_M loci are not understood.

Differences between the tetramers

There are differences between the two homotetramers (LDH-1 and LDH-5) and lesser differences between the heterotetramers which are of interest.

The LDH isoenzymes rich in H subunits (e.g. LDH-1 and LDH-2) are inhibited by high concentrations of lactate. Heart tissue, which is rich in these isoenzymes, does not accumulate lactate, and it has been argued that this is because any tendency for lactate to build up would block the further production of lactate from pyruvate, so the latter must, perforce, be oxidatively metabolized. On the other hand, skeletal muscle is sometimes called upon to do intensive anaerobic work, under which circumstances it must accumulate lactate. This it is perfectly well able to do since it is rich in the LDH isoenzymes (LDH-5 and LDH-4) that function satisfactorily in the presence of raised lactate. It is interesting to note that the flight muscles of chickens (which are only used intermittently) have LDH components rich in M subunits, whereas the flight muscles of the storm petrel and the humming bird (muscles which need to function continuously for long periods and cannot afford to accumulate lactate) contain LDH isoenzymes rich in H subunits.

The human LDH isoenzymes decrease in stability with increased content of the M subunit. The production of the two LDH subunits is about equal in human reticulocytes. However, the LDH-5 component (of structure MMMM) is usually not detectable after electrophoresis of normal haemolysates: it presumably has become inactivated during ageing of the erythrocyte in the circulation.

Clinical applications

Diseased tissues release enzymes into the plasma, possibly because their cells have leaky membranes. Therefore raised plasma levels of enzymes can give useful evidence of the existence of disease. In cardiac infarction the plasma LDH activity may be raised fivefold. Furthermore the LDH isoenzymes responsible for the raised enzymic activity are those characteristic of the damaged tissue. Raised levels of LDH-1 and LDH-2 suggest damage to cardiac tissues, whereas raised LDH-5 and LDH-4 could indicate liver damage (because liver, like skeletal muscle, has LDH isoenzymes rich in M subunits).

There are alternative procedures to electrophoresis for establishing the types of isoenzymes causing a raised plasma LDH. In one such procedure the LDH activity of a plasma sample is measured before and after incubation (at 60° for 60 min): an elevated LDH level that is not much reduced after incubation indicates the presence of raised amounts of the stable LDH-1 and LDH-2 components and would suggest heart disease, whereas an elevated LDH that was substantially reduced by the incubation procedure would indicate the presence of raised levels of the LDH-5 and LDH-4 components that are known to be unstable and would be consistent with liver disease.

Allelic variations at one or other of the LDH loci

Allelic variants at both the LDH_H and the LDH_M loci have been reported, but they are rare. Each gives rise to a new series of isoenzymic components. Such a series, for a rare variant heterozygous at the LDH_H locus, is set out in Table 5. Four additional closely spaced electrophoretic components appear adjacent to the normal LDH-1 component, equally spaced from each other; three additional components appear adjacent to the LDH-2 isoenzyme; and so on. In all, fifteen different isoenzymic components are possible in such a heterozygous individual.

D. THE HUMAN HAEMOGLOBINS

The haemoglobins have probably been investigated more thoroughly than any other group of proteins. Investigators have been attracted to this molecule not only because of its importance in the carriage of oxygen for respiration, but also for the practical reason that it is relatively easy to isolate in a pure state in large amounts. Erythrocytes contain about one third of their weight of haemoglobin whereas, by contrast, most enzymes are present in tissues in very low concentrations. Inevitably, the haemoglobins have already been mentioned in a variety of contexts in this chapter.

The haemoglobin molecule, like LDH, is a tetramer, but it differs from LDH in many respects. Thus haemoglobin is produced in only one cell type—the erythroblast. The haemoglobin subunits (at least five are known) associate together in certain definable and predictable ways, and not at random, as with LDH. There is a complicated and ill-understood "switching mechanism" for turning on and off the production of certain of the subunits, and consequently different haemoglobins are formed at different stages of development.

Formation of tetramers

The locus for the α-polypeptide chains of haemoglobin is active throughout life, and hence α chains are always being produced. The

α chains readily form dimers with unlike chains (β, γ, δ or ε chains, depending on the type available) but not with themselves. (Haem units are also incorporated, one per polypeptide chain.) The dimers then associate with like dimers to give tetrameric haemoglobin molecules: $\alpha_2\beta_2$ (Hb A); $\alpha_2\gamma_2$ (Hb F); $\alpha_2\delta_2$ (Hb A_2) and $\alpha_2\varepsilon_2$ (Hb Gower II). When α subunits are in short supply small amounts of homotetramers are formed: β_4 (Hb H); γ_4 (Hb Barts) and ε_4 (Hb Gower I).

TABLE 6. Relative rates of synthesis of five different human haemoglobin sub-units with age of development, showing also the different types of haemoglobins formed. The figures represent percentages of the total haemoglobin chain production. Rates of synthesis of less than one per cent are ignored

Gene pairs	$Hb_\alpha Hb_\alpha$	$Hb_\beta Hb_\beta$	$Hb_\gamma Hb_\gamma$	$Hb_\delta Hb_\delta$	$Hb_\varepsilon Hb_\varepsilon$	Haemoglobin types and typical percentage amounts
Gene products	α	β	γ	δ	ϵ	
6 weeks gestation	46	—	27	—	27	Gower II ($\alpha_2\epsilon_2$) 46% F ($\alpha_2\gamma_2$) 46% Gower I (ϵ_4) 3% Barts (γ_4) 3% Gower III ($\gamma_2\epsilon_2$) 2%
2 months gestation	48	1	40	—	11	F 72% Gower II 22% Barts 4% A ($\alpha_2\beta_2$) 2%
3 months gestation	50	2	48	—	—	F 96% A 4%
Birth	50	10	40	—	—	F 80% A 20%
6 months of age onwards (Adult pattern)	50	48	1	1	—	A 96% A_2 ($\alpha_2\delta_2$) 3% F 1%

At six weeks gestation in the human the α chains make up slightly less than half the total haemoglobin globin, the remainder comprising about equal amounts of ε chains and γ chains (see Table 6). By three months of gestation the production of α chains has increased to exactly half the total globin, and this remains so throughout life. The remaining half is made up largely of γ chains, for at three months the synthesis of β chains has only just begun, and the ε chain contribution has fallen nearly to zero. At birth the γ chain production is declining and is being replaced by β chains. By six months of age the amounts of γ chain have fallen to a low level and been replaced largely by β chains, together with a small proportion of δ chains (about 1% of the total polypeptide amount). The mechanisms that operate these intriguing changes are obscure.

Nature of the subunits

Like the two subunit types involved in the production of LDH isoenzymes, the five different subunits concerned in the formation of the various human haemoglobins differ from each other in a number of respects and yet they bear such resemblances to each other (and also to the single polypeptide of muscle myoglobin) that there can be little doubt that they all evolved from a common primordial polypeptide. For instance, fifty of the amino acid residues in α, β, γ and δ chains match, in identity and in sequencial position, and the chains are all of very similar lengths.

Of all the chains, the β and δ varieties bear the greatest similarity to each other, both possessing 146 amino acids and differing in respect of only 10 of them. The loci for these two chains are known to be closely linked (i.e. are sited close to one another on the same chromosome) because in certain rare families where allelic variants occur at both loci, the two alleles do not segregate independently when inherited. There is also evidence that the γ locus is linked to the β and δ loci, but more distantly. The locus for the α chain, which is removed (from an evolutionary point of view) from the other loci by the lapse of a greater period of time and produces a polypeptide that differs from the other polypeptides in respect of more than half the amino acids, is known not to be linked to the β, γ or δ loci. Nothing is known of the linkage of the ε locus.

"Finger-printing" of fetal haemoglobins has shown that some individuals possess two varieties that differ in respect of a single amino acid residue in the γ chain: the particular substitution involved is not accompanied by any charge change. This sort of difference would normally be considered to be typical of alleles, but genetical considerations show that there must be two different γ loci. It would appear that the two loci are very close, and perhaps adjacent, for they do not segregate when inherited. They presumably have arisen by duplication, just as the other haemoglobin subunit loci are thought to have done but the separation must have been comparatively late. Evidence for the existence of two loci for α chains has very recently been put forward.

Allelic variations

Not only are there multiple loci coding for haemoglobin sub-units:

there are also a number of alleles at each of these loci. Over one hundred variant human haemoglobins have now been characterized, mostly accountable for on the basis of allelism. Many of the alleles arise by single amino acid substitutions. About half the known replacements are situated on the surface of the haemoglobin molecule, and are relatively innocuous (Hb S constitutes an exception to this). Replacements around the haem pocket that disturb its hydrophobic nature give rise to the so-called "unstable haemoglobins". Replacements at the points of contact between different polypeptide chains tend to affect the oxygen dissociation curve. The reader must, however, consult other texts for a detailed exposition of the extent and variety of allelic haemoglobin variation.

Each of the two β chains of Hb S contains a single amino acid substitution (glutamate→valine) and this substitution occurs on the surface of the molecule. Hb S has lowered solubility, especially when in the reduced form. It has been suggested that in Hb S the two valine residues adhere to binding sites on adjacent haemoglobin molecules, especially when in the reduced form, and build up long chains of haemoglobin moleucles, thus accounting for the lowered solubility and for the sickling phenomenon. In the heterozygous state (sickle-cell trait) the erythrocytes contain both Hb A and Hb S, and the subject shows no particular disability under normal living conditions, but minor complaints (e.g. joint pains) may occur when the oxygen tension of the air is reduced (e.g. at high altitudes, or in aircraft). In the homozygous state (sickle cell anaemia) the clinical consequences are serious and often fatal.

All populations harbour rare haemoglobin variants. In certain areas, however, a variant occurs with a particularly high frequency, and undoubtedly the most striking example of this is in the occurrence of Hb S over a substantial part of the African continent (the sickle cell trait attains a frequency of as high as 40% in some parts). Hb C (see pp. 8 & 9) is another example of a haemoglobin variant occurring with an appreciable frequency in a particular geographical region.

"Balanced polymorphism"

There is no doubt about the fact that homozygous sickle cell disease is serious: indeed it is often fatal in early life, and is held to be responsible for 80 000 deaths per year in Africa. Thus something like 160 000 sickle cell genes are eliminated annually, and furthermore they are usually eliminated before reproductive age is reached, so that the deleterious genes do not get passed on to the next generation. On this count it would be expected that the frequency of the sickling gene would rapidly decline. This is especially so since it does not seem likely that the gene can enter a population anew at any significant rate, by fresh mutation, for the sickling gene is practically unknown over large areas of the world. Yet 1% of all children born in tropical Africa have sickle cell disease; they are most usually the offspring of unions between two heterozygous parents, who represent about 20% of the population. One is left with the conclusion that heterozygous individuals have greater biological fitness than their non-affected fellows, i.e. in some way they possess an advantage that enables them to produce more children than average. so that there will be an increased frequency of the sickle cell gene in the new born of the next generation such that, allowing for the early death of the homozygotes with sickle cell disease, the frequency of the gene in the population as a whole remains static.

A hypothesis about the nature of this heterozygous advantage was put forward by Allison in 1954. He drew attention to the fact that the world distribution of falciparum malaria and of the sickle cell gene closely paralleled each other, and he adduced evidence demonstrating that in young children (in whom an immunological defence against malaria has not had time to develop) those with the sickle cell trait were less prone to develop malaria, and had lower parasite counts when they did succumb, than non-sickling individuals. Thus individuals with the sickle cell trait would have an increased survival rate to the reproductive years. This explanation for the high frequency of the sickle cell gene in Africans is now generally accepted. The nature of the increased resistance to malaria in heterozygotes is not yet fully resolved: one theory is that when an erythrocyte is parasitized it becomes deoxygenated and sickles, whereupon it is promptly destroyed.

It has been proposed that other genes occurring with a relatively high frequency in certain areas of the world (e.g. the genes for thalassaemia and glucose-6-phosphate dehydrogenase deficiency) also confer some protection against malaria, but in these instances the evidence is equivocal.

The thalassaemias

The thalassaemias are a group of inherited conditions in which the synthesis of haemoglobin is depressed, resulting in anaemia. The haemoglobin that is produced appears perfectly normal in structure, suggesting that the defect does not concern a structural locus, and some malfunctioning of the regulatory mechanisms controlling the synthesis of the polypeptide chains is implicated. In one type there is depressed production of α-chains (α-thalassaemia). This is accompanied by raised amounts of Hb Barts (γ_4) during the fetal and neonatal period, and significant amounts of Hb H (β_4) during adult life. The other main type is caused by the lack of adequate amounts of β-chains (β-thelassaemia). It comes on at about birth, when the β-chain production should be rapidly rising, and is accompanied by persistence of Hb F into adult life, and raised production of Hb A_2.

E. GLUCOSE-6-PHOSPHATE DEHYDROGENASE

More than fifty inherited variant forms of human erythrocyte glucose-6-phosphate dehydrogenase (G6PD) have been described. Several are deficient in catalytic activity and lead to haemolytic episodes under some circumstances. One interesting variant shows four times the normal level of activity. There are also a number of electrophoretic variants. There appears to be a single locus for human G6PD, and it is X-linked. Since G6PD can be measured quantitatively some important genetic concepts relating to the dosage of X-linked characters can be examined (see p. 20).

Primaquine sensitivity and haemolytic anaemia in Africans

An improved synthetic antimalarial drug, primaquine, became available in 1944, but a disadvantage was that it induced haemolytic episodes in certain susceptible people. About one male American Negro in ten gave a haemolytic response, and this susceptibility appeared to be inherited.

The haemolysis induced in susceptible people by a daily dose of 30 mg of primaquine is dramatic. Two or three days of treatment elapse without untoward symptoms. During the next week severe intravascular haemolysis occurs, involving the destruction of up to half the total erythrocytes, and constituting a serious clinical problem. However at this stage the haemolysis ceases, even although the drug treatment is continued. The haemoglobin and haematocrit levels then rise slowly, returning to normal after about two months. (Fortunately, it was found that reducing the dose to 15 mg per day eliminated the likelihood of severe haemolysis and yet provided an acceptable level of antimalarial activity.)

It was shown that it was the red cells of susceptible individuals that were at fault (i.e. the defect had nothing to do with the plasma) because susceptible cells also became haemolysed when transfused into the circulations of non-susceptible individuals under treatment with primaquine. It was also shown (by tagging the red cells with ^{59}Fe) that haemolysis is related to cell age: only erythrocytes of more than about 40 days of age are haemolysed, whereas the young cells are resistant. During the period of resistance after a haemolytic incident a new population of young red cells becomes released into the circulation from the erythropoietic tissues.

Red cells contain about 50 mg per 100 ml of reduced glutathione (GSH). Although its function is still not clear, it appears that GSH exercises a protective rôle on the integrity of the erythrocyte by maintaining a reducing environment. GSH levels in the red cells of normal and primaquine-sensitive individuals are not very different. However, when sensitive individuals are treated with primaquine a significantly reduced GSH level is detectable in the red cells twenty-four hours before the onset of haemolysis, suggesting that the haemolysis is causally related to the decline in GSH. Indeed, an early diagnostic test for drug sensitive individuals involved the *in vitro* incubation of blood with acetylphenylhydrazine: a fall in the level of red cell GSH indicated a sensitive individual.

In the ordinary course of metabolism in the red cell there is a slow oxidation of GSH to its oxidized form, GSSG, but since the latter is promptly reduced again by the enzyme glutathione reductase:

$$\text{GSSG}+\text{NADPH}_2 \longrightarrow 2\ \text{GSH}+\text{NADP},$$

the erythrocyte GSH levels remain more or less constant. Primaquine greatly increases the rate of GSH oxidation, but in normal individuals the rate of reduction of GSSG by the above reaction is correspondingly stepped up so that the GSH levels remain unchanged. In primaquine sensitive individuals however, the reaction cannot be stepped up to cope with the increased demand. Since there is no evidence for any impairment of the catalytic activity of the glutathione reductase in sensitive erythroctyes, one has to conclude that there must be a lack of available NADPH_2.

The maintenance of the levels of NADPH_2 in red cells is brought about by two successive enzymic steps in the phosphogluconate pathway:

$$\text{Glucose-6-phosphate}+\text{NADP} \xrightarrow{\text{G6PD}} \text{6-Phosphogluconate}+\text{NADPH}_2$$

$$\text{6-Phosphogluconate}+\text{NADP} \xrightarrow{\text{PGD}} \text{Ribulose-5-phosphate}+\text{NADPH}_2+\text{CO}_2$$

However, since the substrate for the PGD reaction is the product of the G6PD reaction, the control of NADPH_2 production is effectively vested in the former reaction. In 1956 Carson and colleagues found that the activity of the G6PD in primaquine-sensitive erythrocytes was severely reduced and constituted the basic cause of primaquine sensitivity.

There is no evidence for the existence of a second G6PD locus. Yet cells other than erythrocytes (e.g. white cells) from subjects with the African variety of G6PD deficiency show no significant degree of deficiency. The reason for this appears to be that, like the Whitechapel allele for PGD discussed earlier, the allele for the African variety of G6PD deficiency is very labile. In erythrocytes, which remain functional in the circulation for several months after they have lost their nuclei and their capability for protein and enzyme synthesis, the G6PD level in deficient individuals soon falls, and the mean level settles at about 10% of normal, most of this being contributed by the younger cells. On the other hand, white cells possess a functioning mechanism for protein biosynthesis and are able to replace G6PD that has become inactivated, so the G6PD levels in these cells are normal or near normal. Thus, the African variety of G6PD deficiency might more properly be classified as an example of G6PD lability.

It has been shown that even in the absence of primaquine, erythrocytes from G6PD deficient individuals of the African type have a reduced cell life (of about 90 days). Presumably after this period the red cells have lost their G6PD activity altogether, and with it their ability to maintain their levels of $NADPH_2$ and GSH, even though they are not under stress. The normal erythrocyte, it will be recalled, is removed from the circulation at about 120 days (and there is no reason to believe that this removal is attributable to a lack of G6PD activity).

The Mediterranean variety of G6PD deficiency

Whereas erythrocyte G6PD levels of 5–15% of normal are found in the African type of deficiency, there is another variety of G6PD deficiency, occurring in Mediterranean regions, where the level of activity in the red cells of fully expressed individuals is only 1–3% of normal, and where there is also some reduction in the G6PD activity of the white cells.

The findings are consistent with the presence of a G6PD allele which produces an enzyme even more labile than that produced by the African allele, but which nevertheless has catalytic activity when freshly synthesized. Fully affected individuals are subject not only to drug induced haemolytic anaemia, but also to haemolytic episodes brought on by other causes such as viral infections, and during the neo-natal period. A particularly intriguing cause of haemolytic incidents in some (not all) of Mediterranean type G6PD deficient individuals is exposure to the broad bean (or fava bean), a condition known as favism. Favism may be precipitated by eating raw or cooked beans, or even by walking through fields of growing beans and inhaling the pollen. The agent in broad beans that causes the

Stanbury, A. B., Wyngaarden, J. B. and Fredrickson, D. S., Editors (1966). "The metabolic basis of inherited disease". 2nd Ed. McGraw-Hill, New York.
Strickberger, M. W. (1968). "Genetics". Macmillan, New York.
Wilkinson, J. H. (1970). "Isoenzymes". 2nd Ed. Chapman and Hall, London.
Yunis, J. J., Editor (1969). "Biochemical Methods in Red Cell Genetics". Academic Press, New York and London.

10. General Mammalian Cytogenetics

C. E. FORD
Sir William Dunn School of Pathology,
South Parks Road, Oxford

I. Introduction

Cytogenetics deals with the properties of chromosomes, their transmission from mother cell to daughter cells in development and from parents to offspring in heredity, and their influence on the expression of the phenotype. Its associations are therefore with genetics and embryology rather than cytology. It was founded on observations made on plants, invertebrates and to a lesser extent, lower vertebrates and now has a considerable history. Mammalian cytogenetics is a development of the last 15 years.

The remarkable uniformity of the essential events of mitosis and meiosis throughout the eukaryotes allows generalisations of very wide applicability. Observations made on a lily may illuminate a problem in man and vice versa. In the text that follows it has consequently sometimes been necessary and often more appropriate to illustrate principles by reference to information obtained from non-mammalian species.

Human cytogenetics, though by definition a part of mammalian cytogenetics, has a special place and a chapter to itself. But though its history is equally brief, it has already made unique contributions to general cytogenetics and it will be necessary to refer to some of them to ensure a balanced and, it is hoped, reasonably coherent account.

II. Mitotic Chromosomes and the Mitotic Cycle

A. Organisation of the Nucleus

The chromosomes are clearly visible by conventional microscopy only during cell division. Mitosis, the division of the nucleus, precedes cytokinesis, the division of the cytoplasm. When not in mitosis, that is during the greater part of its life, the typical cell contains a single approximately spherical nucleus, bounded by a nuclear membrane. Though invisible, the chromosomes are inferred to maintain their identity in an extended, hydrated form throughout the interphase between successive mitoses. The interphase nucleus may seem almost structureless but usually contains one or more darkly-staining bodies of varying size called *chromocentres*. The size of the nucleus and the number and character of the chromocentres vary considerably between the cells of different tissues. A single larger body, the nucleolus, is also usually visible. Some nuclei may contain more than one nucleolus. Histochemical tests show that the chromocentres contain deoxyribose nucleic acid (DNA) and the nucleoli, ribose nucleic acid (RNA).

B. MITOSIS

As mitosis commences the chromosomes become visible as long, narrow, loosely coiled threads which, in favourable preparations, are seen to consist of two strands. This is the prophase stage. As the cycle advances the chromosomes become progressively shorter and thicker until they appear as discrete, rod-like structures. At this stage the nuclear membrane disappears and a specially differentiated region of the cytoplasm, the spindle, develops. The chromosomes, now at their shortest, congress in the equator of the spindle. This is the metaphase stage. The two half chromosomes, (the future daughter chromosomes), are called *chromatids*: each consists of a closely-coiled, spiral *chromonema*. As soon as all are assembled, the two chromatids of each chromosome separate from one another and move to opposite poles of the spindle. This phase of polewards movement is the anaphase. It is followed by telophase, in which the chromosome coils relax, the chromonemata become more faintly stained and diffuse and finally attenuated, disappear, and a new nuclear membrane is formed. No description of mitosis and no still photographs can be adequate substitutes for a time-lapse film of the process.

To display the chromosomes of mammalian cells clearly special methods have to be used. It is true that much was learned of the structure of the large chromosomes of some plant and invertebrate species by the classical cytological procedures of fixing, embedding, sectioning and staining. The chromosomes of mammals, by contrast, are frequently numerous and rarely large, so that when assembled close together at metaphase it was difficult even to get correct counts, let alone observe details of individual morphology. The introduction some 15 years ago of three simple technical devices, two of them borrowed from plant cytogenetics, revolutionized the position and mammalian chromosomes suddenly became technically favourable material.

C. PREPARATIVE METHODS

A contemporary procedure would begin with the exposure of the cells to colchicine or a similar drug by injection into a living animal or by addition to the culture medium in which cells are growing *in vitro*. Colchicine prevents the formation of the spindle. This has three important technical consequences. First, the chromosomes become widely distributed in the cytoplasm instead of congregating closely together in the middle of the cell. Second, the chromosomes continue to contract, making them easier to count and observe. Third, since there is no spindle, the anaphase stage does not take place and the cells persist in metaphase, accumulating progressively as the duration of exposure increases.

The next step is to expose the cells to hypotonic fluid. If the cells to be studied are in a living animal, the animal must first be killed and dissected, and a suspension of free cells prepared from the chosen tissue, such as bone marrow. Hypotonic solutions of tri-sodium citrate or potassium chloride are commonly used, at about one quarter of normal tonicity. The cells imbibe water and the cytoplasm becomes less viscous. Next, the cells are fixed, usually in a mixture of three parts absolute ethanol to one part glacial acetic acid. The final step is to obtain good spreads of mitotic chromosomes. One method is to take cells dispersed in an aqueous medium, put a droplet between cover slip and slide and then squash by direct manual pressure between layers of filter paper. The alternative, and now more usual method, is to transfer droplets of volatile fixative containing suspended cells to a clean slide and simply allow them to dry in air. Much of the art of obtaining good preparations depends on the details of fixation and air-drying. Staining of the cells is usually done before squashing but necessarily has to follow the air-drying procedure. A mitotic cell at metaphase photographed by a 100× objective oil-immersion of a standard bright-field microscope is shown in Fig. 1.

Direct preparations can be obtained from bone marrow, spleen, lymph nodes, thymus, cornea, the basal layer of the skin, intestinal epithelium, ovary (granulosa cells) and testis. However, cells grown from small tissue explants in monolayer culture or from short term cultures of peripheral blood lymphocytes usually give rather more favourable results. To obtain mitoses in cultures of peripheral blood it is necessary to incubate the cells in the presence of phytohaemagglutinin (PHA) (Moorhead *et al.*, 1960). This is a substance extracted from the seeds of the bean, *Phaseolus vulgaris,* which has the remarkable property of stimulating many small lymphocytes to transform into large mononuclear cells which have the capacity to enter mitosis. Other *mitogens* which may stimulate different classes of lymphocytes, may be substituted for PHA.

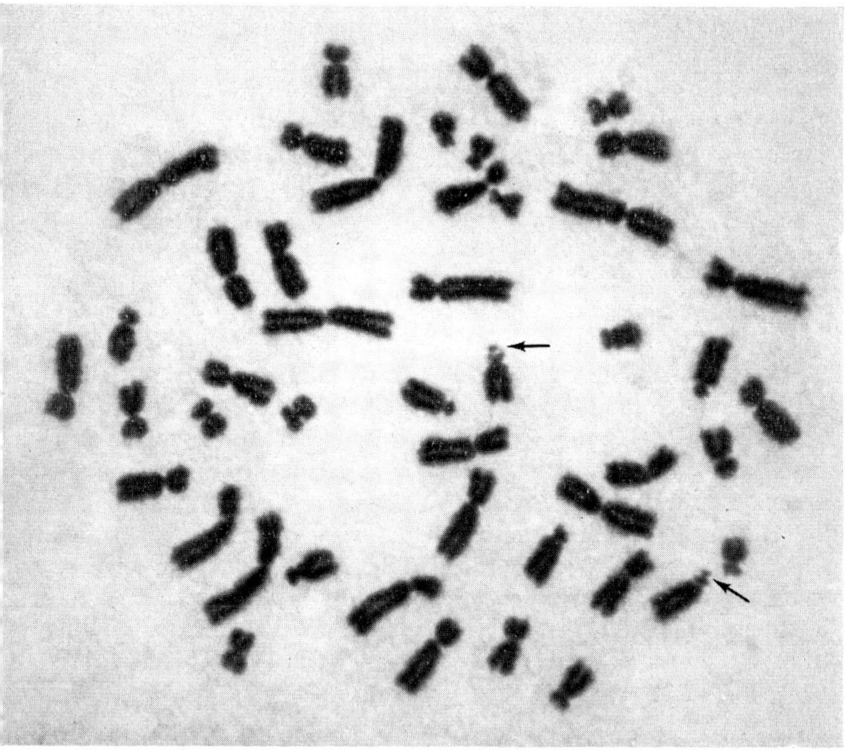

Fig. 1. Chromosome spread from a normal human male. Blood culture. Arrows indicate satellites. × 2 100.

D. NORMAL SOMATIC CHROMOSOMES

Figure 1 shows the chromosomes from a human male mitotic cell at metaphase. In a typical spread the chromosomes range from about 1·5μm to about 7μm in length. At one point in each chromosome the two chromatids appear to be constricted and attached together. This constricted region is the *centric constriction*. It bears an organelle, the *centromere,* which appears to divide as anaphase is initiated, the daughter centromeres then leading the way in poleward movement and drawing the arms of the daughter chromosomes behind them. The centromere separates the chromosome into two arms, and its relative position can be indicated numerically by the *arm-ratio,* the ratio of the shorter of the two arms to the longer. The term *telomere,* was introduced as a name for hypothetical bodies postulated to exist at the end of each chromatid. Accordingly, a chromosome at metaphase has one centromere and four telomeres. Individual chromosomes are distinguished primarily by relative length and the position of the centromere. When the centromere lies in the middle, or close to the middle, the chromosome is said to be *metacentric*. The centromeres of *acrocentric* chromosomes lie very close to one end; those of *sub-metacentric* and *sub-acrocentric* chromosomes lie in intermediate positions.

Two of the acrocentric chromosomes in Fig. 1 have pairs of small, rounded bodies attached to the short arm. These are *satellites*. Each is connected to its chromatid by a narrow filament. The filaments are thought to be *nucleolar organizers,* the sites at which nucleoli are formed in telophase. Points on the chromosomes (other than the centromeres) where both chromatids are narrowed are called *secondary constrictions*. Though they vary greatly in expression between one cell and another they tend to recur at the same positions in individual chromosomes and are therefore of value for chromosome identification.

The array of chromosomes shown in Fig. 2 was prepared from a photograph of the cell shown in Fig. 1 simply by cutting round the individual chromosomes, matching them in pairs and arranging them in order of diminishing length. Such a *karyotype* conveniently displays the chromosomes of a single cell. The number of chromosomes and their morphological features are normally constant from cell to cell within an individual and from individual to individual within a species. One karyotype may therefore be sufficient to demonstrate the chromosomes of an individual and one of each sex sufficient for the species. The term, *genome,* is used to connote the total of genetic material contained within the cell. *Karyotype* gives emphasis to number and morphology, *genome* to function.

According to a convention established at a meeting held in Denver in 1960 the human chromosome set is divided into seven groups now designated by the letters A–G (Fig. 2). the 23 pairs include 22 pairs of *autosomes* and a pair of *sex chromosomes,* X and Y in the male, two X chromosomes in the female. By convention also, the autosome pairs are numbered serially from 1–22 in order of length. At a further meeting (Chicago Conference, 1966) a shorthand system for the designation of human karyotypes was adopted (see also Chapter 11). Thus, 46, XY is to be read as 46 chromosomes with sex chromosomes X and Y, i.e. a presumptively normal male karyotype. The Chicago system also included methods for designating most classes of abnormal karyotypes. It has been widely accepted and is suitable for use in other mammalian species.

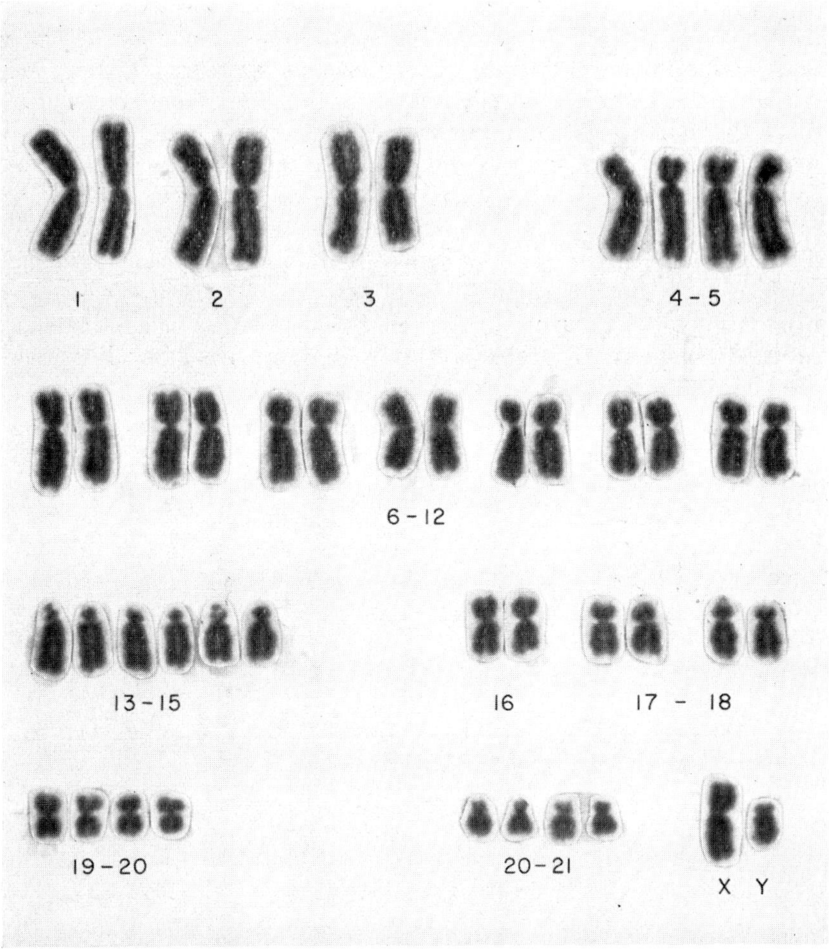

Fig. 2. Karyotype of a normal human male prepared from the spread shown in Fig. 1. Chromosome pairs 1, 2 and 3 form the A group; pairs 4 and 5 the B group; pairs 6–12 and X the C group; pairs 13, 14 and 15 the D group; pairs 16, 17 and 18 the E group; pairs 19 and 20 the F group; pairs 21 and 22 plus the Y chromosome the G group. X is not confidently identified.

E. BAND SEQUENCES

Only four of the human autosome pairs, numbers 1, 2, 3, and 16, can be distinguished individually with complete confidence by the simple criteria of length and morphology. However, newly introduced staining procedures have made it possible to distinguish all the individual pairs. The first of the new methods was to stain a standard air-dried preparation with an aqueous solution of a fluorochrome and then examine it with an ultraviolet microscope. It was observed that some chromosome regions fluoresced brightly, others less so. Each chromosome type was found to have its own specific pattern of brighter and darker fluorescent bands (Caspersson *et al.*, 1971; Paris Conference, 1971). Most distinctive of all is the Y chromosome, which has a particularly brilliant fluorescing region located distally in the long arm. The fluorescence staining method has also revealed a new form of polymorphism. For example, some individual No. 3 chromosomes have a very brightly fluorescing spot close to the centromere. This feature seems to be inherited in a normal Mendelian manner, so that some human subjects have two No. 3 chromosomes with bright spots, some have only one and some none at all. A similar property is shown by the short arms of some, possibly all, of the acrocentric chromosomes of the D and G groups.

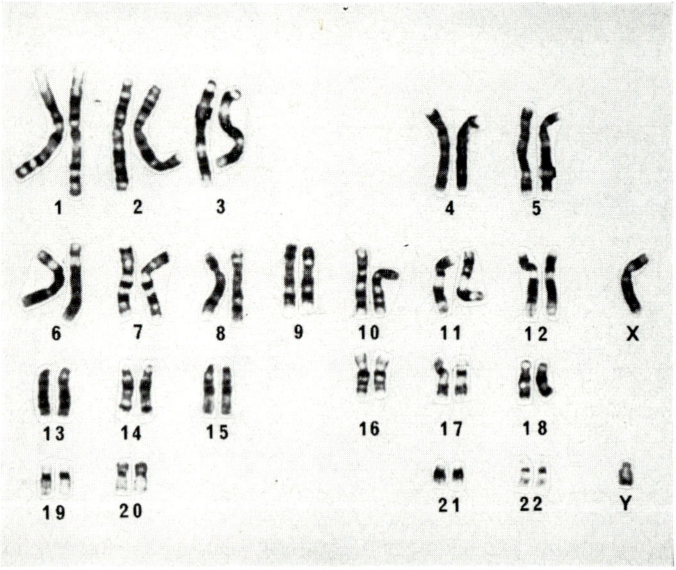

Fig. 3. Karyotype of a normal human male prepared from a chromosome spread "banded" by the trypsin method. (Prepared by Mrs. M. Seabright.)

More recently it has been found that apparently the same patterns of darker and lighter bands can be demonstrated by staining with Romanowsky-type stains following certain pretreatments. A simple method is to expose a standard air-dried preparation briefly to the digestive action of trypsin, and then to stain (Seabright 1971) with Leishman's stain. A karyotype prepared from a cell treated in this way is shown in Fig. 3.

Specific banding patterns are seen in the chromosomes of other mammalian species, both by fluorescence microscopy and after use of the special staining methods. The new methods are of particular value in the mouse, where the chromosomes are all acrocentric and at most three of the 19 autosome pairs can be distinguished at mitosis by standard staining procedures. All 19 pairs, as well as the X and Y chromosomes, can now be identified by their banding patterns and 14 of the 18 known autosomal linkage groups have been assigned to the chromosomes that bear them.

Nothing can yet be said about the nature of the bands revealed in this way, about what they represent in genetical or cytochemical terms, nor indeed whether identical patterns are produced when different methods are employed, (though it is already evident that there is a considerable degree of concordance). But their value for chromosome identification is unquestioned and in many instances more accurate definition of the segments involved in structural rearrangements of the chromosomes (to be mentioned later) should be possible.

F. POLYTENE AND LAMPBRUSH CHROMOSOMES

Evidence of the longitudinal differentiation of chromosomes is also provided by the specific frequencies of variable morphological features seen in the *polytene* chromosomes of dipteran larvae, in the *lampbrush* chromosomes of vertebrate (particularly urodele) oocytes, and commonly in bivalents at the pachytene stage of meiosis (see later). Polytene chromosomes are well known in species of the genus *Drosophila*. They are most prominent in the salivary glands but are also present in some other tissues. They are greatly extended, multistranded structures produced by the close lateral association of pairs of homologous chromosomes, in a manner comparable with the pairing that occurs in the prophase of meiosis (see later). But their most striking feature is the sequence of narrow, transverse, darker and brighter bands which in several species (notably *D. melanogaster*, the fruit fly so important in the early development of genetics) have been mapped in detail.

G. CHROMSOME STRUCTURE

The spirally coiled chromonemata making up the chromatids of normal metaphase chromosomes can be seen clearly in favourable preparations. Below this level of organization the structure of the chromatid has been a matter of controversy, some contending that it is multistranded, others that it is a single strand, though a single-strand model is more readily reconciled with the basic facts regarding recombination and mutation established by genetics. It might be thought that all questions of fine structure of chromosome would have been resolved by the application of electron microscopy. It is true that many studies of the cell nucleus and of individual chromosomes have been made by electron microscopy but it has proved extraordinarily difficult to arrive at an integrated interpretation of the photographs obtained. Valuable information has been secured by spreading individual interphase nuclei on surface-active films and picking them up from below on suitable grids and there seems to be general agreement that the chromosomes in the interphase nucleus consist largely, if not entirely, of fibres of approximately 20nm diameter. The central problem now is how these fibres are packed into the chromosomes as they are seen at mitotic metaphase by conventional optical microscopy.

H. CHROMOSOME CHEMISTRY

Chemically, the chromosomes consist largely of DNA together with basic proteins of relatively low molecular weight called histones, more complex acidic proteins and a certain amount of RNA. Whichever model of the structure of the chromonema proves to be correct it is the long, fibre-like molecules of DNA that endow it with structural continuity and carry the genetic message. The DNA molecules are spirally intertwined twin helices, each helix being a linear copolymer of mononucleotides. Each mononucleotide consists of three parts, a heterocyclic base, a sugar group (deoxyribose) and an acid group (orthophosphate). Four kinds of heterocyclic bases are included : adenine and guanine are purines, thymine and cytosine are pyrimidines. The discovery, approximately 20 years ago, that adenine and thymine occurred in DNA in nearly equal molar proportions and likewise guanine and cytosine provided one of the important clues that led to the eventual solution of the structure of the DNA molecule. It is now known that the mononucleotides are linked together by their phosphate groups to form a backbone for each strand of the double helix. The base group of each mononucleotide residue faces inwards and is linked by hydrogen bonds to the base group of the corresponding mononucleotide in the other strand. Adenine residues in the one strand are linked to thymine residues in the other and guanine residues are linked to cytosine residues. Twin-helix DNA molecules are therefore linear sequences of mononucleotide pairs, AT and GC. Genetic experiments with microorganisms have now established that the genetic information carried by DNA molecules resides in the specific linear sequences of these base pairs.

It is now generally believed that individual segments of the DNA strands are *transcribed* into high-molecular-weight messenger RNA which then passes through the nuclear membrane into the cytoplasm, proceeds to small bodies called ribosomes and is there *translated* into polypeptide chains which fold in specific three dimensional forms to produce the proteins of the cell. (See Chapter 4 for a detailed account of the chemistry of chromosomes.)

The reproduction of the chromosomes is related to the replication of DNA, which takes place during part of interphase referred to as S (synthesis phase). Interphase is therefore divided into three periods. The first, which succeeds telophase, is designated G1; S and G2 follow, after which prophase of the next mitosis begins. The times occupied by G1, S and G2 vary greatly according to cell type, to tissue site within the animal body and to conditions of growth if in culture. Rapidly dividing cells, like many in bone marrow of the living animal and in culture, may have a total cycle time of 24 hours or less. (See also Chapter 13.)

I. AUTORADIOGRAPHY OF DNA REPLICATION

In DNA replication the twin-helices separate into their two constituent strands and a new complementary strand is assembled for each, possibly by the successive addition of single nucleotides with the old strand serving as a template (Chapter 4). Much has been learnt about the time and rate of this process by the use of radioactive isotopes, originally ^{32}P and ^{14}C, more recently tritium (^{3}H). A powerful method of investigation depends on the synthesis of DNA containing tritiated thymidine. The basic procedure is to expose living cells to tritiated thymidine, either in the animal itself or in culture, for periods of up to 8 hr, during which time radioactive thymidine molecules are incorporated direct into chromosomes at sites where DNA synthesis is in progress. Preparations are made in the normal way, coated with an appropriate photographic emulsion, stored in the dark for from two to three days to several months and then developed. The atomic disintegrations are associated with the emission of soft β rays which only travel 1 or 2 nm and therefore leave latent images in the close vicinity of the atoms that disintegrated. These are converted into silver grains on development and their positions can be seen by direct microscopical observation.

The first cells to reach metaphase after exposure to tritiated thymidine will have been in mitotic prophase or G2 when the exposure began and will not show in the autoradiograph. The next cells to arrive would have been in late S phase and attention has been particularly concentrated upon them. Some chromosome segments have associated silver grains. The positions of these segments constitute the so-called late-labelling pattern, which is specific for each individual chromosome type, at least in the human karyotype. Autoradiography with tritiated thymidine therefore provided a means of specific identification of individual chromosomes before the new banding techniques were introduced. One of the two X chromosomes in females labels particularly heavily (Fig. 4) and is referred to as the late-labelling or "hot" X chromosome (see also Chapter 12).

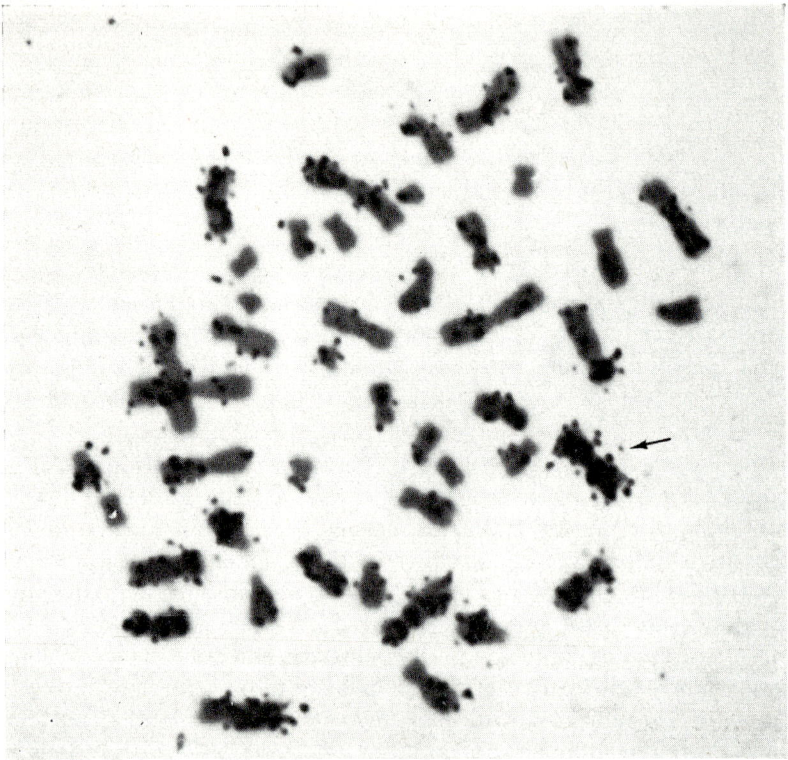

Fig. 4. Autoradiograph of a chromosome spread from a normal human female. Blood culture harvested six hours after addition of tritiated thymidine. Arrow points to "hot" X chromosome. × 2 300.

Two other special properties of X chromosomes in females require mention. One of the two is often *heteropycnotic* in late prophase, that is, it is more strongly condensed than the other chromosomes and more heavily stained. This chromosome and the late-labelling chromosome are believed to be identical and to be represented in many interphase nuclei by a large, darkly-staining chromocentre, unique to the normal female and called the *sex chromatin*. All three properties are thought to be different visible manifestations of the genetic inactivity postulated by Lyon for one of the two X chromosomes in normal mammalian females. (See Chapter 12.)

J. DNA FIBRE AUTORADIOGRAPHY

A new technique that employs autoradiography in a different way promises to reveal much about absolute rates of DNA synthesis and about the number and relative locations of the sites at which DNA synthesis is initiated (Huberman and Riggs 1966). DNA fibres are prepared from cells that were killed a known time after they had been supplied with tritiated thymidine. Autoradiographs then reveal the points in the fibres at which DNA synthesis was in progress when the cell was fixed. If DNA synthesis had already commenced at the time the tritiated thymidine was added, each site of initiation of synthesis on the fibre is marked by two regions of autoradiograph separated by a gap, the size of the gap being a function of the time interval between the commencement of synthesis and the addition of tritiated thymidine. But if synthesis began after the time when tritiated thymidine became available then each initiation site is marked by a single linear autoradiograph. That DNA synthesis procedes outwards from these points of initiation at equal rates in both directions is demonstrated by using such devices as dilution of the label at a certain point in time, the time of dilution then being marked by a sudden change to a less intense autoradiograph (Callan 1972). The first results show that points of initiation are evenly distributed along the DNA fibres, that specific rates of synthesis are effectively constant at constant temperature, and that differences in overall rates of DNA synthesis are the consequence of change in the number of sites of initiation.

K. *In situ* AUTORADIOGRAPHY

Another powerful new technique arises out of the steady development of methods for the extraction and fractionation of DNA from cell nuclei. Density gradient centrifugation of mouse DNA, for example, has revealed a band of somewhat lesser density which can be separated off from the main band. This is known as *satellite* DNA The idea apparently occurred to several investigators simultaneously that if the DNA in the chromosomes of conventional preparations could be separated into its two constituent strands *in situ*, then it might be possible to react the preparations with single-stranded DNA of specific types in such a way that restoration of the twin-strand structure might occur through the union of single strands of the applied DNA with single strands of the chromosomal DNA at specific sites of homology. Such *in situ* hybridization between applied DNA and chromosomal DNA was first achieved with mouse satellite DNA. The satellite DNA was radioactive, having been prepared from mice previously injected with tritiated thymidine. The chromosomes on the preparation were non-radioactive. Proof of *in situ* hybridization was provided when autoradiographs were made and silver grain were found over the centromere regions of all the chromosomes except the Y chromosome (Jones 1970; Pardue and Gall 1970). Evidently the satellite DNA of the mouse comes from the regions of the chromosomes around the centromeres of the autosomes and the X chromosome. Doubtless much more will soon be learned about the location of specific kinds of DNA on chromosomes by the use of this method.

L. "HETEROCHROMATIN"

The student may encounter the term "heterochromatin" in his reading and be puzzled. The natural meaning of the word is "chromatin that is different from the normal chromatin" (euchromatin), but it has been employed for so many differential properties of whole chromosomes or chromosome segments that the precise meaning intended invariably has to be inferred from the context if it can be understood at all. There is less objection to the adjectival form "heterochromatic".

III. Meiosis

A. GERM CELLS

The germ cells of mammals first appear in the yolk sac endoderm and migrate from their sites of origin to the gonad rudiments, the gonadal ridges, which lie in the dorsal mesoderm close to the mesonephroi. For a time the primitive germ cells multiply by mitotic division in the still indifferent gonads. Subsequent to the development of specific testicular and ovarian architecture the germ cells are known as *spermatogonia* and *oogonia* respectively. Increase in the number of oogonia ceases, perhaps in all mammals, before birth. Spermatogonial multiplication however, may continue throughout life. Eventually spermatogonia and oogonia transform into *spermatocytes* and *oocytes* and the first of two successive nuclear divisions, both of which are exceptional, is initiated. The two divisions jointly constitute the process of meiosis and have the effect of reducing the somatic, *diploid* number of chromosome to the gametic, or *haploid* number. The most distinctive meiotic stage is *diakinesis* (Fig. 5).

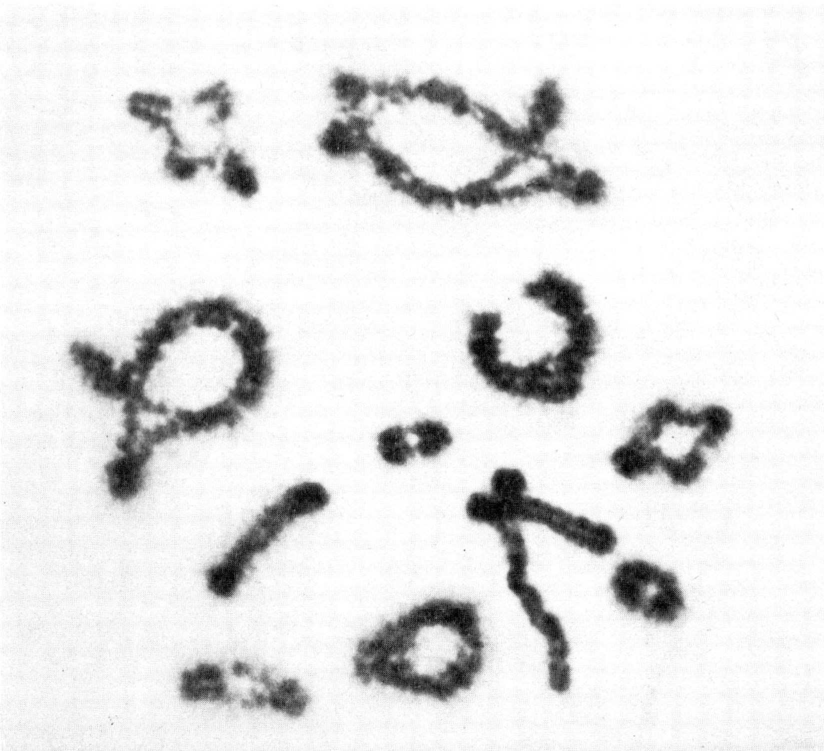

Fig. 5. Primary spermatocyte of a Chinese Hamster at diakinesis. Some of the chiasmata show clear exchanges of partner chromatids. The unequal bivalent is the sex chromosome bivalent, XY. (Preparation by Dr. E. P. Evans.) × 2 400.

B. THE COURSE OF MEIOSIS

The chromosomes of the nucleus entering first meiotic prophase first appear as long, slender, apparently single strands (*leptotene* stage). The thin leptotene strands come to lie closely side-by-side in homologous pairs (*zygotene* stage), the homology sometimes being indicated by the presence of matching couplets of dark granules varying in size and degree of linear separation, the *chromomeres.* The process of pairing (*synapsis*) is inferred to involve specific attractions operating at a distance. These two earliest stages of meiotic prophase were originally described from plant and invertebrate material; they have not been defined in mammalian species, though the approach and eventual close association of pairs of darkly-staining bodies, presumptively homologous centric regions, can be seen in spermatocytes of the mouse (Fig. 6).

The pairs of homologues, called *bivalents,* become progressively shorter and thicker. At this stage, *pachytene*, the homologous

6

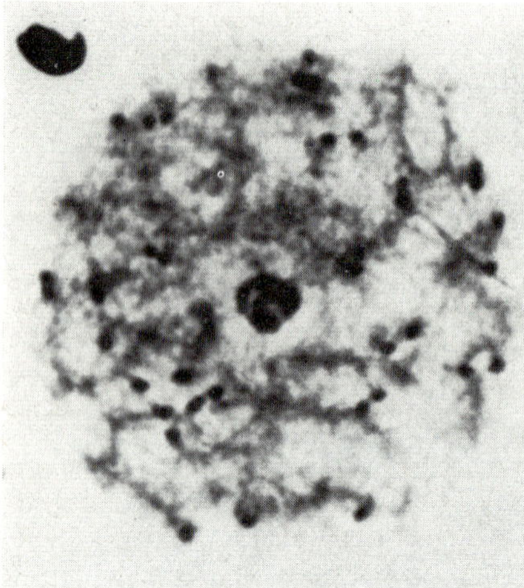

7

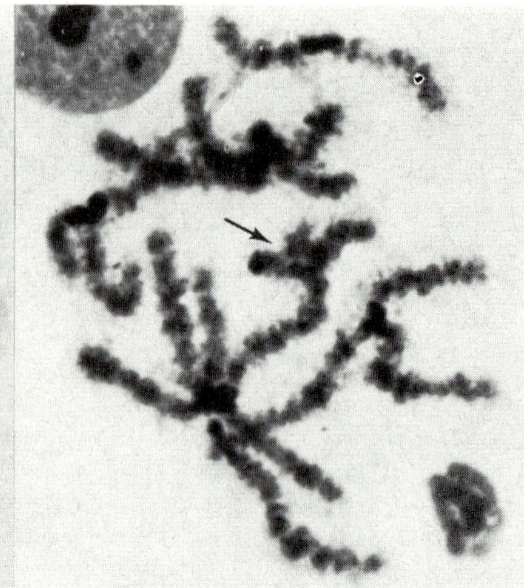

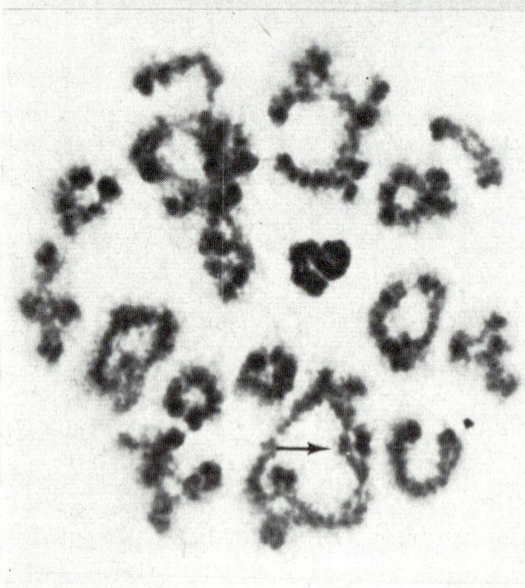

8a

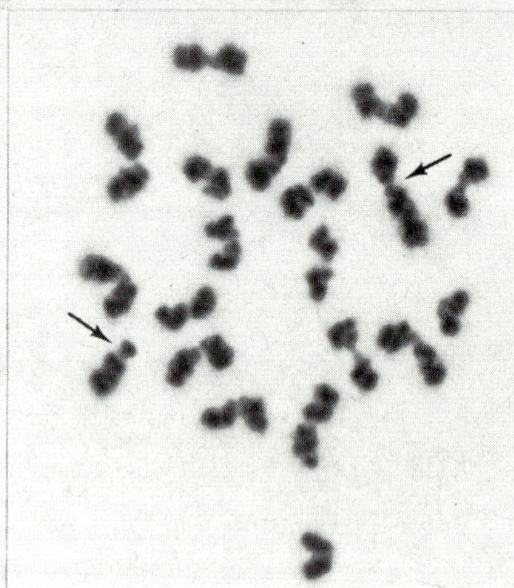

9

Fig. 6. Primary spermatocyte of the mouse at zygotene. Note the pairs of darkly staining bodies. There are 17 pairs and 4 single bodies presumptively representing the centric regions of the 19 autosome pairs. × 1 400.

Fig. 7. Primary spermatocyte of a mouse heterozygous for the translocation T26H at pachytene. The centric regions remain darkly staining but homologues are now so closely associated that they can no longer be resolved into pairs as in Fig. 6. The arrow points to the cross configuration indicative of heterozygosity for a reciprocal translocation. Note the association of bivalents at their centric regions to give clusters of 6, 4, 3, 3 and 2 bivalents. The structure at lower right is the sex chromosome bivalent. × 1 500.

Fig. 8a. Primary spermatocyte from a mouse heterozygous for the T26H translocation at late diplotene. Most of the darkly stained centric regions are still visible and some are now resolvable into pairs of bodies. The translocation ring quadrivalent lies at lower centre. Two interstitial chiasmata lie close to the centric regions and alternate in the ring with two terminal chiasmata. Darkly staining telomeric bodies, similar in appearance to the centric bodies, are present in some chromosomes of some animals; this animal was presumptively heterozygous in respect of the size of the telomeric body on one of the chromosomes involved in the translocation. The arrow points to the structures concerned. Had the terminal attachment at this point arisen as a terminal association *ab initio* telomeric bodies of the same size would have been expected to lie side-by-side and not end-to-end. The end-to-end arrangement is therefore evidence that the association has been produced by terminalisation of a chiasma that arose in an interstitial position. × 1 000. (See pp. 84 and 92 for *translocation*; p. 93 for *quadrivalent*.)

Fig. 9. Secondary spermatocyte from a mouse heterozygous for the T70H translocation at metaphase. Arrows point to the centromeres of two unequal dyads produced as a result of crossing over in the interstitial segments of the translocation quadrivalent. × 1 800.

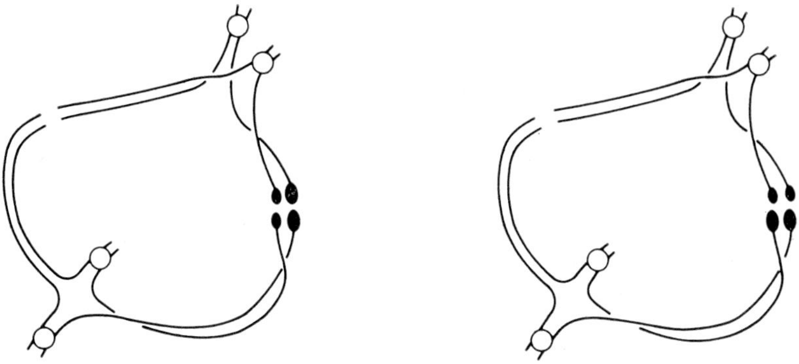

Fig. 8b. Chromatid diagram of the ring quadrivalent shown in Fig. 8*a* (left) and the expected appearance in a similar ring quadrivalent if the association involving the pairs of larger and smaller chromatin bodies had been terminal *ab initio* (right).

chromosomes are so closely paired that the dual nature of the bivalents may not be apparent (Fig. 7). In particularly favourable preparations it is possible to make out a linear pattern of larger and smaller chromomeres that is specific for the individual bivalent. These form the basis of the pachytene maps of the autosomes of the mouse and of some human chromosomes that have been prepared (Hungerford 1971). The centric regions commonly stain more heavily; they also frequently act as foci of association, yielding clusters of bivalents with arms radiating from a common centre like the spokes of a wheel (Fig. 7). Such clusters are specially common in the mouse. Judged by the very large proportion of primary spermatocytes seen in pachytene, this stage is a relatively long one. Eventually the pachytene bivalents appear to separate, though incompletely, into their constituent chromosomes and are now seen to be composed of pairs of *double* strands (*diplotene* stage). At the points of incomplete separation the unit strands exchange partners (Fig. 5). These points of exchange are called *chiasmata*. In the mouse there are usually one or two per bivalent but the numbers are generally much greater in man and there may be five or more in the larger human bivalents. Where there is more than one, the loops between successive chiasmata rapidly expand to almost circular form and lie in planes at right angles to one another. For a long time it was thought that the reproduction of the chromosomes occurred at the end of pachytene and was the signal for the initiation of diplotene "looping". But it is now known that DNA synthesis, and therefore presumptively, chromosome reproduction, occurs before the onset of leptotene.

Even in early diplotene some bivalents may show one or more terminal associations in the place of, or in addition to, interstitial chiasmata. It is common to refer to such terminal associations as terminal chiasmata. This is because there is reason to believe they arise from previously existing interstitial chiasmata by progressive movement of the pairs of strands in relation to one another, a process called *terminalization*. Qualitative evidence in favour of the interpretation of the terminal associations as terminal chiasmata is provided by Figs. 8*a* and *b*.

The bivalents become more contracted and more heavily staining. This stage, diakinesis, the final stage of meiotic prophase, is not distinguished from diplotene by any distinctive change in bivalent structure. Contraction continues further and reaches its maximum at metaphase. Further evidence of terminalization is provided in some species by reduction in the total number of chiasmata from diplotene to diakinesis and again from diakinesis to metaphase, coupled with an increase in the proportion of chiasmata that are terminal. The stage of metaphase is normally defined by the aggregation of the bivalents at the centre of the spindle. This is rarely seen in mammalian cells because the employment of hypotonic pretreatment prevents the formation of the spindle. Metaphase therefore has to be judged arbitrarily by the degree of contraction of the bivalents.

First meiotic metaphase-anaphase is quite distinctive. Whereas all the centromeres at mitotic metaphase are located in the *equatorial plane* before anaphase separation of the daughter chromosomes or *disjunction* begins, the two centromeres of each bivalent at first meiotic metaphase are positioned symmetrically above and below the equatorial plane. Moreover the disjoining chromosomes are composed of *two* chromatids and for that reason are sometimes called *dyads*. There is little doubt that irregularities of disjunction are more frequent at first meiotic anaphase than at mitotic anaphase. The interphase that follows, if it is rightly called interphase at all, is brief and atypical: the chromosomes may not relax completely before the onset of the ensuing prophase and there is no S phase. The second meiotic division presents two unique features: the spiral chromonemata tend to be particularly prominent; and the chromatids diverge from one another so strongly that in an acrocentric chromosome they may lie in nearly a straight line and in a metacentric chromosome take the form of a cross.

C. SPERMATOGENESIS AND OOGENESIS

In the tubules of the adult testis a series of mitotic divisions of spermatogonia gives rise to primary spermatocytes in which the two meiotic divisions occur, yielding first two equal secondary spermatocytes and finally four equal spermatids, that differentiate into spermatozoa (spermiogenesis). The primordial germ cells in the ovary are oogonia. During foetal life they proliferate and form primary oocytes. These are surrounded by flattened cells forming the so-called primordial follicles lying evenly scattered just below the ovarian epithelium.

Whereas the male mammal has a continuously replenished supply of spermatogonia and spermatozoa, the female's store of oocytes diminishes progressively throughout reproductive life. Multiplication of oogonia and the initiation of meiosis is completed during fetal life in the mouse but may be delayed until about the time of birth in some mammalian species.

Meiosis thus commences early in primary oocytes but it is later arrested and resumed (if at all) only during reproductive life. It is said to reach diplotene before it is interrupted but convincing evidence is still lacking. Whatever the stage of interruption, the chromosomes relax again and give an appearance resembling an interphase nucleus (the *dictyate* stage), though the degree of relaxation seems to differ considerably between different species. The oocyte remains in this state until (should it be one of the chosen few) the meiotic process is resumed in adult life. In the human female the dictyate stage may persist for 40 years or even more. Most oocytes, however, eventually undergo atresia and disappear. This is a progressive process, commencing at birth and continuing throughout life.

The first meiotic division takes place about the time the oocyte is released and starts to find its way into the oviduct. The products of this division are unequal, a large secondary oocyte and a minute first polar body. This consists of a nucleus, genetically equivalent to the nucleus of the secondary oocyte, and a negligible amount of associated cytoplasm. The second meiotic division in the secondary oocyte commences almost immediately and there is no true interphase. In most mammals it proceeds as far as metaphase, at which point it is arrested and is only resumed after the penetration of the sperm. After the completion of the second meiotic division one daughter nucleus remains in the egg and the other becomes eliminated with a negligible amount of cytoplasm as the second polar body. The nucleus of the first polar body may also divide, but usually does not. In any case the polar bodies have no genetic future and soon break down and disappear. The final product of oocyte meiosis is therefore a single haploid gametic nucleus comparable with the haploid spermatid nuclei of the male.

D. THE CHIASMATYPE HYPOTHESIS

The parallelism between chiasmata seen in meiosis and crossovers inferred in genetic experiments was recognized long ago and the hypothesis that each chiasma represents a single genetic crossover was put forward. It was supposed that one of each pair of sister strands breaks at homologous loci and that cross reunion of the broken ends follows. In view of the terminal movement that chiasmata are believed to undergo (see above) the positions of the chiasmata as seen in diplotene cannot be assumed to be the actual sites at which crossing-over has occurred. That at least some chiasmata must represent true physical crossovers can be inferred directly from observations at second metaphase in some types of structural heterozygotes (see below). The unequal chromosomes in the second metaphase shown in Fig. 9, for example, must have arisen from crossing over in homologous proximal segments of partially homologous longer and shorter chromosomes in the preceeding first meiotic division.

The amount of recombination observed between pairs of linked genes is the basis for the construction of linkage maps. In the simplest type of experiment an individual with two dominant genes, *AB*, on one chromosome and the corresponding recessive alleles, *ab*, on its homologue would be crossed with individuals that are homozygous recessive, *aa bb*, at both loci. In the absence of viability differences between the different genotypes the progeny would be expected to consist of equal numbers of non-recombinant types, *Aa Bb* and *aa bb*, and equal, but smaller, numbers of recombinant types *Aa bb* and *aa Bb*. The simple empirical measure of recombination is then taken as the ratio of both classes of recombinant progeny combined to the total number of progeny observed, expressed as percent.

The loci are said to be closely linked if there is relatively little recombination between them and to be loosely linked if the amount of recombination approaches 50%, in other words, independent segregation. The greater the recombination frequency, the larger the body of data required to establish whether two genes are linked or not and some reports of loose linkage were subsequently shown to be spurious. One of the triumphs of classical formal genetics was to show that linkage data were consistent with a series of linear, unbranched gene sequences, the number of sequences (linkage groups) maximally equalling the haploid number of chromosomes.

When the data for three successive loci are considered it is almost invariably found that the frequency of recombination between the two outer loci is less than the sum of the recombination frequencies between each outer locus and the middle locus. This effect was attributed to the occasional occurrence of double crossovers, the second nullifying the first and resulting in no net recombination between the two outer loci, and was given the name *coincidence*. The assumption of double crossing over was verified when recombination between three or more loci was investigated simultaneously.

If crossing over occurs independently in adjoining chromosome segments the frequency of simultaneous occurrence in both segments (i.e. of double cross overs between the two outer loci) would be expected to be equal to the product of the frequencies of recombination in the two segments considered separately. In fact, the observed frequency of double crossovers in three-point and multi-point experiments is almost invariably found to be less than this. One crossover evidently reduces the likelihood of occurrence of another crossover in its neighbourhood. This phenomenon is known as interference and is measured by the coefficient of coincidence, namely, the ratio of observed double crossovers to expected double crossovers on the assumption of independence. The numerical value of the coefficient of coincidence may vary between 0 and 1, the value 0 corresponding with total interference and the value 1 with absence of interference. It will now be evident that the observed recombination frequency should be less than the true map distance, as measured by the total frequency of crossovers. Several mathematical mapping functions, differing slightly from one another according to the assumptions made regarding the nature of interference, have been put forward for the conversion of observed recombination frequencies into map distances. Such transformations may be useful for some comparative purposes but linkage maps are commonly constructed on the basis of empirical recombination frequencies observed between loci that provide good phenotypic marker features, using closely linked loci as far as possible to minimize the effect of interference. The unit of genetic map distance is the centimorgan (cM) and is defined as 1% true frequency of crossing over. The total genetic length of a whole genome may be expressed in the larger unit, the morgan (M).

Bias associated with ascertainment and sampling, the problem of combination of small bodies of data and probably the further complication of heterogeneity between different families have combined to make the construction of linkage maps of the human genome extraordinarily difficult. Nevertheless, considerable progress is now being

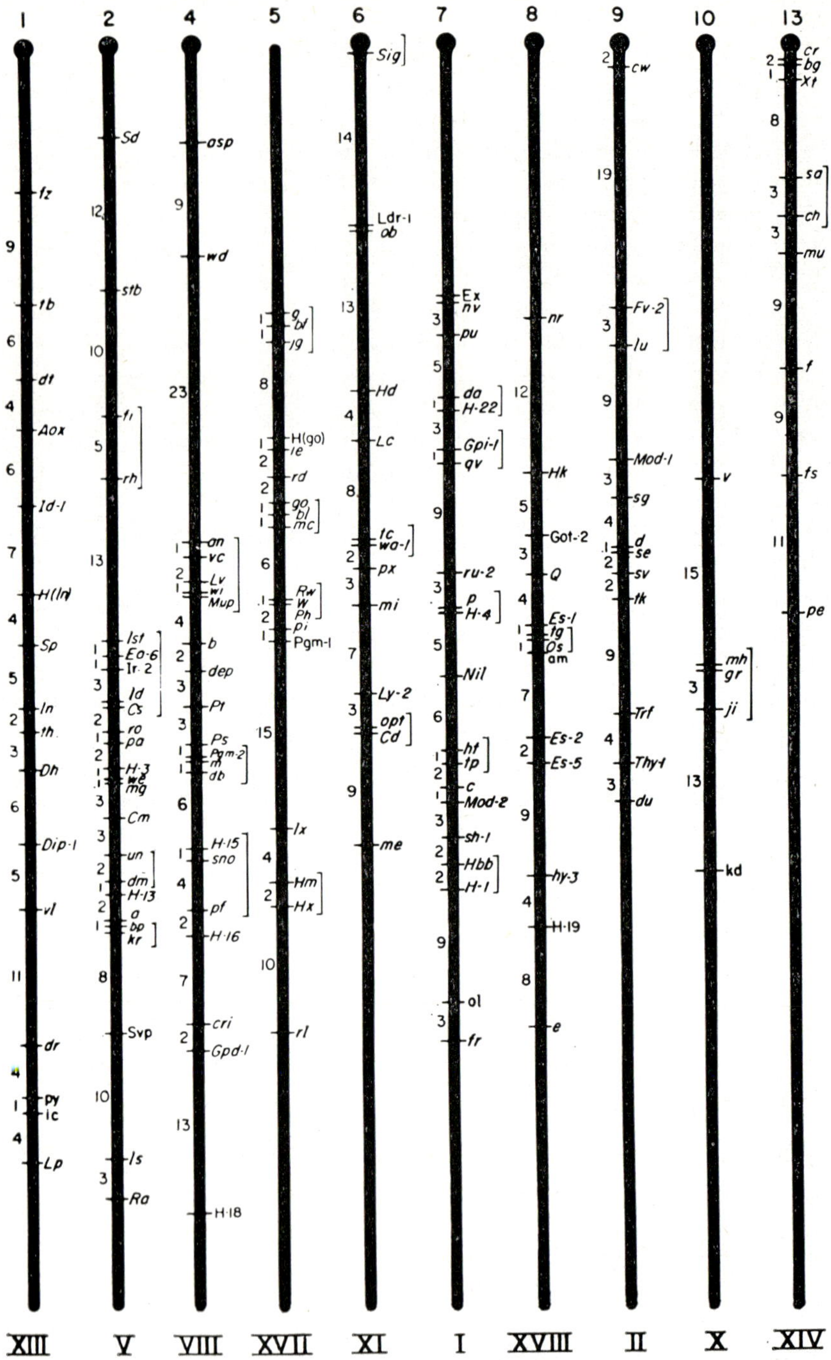

Fig. 10. Linkage map of the mouse prepared by Dr. M. C. Green (1972). Arabic numerals at top are chromosome numbers. Roman numerals at bottom are linkage group numbers. Map

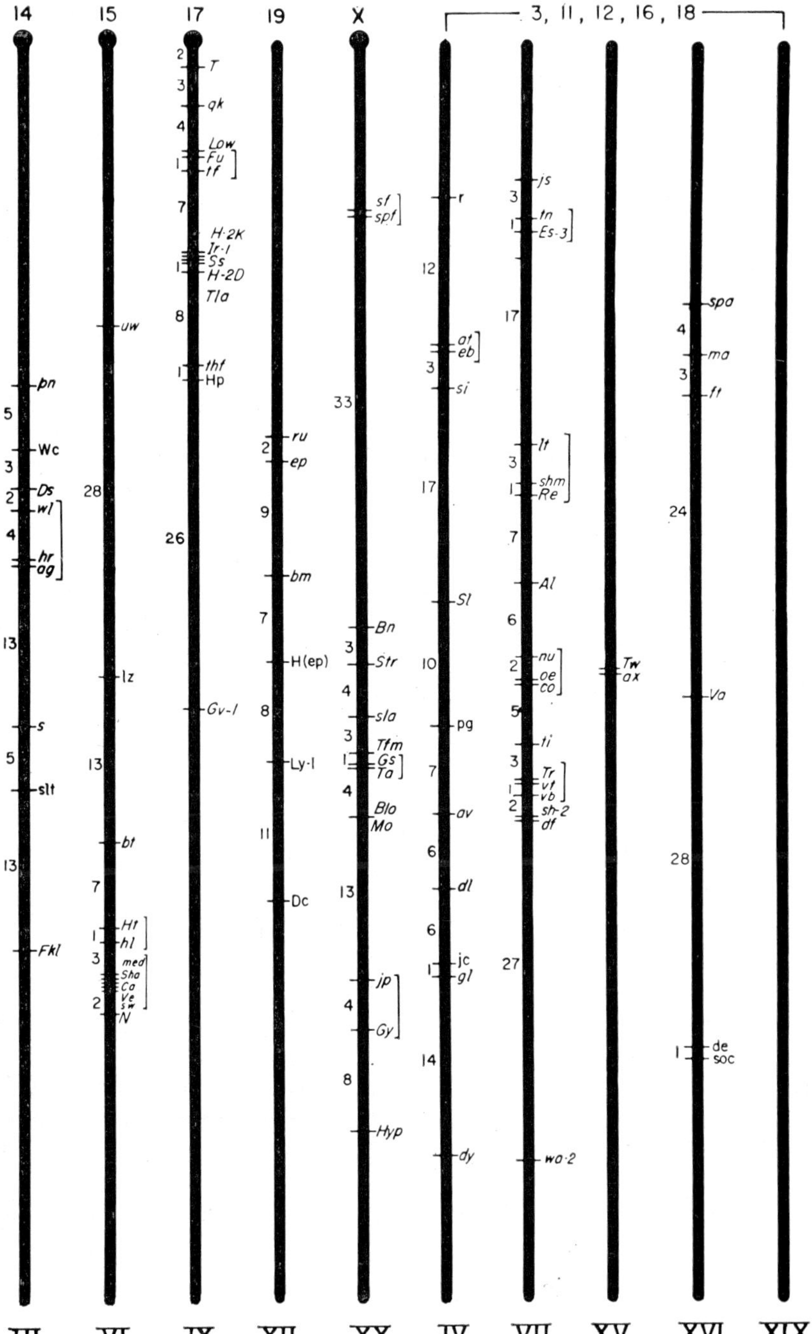

distances in recombination units are shown on the left of each chromosome. Brackets indicate that the order within the bracketed groups has not been established.

made. Two groups of X-linked genes, centered round the *Xg* blood group locus and the two closely linked colour-blindness loci have now been mapped and 10 autosomal linkage groups are known (Renwick, 1969).

Quantitative data bearing on the relationship between chiasmata and genetic crossovers have not yet been published for any mammalian species. In the mouse, however, circumstantial evidence favouring a one-to-one relationship is strong. The mean number of chiasmata in the autosomes at diplotene is approximately 30 in oocytes and approximately 25 in spermatocytes, counting the terminal associations as chiasmata. Each chiasma corresponds with 50 cM of map distance (since the exchange is between two of the four strands) so the total expected autosomal map length in the mouse would be approximately 15 M in the female and 12·5 M in the male. Eighteen autosomal linkage groups have now been established and the total length mapped amounts to approximately 9·5 M (Fig. 10) (Green, 1972). A recent tabulation of data has shown that when recombination in defined segments of the linkage groups in the two sexes is compared, there are 19 segments with significantly more recombination in the female than in the male, six segments where the difference is not significant and only three in which recombination is more frequent in the male than in the female (Dunn and Bennet, 1967). The longest linkage group (83 cM) in the present map is group XIII and it has recently been established that this group is carried by the longest autosome. The large bivalents at male meiosis commonly have two chiasmata, sometimes one and occasionally three. The expected maximum map length would therefore be approximately 100 cM for the longest chromosomes of the male mouse.

The frequency of chiasmata per bivalent in the human male is much greater than in any other mammalian species yet studied. Five are frequently present in the larger bivalents and more have been seen. The mean total frequency per spermatocyte is approximately 55, corresponding to a map distance of 27·5 M for the whole genome. Although the amount of DNA in the human diploid nucleus (6×10^{-12}gm) is almost the same as in the mouse (and mammals generally) the total genetic length of the genome is about twice as great in man as in the mouse. Linkages would therefore be expected to be relatively looser in man, increasing still further the difficulties of mapping already mentioned. No satisfactory counts of chiasmata in human oocyte meiosis have yet been obtained but recombination is considerably greater in the female wherever sufficient data has been assembled to permit an efficient comparison between the sexes.

IV. Chromosome Abnormalities—Numerical

A. POLYPLOIDY

Mitosis and meiosis are normally remarkably regular processes and the chromosome complement shows a high degree of constancy. Nevertheless abnormalities do occur from time to time and are sometimes perpetuated. They are conveniently divided into numerical abnormalities, arising very largely as a consequence of errors at anaphase, and structural abnormalities, arising almost exclusively during interphase.

The first chromosome abnormalities were identified in the early years of the century through the examination of phenotypically abnormal plants of the species *Oenothera Lamarckiana* (evening primrose), already famous as the species whose capacity to produce abnormal individuals (mutants) provoked de Vries to write his classic work, Die Mutationstheorie. De Vries thought he had seen the essential step in the origin of new species, not by the Darwinian accumulation of small changes but by a sudden, one-step change involving many characters simultaneously, though many of the new forms showed reduced fertility and most did not breed true.

Among the abnormal plants some were taller and altogether more robust than normal individuals of the species. Eventually it was found that these plants had 28 chromosomes instead of the normal diploid number of 14. Crossing such a *tetraploid* mutant with normal plants gave *triploids* with 21 chromosomes. Plants or animals (or single cells) with exactly three or more times the normal haploid number of chromosomes are said to be *polyploid*. The symbol n has been adopted to designate the haploid number of chromosomes, so that $2n$, $3n$, $4n$, indicate diploid, triploid and tetraploid numbers respectively.

Polyploidy is widespread, particularly among flowering plants, where polyploid series of species within a genus are common, some of the species having up to ten or more times the basic number of the series. Many of these species appear to have originated from natural interspecific hybrids by doubling of the chromosome set. Such *allopolyploids* are usually fully fertile although the hybrids from which they are presumed to have arisen may have had very low fertility. This is because meiotic abnormalities attributable to lack of full homology between the chromosomes of the two parental species are common in hybrids, whereas after doubling, each chromosome has an exact homologue to pair with, segregation is regular and fertility is restored.

Proof of the origin of some new species in this way provides a certain vindication for the views of de Vries, despite the fact that the principle is of limited applicability. A striking example is the bread wheat to one, though the numbers recorded are small and normals are in excess. These observations could have served as a very strong hint of primary autosomal trisomy before direct chromosome observations in Down's syndrome were made.

Children with Down's syndrome now have an expectation of life at birth which approaches twenty years. Two other types of primary autosomal trisomy in man are also compatible with live birth but both are associated with severe multiple congenital abnormalities and survival is short (Chapter 11). Few examples of liveborn autosomal trisomics have been claimed in other mammalian species. A type characterized by a short lower jaw frequently associated with other congenital abnormalities occurs in cattle but reports of liveborn primary autosomal trisomics in the mouse have not been confirmed.

If liveborn autosomal trisomics are few, numerical abnormalities of the sex chromosomes are compatible with live birth. XXY, XYY, XO, and XXX types are now well known in man and all have a reasonable expectation of continued post-natal life. One or more of these types has also been identified in the mouse, pig, cattle, cat, dog and kangaroo and doubtless the list will be extended. The effect on the phenotype is severe only in the human XO female, who exhibits a condition known as Turner's syndrome that includes dwarf stature, various congenital malformations, failure of secondary sex development and infertility (see Chapter 11). In contrast, the XO female mouse is phenotypically normal and fertile.

Monosomics, trisomics and all individuals with chromosome numbers deviating from the normal by less than the full haploid set are known collectively as *aneuploids*. *Euploids* have a chromosome set that is an integral multiple of the haploid number, so polyploids are euploids, though the converse statement is not true. Three other terms of particular relevance to the cytogenetics of neoplasia and of cells in culture are *hypodiploid* (less than the diploid number of chromosomes) *hyperdiploid* (more than the diploid number of chromosomes) and *pseudodiploid* (diploid number, with addition of one or more chromosomes compensated by an equal loss).

C. DISJUNCTIONAL ERRORS

As already said, most numerical abnormalities arise as a result of errors at mitotic or meiotic anaphase. *Non-disjunction* is the failure of daughter chromosomes to separate from one another in mitosis, or of bivalents to disjoin in meiosis, so that the one polar group receives both and the other neither. *Lagging* is the failure of one or both chromatids at mitosis, or one or both chromosomes of a bivalent at meiosis, to reach the pole; they are left in the cytoplasm and are likely to form micronuclei that eventually disappear. An abnormality that is probably very rare in normal tissue but which has been seen in regenerating and neoplastic tissue and also in cultured cells is the formation of a *multipolar spindle*. The daughter chromosomes are then distributed into three or four groups at anaphase instead of the normal two, with the consequent generation of reduced and grossly abnormal numbers.

Tetraploid cells can arise in tissue in three known ways. A single nuclear membrane can be formed round both daughter groups of chromosomes in telophase, particularly where errors of disjunction have interfered with the normal full separation of the daughter groups. A tetraploid nucleus generated in this way is known as a *restitution nucleus*. Also, in some nuclei, there can be two S phases and two duplications of the chromosomes. At least this is the interpretation placed on the observation of somatic metaphases with a tetraploid set of chromosomes arrayed as a diploid number of closely aligned pairs. Such nuclei are said to be *endoreduplicated,* though there is no proof of the proposed interpretation. Simple fusion of two diploid nuclei is also a possible mechanism, but direct evidence of this is very difficult to obtain.

D. ZYGOTIC LETHALITY

Though there are few types of liveborn autosomal primary trisomics in man, and probably in mammals generally, other types are known to occur in man that are incompatible with survival to birth and the same is known to be true of the mouse. The human evidence has been obtained by the study of the chromosomes in cultures set up from spontaneously aborted fetuses (Carr, 1971). Owing to the difficulty of carrying out a systematic study with complete ascertainment it is difficult to determine the true proportion of aborted foetuses with abnormal karyotypes but it seems that the figure must lie between one-third and one-half. The great majority of the abnormal karyotypes are numerical deviants; approximately 50% are trisomics, 25% polyploids (triploids plus a few tetraploids), 15% X monosomic and 10% others (monosomics, double trisomics, mosaics and structural abnormalities). It is likely that all the 22 possible different human primary autosomal trisomics are represented, though specific identification so far has been confined to the four individual chromosome types that can be recognized with confidence. Use of the new banding techniques should eventually establish the frequencies with which the different possible types occur. Monosomics would be expected at conception with approximately the same frequency as trisomics, yet only a negligibly small proportion of the aborted foetuses examined

B. INVERSION

Most chromosome rearrangements are presumptively *balanced*, that is to say, they provide no evidence of either gain or loss of chromosome material as a result of rearrangement, whether from direct chromosome examination or from genetic analysis when that has been possible. The simplest types require two breaks and two rejoins. When the two breaks are within one chromosome and the segment between them is, in effect, rotated through 180 degrees before rejoining takes place, an *inversion* is produced. If both breaks are in one arm, a *paracentric* inversion results; if the breaks are on opposite sides of the centromere, the result is a *pericentric* inversion. A paracentric inversion was among the very first of the rearrangements to be recognized in *Drosophila* as a result of a changed order of the genes in a chromosome segment identified by genetic analysis. Several presumptive pericentric inversions have been recognized in human somatic chromosomes as a result of a change in position of the centromere but there have been few reports in other mammalian species. In the mouse, where all the chromosomes are acrocentric, it would appear that paracentric inversions only are possible. Several have been recognized by their property of forming *bridges* in first meiotic anaphase and some have subsequently been confirmed by genetic analysis (Roderick and Hawes, 1970).

Balanced polymorphic systems in which there is a standard gene sequence and one or more inverted sequences are widespread among natural populations of *Drosophila* species and other species of Diptera. The presence of inversion heterozygosity is easily recognized by examination of the polytene chromosomes in the salivary glands of the larvae and the analysis of the behaviour in these systems in nature has contributed in an important way to the theories of evolution and speciation.

C. TRANSLOCATION

When single breaks occur in two different chromosomes and the segments distal to the break points are exchanged, a *reciprocal translocation* or *interchange* results. Rearrangements of this class were first recognized in plants and immediately provided the solution to a genetic problem that had persisted unsolved for nearly twenty years. In certain inter-strain crosses, the F_1 hybrid plants had been found to be characterized by a reduction in fertility (seed viability) to about 50% of the normal level and by transmission of this property to about half their offspring when backcrossed to normal plants, (*dominant semi-sterility*). Following the discovery of reciprocal translocation it was immediately realized that individuals heterozygous for such translocations could produce about 50% of gametes with grossly unbalanced genomes as a consequence of irregularities of pairing and segregation in meiosis, and, if unbalanced gametes were able to take part in fertilization, would exhibit a corresponding reduction in fertility.

Many reciprocal translocations have been produced in the mouse, largely by the irradiation of adult males, and they have been studied extensively by both genetic and cytological methods. Reciprocal translocation heterozygotes, regardless of sex, exhibit the property of dominant semi-sterility as in the plants just mentioned. If the female partner of a cross between a heterozygote and a normal mouse is killed and dissected midway through pregnancy, about half the implantation sites in her uteri are found to be occupied by normally developing embryos and the remainder by dead embryos or resorption scars. Counts of the number of corpora lutea in the ovaries will show that approximately the normally expected number of oocytes has been shed. Systematic examination of the uterine contents in such females, coupled with breeding experiments and chromosome analysis in male meiosis, indicates that unbalanced sperm and ova are capable of completing normal maturation and effecting fertilization; and furthermore, that the fertilized egg with an unbalanced diploid genome is able to develop at least to the point of evoking a decidual reaction in the uterine wall (Ford, 1972). Only in three of the many translocation stocks studied is it known that an embryo with an unbalanced genome can continue development to term and be born alive. But survival is usually brief. There is reason to believe that at least one of the two segments exchanged in these translocations was small and the unbalance of the embryonic genome minimal.

Translocations of a different class that involve the generation of one large metacentric chromosome from two smaller acrocentric chromosomes are common in man and widespread among mammals. They are called *Robertsonian* translocations (after their discoverer), or alternatively, *centric-fusion* translocations, though the latter term begs the question of the nature of the rearrangement. This is not yet known, though some have assumed it to be a special case of reciprocal translocation with breaks very close to the centromere in the long arm of one chromosome and in the short arm of the other. Whatever the nature of the mechanism producing the rearrangement, the expected short complementary chromosome (representing, in effect, the association of the short arms of the two acrocentric chromosomes involved) has only been seen in a few exceptional individuals, all of them human subjects. The absence of the expected complementary chromosome appears to be without effect on the phenotype, and it is supposed that the small amount of chromosome material lost carried no important

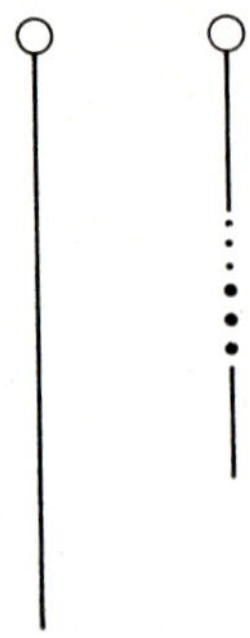

Original chromosomes

Rearranged chromosomes

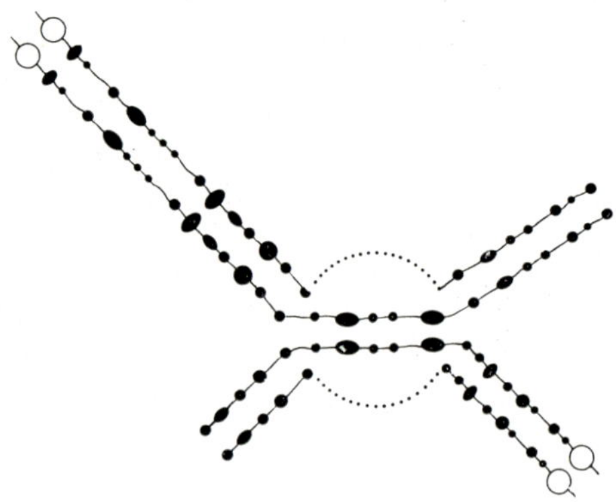

Pachytene

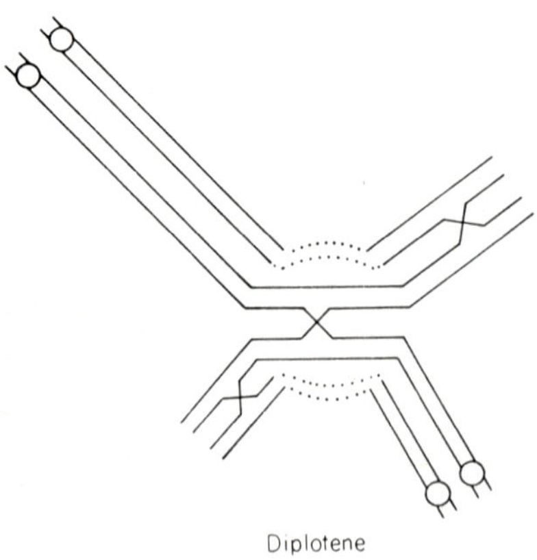

Diplotene

Anaphase

genetic information. Robertsonian translocations are therefore regarded as balanced rearrangements.

In several mammalian species, notably the Common Shrew (*Sorex araneus*), balanced polymorphic systems exist in which the alternative forms are a Robertsonian metacentric chromosome and the two acrocentric chromosomes from which it is presumptively derived. Some shrew populations in Britain may carry as many as four of these translocations and since each has three possible states (homozygous metacentric, heterozygous and homozygous acrocentric) there are $3^4 = 81$ different possible karyotypes. Robertsonian translocation seems to have played a prominent part in the evolution of mammalian karyotypes and such systems may possibly represent steps in the change from one stable karyotype to another. But it is not known what advantage they may confer upon the populations that carry them.

More complex balanced rearrangements involve a minimum of three breaks and rejoins. If three breaks occur in one chromosome and all broken ends rejoin in new ways, the segment released by two of the breaks may be replaced between the two broken ends generated by the third break. This type of change is technically known as *shift*. No examples have yet been recorded in man or other mammals. When two breaks occur in one chromosome arm and one in another chromosome, new rejoins can result in the transfer of a segment from the one chromosome to an intercalary position in the other (*insertion*). An example in the mouse involves the insertion of a segment derived from the chromosome carrying linkage group I distally into the X chromosome. The segments transposed in shifts and insertions can be reattached with the order of the genes unchanged from the original relation to the centromere, or with the order inverted. The mouse insertion just mentioned is inverted, so that on the rare occasions when the inserted segment pairs with its homologue in meiosis and under-

---

Fig. 11. Diagrams to show how a dicentric chromatid bridge and complementary acentric chromatid fragment are produced at first anaphase as a result of formation of a single chiasma between the displaced segment and its normally located homologous segment in an animal heterozygous for an inverted insertion. The two remaining chromosomes of the complex are interrupted in the pachytene and diplotene diagrams to clarify the chromatid relationships in the neighbourhood of the critical chiasma. Continuity is indicated by the thin dotted lines connecting the proximal and distal segments. The natural configuration would appear as a loop resembling an inversion loop but with four radiating arms rather than two (cf. Fig. 17). The centromeres are shown as open circles. At diplotene, chiasmata could be present in any one, or any combination of the arms of the configuration; only two are shown additional to the one in the insertion region, but the number and positions of chiasmata outside the insertion region are irrelevant to the disjunctional consequences. At anaphase the chromatid bridge lies nearly parallel to the axis of the spindle; the acentric fragment (normally) remains at the equator.

goes crossing over, a dicentric chromatid is formed which reveals itself as a bridge in first anaphase (Fig. 11).

The unqualified word translocation is sometimes employed as a generic term to include all balanced rearrangements other than inversions, and sometimes for any one of the four classes of rearrangement just considered. The mouse insertion, for example, is known as Cattanach's translocation.

Some recessive sex-linked lethal gene mutations in *Drosophila* are associated with apparently balanced inversions or reciprocal translocations. In many instances it has been impossible to separate the lethal mutation from the rearrangement by crossing-over on continued breeding, so the lethal effect must be associated with the rearrangement itself. Formally, there are three ways in which this association can be explained, namely an independent gene mutation so close to the point of exchange that they cannot be separated by crossing over, a small duplication or deletion located at the point of rearrangement, or a position effect. By position effect is meant an effect on the phenotype arising directly from the close physical proximity of gene loci that were formerly located in more distant parts of the genome. Some reciprocal translocations in the mouse act as recessive lethals, i.e. the homozygotes, if they are viable at all, are not found among the live-born progeny of an intercross between heterozygotes.

D. DUPLICATION AND DELETION

Duplications and *deletions* are the two basic types of unbalanced rearrangements. The term deficiency has been widely used as a partial synonym of deletion and tends to be used for the smaller losses, for interstitial losses rather than terminal ones and to be favoured by geneticists.

Duplication and deficiency can be generated simultaneously as complementary products of a single event if breaks occur at slightly different positions in the two chromatids of one chromosome and then the broken ends undergo exchange reunion (Fig. 12). One resultant chromatid would be a little longer than normal and have one short segment represented twice (a *repeat,* or *tandem duplication*); the other chromatid would lack this segment. This type of event is believed to have taken an important part in genome evolution. For example, the genes producing α and β haemoglobin in man are very closely linked and their amino-acid sequences are very similar. The original postulate was that they represent a small duplication that had arisen through unequal crossing over between homologues at meiosis. But the assumption does not have to be so restrictive; the required event could occur within a single chromosome and at any time of the life cycle.

In view of the resultant change in balance, duplications and deletions would be expected to have a more or less severe effect on the phenotype dependent upon the amount of genetic material involved. Some very small deficiences have been identified in the mouse by genetic analysis but none of them has been large enough to offer hope of visual detection by present methods. Large deletions, clearly visible in the somatic chromosomes, have been recognized in the human species. At least three types and possibly more (Table III Chapter 11), are known, each characterized by a specific array of congenital malformations.

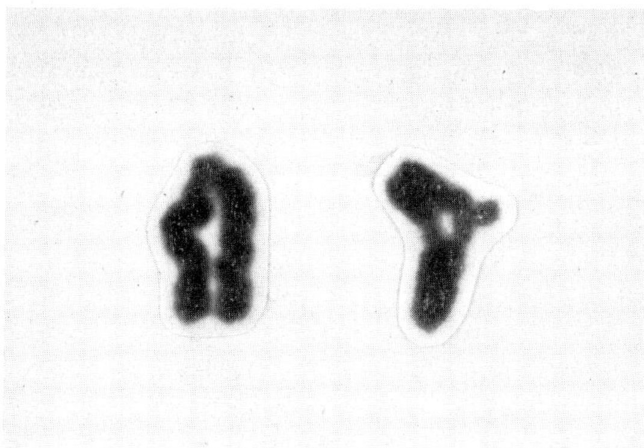

Fig. 12. Chromatid intrachanges in chromosomes from a culture of irradiated mouse blood. The structure on the left is inferred to have arisen as a result of breakage of both chromatids at slightly different positions and subsequent cross reunion to give one longer, duplicated chromatid (with the kink) and one shorter deficient chromatid. The structure on the right is inferred to have arisen as a result of two breaks close together in the one chromatid followed by reunion to give one deficient chromatid and one normal chromatid with associated minute ring fragment. × 4000.

E. ISOCHROMOSOMES

Isochromosomes are chromosomes with an exactly median centromere and two apparently identical arms. Unlike all other structural rearrangements of the chromosomes, the evidence suggests that they are formed at anaphase as a result of a rare abnormality known as *misdivision* of the centromere. This seems to occur with a frequency sufficient for study only when meiosis is disturbed by the presence of one or more univalents, the classic material for its demonstration being monosomics of wheat. It appears that the centromere, instead of dividing as it does normally in a transverse plane, divides in a vertical plane, so separating the two arms of the chromosome. If one arm is incorporated into one of the daughter nuclei and if the normal events of the next S phase are not disturbed, a new chromosome with two identical arms will be present at the next metaphase (Fig. 13).

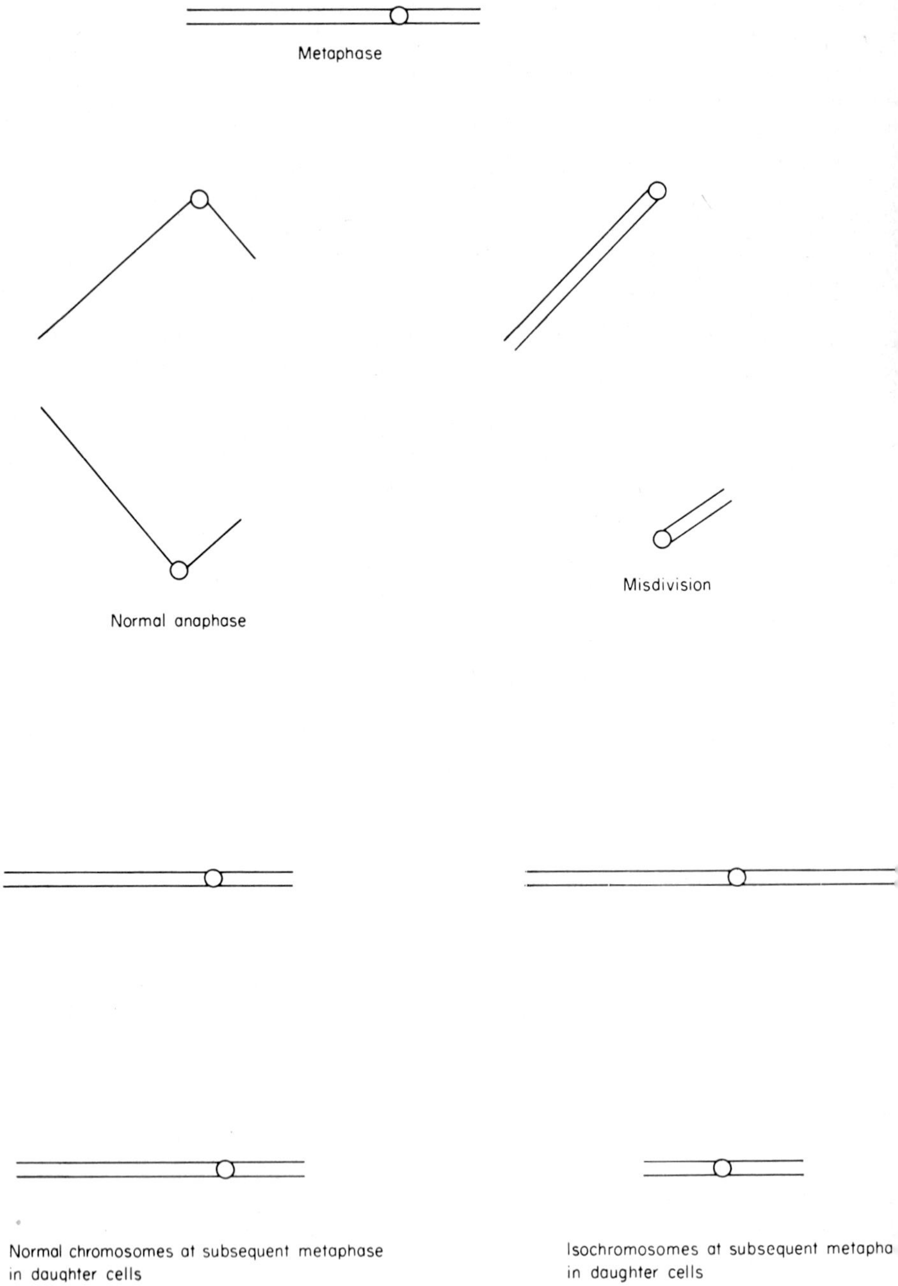

Fig. 13. Diagram to show the origin of isochromosomes as a result of misdivision at anaphase of a somatic mitosis or the second meiotic division.

Isochromosomes derived from the long arm of an X chromosome are found commonly in cases of Turner's syndrome in man but as yet no examples have been described in other mammalian species.

Much can be learned about the nature of chromosome rearrangements by studying the chromosomes at meiosis in heterozygotes. Probably the most helpful generalization is that homologous segments are potentially capable of pairing whether they lie in the same chromosome or in different chromosomes. A likely, but unproven, limitation is segment size. For instance, it is possible that a small inserted segment may not find its normal homologue before the phase of zygotene pairing ends. Whatever the limitations on the pairing process, the configurations observed in pachytene are those expected from the generalization just given. A cross-shaped configuration is formed in a reciprocal translocation heterozygote (Figs. 7, 14a) and a linear bivalent with an attached loop in an inversion heterozygote (Fig. 14b), the loop being formed by the inverted segment and its normal homologue, paired specific locus by specific locus. Duplications

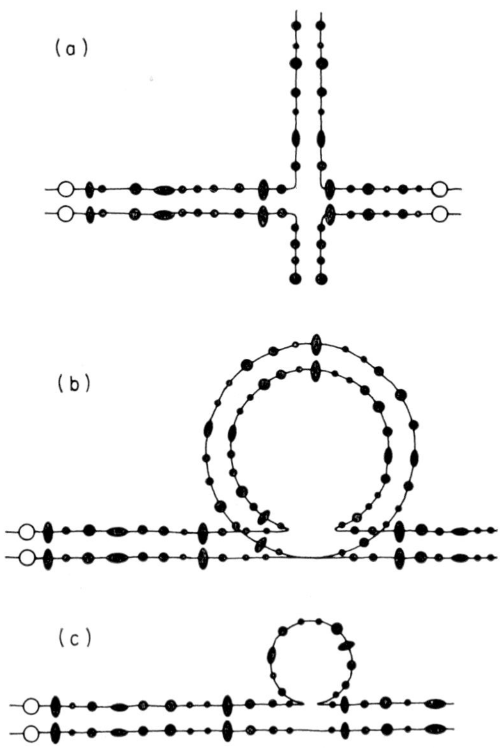

Fig. 14. Diagrams to show pairing at pachytene in animals heterozygous for a reciprocal translocation (*a*) a paracentric inversion (*b*) and a duplication or deficiency (*c*).

and deletions appear in the heterozygous state at pachytene as otherwise normal bivalents with small intercalary unpaired loops (Fig. 14c).

F. MEIOSIS IN STRUCTURAL HETEROZYGOTES

Chiasma formation between homologous but relationally displaced segments leads to the appearance of an equally characteristic series of abnormal configurations in diplotene, diakinesis and first metaphase.

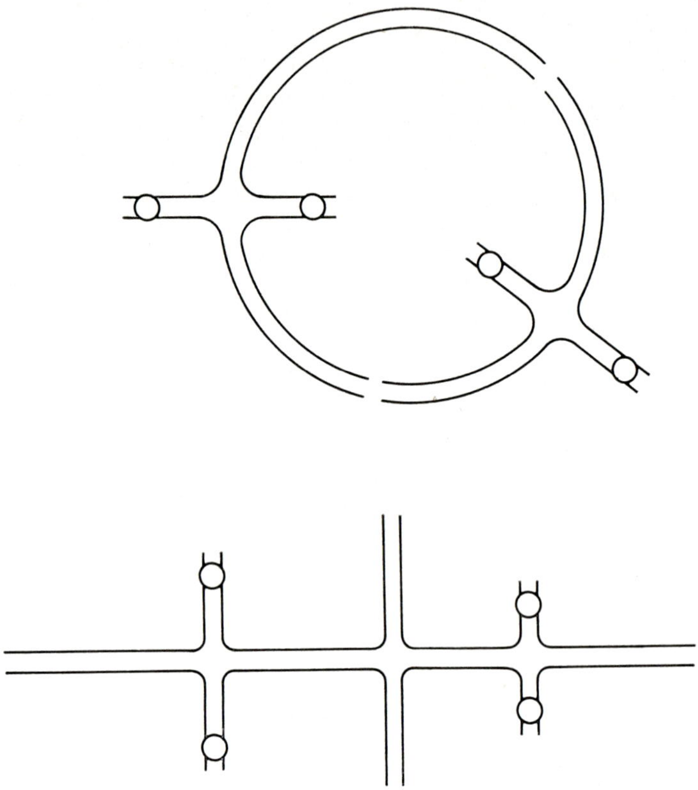

Fig. 15. Chromatid diagrams of a ring quadrivalent (above) and a chain quadrivalent (below) as formed in an animal heterozygous for a reciprocal translocation between two acrocentric chromosomes. A translocation between acrocentric chromosomes has been chosen deliberately because the corresponding configurations when the translocation is between metacentric chromosomes are more complex, though the principles remain the same. The relative lengths of the chromosomes involved and of the segments exchanged are the same as in the pachytene cross shown in Fig. 14.

The possibilities can be illustrated by reference to the reciprocal translocation heterozygote. If one or more chiasmata are formed in each arm of the pachytene cross referred to above, then a ring-like association of four chromosomes, each chromosome attached to the next by at least one interstitial or terminal chiasma, takes the place of the pachytene cross configuration in diplotene and later stages. Such an association is usually referred to as a *ring quadrivalent*. If a chiasma fails to form in one arm of the pachytene cross a *chain quadrivalent* (Fig. 15) results. Failure of chiasma formation in two arms results in a *trivalent* plus a *univalent*, or two bivalents, according to whether adjoining or opposite arms of the pachytene cross remain associated. Many different diplotene-diakinesis configurations may thus result from the same configuration at pachytene depending on the number of chiasmata formed, their positions of origin and the positions to which they may move.

At first meiotic anaphase a quadrivalent configuration can disjoin two-and-two in three ways. Alternate chromosomes may go to the same pole (*alternate disjunction*); or adjacent chromosomes with non-homologous contromeres (*adjacent–1 disjunction*); or adjacent chromosomes with homologous centromeres (*adjacent–2 disjunction*). When no chiasmata are formed in either of the *interstitial segments* that lie between the centromeres and point of exchange, all the products of alternate segregation have balanced haploid genomes and all the products of both types of adjacent segregation have unbalanced haploid genomes. The consequences are different when one or more chiasmata are formed in one or both of the interstitial segments, which happens regularly in mouse translocation heterozygotes and probably also in man. In such cases alternate and adjacent–1 segregations become geometrically indistinguishable and give rise to equal numbers of balanced and unbalanced haploid genomes. Adjacent–2 disjunctions still yield unbalanced haploid genomes only, though there are now potentially six different types instead of two. For any given reciprocal translocation heterozygote then, the proportion of all gametes produced that are unbalanced (and the consequent degree of infertility) is a direct function of the types of anaphase disjunction and of the frequency with which chiasmata are formed in the interstitial segments. If AB, CD are the two original chromosomes and AD, CB the two translocation chromosomes, with the centromeres lying in segments A and C, the principle types of unbalanced gametes are AB, CB, and AD, CD. Each has an extra segment and a missing segment, so both unbalanced gametes and unbalanced zygotes are simultaneously duplicated and deficient (Fig. 16).

The production of unbalanced gametes by inversion heterozygotes (Fig. 17) is also variable but generally much less. It is dependent upon the frequency of crossing over in the inversion loop. One crossover results in a single chromatid bridge and associated fragment at anaphase and leads to the formation of two unbalanced gametes and

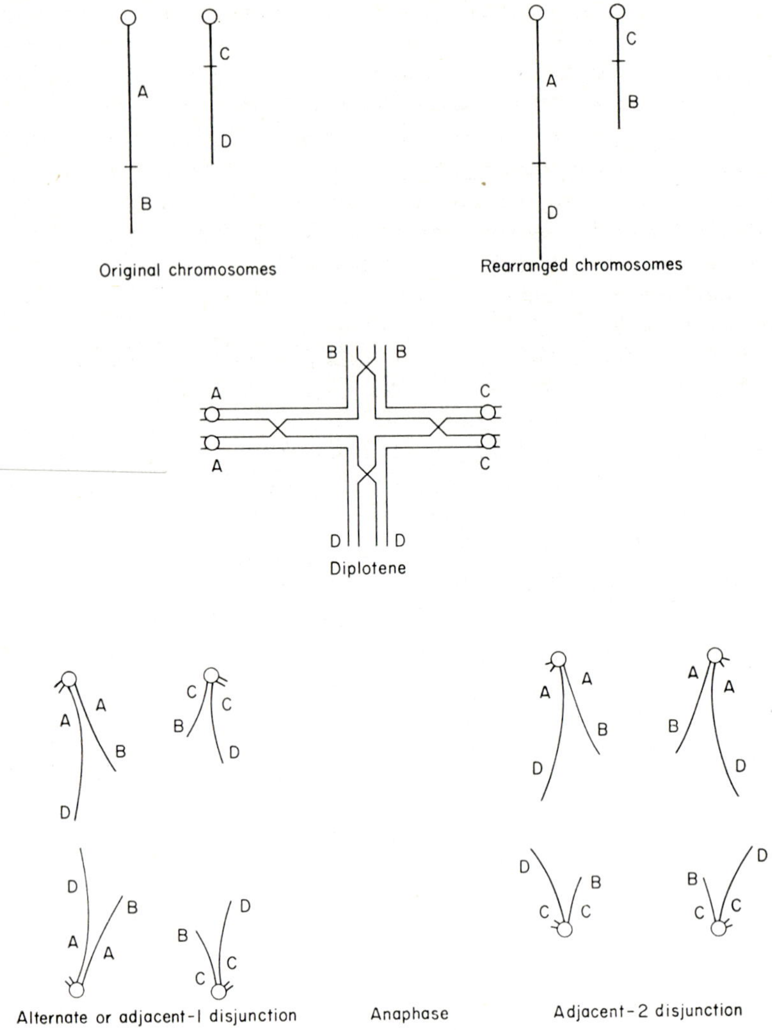

1 AB, CD
1 AD, CB
1 AB, CB
1 AD, CD

1 AB, AB
2 AB, AD
1 AD, AD
1 CB, CB
2 CB, CD
1 CD, CD

Gametic combinations resulting from second metaphase segregation.

Fig. 16. Shows the disjunctional consequences at first and second anaphase in an animal heterozygous for a reciprocal translocation between acrocentric chromosomes, following chiasma formation in all four arms of the pachytene cross. Compare figures 13 and 14.

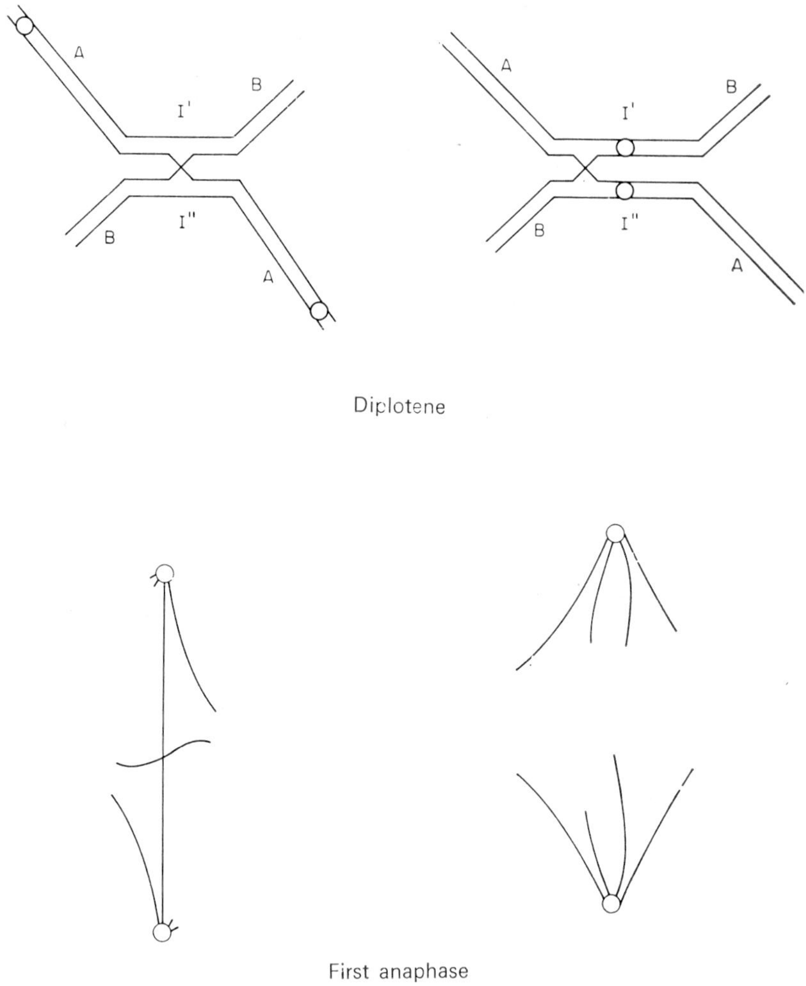

Fig. 17. Diagrams to show the disjunctional consequences of crossing-over in the inverted segments of animals heterozygous for a paracentric inversion (left) and a pericentric inversion (right). The configurations at diplotene are deliberately "opened" so that the paths of the chromatids through the critical chiasmata can be followed more readily. In the natural configuration the inverted segment (I′) and its normally located homologue (I′′) would form a loop and the homologous segments AA, BB would be paired. It will be readily seen that the exchange of partner chromatids at chiasmata between the two segments A or between the two segments B would not change the course of the chromatids through the critical chiasmata. Note that at first anaphase the chromatid bridge (left) lies in the axis of the spindle (contrast the inverted insertion bridge shown in Fig. 11). Note also that segregation at second anaphase following crossing-over in the inversion loop of the pericentric inversion would give rise to a structurally normal chromosome, AI″B, an inverted chromosome AI′B, and two structurally unbalanced chromosomes, AIA and BIB.

two balanced gametes (unless the bridge fails to break and causes mechanical disturbance of the second meiotic division). Should a second chiasma be formed in the loop, which is only likely in exceptionally large inversions, it could cancel the first and lead to normal disjunction at anaphase without bridge or fragment, or there could be a single bridge and single fragment, or a double bridge and two associated fragments, depending upon whether two, three or four strands were involved in the chiasmata.

In Robertsonian translocation heterozygotes there is normal pairing of the two homologous acrocentric chromosomes with the single metacentric chromosome at zygotene-pachytene. This configuration gives place to a characteristic chain trivalent in stages from diplotene to first metaphase. There appears to be no obvious mechanical restraint upon normal segregation at anaphase, with the metacentric chromosome going to one pole and the two acrocentric chromosomes to the other with consequent production of four balanced haploid genomes. However one of the two associated acrocentric chromosomes may proceed to the same pole as the metacentric chromosome and lead to the formation of duplicated and deficient gametic genomes in equal proportions. The frequency with which this occurs varies between one Robertsonian metacentric and another. Heterozygosity for a Robertsonian translocation is therefore associated with an unpredictable degree of zygotic unbalance and consequent infertility.

Heterozygosity for a simple duplication or deficiency, if the segment concerned were large enough, would be expected to lead to the appearance of visibly asymmetrical bivalents at diplotene and later stages. Such *unequal bivalents* should segregate normally at first anaphase and would therefore not be expected to influence fertility if the heterozygote itself were fully viable.

The greater part of present knowledge about the behaviour of structural heterozygotes at meiosis has been derived from the study of plant species, notably Indian corn (*Zea Mays*). Studies of meiosis in human and mammalian structural heterozygotes are still few, but the observations made are consistent with the conclusions drawn from plant cytogenetics outlined above.

Figures, 7, 8a and 9 illustrate this point. Figure 7 is of a pachytene cross in a male mouse heterozygous for a reciprocal translocation. Figure 8a shows a ring quadrivalent at diakinesis from the same translocation. Figure 9 is of a similar translocation and shows a second metaphase with two unequal chromosomes. The translocation concerned is a grossly unequal one, a very long segment having been exchanged for a very short one. A single crossover in an interstitial segment therefore leads to the exchange of a connection to the long distal segment for a connection with the short distal segment. Since crossovers are symmetrical exchanges, the two chromosomes associated at the chiasma are structurally identical, each having a longer and a shorter chromatid attached to its centromere. Segregation of the centromeres at first anaphase then results in the appearance of unequal chromosomes at second metaphase. When a single chiasma is formed in both interstitial segments, two different unequal-armed chromosomes enter each of the two daughter nuclei. Comparison of the frequency of unequal chromosomes at second metaphase with the frequency of chiasmata in the interstitial segments at first metaphase shows good agreement and supports the hypothesis of a one-to-one relationship between chiasmata observed in meiotic prophase and physical crossovers.

VI. Mosaics and Chimaeras

A. DEFINITIONS AND DIFFERENCES

Mosaics and chimaeras are rare curiosities which, like other experiments of nature, are often of considerable investigational value. In the simplest terms, they are animals compounded of cells of two or more different genotypes. The two words have been used in different ways in different biological contexts for over fifty years. Two quite distinct classes of biological events are involved in their formation and it is therefore logical to use "mosaic" for the one and "chimaera" for the other. Accordingly, *mosaics* have been defined as individuals with two or more distinct cell-genotypes that have arisen *from a single zygote*; and *chimaeras* as individuals with two or more distinct cell-genotypes contributed by *two or more separate zygotes.* If this distinction is accepted two further differences follow: a mosaic arises as a consequence of one or more mutational (in the wide sense) events, no other abnormal or unusual biological process being involved, and, discounting the mutational change (or changes), the residual genotype is the same in all cells of the body; chimaeras are formed as the result of an exceptional biological event (other than mutation) and the component cell lines are likely to have completely different genotypes.

B. CHROMOSOMAL MOSAICISM

Many examples of chromosomal mosaicism, particularly involving the sex chromosomes, have been recorded in human subjects. For example, some patients with Turner's syndrome have been found to have two different cell lines, 45,X/47,XXX, others have three cell lines, 45,X/46,XX/47,XXX and still others have two cell lines 45,X/46,XX. Since it is usual to find that mosaicism, if it is detected at all, is widespread over all the tissues of the body that can be investigated, the generating event or events must have occurred very early in development. It is plausible to suppose that mosaics of the first type mentioned arise as a result of non-disjunction of one of the two X chromosomes at the first division of an XX zygote. If a similar non-disjunctional event should occur at the second or later cleavage division a mosaic of the second type could be formed. For the third type of mosaic there are two formal possibilities. One is a non-disjunctional event at the second or a later cleavage division, with accidental exclusion of cells with three X chromosomes from the group of cells that give rise to the embryo proper. The second and more likely alternative is the lagging of one chromatid of one of the two X chromosomes at the anaphase of the first cleavage division and its exclusion from one of the daughter nuclei at telophase.

C. MONOZYGOTIC TWINS WITH DISCORDANT PHENOTYPES

Whatever the nature of the events that generate the founder cells of the different cell lines there is little doubt that chance exclusion of a variant cell type from the inner cell mass can occur. The evidence is provided by a remarkable phenomenon, the occurrence of indubitably monozygotic twins with discordant phenotypes, for example, a normal male and a female with Turner's syndrome. In the first such pair examined the cells karyotyped were all normal 46,XY in the male and all 45,X in his sister (Turpin *et al.*, 1961). The evidence of monozygotic origin was provided by complete concordance for an extensive series of genetic markers, close similarity of palm prints and general phenotype as well as supporting obstetrical records. The explanation can only be that a 45,X cell was formed during cleavage of a 46,XY zygote that subsequently divided to give two embryos, one of which received normal 46,XY cells only and the other 45,X cells only.

D. NATURAL CHIMAERISM

A striking type of natural chimaerism has been identified in man by combined cytogenetic and genetic analysis and in the mouse by genetic analysis alone. This is the participation of two independent spermatozoan nuclei in syngamy with two normal haploid female pronuclei. All the human chimaeras of this class so far identified have been 46,XX/46XY mixtures and have mostly been studied because of an abnormality of the sex phenotype.

Other natural human chimaeras have been identified in the course of blood grouping. Several pairs of twins have been found with red cells of two kinds, one identifiable as their own, the other from the twin partner. Chimaerism has subsequently been confirmed in heterosexual pairs by cytogenic identification of 46,XX and 46,XY cell lines in lymphocyte cultures. Evidently there must have been an exchange of haemopoietic stem cells between the twins during fetal life. Though rare in man, this phenomenon is common in cattle and a normal feature of gestation in monkeys of the marmoset family. In cattle the female member of heterosexual twin pairs invariably develops into an intersex long known to farmers as a "freemartin".

E. EXPERIMENTAL AND GRAFT CHIMAERISM

Two further kinds of chimaeras are mentioned for their great experimental value. One kind is made artificially by manipulation of early embryos. This was first achieved in the mouse by Tarkowski (1961), who succeeded in fusing together mouse eggs at the eight-cell stage, incubating the resultant mass *in vitro* until it became a blastocyst and then injecting it into the uterus of a pseudo-pregnant female. Another method, due to Gardner (1968), is to transfer cells by micromanipulation from one blastocyst into the blastocoel of another and then to proceed in the same way as Tarkowski. The two components of such chimaeras can be "marked" by suitable genetic tags or marker chromosomes and the derived cells subsequently identified in the adult animals.

The second type is made by exposing a young adult mouse to a just sub-lethal dose of X-rays and then injecting it with cells from bone marrow or other suitable source, such as fetal liver. The injected cells reconstitute the radiation-depleted tissues of the lympho-myeloid complex and if they are tagged by a suitable marker chromosome their persistence and transformations can be followed by a combination of cytogentic methods with other experimental techniques. Such "radiation chimaeras" have been used extensively in experimental haematology and immunology.

Human patients who have received a transplant of an organ or other tissue from another person could logically be called "graft chimaeras". The term is perhaps particularly appropriate when cells of the lympho-myeloid complex are partly or wholly replaced, following injection of bone marrow cells from a suitable donor. This has occasionally been practised with success in the treatment of diseases of the immune system.

VII. Sex Chromosomes and Sex Determination

Inferences about the part played by the sex chromosomes in sex determination can be drawn from the comparison of the genital anatomy of individuals with abnormal sex chromosomes. An enormous number of different types of abnormality have now been described in man (see Chapter 11) and several are known in other mammalian species. The data permit the broad generalization that in man, and mammals generally, the presence of a single Y chromosome is normally sufficient to determine that the primordial gonads shall become testes and that masculine development of the accessory genitalia will follow. In man, individuals with a normal set of 22 autosome pairs and one to five X chromosomes develop a basically female phenotype, whereas the testis-inducing function of the Y chromosome is not impaired even when it is accompanied by four X chromosomes, as in the constitution 49,XXXXY.

The developmental significance of the mammalian testis was first demonstrated by A. Jost (1947). He succeeded in ablating the gonads of 19–day old rabbit embryos *in utero* and showed that subsequent development of both the internal and external genitalia followed the female patern, regardless of whether the gonads removed were destined to become ovaries or testes. This experiment shows that the rabbit testis has a morphogenetic function and is essential for normal development of the accessory male genitalia.

There is a condition in man which probably represents a natural parallel to Jost's experimental rabbits and which at first seems to conflict with the generalization just given. Some women who present with primary amenorrhoea show no secondary sex development, are usually tall and eunuchoidal and, if a laparotomy is performed, are found to have their gonads replaced by a streak of connective tissue. The condition is known as *pure gonadal dysgenesis* and is of particular interest because the karyotype may be either normal female or normal male. The reasonable explanation is that gonad development must be arrested before the time that a normal testis would have begun to influence the development of the accessory genitalia. Furthermore, the XX cases and XY cases are so similar in phenotype that the influence of the Y chromosome on the soma must be negligible.

A second apparent exception to the principle that the mammalian Y chromosome induces testicular formation and male development is provided by *testicular feminization,* a condition known in man and mouse and suspected in cattle. The external phenotype is female, testes and male ducts are present internally, the karyotype is that of a normal male and the condition is determined by a single gene that is sex-linked in the mouse and probably also sex-linked in man. Analysis of the steroid sex hormones indicates that their concentration in urine and serum is approximately the same as in a normal male. Present information is therefore consistent with the assumption of failure of end-organ response. Ohno (1971) has recently given support for this assumption and a plausible hypothesis to explain the action of the gene at the molecular level.

A third group of apparent exceptions to the general principle is provided by mosaics and chimaeras that are compounded of one or more cell lines that contain a Y chromosome and one or more cell lines that do not. The sex phenotype presented by the human cases ranges from apparently normal female through to apparently normal male and includes some of the class of true hermaphrodites, bearing both testicular and ovarian tissue. All the data gathered so far are consistent with the supposition that the critical factor in the eventual determination of the sex phenotype is the proportion of Y-bearing to non-Y-bearing cells in the gonadal primordia. The cytogenetic evidence suggests that a relatively small proportion of cells containing a Y chromosome, perhaps only one in five, is sufficient to determine that the whole organ shall become a testis. The critical proportions may well differ between one species and another and even between one individual and another of the same species, depending upon the residual genotypes of the cell lines concerned. Much of the limited information that bears on this point has come from the study of XX/XY chimeras, spontaneous cases in man and experimentally produced chimaeras in the mouse (Mystkowska and Tarkowski, 1968).

None of the apparent exceptions are therefore contrary to the general principle that it is the presence of the Y chromosome that determines the formation of testes. Secondary factors presumptively interfere with primary gonad formation in cases of pure gonadal dysgenesis and with events subsequent to testis formation in testicular feminization; the analysis of mosaics and chimaeras takes the principle a stage further and indicates a quantitative aspect of the process of gonadal determination.

Jacobs and Ross (1966) gave reasons for supposing that the essential genetic information concerned with masculinization carried by the human Y chromosome was located in the miniscule short arm. The evidence was provided by the comparison of individuals with different abnormalities of the Y chromosome. Two had an isochromosome of the long arm of the Y, that is to say, a chromosome symmetrically duplicated about the single centromere and with the short arm totally absent; two had somewhat similar, but shorter, Y chromosomes interpreted to have arisen by pericentric inversion. The first two patients with no evident short arm material, were female; though they had never menstruated. The other patients, also with rearranged Y chromosomes but with the short arm presumptively retained, were male. Apart from the unlikely possibility that the constitution of these abnormal chromosomes was misinterpreted, the only objection that can be made to the inference that the short arm of the Y must therefore carry the masculinizing factor or factors would appear to be the possibility that the patients with the long arm isochromosomes may have been undisclosed mosaics, each with an XY cell line that had been insufficiently represented in the primordial gonads to ensure testis formation. This is also unlikely and would be effectively excluded if the phenotypic distinction should be confirmed on a sufficient number of further cases.

In view of this strong evidence that the part of the human Y chromosome that carries the essential masculinizing factor or factors is so small, it is of interest to note that the Y chromosome in mammals is nearly always the smallest member of the complement, and that in all groups that have been sufficiently studied, at least one species has been recorded in which the diameter of the Y chromosome is about equal to the width of the autosomal chromatid or even less. The reduction in size of the Y chromosome reaches its ultimate in some marsupial species, where it is only just above the level of visible resolution by the light microscope. This is further evidence that the segment of the Y chromosome carrying the masculizing gene or genes may be very small indeed.

Formally, there are two ways in which the Y chromosome could influence the developing gonad. It could itself bear the essential structural genetic information for testis formation; or it could have a regulatory function, serving as an activator (or derepressor) of a structural gene (or genes) located elsewhere in the genome. It has not yet been possible to discriminate between these two possibilities though the balance of the evidence is now rather in favour of the latter. This evidence comes from the consideration of "XX males", namely individuals with a normal female chromosome complement but which develop testes and anatomically normal male genitalia. Such individuals are known in man, mouse, goat and pig and are always sterile. A recently discovered autosomal mutant gene in the mouse causes XX zygotes to develop into sterile males. This would, at first sight, appear to demonstrate that there is at least one autosomal gene locus involved in testicular development. However, in view of the fact that the segment of the Y chromosome concerned with masculinization can be so small, the possibility of translocation of this segment from the Y chromosome to an autosomal site cannot yet be excluded.

A further requirement of the efficient Y chromosome is that it should pair and disjoin regularly from the corresponding X chromosome at meiosis. This holds true in all mammalian species yet studied. In most of them X and Y are associated terminally from the earliest stage (diplotene) at which they can be seen as a separate bivalent. It has been contended that this is sufficient evidence for supposing that the mechanism of chromosome pairing is different from the one responsible for the synapsis of the autosomes. However the observations do not exclude the possibility of formation of an interstitial chiasma and exceptionally early terminalization. Furthermore, in the zygotene phase, X and Y must attract one another just as the pairs of homologous autosomes do. It seems rather unlikely that different molecular mechanisms of attraction-at-a-distance should operate between sex chromosomes and between autosomes, particularly as the sex chromosomes must have evolved originally from an autosome pair. Whatever the nature of the mechanisms of attraction and pairing it is convenient to consider all sex chromosomes in XX:XY systems as being compounded of a *pairing segment* homologous with a corresponding segment in the other member of the sex chromosome pair, and a *differential segment* unique to the chromosome concerned. In man, mouse and most mammalian species the pairing segment may be reduced to the shortest length consistent with the maintenance of regular association and disjunction. Although abnormal sex chromosome systems of several different types have been identified in other mammalian species (Fredga, 1970), their characteristics, so far as they are known, do not conflict with the principles stated above.

VIII. The Cytogenetic Effects of Ionising Irradiation

A. CHROMOSOME ABERRATIONS INDUCED BY RADIATION

It has already been said that chromosome aberrations are induced in dividing cells by exposure to ionizing radiation. This statement applies to all dividing cells in all species, plant or animal, that have been investigated and irrespective of whether the cells are in their natural sites or explanted and growing in culture. The effect is first observed in the metaphase immediately following exposure and the proportion of cells affected as well as the mean number of aberrations per cell is primarily dependent upon the total dose delivered, the rate of delivery and the nature of the irradiation (X-rays, gamma-rays, neutrons etc.). Exposure during the G1 period of interphase results in the formation of chromosome-type aberrations (see above), exposure during G2 results in chromatid-type aberrations and exposure during the S period may result in a mixture of both.

A special and characteristic consequence of irradiation is that the production of dicentric chromosomes or chromatids leads to the occurrence of bridges in anaphase. These may sometimes impede normal anaphase disjunction with the result that a single nuclear membrane is formed round the two sets of incompletely separated chromosomes at telophase. If the resultant restitution nucleus should enter metaphase again, a tetraploid complement would be expected with each abnormal chromatid or chromosome represented twice.

B. DOSE-YIELD RELATIONSHIP

When the number per cell of a particular class of aberration, such as dicentric chromosomes, is studied in relation to the total delivered dose of X-rays or gamma-rays, statistical analysis shows that the total yield is the sum of two components, one that increases in direct proportion to the dose (i.e. linearly) and the other which increases as the square of the dose. The dose-yield relationship for simple breaks is influenced almost exclusively by the first of these components, a result to be expected if the simple breaks are normally the consequence of a single intra-nuclear event. But the yield of *exchange aberrations*, such as the dicentric chromosomes just mentioned, is determined very largely by the second component. This is explained if such aberrations arise largely as the consequence of the interaction of two independent intra-nuclear events.

C. TARGET THEORY

These intra-nuclear events are readily envisaged as the passage of an ionizing particle through the chromosome. Taking the image of the ionizing particles generated by the absorption of X-ray or gamma-ray quanta as shots fired, and the points of passage through the chromosomes as targets hit, this model of the effect of ionizing radiation on chromosomes has been called the "Target Theory". Accordingly the expressions, "single-hit aberrations" and "two-hit aberrations" are often used for types where the linear component and the square law component of the dose-yield relationship are dominant respectively. The relationships between the yield of different classes of aberrations and the basic physical variables were established by studies with plant cells, notably pollen grains of *Tradescantia* species and cells from the root tips of broad bean (*Vicia Faba*) seedlings growing in water culture under constant conditions. More recently they have been confirmed by study of irradiated cultures of blood from human volunteers and from various mammalian species.

D. RESPONSE OF THE SMALL LYMPHOCYTE

The small lymphocytes which transform under the influence of PHA into large mononuclear cells and then enter mitosis in blood cultures are in many ways ideal for studies of radiation-induced chromosome breakage. They are cells of a single, sharply circumscribed morphological type and effectively all of them are in the same (G1) phase of the mitotic cycle at the time of removal from the body. Within the mammalian body they are distributed to all organs and tissues by the blood and lymph. Furthermore, it is now known that they are continually being recirculated, into the tissues and back again into the blood stream. It has therefore been suggested that the determination of the frequency of aberrations in blood cultures established from irradiated human subjects could serve as a form of biological dosimetry.

E. VARIATION IN SENSITIVITY

Statistical analysis of the variation between cells in the numbers of chromosome aberrations observed shows that the small lymphocytes are very uniform in their response to irradiation, the results being consistent with uniform sensitivity and a random distribution of aberrations between cells. The results obtained from the irradiation of a complex cell population, like that of bone-marrow, are very different. If a mouse is exposed to a whole-body dose of 1 000 rad of 250 kvp X-rays and preparations are made from the bone-marrow between one and two days later when the first mitoses after exposure are beginning to appear, the chromosomes in the great majority of mitoses are broken into so many small fragments as to be unanalysable; in some cells the number of aberrations is small enough for them to be identified and counted; and in a very few cells there are no aberrations at all. This great variation in response of individual cells indicates that they must have differed considerably in radiosensitivity. Inherent variation in sensitivity between the different species of dividing cells making up the bone-marrow and variations attributable to the particular stage of the cell cycle at which the cells were exposed are both likely to have contributed to the total variation observed.

F. GENETIC DEFICIENCY AND CELL DEATH

It may be possible for a small lymphocyte to persist in the body of an irradiated human patient for ten years or even longer and then enter mitosis in culture (Buckton *et al.*, 1967). So much can be inferred from the presence, many years after exposure to irradiation, of mitoses that include one or more acentric fragments and in which analysis of the karyotype is consistent with conservation of the whole genome. At the first anaphase following irradiation, acentric fragments are unable to move on the spindle and are usually left out from the telophase groups to form micronuclei which eventually degenerate and disappear. If a daughter cell from such a division should enter mitosis again it could be recognized by an abnormal karotype in which the number of centromeres was conserved but in which the genome was clearly deficient. Such X2 nuclei have sometimes been recognized in cultures of blood from irradiated patients though not in the irradiated tissues of small laboratory animals. It seems that deficient cells in general must either die, or at least be incapable of further proliferation.

Depletion of a proliferating cell population in this way would doubtless be made good by a temporary greater call on the stem cells if the loss were not too severe (see also Chapter 13). But for each species and each tissue there must be a point beyond which cellular restoration does not occur and it is therefore not surprising that among the prominent effects of over exposure to ionizing radiation are destruction of lymphoid tissue, bone marrow, intestinal epithelium and seminiferous epithelium of the testis tubules.

G. CLONES

Should a heavily irradiated mouse survive, the cell populations of its depleted tissues may eventually be restored to nearly their normal numbers. And if chromosome preparations are made from, say bone marrow, some weeks or months after the irradiation was delivered, cells with abnormal karyotypes may indeed be found. The abnormalities, however, are restricted in two ways; the altered chromosomes all have single centromeres and the total genome appears to be conserved. Since the normal chromosomes of the mouse are all acrocentric and vary continuously in length, the abnormal (*marker*) chromosomes are metacentrics (few) and acrocentrics that are either clearly longer or clearly shorter than any normal chromosome.

Very frequently, apparently the same marker chromosome, or set of marker chromosomes, is present in many mitoses on the same preparation. Now a single longer or shorter acrocentric chromosome is not specially distinctive and might have originated independently in two or more different ways, but experience has shown that the sets of two, three or more marker chromosomes are unique. The inference is that all cells that have the same set of marker chromosomes are derived by division of a single ancestral cell in which the rearrangements were induced by the irradiation (Ford, 1964). The term *clone* (derived from a Greek word meaning a twig) has been borrowed from plant genetics for such a group of cells with a common ancestor. The dominant clone often constitutes 50% or more of the mitoses in the preparation and the degree of dominance tends to get even greater as the time elapsed since irradiation increases. Furthermore, cells of the clone become disseminated to other sites of the lymphomyeloid complex: spleen, lymph nodes, thymus and Peyer's patches. The implication that the clones compete with one another and that one tends to gain increasing ascendancy over all the others has been confirmed by experiment. This is not a surprising result since the cells that survive exposure to such a high dose must be heterozygous for many different gene mutations which could modify the normal course of as many different biosynthetic systems. In man, as yet, such clones have only been recognized in cultures established from irradiated skin.

Apart from the phenomenon of clonal proliferation described in the last paragraph little attention has been given to the long-term consequences of chromosome breakage and rearrangement. They have been associated speculatively with both carcinogenesis and accelerated ageing but there is no direct evidence of a connection with either and only limited circumstantial support.

IX. The Cytogenetics of Cells in Culture

A. TECHNICAL VALUE

Cytogenetics and tissue culture complement one another. It is relatively easy to obtain preparations of high technical quality from cells growing in culture and they have often been preferred for establishing karotypes for this reason; and since, from the late fetal stage on, mitosis is too infrequent in most tissues for cell karyotypes to be established by direct methods, tissue culture is particularly valuable for the investigation of mosaics and chimaeras. To minimize the nagging possibility that karyotypic changes may have taken place between removal of tissue from the body and preparation of slides for examination, it is obviously best to use the primary culture itself, and failing that, the earliest sub-culture possible. Setting up duplicate cultures from the same biopsy specimen provides an additional safeguard, since the likelihood of the same karyotypic change taking place in both cultures must be negligibly low.

For his part the cytogeneticist can serve the experimentalist working with long-established cell lines by revealing variability, verifying identities and pointing out relationships that might otherwise have been concealed. Many instances of one cell line masquerading under the name of another have been uncovered by the simple expedient of taking a look at its chromosomes.

B. CELL STRAINS AND CELL LINES

The natural history of a culture may take the following course. Outgrowth from the primary tissue explant is slow. A period of adaptation to conditions of growth in culture follows. As division gathers momentum and the culture enters a period of exponential growth frequent sub-culture becomes necessary (cultures in this stage of their history can usually be preserved, perhaps indefinitely, by careful cooling and storage in liquid nitrogen). Eventually, growth slows down and after at most some forty to sixty cell generations ceases entirely. Occasionally, however, if such static cultures are nevertheless maintained by regular changes of medium, a new focus of growth may appear from which further sub-culture becomes possible. From this point on the culture is potentially immortal. There is therefore a vital difference in the expectation of further life before and after the period of growth stagnation. Cultures prior to this point in their history have been known as *cell strains;* after this point they are known as *cell lines.*

C. KARYOTYPIC COMPLEXITY

The karyotypes of cell strains are predominently normal though aneuploidy and structural changes are much more frequent than in cells taken directly from somatic tissues. Cell lines however, present a striking contrast. The cells exhibit a great range of variant karyotypes round a mode that often differs greatly from the normal karyotype of the species from which the culture was established. The term, *heteroploid,* is now often used to specify cell populations, like these, that show greater or lesser degrees of karyotypic diversity. In many instances the degree of variability is so great that no two cells are found with identical karyotypes in the samples examined. Despite the diversity, each cell line has its own unique cytogenetic features, a *modal number* of chromosomes, a specific range (of chromosome number) and often one or more marker chromosomes, not all of which are recognizable in all cells. The modal number is invariably hyperdiploid, often approximately triploid.

The same modal karyotype is normally maintained on sub-culture under constant conditions, but if the conditions of growth are varied and particularly if deliberate selection is practised (such as growth in a medium containing a particular antimetabolite) a change may take place and a new modal karyotype emerge. But if so, the karyotypic features of the derived cell population remain clearly, and usually closely, related to those existing before selection began.

New modal karyotypes may also be produced by "cloning", i.e. growing up new populations from single cells. Clones also quickly become complex mixtures of cells with different karyotypes. This observation shows that new karyotypic variation is continuously being generated. The new variation is mostly numerical but includes some that is structural as shown by the occasional appearance of new marker chromosomes.

D. EVIDENCE OF CONTINUOUS SELECTION

Current knowledge of genetics permits the formal subdivision of the chromosomes of any cell into segments of three kinds: segments consisting of "active" loci, whether structural or regulatory; genetically "inactive" segments containing structural and/or regulatory loci; and genetically "silent" segments subserving a mechanical function only, such as the centromeres and their associated chromatin (and there might be others concerned with maintaining the physical integrity of the chromosome).

It is highly unlikely that the additional chromatin acquired during the origin of the modal karyotype of the cell line should consist solely of segments of the last two types. Nor is it likely that the difference between modal karyotype and a variant karyotype should always be limited in the same way, though some may be. In short it *is* likely *a priori*, that the genomes represented by the different karyotypes of a cell line differ from one another in their content of active genetic loci. The cells of such a population would therefore be expected to differ among themselves in respect of many different biochemical properties which would be expressed collectively as hereditary differences in capacity for growth and proliferation, differences on which selection would inevitably act.

This view is supported by the following simple argument. If all cells in a culture were essentially equal in their capacity to proliferate under given conditions of growth, the appearance of new karyotypic variability should bring about a progressive increase in the total range of variability on continued sub-culture. This does not occur. Instead, both mode and range are normally conserved. If we can be absolutely sure of the relevant facts, namely that new karyotypic variation is continuously arising and that total variability is conserved, the inference is inescapable: the more extreme variants do not contribute proportionally to the perpetuation of the culture; they must be at a selective disadvantage and be progressively eliminated (Ford, 1964).

X. Cytogenetics and Neoplasia

A. Ascites Tumours of Rodents

The transmissible ascites tumours of rats and mice grow as populations of free cells in the peritoneal fluid and can be perpetuated indefinitely by serial innoculation into new recipients of the same inbred line. The cytogenetic features of these tumours have much in common with those of the cell lines just described. Each tumour has its own specific modal number and specific range of variation around the mode. Many have one or more marker chromosomes and in all there is continuous generation of fresh numerical and structural variation. The degree of karyotypic diversity within the population may in general be less than in the cell cultures and the modal karyotype the more readily discerned. But differences in the extent of variation and the rate of accretion of new variation are not important in themselves. What is important is the often striking deviation of the modal karyotype of the tumour from the normal karyotype of the species and the contrast between the karyotypic variability of the tumour cell population and the near constancy of the karyotype in normal healthy tissues of the animal.

B. STEM LINES

Cells with the modal karyotype are known collectively as the stem line. It was postulated that the tumour maintained itself by proliferation of the stem line and that cells of the other karyotypic classes arose from it as variants. A corollary of this proposition is that the total karyotypic variability of the whole population would increase in the absence of selection against the variant forms. But since the total range of variation remains approximately the same from one transplant generation to another we have the same argument as that just given for cell lines and an equivalent conclusion, namely that cells with the more extreme karyotypes must contribute less than the cells with the modal and less extreme karyotypes to the continued growth of the tumour cell population.

If continuous origin of new karyotypic variability and equally continuous selective elimination of cells with less successful genomes occurs in transplanted ascites tumours, could the same process take place in other types of transplanted tumour, and, more important, in primary neoplasms growing in their original hosts? The answer appears to be, "yes"; at the very least in many neoplasms and perhaps in all that are truly malignant.

Many primary neoplasms, most human carcinomas and sarcomas among them, show a greater degree of karyotypic variability than even the most variable cultured cell populations. It is therefore not surprising that for most of the early investigators of chromosomes in cancer cells their observations had no genetic significance. However, even in the most difficult cases, with good technique and careful analysis it may be possible to discern some element of order in the apparent chaos and obtain a hint of the existence of a modal karyotype. Unless *some* element of repeatability can be detected in the observations, there is nothing to show that an abnormal karyotype is other than the product of disordered growth itself and therefore, as many have argued, epiphenomenal.

C. CLONAL PROLIFERATION

Fortunately for the purposes of cytogenetic analysis, not all primary neoplasms present such a disordered picture as this. In many cases of acute leukaemia in man no trace of karyotypic abnormality has been found at all, though in these cases there is no assurance that the neoplastic cells themselves had entered mitosis in the PHA-stimulated blood cultures used for study. When abnormality *is* detected there is usually little variation between one cell and another: the abnormal cells may all have the same karyotype or a closely related one. The inferences are that the abnormal cells are all members of the same clone and that the variant karyotypes have arisen during proliferation of the clone. The modal karyotype is usually hyper-diploid, commonly with 47 or 48 chromosomes, but some hypodiploid cases have also been recorded.

Many tumours in the Chinese hamster also have abnormal clonal karyotypes with a modal chromosome number in the diploid range and little or no detected karotypic variation in the necessarily limited samples of cells that can be analysed (Ford and Clarke, 1963). The karyotype is often pseudo-diploid, with a chromosome number of 22 as in normal somatic cells of the species, but including one or more abnormal marker chromosomes. The cell populations in the Chinese hamster tumours and in the karyotypically abnormal cases of human leukaemia are comparable in cytogenetic structure with the clones of cells observed in normal tissue of the mouse and Chinese hamster following recovery from irradiation. But whereas the clones of cells in irradiated tissue are no more variable than normal somatic cells and appear to have conserved a normal, though rearranged, genome, the leukaemic and tumour cell populations invariably show at least some degree of variability and the modal karyotype often suggests that a quantitative change has taken place in the genome. This comparison leads on to an apparent paradox. In irradiated normal tissue, cells with quantitatively abnormal genomes are produced as a consequence of chromosome breakage and rearrangement yet do not proliferate; in neoplastic tissue, many cells with grossly abnormal genomes evidently proliferate without restraint. A reasonable solution to the paradox would be that the essential first step in neoplastic transformation occurs in a cell with a normal genome (or a cluster of such cells) and that the observed karyotypic changes arise secondarily and are incorporated into the expanding cell population when the physiological conditions have become in some way different from the conditions in normal somatic tissues.

D. ALTERNATIVE HYPOTHESES

It is therefore not necessary to suppose that karyotypic abnormality, when it is observed in a neoplasm, is to be referred back to changes that were part of the process of carcinogenesis itself. How then does the observed karyotypic complexity of so many neoplastic cell populations arise? Two formal hypotheses are possible. According to the one, a sudden, cataclysmic repatterning of the genome would produce the observed modal karyotype and subsequent disjunctional errors and structural changes would produce the remaining observed variation about the mode. This hypothesis does not require the assumption of selective differentials between cells with different karotypes. According to the other hypothesis, the changes would be accumulated, single step by single step, each in turn conferring a selective advantage on the cell in which it had occurred and its descendants. It is, of course, not necessary for this hypothesis that *all* karyotypic changes should be of selective value. On the contrary, since disjunctional irregularity and structural change are essentially random processes, it is to be expected that most karyotypic variation would be deleterious to the cell as a neoplastic cell rather than advantageous and lead eventually to the elimination of the descendants of the cells concerned.

It will be evident that primary neoplasms, except perhaps under special experimental conditions, can only be studied after they have given clinical evidence of their presence. Access to the all-important period of early development is therefore denied and it may be impossible to exclude either hypothesis by direct observation. However, it seems most unlikely that a random cataclysmic change could chance to produce an efficiently balanced neoplastic genome, or alternatively, that a cataclysmic change could be at the same time a directed change. Moreover, at least in the transplanted rodent ascites tumours, there is positive evidence of a selective process related to karyotypic differences. The second hypothesis is therefore preferred. It has the advantage of accounting for two outstanding features in the biology of malignant neoplasms. One is the uniqueness of individual neoplasms, even within narrowly defined pathological classes, when studied in detail by histological, biochemical or immunological methods. The second is the progressive change through the course of the disease from cells that must originally have been not very different from those of the normal tissue origin to the rapidly dividing, anaplastic cells of later stages. Cataclysmic restructuring of the genome could formally account for neoplastic uniqueness, but not for progressive change in the histological (and presumptively, biochemical) character of the cells.

XI. Cell Hybrids

A. Induction of Cell Fusion

Although some cell hybrids had been obtained before 1965 as a result of natural cell fusions in mixed cell cultures it was not until that year that a general method for producing them was introduced by Harris and Watkins (1965). A true cell hybrid involves the fusion of first the cytoplasm and then the nuclei of cells of different origins.

The intermediate stage, when the cytoplasm has fused but the nuclei have not, is called a *heterokaryon*.

The essential step in the method of Harris and Watkins was to add particles of Sendai virus that had been inactivated by exposure to ultraviolet light to a mixture of two types of cell growing together in culture. Evidently it is the virus particles that promote fusion and Harris has pointed out that there had been many hints in the older literature that sometimes a virus infection might promote natural fusions of cells in certain human diseases. Electron microscope studies have shown that the virus particles first attach to the cell membrane and that subsequently fine cytoplasmic bridges form between cells in close contact which widen until the whole cell membrane in the region of contact disappears and the cytoplasm becomes completely confluent. In some instances the hybrid cells have continued to proliferate and have given rise to hybrid cell lines. Cell hybridization is now a widely practiced technique, valuable in many different kinds of investigation, notable among them being questions concerning the transmission of information from nucleus to cytoplasm and from cytoplasm to nucleus.

Hybrid cells can be obtained not only between rapidly proliferating cells of different origin growing in culture, but also between cultured cells as one component and cells taken directly from the body as the other. The latter have included small lymphocytes, macrophages and the nucleated erythrocytes of birds. Hybridization is now regularly achieved between cultured cells of different mammalian species and some have even been obtained between a mammalian cell line and cells from a non-mammalian vertebrate species such as between human HeLa cells and the nucleated erythrocytes of the domestic hen.

Heterokaryons can often be positively identified by the microscopic character of the nuclei contributed by the two parental cell types. Subsequent to nuclear fusion other methods are necessary to demonstrate hybridity. If the cell continues to proliferate, direct examination of the chromosomes can provide proof (Harris *et al.*, 1966). When cell lines derived from the same species are hybridized the presence of distinct marker chromosomes contributed by the two parental types is decisive. Failing karyotypic evidence, immunological methods can be used.

B. SELECTIVE PROCEDURES

When the hybridization procedure is applied to a mixture of cells from two different actively proliferating cultures it might be thought that an insufficient number of hybrids would be formed to prevent their being outgrown by descendants of the remaining unfused cells.

This is a real difficulty and several different selective procedures have been devised to overcome it. Perhaps the most elegant method is to use cells that lack a vital enzyme and only grow when its product is supplied in the culture medium. If the hybridization procedure is applied to a mixture of two such cultures, each lacking a different enzyme, and the cells subsequently transferred to minimal (i.e. unsupplemented) culture medium, the only cells able to grow and multiply are the hybrids, which have both enzymes and can make both products for themselves. This method is an exact formal parallel to the one first used to demonstrate genetic recombination in bacteria.

C. CHROMOSOME LOSS AND THE DETECTION OF LINKAGE

In the first inter-species hybrid cell lines studied the initial hybrid karyotype was found, as expected, to be approximately the sum of the two parental karyotypes. But after several generations of subculture it was discovered that the total number of chromosomes tended to decrease. This process of chromosome loss was found to be generally much more extreme in inter-species hybrids than intra-species hybrids. Furthermore, the chromosomes lost were largely, if not quite exclusively, from one of the two parents. This phenomenon has now been generally confirmed and promises to provide a powerful general method for identifying genetic linkages and for locating genes on the chromosomes that carry them. For example, Weiss and Green (1967) produced a hybrid cell line between metabolically normal human cells and mouse cells that lacked the enzyme thymidine kinase. After first isolation of the hybrid cell line growth was continued in medium lacking thymidine. Progressive loss of the human chromosomes took place until the complement contained only a single human chromosome. It was therefore concluded that this chromosome, a no. 17 or no. 18, carried the structural locus for the enzyme thymidine kinase. The limitations of this method appear to be set only by the availability of cell lines lacking the capacity to synthesize specific metabolites.

D. GENETIC ANALYSIS OF THE NEOPLASTIC STATE

But perhaps the most outstanding achievement of the cell hybridization technique has been the demonstration that the property of malignancy can be suppressed by hybridization of malignant cells with cells from a non-malignant cell line (Harris, 1971). This observation has now been confirmed in several combinations of different mouse tumour cells with non-malignant cells, including normal diploid fibroblasts. The criterion of malignancy in these experiments was the formation of a tumour following injection of standard recipient mice by a standard cell inoculum and progressive growth of the tumour to death of the host. The detailed results indicate that the effect is better described as change from a high level of malignancy to a very low one rather than absolute suppression.

Even more remarkable, in several instances, when cells from two highly malignant cell lines were fused together, the hybrid product proved to be non-malignant. Whatever is responsible for the suppression of malignancy must therefore be different in the two cell lines which, in a sense, complement one another to restore "normality". Relatively few experiments of this kind have yet been done and it may be that more than two "complementation groups", possibly many more, will eventually be found, implying as many different ways of "suppression" of malignancy.

A further result completes a very striking trio. When non-malignant hybrid cells with at least one malignant parent are cultured *in vitro* the usual steady reduction in number of chromosomes takes place and tests of the derived cells have shown that in some cases the property of malignancy has re-emerged. This result raises the possibility that malignancy was suppressed by a particular chromosome and that subsequent loss of this chromosome led to restoration of malignancy. If this should be true, the complementation experiment would imply that at least two different chromosomes (perhaps specific loci on these chromosomes) have the capacity to suppress malignancy. It may be that the eventual solution will not be quite so simple as this, but whatever the explanation, the hybridization method offers the possibility of gaining a new insight into the nature of the genetic factors involved in neoplasia.

Further Reading

Darlington, C. D. (1965). "Cytology". J. and A. Churchill Ltd., London.

Darlington, C. D. (1958). "The Evolution of Genetic Systems", 2nd edn. Oliver and Boyd, Edinburgh and London.

DuPraw, E. J. (1970). "DNA and Chromosomes". Holt, Rinehart and Winston, Inc., New York.

Evans, H. J., Court-Brown, W. M., and McLean, A. S. (1967). Editors, "Human Radiation Cytogenetics". North Holland Publishing Company, Amsterdam.

Ford, C. E. (1970). "Cytogenetics and Sex Determination in Man and Mammals". *J. biosoc. Sci. Suppl.* **2**, 7–30.

Hamerton, J. L. (1971). "Human Cytogenetics", Vol. I "General Cytogenetics". Vol. II. "Clinical Cytogenetics". Academic Press, New York and London.

Harris, Henry (1968). "Nucleus and Cytoplasm". Clarendon Press, Oxford.

Harris, Henry (1970). "Cell Fusion". Clarendon Press, Oxford.

Lewis, K. R. and John, B. (1963). "Chromosome Marker". J. and A. Churchill Ltd., London.

White, M. J. D. (1973). "Animal Cytology and Evolution". 3rd edn. Cambridge University Press, Cambridge.
Whitehouse, H. L. K. (1965). "The Mechanism of Heredity". Edward Arnold Ltd., London.

References

The references were chosen either as classic sources or as giving new information of particular current interest; they are necessarily incomplete. Original publication in general mammalian cytogenetics is principally in the journals "Chromosoma" and "Cytogenetics and Cell Genetics".

Buckton, K. E., Court Brown, W. M. and Smith, P. G. (1967). *Nature, Lond.* **214**, 470–473.
Callan, H. G. (1972). *Proc. Roy. Soc. B.*, **181**, 19–41.
Carr, D. H. (1971). Advances in Human Genetics, **2**, 201–257.
Caspersson, T., Lomakka, G. and Zech, L. (1971). *Hereditas* (*Lund*), **67**, 89–102.
Chicago Conference. (1966). Birth Defects, Original Article Series, **2** (2), 1–21. The National Foundation—March of Dimes, New York.
Dunn, L. C. and Bennett, D. (1967). *Genet. Res.*, **9**, 211–220.
Ford, C. E. and Clarke, C. M. (1963). *In* Canadian Cancer Conference, **5**, 129–146, (Begg, R. W., ed.) Academic Press, New York and London.
Ford, C. E. (1964). *In* Cytogenetics of Cells in Culture, Symposium of the International Society for Cell Biology, **3**, 27–45. Academic Press, New York and London.
Ford, C. E. (1972). *Proc. Int. Symp.* "The Genetics of the Spermatozoan" (Beatty, R. A. and Gluecksohn-Waelsch, S., eds.), 359–369 (Edinburgh and New York).
Gardner, R. L. (1968). *Nature, Lond.*, **270**, 596–597.
Green, M. C. (1966). *In* "The Biology of the Laboratory Mouse", 2nd edition, (Green, E. L., ed.) McGraw Hill, New York.
Green, M. C. (1972). Jackson Laboratory Annual Report.
Harris, H. and Watkins, J. F. (1965). *Nature, Lond.*, **205**, 640–646.
Harris, H., Watkins, J. F., Ford, C. E. and Schoefl, G. I. (1966). *J. Cell Sci.*, **1**, 1–30.
Harris, H. (1971). *Proc. Roy. Soc. B.*, **179**, 1–20.
Huberman, J. A. and Riggs, A. D. (1966). *Proc. Nat. Acad. Sci.* (*Wash.*), **55**, 599–606.
Hungerford, D. A. (1971). *Cytogenetics*, **10**, 23–32.
Jacobs, P. A. and Ross, A. (1966). *Nature, Lond.*, **210**, 352–354.
Jones, K. W. (1970). *Nature, Lond.*, **225**, 912–915.
Jost, A. (1947). *Archs. Anat. microsc. Morph. exp.*, **36**, 271–315.
Moorhead, P. S., Nowell, P. C., Mellman, W. J., Battips, D. M. and Hungerford D. A. (1960), *Expl. Cell. Res.*, **20**, 683–000.
Mystkowska, E. T. and Tarkowski, A. J. (1968). *J. Embryol. exp. Morph.*, **20**, 33–13.
Ohno, S. (1971). *Nature, Lond.*, **234**, 134–137.
Pardue, M. L. and Gall, J. G. (1970), *Science N.Y.*, **168**, 1356–1358.
Paris Conference (1971): Standardisation in Human Cytogenetics. *Cytogenetics*, **11**, 313–362 (1972).
Renwick, J. H. (1969). *Brit. Med. Bull.*, **25**, 65–73.
Roderick, T. H. and Hawes, N. L. (1970). *Proc. Nat. Acad. Sci.* (*Wash.*), **67**, 961–967.

Seabright, M. (1971). *Lancet*, **ii**, 971–972.
Tarkowski, A. J. (1961). *Nature, Lond.*, **190**, 857–860.
Turpin, R., Lejeune, J., Lafourcade, J., Chigot, P. L. and Salmon, C. (1961). *C. r. hebd. Seanc. Acad. Sci., Paris*, **252**, 2945–000.
Weiss, M. C. and Green, H. (1967). *Proc. Nat. Acad. Sci. (Wash.)*, **58**, 1104–1111.

11. An Introduction to Human Cytogenetics

C. E. BLANK

Sub-Department of Medical Genetics, University of Sheffield, England

I. Introduction

A. Historical Notes

Early twentieth century cytogeneticists, mainly working with sectioned testicular tissue, concluded that man probably had a normal chromosome complement of 48. However, in 1956, Tjio and Levan, two Swedish botanical cytologists, using more advanced techniques of chromosome analysis, demonstrated that man normally has 46 chromosomes. This observation was quickly confirmed by Ford and Hamerton (1956) working in England. They further demonstrated that the human chromosome complement consisted of 22 pairs of autosomes and a pair of sex chromosomes, XX in the female and XY in the male. Tjio and Levan used tissue cultures of embryonic lung fibroblasts and accumulated cells in the metaphase of mitosis by using colchicine. They studied the chromosomes in unsectioned cells and assisted separation of the chromosomes by swelling the cells in hypotonic solution before fixation. These technical advances resulted in preparations that were relatively easy to interpret and stimulated the development of techniques promoting the growth and division of human cells *in vitro*. In 1958 Ford *et al.*, developed a method for obtaining cells suitable for chromosome analysis from bone marrow and Harnden, a fourth member of this group of workers at Harwell, England, successfully cultured fibroblasts from skin biopsy. It was thus no longer necessary to confine cytogenetic investigations to testicular tissue. Chromosome make-up could now also be studied in females and in azoospermic males. A great impetus had been given to the discipline of human cytogenetics and in 1959 Hungerford *et al.*, in America developed a technique for the preparation of cultured white blood cells for chromosome analysis using a bean extract, phytohaemagglutinin, as a mitotic stimulator. Lymphocyte culture has now become the standard method of obtaining cells for chromosome analysis.

In 1959 Lejeune *et al.* in Paris showed that children with mongolism had 47 chromosomes; moreover the extra chromosome was always a member of a particular group of chromosomes. This chromosome has been allocated a number (21) and the cytogenetic state referred to as trisomy 21. Later that year it was shown that two abnormalities of sexual development, Turner's syndrome in the female and Klinefelter's syndrome in the male, were associated with specific abnormality of the chromosome complement. Ford and his colleagues at Harwell showed that patients with Turner's syndrome had 45 chromosomes with only one sex chromosome, an X. Jacobs and Strong in Edinburgh described an XXY sex chromosome complement in a patient with

Klinefelter's syndrome. Early in 1960 Polani *et al.*, at Guy's Hospital, London, discovered a mongol with 46 chromosomes but an abnormal chromosome complement. Besides two normal 21 chromosomes, the subject had a structurally abnormal chromosome belonging to group D that incorporated most of yet another chromosome number 21. She therefore had the greater part of chromosome 21 present in triplicate; and was in effect trisomic for chromosome 21. This type of abnormality (translocation) suggested a mechanism (see p. 162) for details) explaining the occurrence of mongolism in several members of the same family (Fig. 28). Later that year Penrose *et al.*, at University College, London reported a family where the translocation type of chromosomal abnormality of chromosome 21 was present in genetically balanced (i.e. with an essentially normal gene content) and genetically unbalanced (mongol) form. Since then many different chromosome abnormalities have been observed as a cause of disorder of physical and/or mental development and several new clinical syndromes defined. The pathogenesis of many disorders of development, particularly sexual development, have become much better understood as a result of cytogenetic study. It is now clear that chromosome abnormalities are a common cause of developmental disorder in man.

B. CHROMOSOME COMPLEMENT

Human chromosomes at metaphase of mitosis (Fig. 1) consist of pairs of chromatids joined at the centromere. They may be classified with respect to length and with respect to the position of the centromere. Where the centromere is not near the middle of the chromosome arms of unequal length are observed. Particular chromosome pairs may have distinctive arm ratios. If chromosomes in a metaphase preparation (metaphase plate) are arranged with regard to length and centromere position seven groups of chromosomes, designated A–G in descending order of length (Fig. 2), can be distinguished with confidence. Group A may usually be subdivided into autosome pairs 1, 2 and 3 and pair 16 may usually be distinguished from the other autosomes of Group E. The X chromosome is one of the largest chromosomes of Group C, but is not recognizable using these simple morphological criteria. The Y chromosome often has parallel chromatids and sometimes a constriction on the long arm. It is usually distinguishable from the autosomes of the G group. Certain autosomes also have distinguishing features other than length and centromere position: for example the chromosome of groups D and G may have "satellites" on the short arms and a particular chromosome (designated No. 9) of the C group may have a constriction on the long arm.

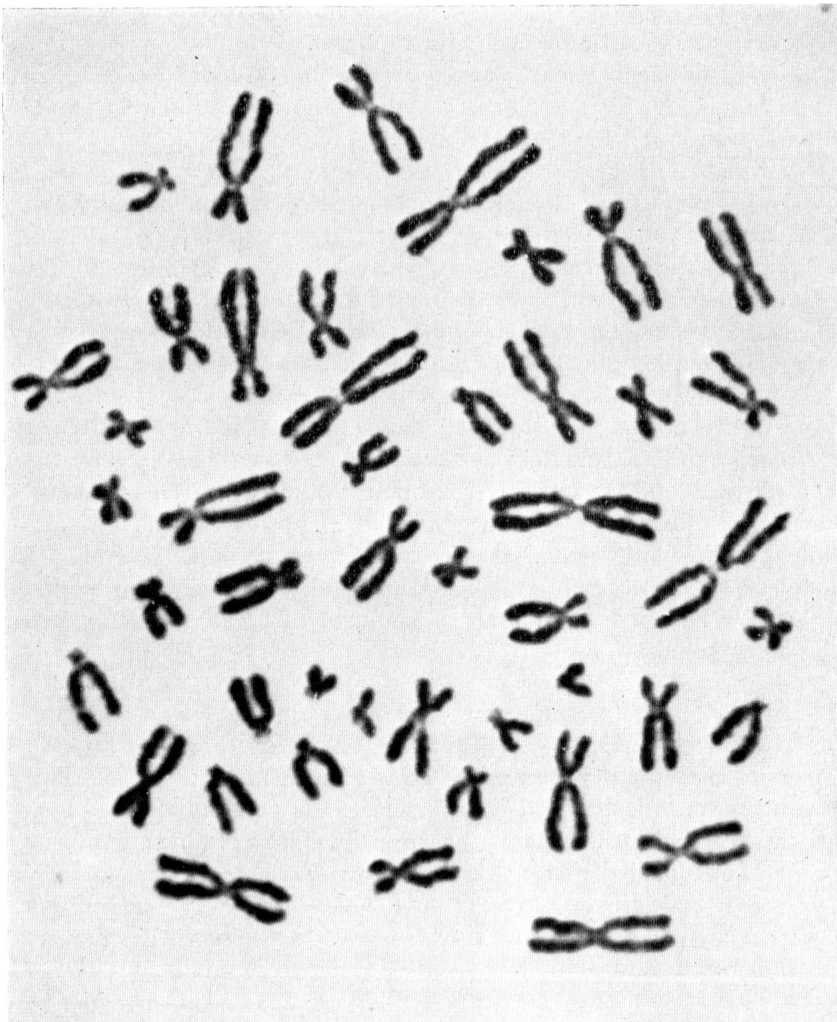

Fig. 1. A metaphase plate preparation as seen under the microscope. The longest chromosome is about 10 μm in length. Lymphocyte culture.

Duplication or deficiency of chromosomal material is suspected when the number or morphology of chromosomes in a group is different from the normal. Chromosome abnormality may cause gene imbalance severe enough to prevent normal development, or may even be incompatible with life.

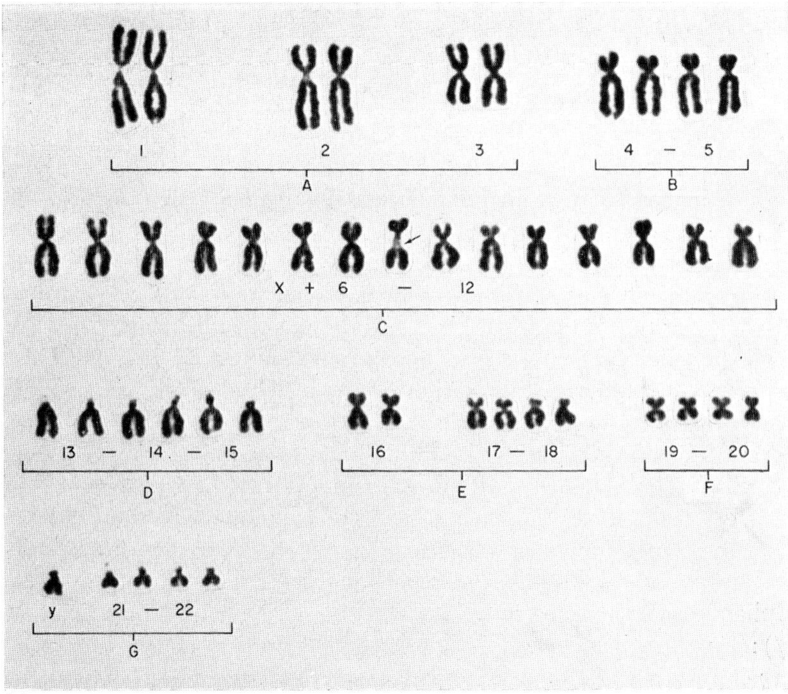

(a)

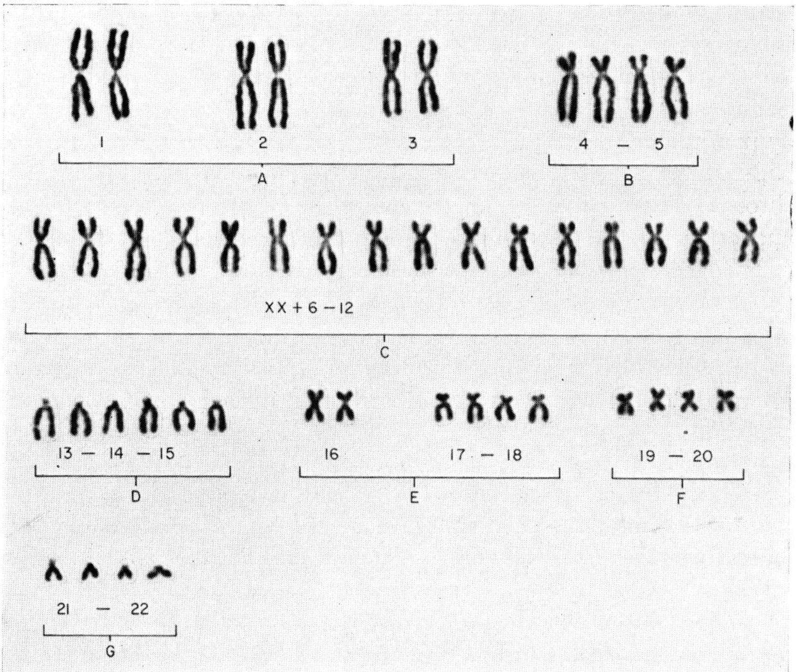

(b)

Fig. 2. (a) Normal male chromosome complement. Secondary constriction (characteristic of chromosome No. 9) is arrowed. (b) Normal female chromosome complement.

In order that the reader may be able to interpret the nomenclature used in current papers on human cytogenetics the chromosome complements discussed in this chapter are formally described in the manner recommended by the Chicago Conference (1966). The first item to be recorded is the total number of chromosomes. The sex chromosome constitution is given next: for example 47,XXY indicates 47 chromosomes with an XXY sex chromosome complement. The autosomes are specified only when there is an abnormality present. Thus, if there is a numerical aberration of the autosomes, the group letter of the extra or missing autosome, followed by a plus (+) or minus (−) sign, succeeds the sex chromosome designation: for example 48, XXY,21 + indicates 48 chromosomes, an XXY sex chromosome complement and an additional No. 21 chromosome. Where two or more cell lines are observed (chromosome mosaics) the chromosome constitutions of the different cell lines are listed in numerical or alphabetical order. The different cell types are separated by a diagonal (/): for example 45,X/46,XY indicates a chromosome mosaic differing in sex chromosome complement. Where structural abnormality is described the short arm of a chromosome is designated by the letter "p" and the long arm by the letter "q". Increase in length of a chromosome is indicated by placing a plus sign (+) and decrease in length by placing a minus sign (−) after the arm designation: for example Bp−. A translocation is indicated by the letter "t" followed by parenthesis which includes the chromosome involved: for example 45,XX,D−,G−, t(DqGq)+ indicates 45 chromosomes, an XX sex chromosome complement, one chromosome missing from the D group and one from the G group, their long arms having united to form a DG translocation chromosome. Isochromosomes, a situation in which one chromosomal arm is duplicated (see also Chapter 10), are designated by the letter "i" placed after chromosome arm involved: for example 46,XXqi. Ring chromosomes (Chapter 10) are indicated by the letter "r" placed after the chromosome involved: for example 46,XXr would indicate a ring X. The term karotype should be applied to a systematized array of the chromosomes from a single cell prepared either by drawing or by photography, with the extension in meaning that the chromosomes of a single cell can typify the chromosomes of an individual. The term idiogram is reserved for the diagramatic representation of a karyotype. Further conventions of use, when formally describing chromosome complements not referred to in this chapter, are to be found in the Chicago Conference (1966).

C. DNA REPLICATION PATTERN

Some chromosomes undergo DNA replication out of phase with the rest of the chromosome complement: a notable example is the late replicating X chromosome in the female (see Chapter 12). Replication pattern can be studied (Fig. 3) using autoradiographic techniques.

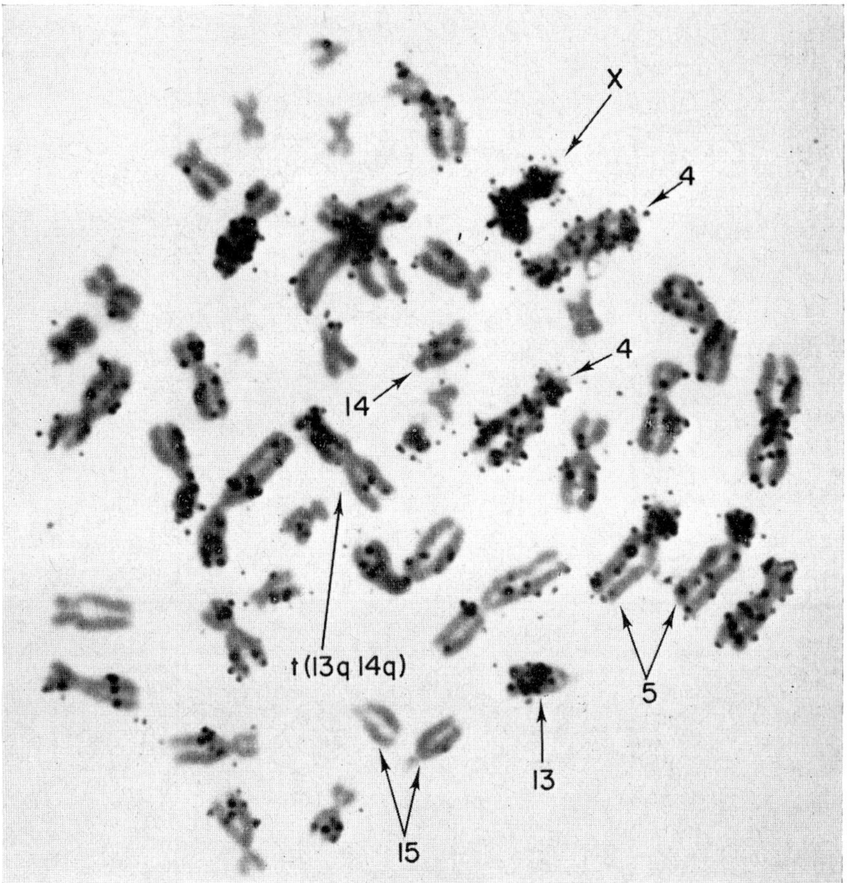

Fig. 3. Autoradiography showing the pattern of tritiated (^3H) thymidine uptake at *end* of DNA replication. Heavily labelled chromosomes or chromosome segments have replicated later than the remainder. They include one of the two X chromosomes (arrowed). A characteristic pattern of labelling is seen on chromosomes No. 4, 5, 13, 14 and 15. Note: t (13q14q).

Combining morphological characteristics and DNA replication peculiarities it is possible to identify individual pairs of chromosomes in groups B, D and E. Autoradiography is a useful aid to the identification of extra or abnormal chromosome material.

TABLE 1. Frequency of chromosome abnormality among liveborn infants

Chromosome abnormality (per 1 000 live births)		Associated developmental abnormality
Trisomy 21	1·5	Mongolism
Trisomy E	0·2	Edwards' syndrome
Trisomy D	0·1	Patau's syndrome
Deletion Short Arm B	0·05	Cri-du-chat
Miscellaneous Autosome Abnormality (unbalanced)	0·2	Physical and mental maldevelopment of varying degree.
Sex Chromosome Abnormality	2·5	About one half will show development anomaly of sex and/or mental disorder. The rest are clinically normal.
Autosome Rearrangements (balanced)	5·0	Clinically normal, but may be infertile.
"Normal" Autosomal Variants	25·0	Clinically normal.
"Normal" Variations of the Y Chromosome	15·0	Clinically normal.
All types: approximately	50·0	Nine out of 10 individuals with a chromosome complement distinguishable from the normal are physically and mentally normal. Nevertheless, at least one in 250 liveborn infants show severe developmental defect of chromosomal origin.

D. NORMAL VARIANTS

Developmental abnormality is not a necessary consequence of an unusual chromosome configuration (Table 1 and Fig. 4). Many variants are probably produced by irregular coiling of chromosomes or by minor duplications or deletions without pathological significance. Consequently, it is sometimes difficult to be confident that a chromosome abnormality observed in patients with physical or mental maldevelopment is not an incidental finding. A chromosomal variant may often be seen in several members of a family, all of whom are phenotypically normal.

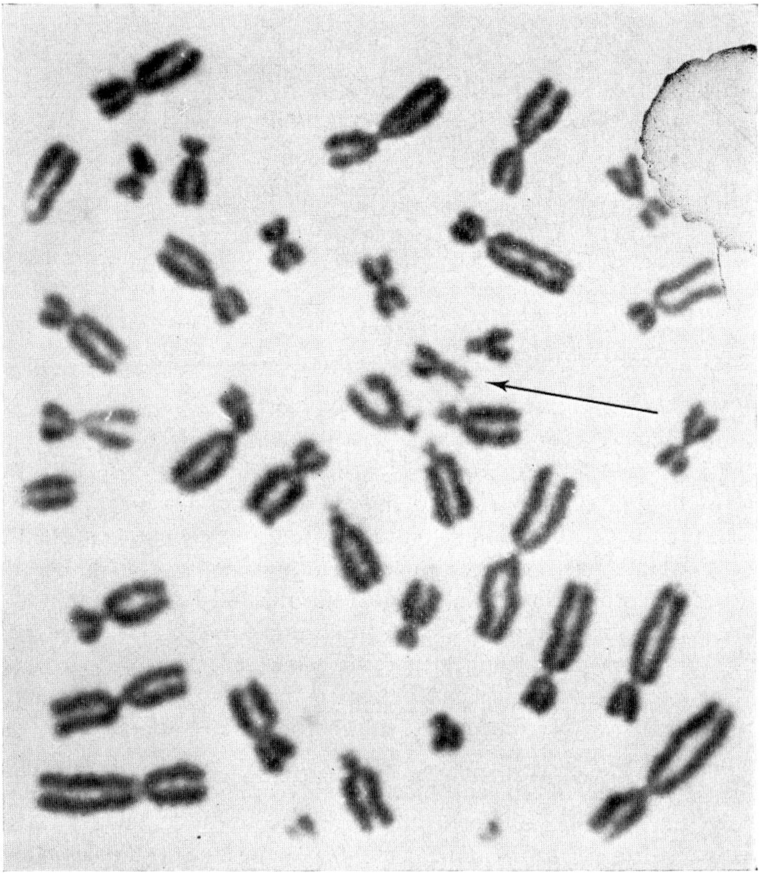

Fig. 4. Normal variant (arrowed). This same "marker" chromosome was also observed in the mother of this child. Both mother and child were clinically quite normal.

E. SEX CHROMATIN

A valuable auxiliary to chromosome analysis in the investigation of development anomalies of sex is the determination of the sex chromatin state (see Chapter 12). A convenient tissue in which to study sex chromatin is the buccal mucosa (Fig. 5). The sex chromatin body (Barr body) is derived from a single X chromosome in a compact (heteropyknotic) form. In a normal male a Barr body is never seen; in a normal female about 30% of nuclei show a single Barr body, the actual count depending to some extent on the staining technique and method of counting.

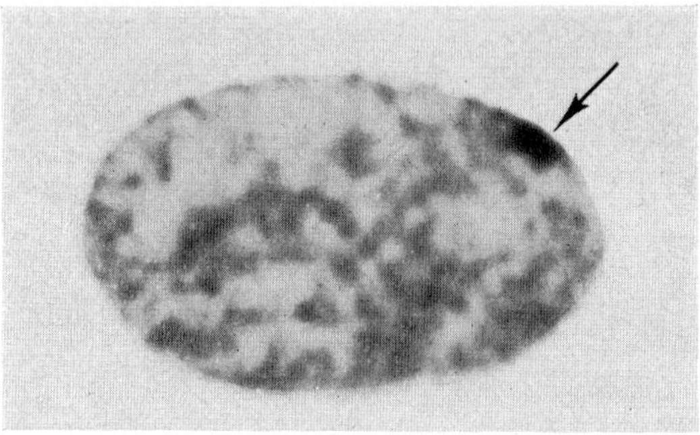

Fig. 5. Sex chromatin in buccal mucosa. A Barr body, about 1 μm in diameter, is seen (arrowed) under the nuclear membrane.

Individuals with a single Barr body in a proportion of buccal smear nuclei have two X chromosomes: individuals with two, three or four Barr bodies have three, four or five X chromosomes respectively. If sex chromatin is absent then only one X chromosome is present. A large, or small, Barr body may indicate abnormality of an X chromosome. If the proportion of cells showing a Barr body is outside normal limits, or the distribution of cells containing Barr bodies is uneven, a sex chromosome abnormality may be suspected, for example a low sex chromatin count sometimes indicated that the patient is a mixture of cell types with respect to sex chromosome complement (sex chromosome mosaicism).

A second type of sex chromatin body is the drumstick, observed in polymorphonuclear leucocytes (see Chapter 12). In a normal female six drumsticks or more per 500 polymorphonuclear leucocytes are usually observed. A normal male has very few. Like the Barr body, the drumstick is a heteropyknotic X chromosome. Drumsticks are not easy to score, even by the expert, and not as much information is to be obtained by examination of a blood smear for sex chromatin as examination of a buccal smear.

F. THE FLUORESCENT Y CHROMOSOME

While sex chromatin studies can tell us a lot about the number, distribution and structure of the X chromosomes present they can tell us nothing about the Y chromosomes. It is the presence of a Y chromosome that determines the development of embryonic testes. Even where two, three or four X chromosomes are present, a single Y chromosome is sufficient to determine that the bipotential gonad shall become a testis: albeit a poorly differentiated and inadequate one.

Recently a technique for the identification of the Y chromosome in interphase and metaphase nuclei, using a fluorescent stain, has been developed. At present this technique is experimental, but undoubtedly will eventually provide a useful additional means of studying the sex chromosome complement and might also aid the identification of pairs of autosomes (see also Chapter 12).

G. TISSUES THAT MAY BE EXAMINED FOR CHROMOSOME COMPLEMENT

In clinical practice chromosome analysis is carried out on the leucocytes of blood, the fibroblasts of the dermis and other connective tissues, the cells of the bone marrow, fetal cells derived from amniotic fluid and, rarely, on cells undergoing meiosis in the gonad. A chromosome analysis may be provided within 4 days of obtaining a specimen for blood culture but if a really quick analysis is required the bone marrow technique can provide metaphase plates within 24 hr of obtaining a sample. A karotype report of fetal cells is possible within a fortnight of obtaining a specimen of cells from the amniotic fluid. Identification of the sex chromatin state by examination of a buccal smear obtained by scraping the buccal mucosa can take less than four hours and is reliable but the demonstration of a fluorescent Y chromosome can not yet be regarded as unequivocal. For a more detailed discussion of technical methods for the display of chromosomes see Chapter 10.

H. FREQUENCY OF CHROMOSOME ABNORMALITY AS A CAUSE OF DEVELOPMENTAL ABNORMALITY

Chromosome abnormality is an important cause of congenital abnormality and abortion. At least one in 250 live born children have, or will manifest by the time they reach puberty, gross defects of physical and/or mental development that are of chromosomal origin

(Table 1). Approximately one in three spontaneous abortions in early pregnancy are of chromosomal origin and the frequency with which chromosome abnormality produces stillbirth is at present unknown. Chromosome abnormality is an occasional cause of mental handicap independent of gross physical abnormality. At least one in six patients who come to the attention of the gynaecologist because of failure to menstruate spontaneously have a sex chromosome abnormality or a male sex chromosome complement.

Chromosome analysis has become an important tool to the physic-

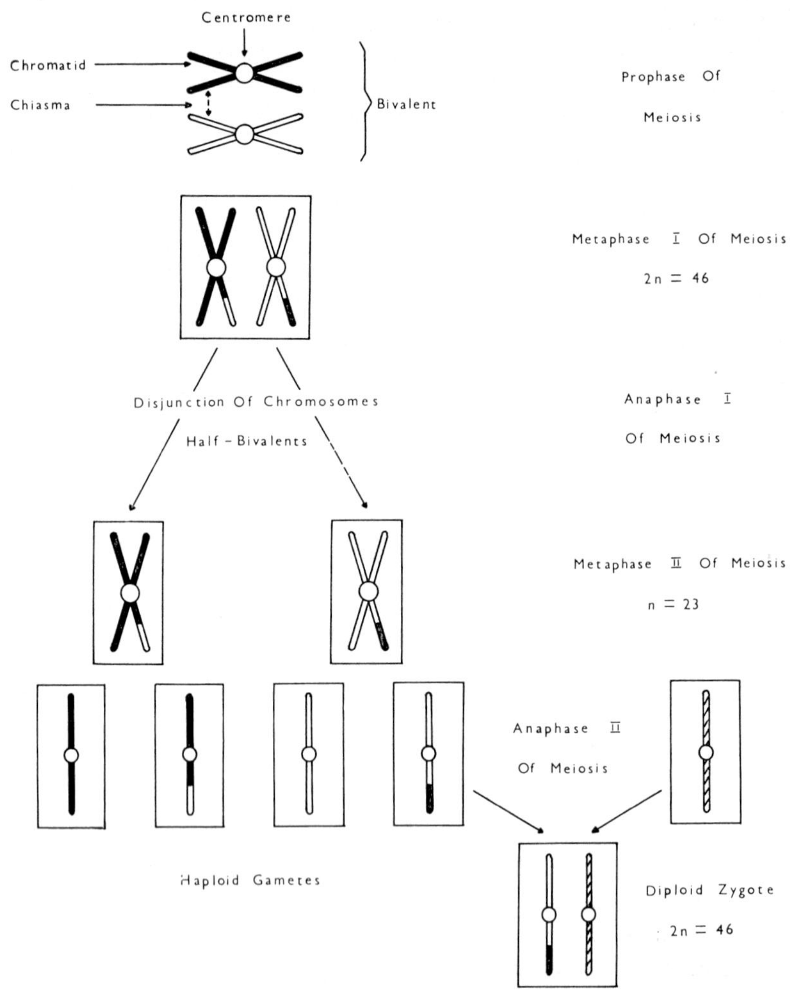

Fig. 6. (a) Normal meiosis; only one pair of chromosomes is represented.

ian; knowledge of the chromosome complement being especially useful in genetic counselling and when a syndrome known to be associated with chromosome abnormality is suspected. The fetal chromosome complement is of fundamental importance where termination of pregnancy is being considered because of an increased risk of a child being born with deformity due to chromosome abnormality.

II. Origin of Chromosome Abnormality

Chromosome abnormality may be divided into two main types:—(a) An error in the number of chromosomes and (b) Structural aberration among the chromosomes. A general description of the origin of chromosome abnormality is to be found in Chapter 10.

A. NUMERICAL ANOMALY

Error in the number of chromosomes present (aneuploidy), usually arises as a result of irregular division in parental gametogenesis. Non-

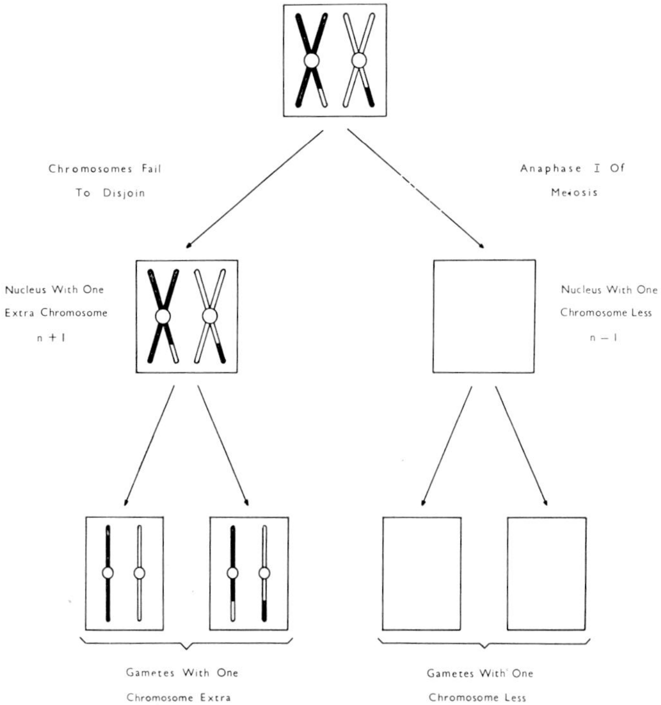

Fig. 6. (b) Non-disjunction; both half bivalents go to same nucleus at first anaphase of meiosis.

disjunction (Fig. 6b), results in the formation of a cell with one chromosome extra and a cell with one chromosome less. At fertilization this will result in a zygote with one chromosome extra (trisomy) or one chromosome less (monosomy). Loss of a chromosome may also result from anaphase lag, in which case a chromosome is simply not included in *either* daughter nucleus. Non-disjunction or anaphase lag can occur at the first or second stage of meiosis. When irregular division occurs during early embryogenesis two or more chromosomally distinct cell lines may be produced, the resultant condition being known as chromosome mosaicism.

When two spermatozoa fuse with a single ovum (diandry) or when the first or second polar body is not excluded from a fertilized ovum (digyny) a zygote with 69 chromosomes and an XXY, XYY or XXX sex chromosome complement is formed. Such triploid zygotes usually abort at an early stage in pregnancy and account for about 5% of spontaneous abortions. It is probable that most instances of triploidy are related to digyny. Simultaneous fertilization of an ovum and a polar body by spermatozoa carrying different sex chromosomes, one an X and the other a Y, can produce a 46,XX/46,XY zygote which may develop into a true hermaphrodite.

B. STRUCTURAL ABERRATION

A chromosome may break at one or more points. In consequence

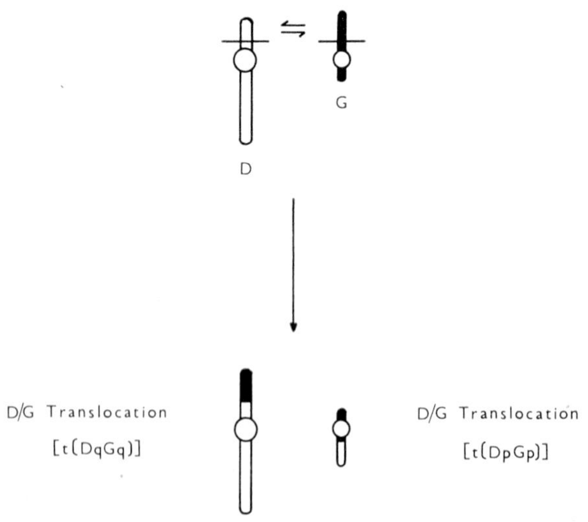

Fig. 7. D/G Translocation (Robertsonian): t(DqGq). The product t(DpGp) is usually lost in subsequent cell division.

a piece of chromosome may be lost (deletion) or transferred to another chromosome (translocation) (Fig. 7). When both ends of a chromosome are lost and fusion occurs between the ends of the remaining arms a ring chromosome is formed. Translocation may occur in a genetically balanced form, in which case the individual is clinically normal, or in a genetically unbalanced form, when the duplication and/or deficiency of genetic material causes developmental abnormality. Structural aberration may be produced during gametogenesis or during embryogenesis.

III. Clinical Syndromes Associated With Autosome Abnormality

Low birth weight for length of gestation, mental subnormality and peculiar, sometimes characteristic, facies are common findings in cases of autosome abnormality that are of clinical significance. Specific cytogenetic states, for example Trisomy E, may produce characteristic clinical syndromes, although some of the development defects that compose these syndromes are sometimes associated with other syndromes as well.

Conditions known to be due to an extra autosome include mongolism (Down's syndrome), Edwards' syndrome and Patau's syndrome, (Table 2). In the Cri-du-chat syndrome a part of a chromosome is absent. Autosome abnormality may sometimes be detectable in non-specific developmental abnormality, although it is probable that some of the latter will with further experience be seen to constitute recognizable clinical syndromes (Table 3).

A. Mongolism (Down's Syndrome)

The clinical syndrome of mongolism (Table 2 and Figs. 8 and 9), is secondary to a specific gene imbalance usually observed as an additional chromosome in the G group. Although it is not possible to distinguish morphologically between chromosome pair 21 and chromosome pair 22, the extra chromosome is traditionally referred to as a 21 and the cytogenic state as trisomy 21; or formally (Chicago Conference, 1966) 47, 21 +. Very recently flourescent microscopy has identified the extra chromosome observed in mongolism as belonging to the shortest pair of the group. *Primary* trisomy 21, (Fig. 10), in which an extra whole chromosome is present, is the usual type of chromosome abnormality seen in mongolism. Less common is translocation trisomy 21, in which transfer (translocation) of a chromosomal segment has occured between a chromo-

TABLE 2. Autosome abnormality associated with established clinical syndromes

Clinical syndrome	Main clinical features	Cytogenetic state invariably or often observed
Mongolism (Down's syndrome)	Characteristic facies (Fig. 8) with small head, flat occiput and low set ears of unusual shape. Palpebral fissures slope upwards and outwards. Vertical folds of skin at the inner canthus of the eye (epicanthic folds) often present and nose often small. Tongue frequently protrudes from a small mouth. Micrognathia. I.Q. between 20 and 60. Poorly developed physically but often overweight. Hypotonia in infancy. Permanent bending (clinodactyly) of little finger with absence of distal flexion crease. Wide gap between big toe and 2nd toe. About 20% have heart defect. Characteristic dermatoglyphics, (Fig. 9). Those surviving infancy have a relatively normal life span.	Trisomy 21; or formally (Chicago Conference, 1966) 47, 21+
Edwards' Syndrome:	Low birth weight. Characteristic facies (Fig. 12a) with small head, prominent occiput, low set and malformed ears. Cleft lip and/or palate sometimes present. Micrognathia. Brain deformity sometimes observed at	Trisomy E (47, ? 13+)

	autopsy. Severe mental subnormality. Hypertonia often marked. Fingers tightly flexed at metacarpophalangeal joints. Often fingers have a characteristic position (Fig. 12b), with index finger overlapping 3rd and 5th overlapping 4th. Absence of distal crease on 5th finger. Heart and/or renal deformity common. Characteristic dermatoglyphics, (Fig. 13). Patients with this syndrome usually die in early infancy. A preponderance of reported cases are females.	
Patau's Syndrome:	Often small baby. Characteristic facies (Fig. 14a), with small head, cleft lip and/or palate and scalp defect. Microphthalmia, anophthalmia or colobomata very common. Brain deformity frequently observed at autopsy. Severe mental subnormality. Post axial polydactyly, (Fig. 14b). Visceral deformity common. Characteristic dermatoglyphics, (Fig. 15). Patients with this syndrome usually die in infancy.	Trisomy D (47, ? 13 +)
Cri-du-chat	Low birth weight. Characteristic round moon-face, (Fig. 16), with small head, broad nasal bridge (ocular hypertelorism), low set and sometimes malformed ears. Oblique, antimongoloid palpebral fissures. Micrognathia. In infancy the cry is diagnostic — like the mewing of a kitten. Characteristic cry and facial appearance tend to disappear in older children. Cardiac anomalies sometimes present. Severe mental subnormality usual. Survival to adult life common.	Deletion short arm B (46, ? 5p-)

TABLE 3. Rare autosome abnormalities for which recognisable clinical syndromes are suspected, and in the process of definition.

Cytogenetic state	Clinical syndrome	Some clinical notes
Deletion long arm E (46,Eq-)	? Carp-mouth syndrome	Mid facial hypoplasia; carp-shaped mouth; long tapering fingers.
Deletion long arm G (46,Gq-)	? Anti-mongolism	Palpebral fissures slope downwards and outwards; big nose; ears big; hypertonia; mental handicap; heart deformities.
Deletion long arm D (46,Dq-)	Un-named	Eye defects, in particular retinoblastoma, frequently observed.
? Deletion short arm 4 (46, ? 4p-)	Un-named	Broad based nose; pre-auricular auricles; midline scalp defects.
Deletion short arm E (46,Ep-)	Un-named	Eyes close together or cyclops may be observed.

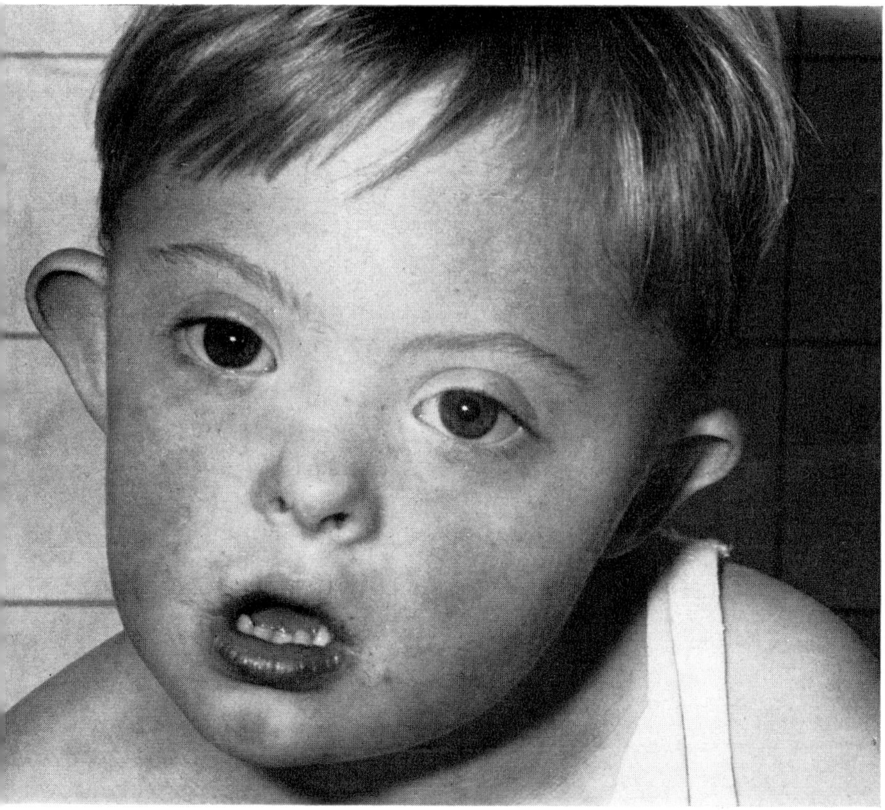

Fig. 8. A four year old boy with mongolism (Down's syndrome).

some 21 and either a chromosome of the D [t(DqGq)] group or a chromosome of the G [t(GqGq)] group. Translocation may occur in a genetically unbalanced form (Fig. 11a), in which case the greater part of chromosome 21 is represented three times and the affected individual is effectively trisomic or it may be present in a genetically balanced form (Fig. 11b), when the individual has virtually a normal gene content. A translocation can be inherited in balanced form for several generations and may be the cause of mongolism in several members of the same family. In some individuals normal/trisomy 21[i.e. 46/47,21+] mosaicism is present and the tissues are mixtures of trisomic and normal cells. Although many normal/trisomy 21 mosaics have the full clinical picture of mongolism, others have abortive features of the condition. Mosaicism is thus a second cause of familial mongolism. Some patients with partial features. of mongolism may have an extra but incomplete (i.e

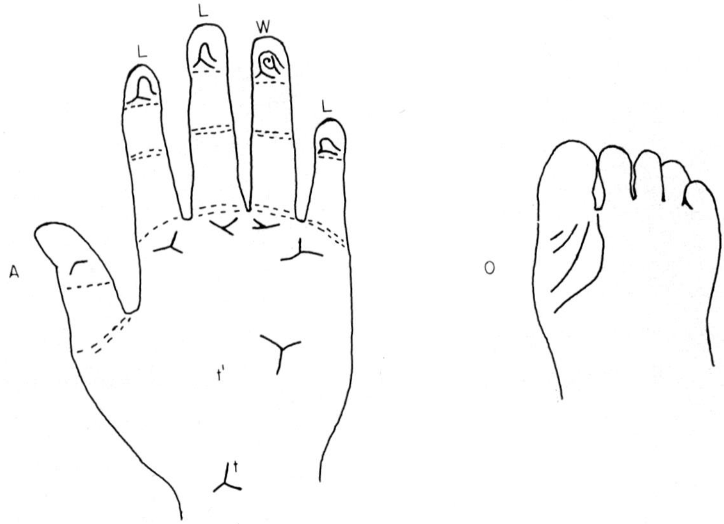

Fig. 9. Dermal ridges in mongolism:- Digits of the hand show an excess of loops (L) opening to the ulnar side. An axis triradius (t′) is commonly found on the hypothenar eminence distal to that usually seen in normal subjects (t). In this patient an arch (A) is depicted on the thumb and a whorl (W) on the ring finger. A characteristic open field (O) without a triradius is shown on the hallucal area of the foot.

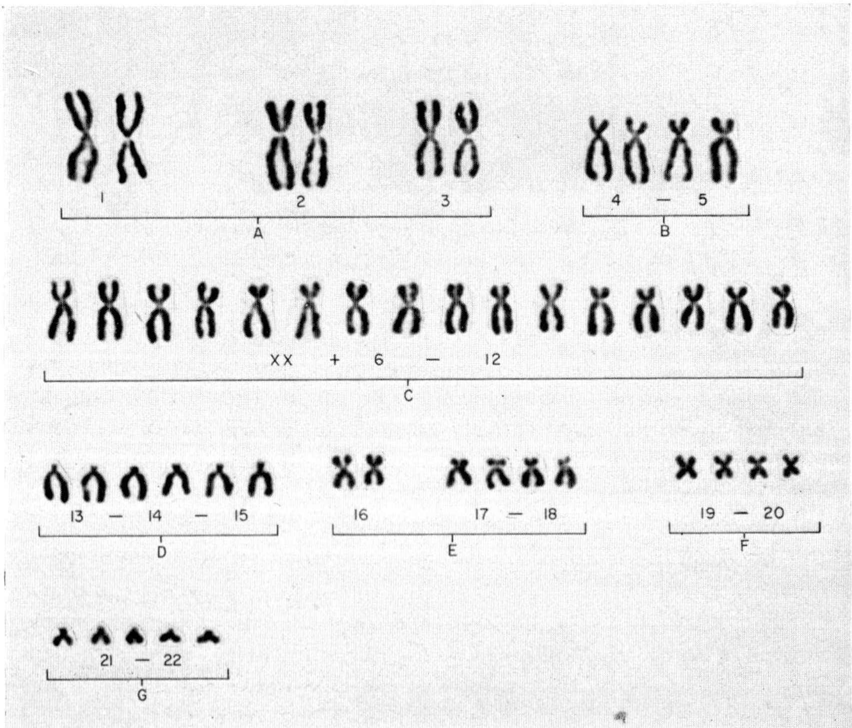

Fig. 10. Primary trisomy 21; there is an extra chromosome in the G group: 47,XX,21+ (Chicago Conference, 1966).

deleted) chromosome 21. It is also possible that clinical features justifying a diagnosis of mongolism may occur in the absence of trisomy 21, but this phenomenon is very rare.

B. OTHER ESTABLISHED CLINICAL SYNDROMES ASSOCIATED WITH AUTOSOME ABNORMALITY

Fairly well defined clinical syndromes are associated with an extra member of the E group of chromosomes (Edwards syndrome—see Table 2 and Figs. 12 and 13), as well as with an extra member of the D group (Patau's syndrome—see Table 2 and Figs. 14 and 15). However, many patients who have what may, on present evidence, be considered the clinical feature of these syndromes, do not have a demonstrable chromosome abnormality. The Cri-du-chat syndrome (Table 2, Fig. 16), is associated with absence of part (deletion) of a member of the B chromosome group (Fig. 17). Unbalanced and balanced translocation as well as mosaisism may be observed in any of these syndromes.

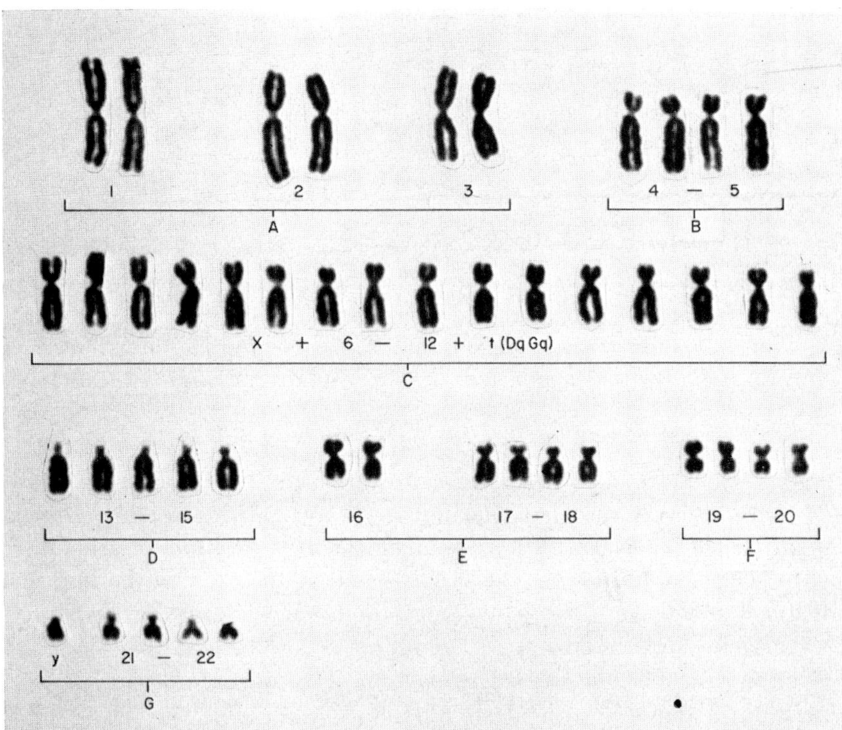

Fig. 11. (a) Translocation trisomy 21 (mongol child): 46,XY,D-,t(DqGq)+. The structurally abnormal t(DqGq) chromosome is one of the smallest of the chromosomes in the C group. It cannot be specifically identified in the karyotype.

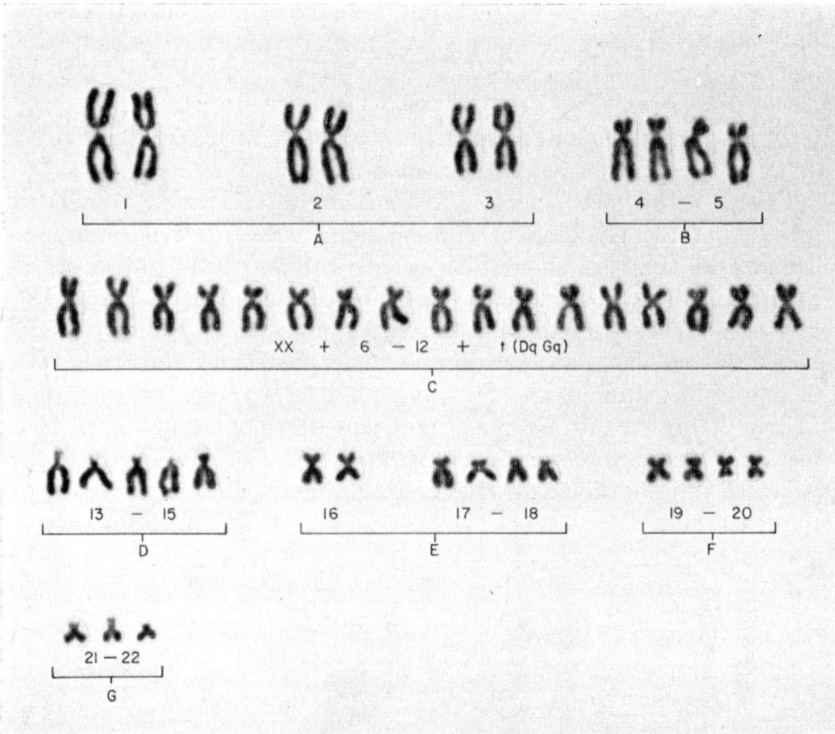

Fig. 11. (b) Translocation in balanced form (mother) 45,XX,D-, G-,t(DqGq)+.

Combined morphological and DNA replication pattern studies involving autoradiography suggest that the particular chromosome involved in Edwards' syndrome, Patau's syndrome and Cri-du-chat are respectively Nos. 18, 13 and 5.

C. OTHER AUTOSOME ABNORMALITY

New, and very rare, irregularities of karyotype are constantly being described in patients with developmental defects. In some patients the chromosome abnormality observed is obviously the cause of the clinical features noted, in others the chromosome irregularity is probably fortuitous. There seems to be good reason to believe that clinical syndromes associated with deletion of a D group chromosome and deletion of the long arm of an E group chromosome will eventually be defined (see Table 3). The occurrence of balanced and unbalanced translocation within a family is occasionally a cause of repeated non-specific subnormality or multiple congenital abnormality.

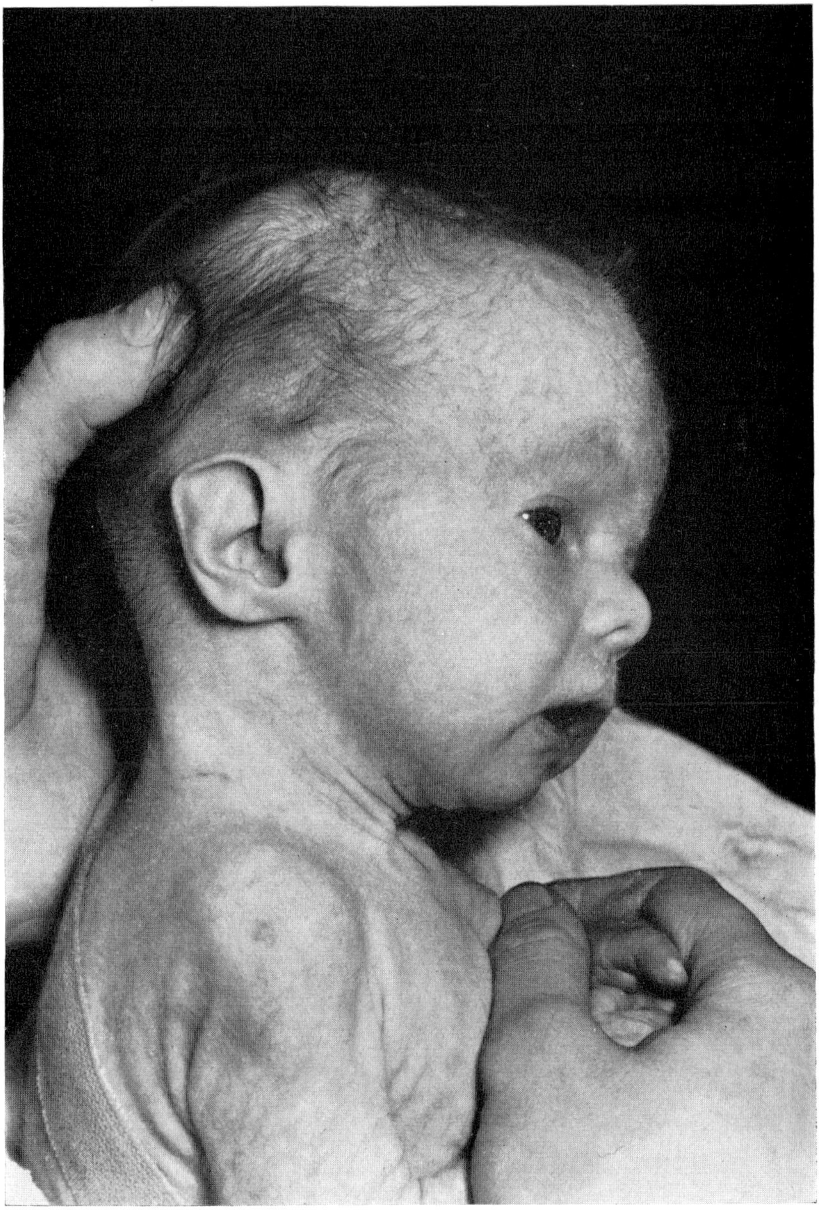

Fig. 12. (a) A six week old girl with Edward's syndrome. Facies.

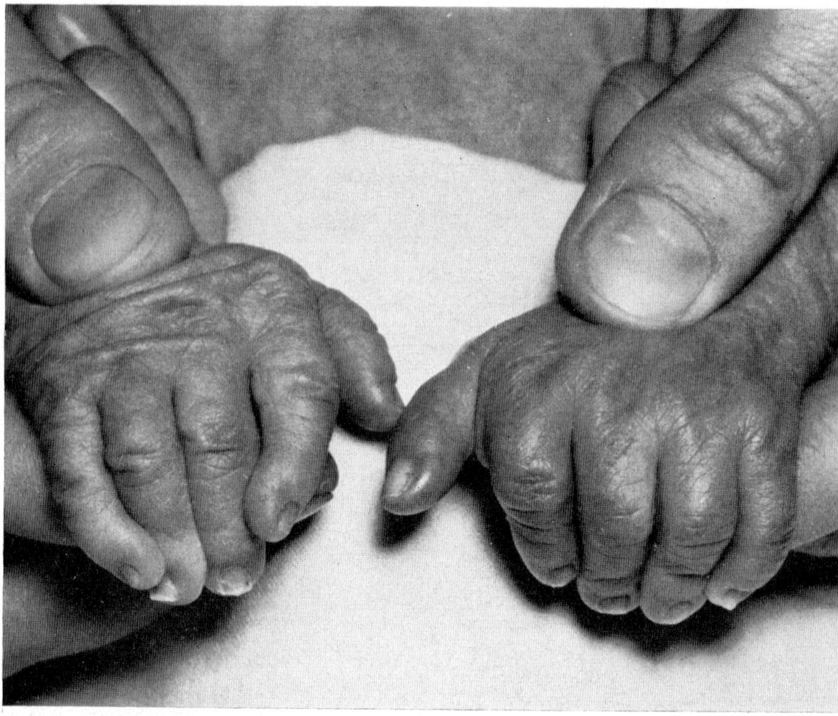

Fig. 12. (b) A six week old girl with Edwards' syndrome. Typical hand position.

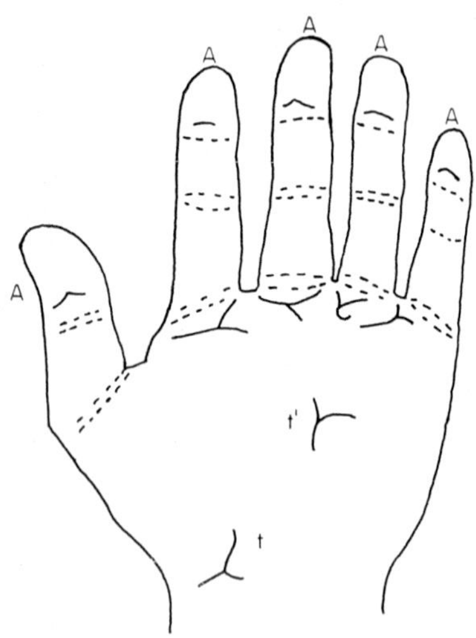

Fig. 13. Dermal ridge patterns in Edwards' syndrome:– Digits of hands show an excess of arches (A). Note additional axis triradius (t').

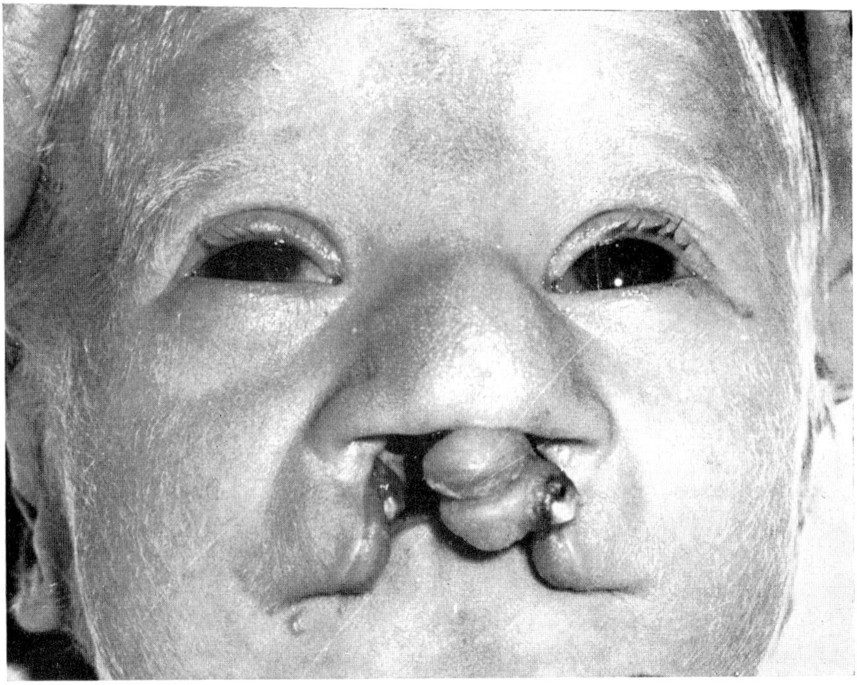

Fig. 14. (a) A seven week old boy with Patau's syndrome. Facies.

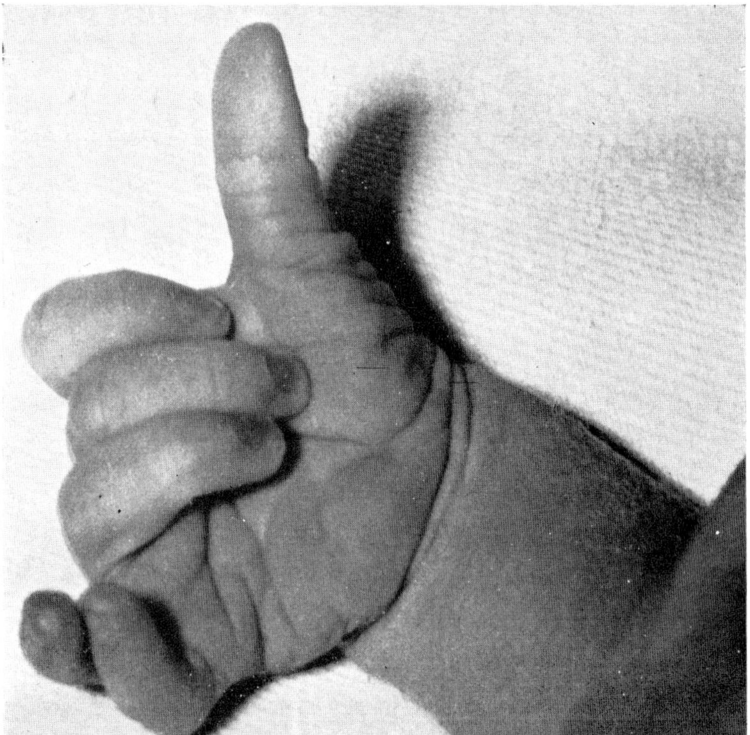

Fig. 14. (b) A seven week old boy with Patau's syndrome. Hand with extra digit.

L

D. CHROMOSOME ABNORMALITY IN NEOPLASIA

Irregularity of chromosome number and structure is frequently observed in malignant tissue. Indeed several distinct abnormal cell lines (clones) may be present. The disappearance and reappearance of cell lines or the formation of new cell lines are sometimes quite clearly related to the progress of the disease, to periods of remission and exacerbation which may be naturally occurring or following treatment. However, with the exception of chronic myeloid leukaemia (see below) the abnormal karyotypes observed are not related to the tissue of origin or the degree of malignancy. It is uncertain whether the malignant change is an effect of the abnormal cell line(s) or whether the production of abnormal chromosome complements is a manifestation of malignancy.

Specific, and virtually constant, abnormality of chromosome complement is, however, observed in chronic myeloid leukaemia in the adult. One of the chromosomes of the G group is consistently small (? deleted) in a proportion of the metaphases obtained from blood culture or bone marrow culture. The abnormal chromosome has been designated the Philadelphia, or Ph^1, chromosome (Fig. 18).

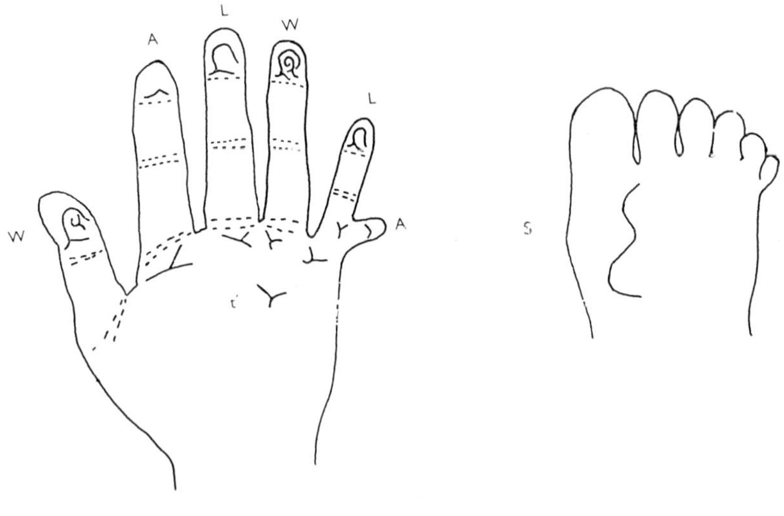

Fig. 15. Dermal ridge patterns in Patau's syndrome:– Very distal axis triradius (t′) S-shaped pattern (s) on hallucal area.

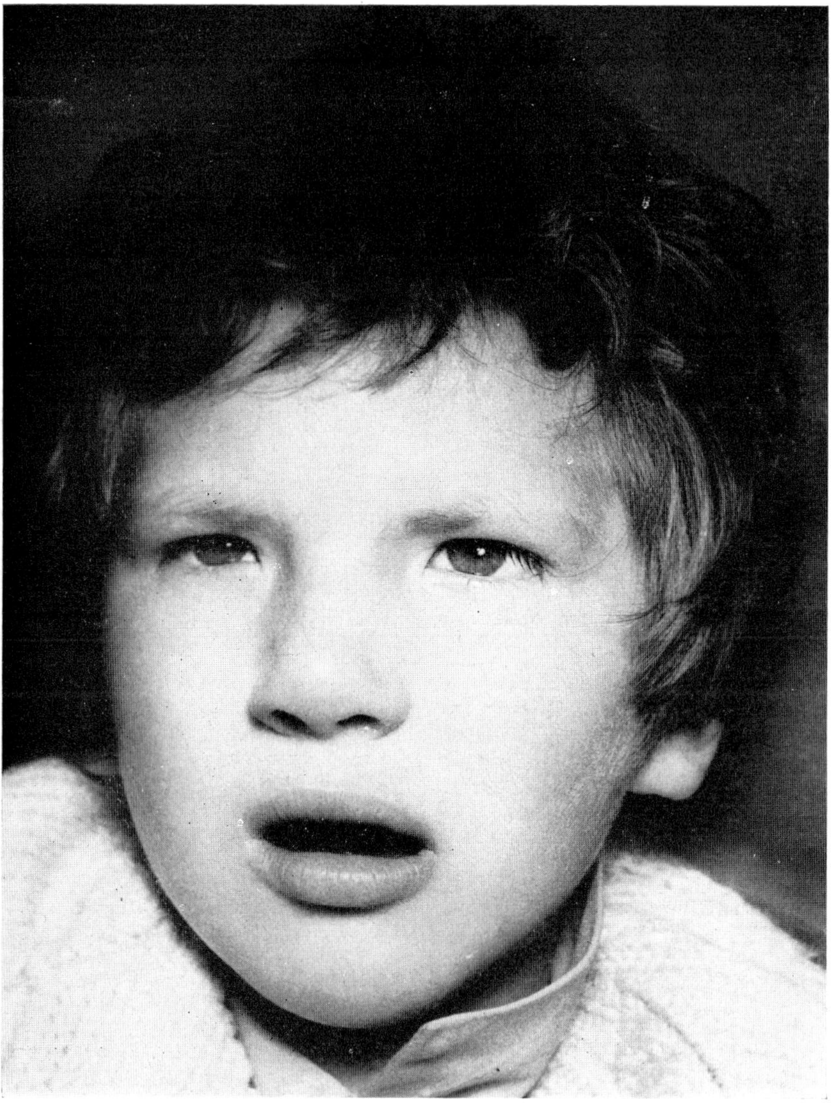

Fig. 16. A four year old girl with the Cri-du-chat syndrome.

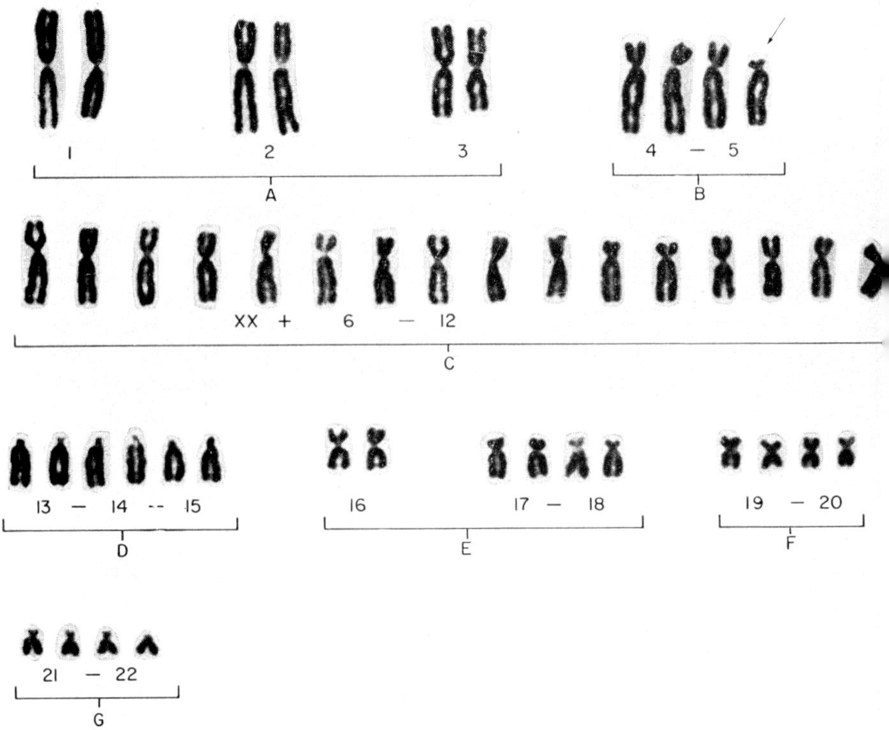

Fig. 17. Chromosome complement of child with the Cri-du-chat syndrome. A member (arrowed) of the B group has a very short, short arm: 46,XX,Bp-.

IV. Chromosome Abnormality in Development Anomalies of Sex

A. Turner's Syndrome

It is convenient to divide cases of Turner's syndrome (Table 4 and Fig. 19), into sex chromatin negative and sex chromatin positive varieties on the basis of buccal smear findings.

Almost all cases of sex chromatin negative Turner's syndrome have a chromosome abnormality (Fig. 20), in which only one sex chromosome, an X, is present though 45,X/46,XY or 45,X/46,XX mosaicism may occasionally be observed. Examples of sex chromatin positive Turner's syndrome can be shown either to have an abnormal second X chromosome (Fig. 21), or to be mosaics of type 45,X/46,XX. Other, more complicated, mosaics have been described: in each case one cell line contained only one complete X chromosome.

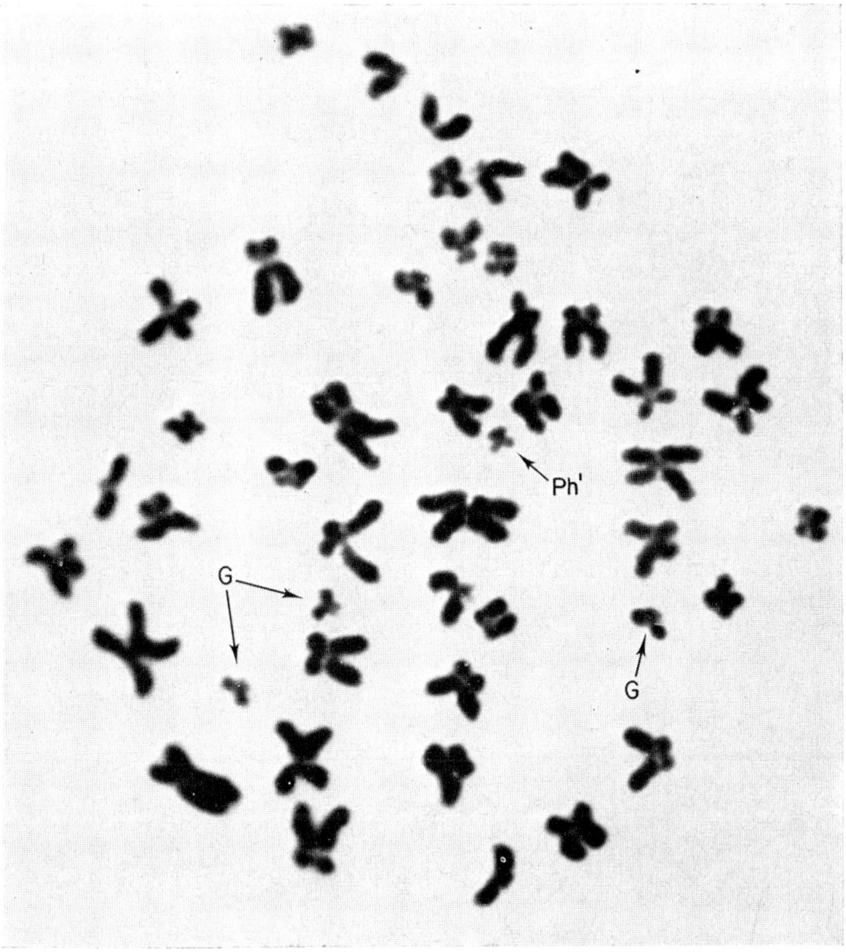

Fig. 18. Chronic Myeloid Leukaemia. The abnormal chromosome Ph[1] and the other G group chromosomes are arrowed: 46,XX,G?q-.

B. KLINEFELTER'S SYNDROME

As in Turner's syndrome, Klinefelter's syndrome is conveniently divided into sex chromatin positive and negative groups. All sex chromatin positive cases (Table 4 and Fig. 22), have a demonstrable chromosome abnormality (Fig. 23), with a cell line containing at least two X chromosomes and a Y chromosome. Sex chromatin negative patients with clinical features of Klinefelter's syndrome have been described and usually have an apparently normal male chromosome complement.

TABLE 4. Sex chromosome abnormality associated with particular developmental anomaly of sex

Clinical syndrome	Main clinical features	Cytogenetic state usually or sometimes observed
Turner's syndrome	Primary amenorrhoea with "streak" (fibrous tissue) gonads. Ovarian follicles present in fetal life but later degenerate. Secondary sex development absent. Somatic stigmata include short stature (usually less than 5ft.), webbing of the neck and minor bony anomalies. Chest is often "shield" shaped and nipples wide apart. Hairline on neck is often low. Coarctation of the aorta is quite common. I. Q. is usually within normal limits. While the somatic stigmata of Turner's syndrome may be seen in a child or infant it is perhaps incorrect to diagnose Turner's syndrome before the age at which puberty would be expected to occur; even where a 45,X karyotype has been identified. In these circumstances Bonnevie-Ullrich syndrome is a convenient designation. Nevertheless it is virtually certain that children with a 45, X, or comparable, karyotype will develop the full clinical features of Turner's syndrome in adult life.	45,X; 46,XXqi; 45,X/46,XX
Klinefelter's syndrome	Eunuchoid body proportions, small testis and poorly developed, or absent, secondary sex characters. Pubic hair of feminine distribution. There may be gynaecomastia. Gonads show degeneration of seminiferous tubules and interstitial cell hyperplasia. Spermatogenesis is usually absent. Slight mental subnormality may be observed in approximately 20% of cases. The clinical features of Klinefelter's syndrome do not become recognisable until after the age at which puberty would occur; although a mentally subnormal boy may occasionally be shown to have an XXY karyotype.	47,XXY: 47,XXY/46,XY; 46,XY
Hermaphroditism:	True hermaphrodites have histological evidence of both ovarian follicles and seminiferous tubules. Development of a Mullerian or Wolfian duct system is usually incomplete. External genitalia are usually ambiguous, e.g. a perineal urethra and one gonad in a scrotal sac. Secondary sex development is usually female in type. I. Q. normal.	Usually 46,XX; occasionally 46,XY/46,XX: occasionally 45,X/46,XY. The last usually lose their ovarian follicles by the time of puberty and are perhaps more properly designated male pseudo-hermaphrodites.

C. TRUE HERMAPHRODITISM

At birth hermaphrodites have both recognizable ovarian and testicular tissue, (Table 4 and Fig. 24). Most are sex chromatin positive and usually have a normal female chromosome complement. Sometimes, however, sex chromosome mosaicism can be demonstrated. Where sex chromosome mosaicism is observed, one cell line has a Y chromosome, and is presumably responsible for the development of testicular tissue and the other cell line lacks a Y chromosome but usually has two X chromosome: it is therefore probably responsible for development of ovarian tissue.

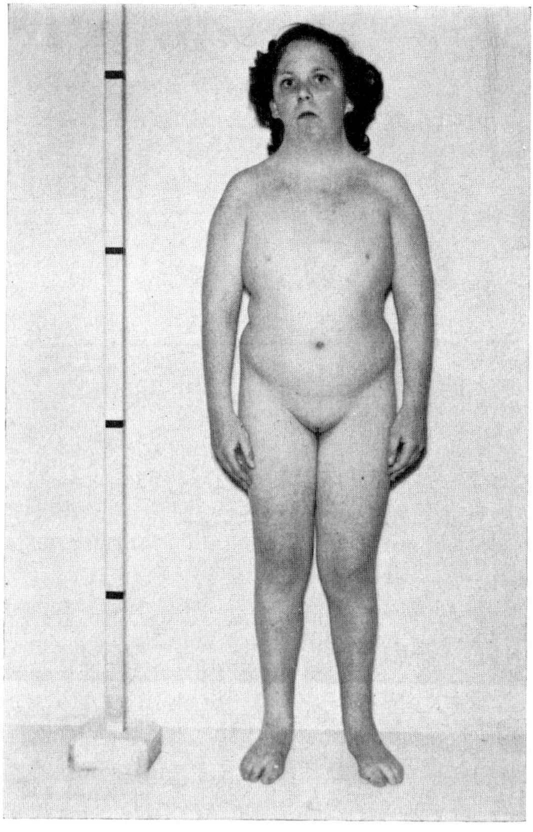

Fig. 19. A 17 year old girl with Turner's syndrome. At operation the ovaries were represented by streaks of whitish tissue: 45,X. (Taken from "Gynaecological Disorders of Infants and Children" by C. J. Dewhurst. Cassel and Co.)

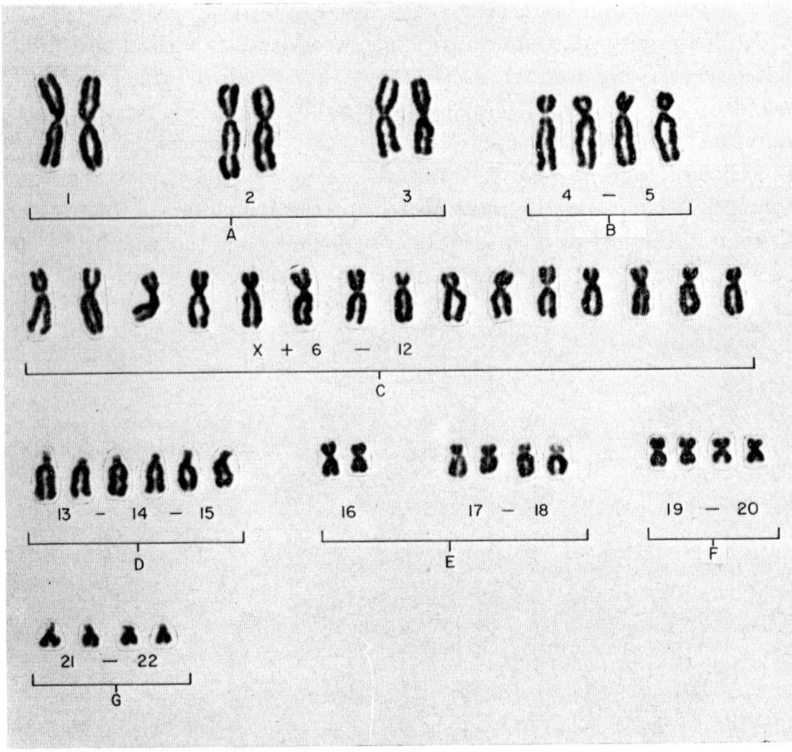

Fig. 20. Chromosome complement of a patient with Turner's syndrome. Only 45 chromosomes are present; a chromosome (an X) is missing from group C: 45,X.

D. 45,X/46,XY MOSAICISM

45,X/46,XY mosaicism has been described in: (1) Examples of gonadal dysgenesis, where the gonads are undifferentiated streaks of fibrous tissue; (2) Examples of male pseudohermaphroditism, where the gonadal tissue is testicular; and (3) Patients in whom there is histological support for a diagnosis of true hermaphroditism i.e. the presence of both ovarian follicles *and* seminiferous tubules. Differences in gonadal pathology observed in 45,X/46,XY mosaicism are probably a consequence of varying distribution of the two cell types in the developing gonads.

E. OTHER SEX CHROMOSOME ABNORMALITIES

An extra X or Y chromosome may be associated with mental subnormality in the absence of overt developmental anomaly of sex. Triple-X females (47,XXX) only rarely have amenorrhoea and XYY males (47,XYY) have normally developed secondary sex characteristics

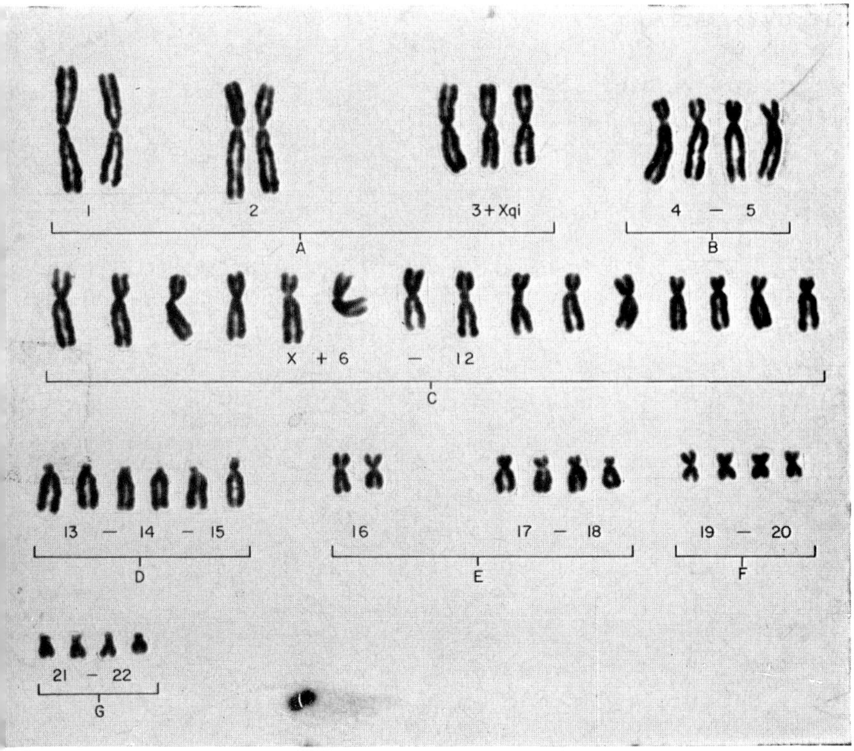

Fig. 21. Chromosome complement of a patient with Turner's syndrome. 46 chromosomes are present, but their distribution is abnormal; an X chromosome is missing from Group C and an abnormal X chromosome (an isochromosome) appears as an extra member of group A: 46,XXqi.

and are usually fertile. XYY males may be unusually tall. There seems to be a rough correlation with the number of extra sex chromosomes and the degree of mental subnormality present. Nevertheless, it is improbable that more than one in five of individuals born with an extra sex chromosome develop a degree of mental subnormality sufficient to require a special schooling or institutionalization.

V. Factors Affecting the Frequency of Chromosome Abnormality

A. MATERNAL AGE

A disproportionate number of patients with chromosome abnormality are born to older parents. Maternal age, paternal age and birth order are interdependent variables—the older a mother is at the birth of her child the older the father will tend to be, and on average the

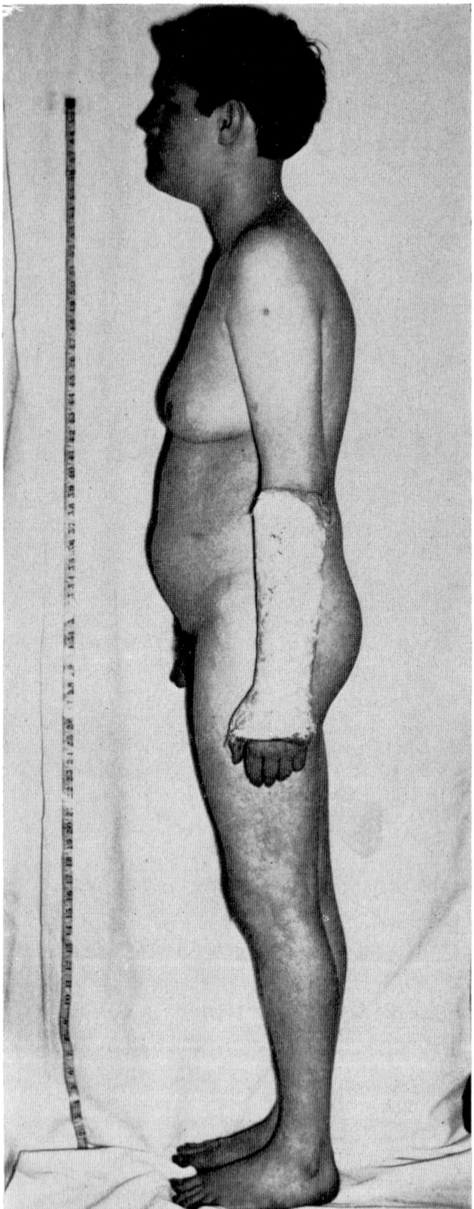

Fig. 22. A case of Klinefelter's syndrome. Both testes are small and on histological examination showed no evidence of sperm maturation: 47,XXY. (Taken from "Gynaecological Disorders of Infants and Children" by C. J. Dewhurst, Cassel and Co.)

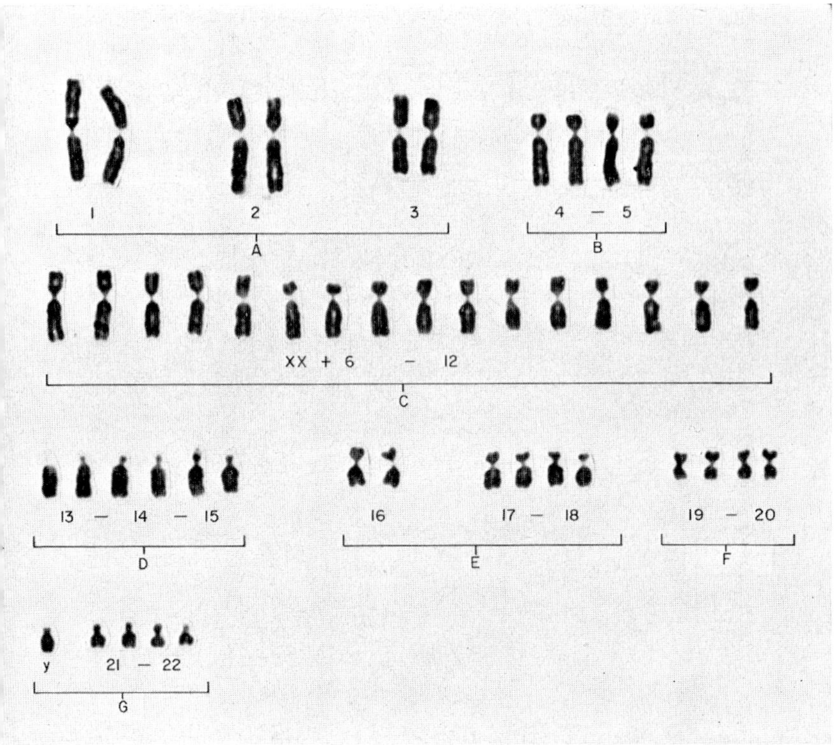

Fig. 23. Chromosome complement of a patient with Klinefelter's syndrome. An extra X is present in group C:47,XXY.

more older sibs the child will tend to have. It is possible, using appropriate statistical techniques, to study the independent effect of each of these variables on the incidence of a particular chromosome abnormality. In general the trisomic conditions are associated with advancing maternal age (Fig. 25), an exception being XYY. On the other hand cases of sex chromatin negative Turner's syndrome (45,X) are more often born to young mothers than chance would lead one to expect. A parental age effect has not been convincingly demonstrated in the case of structural abnormality of the chromosomes. Mosaics have a maternal age effect that is somewhat less than that observed in simple trisomy; perhaps indicating that a proportion are derived from trisomic zygotes.

It has recently been suggested that ova are shed in the sequence in which they are laid down during embryogenesis and that those produced last have fewer chiasmata. Chiasmata are functionally implicated in normal segregation of half bivalents to opposite

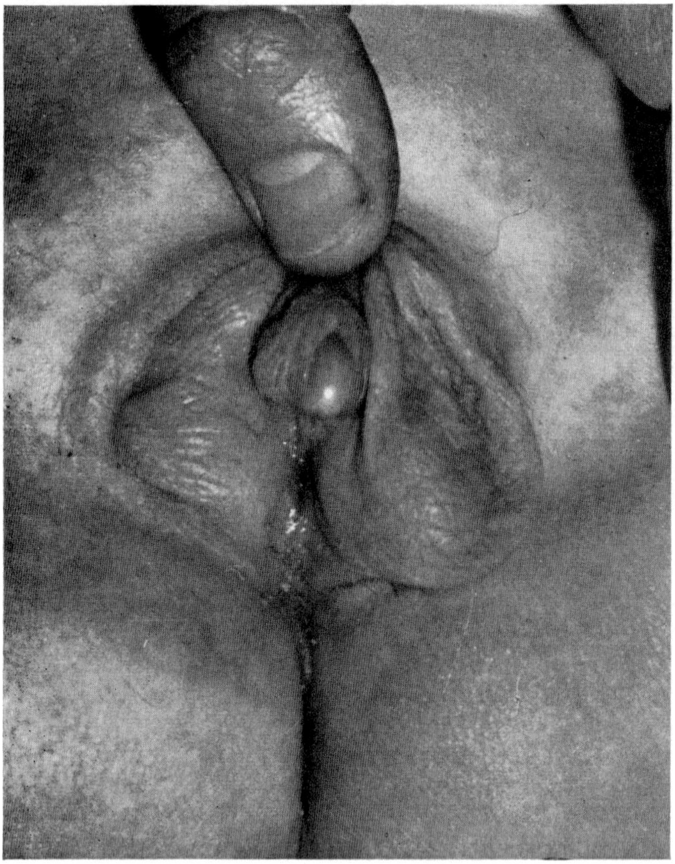

Fig. 24. External genitalia of a 14 months old hermaphrodite. Moderate sized phallus with urethral opening at base. Bifid scrotum with a testicular gonad in left sac only. The right gonad, an ovary, was intra-abdominal. (By kind permission of the Editor of the Journal of Medical Genetics, Dewhurst et al. (1965).)

daughter nuclei. In the absence of chiasmata sister chromosomes might be expected to enter the same daughter nucleus quite frequently. It would therefore be understandable that non-disjunction, leading to trisomy, is seen more often in older ova. The evidence for this explanation of maternal age effect is, however, not proven.

In Table 5 are shown the frequency of mongolism in different maternal age groups, the overall frequency of mongolism in a recent British population, and estimated overall frequencies in two hypothetical populations; one with a mean maternal age five years greater than the recent British population, and one with a mean maternal age

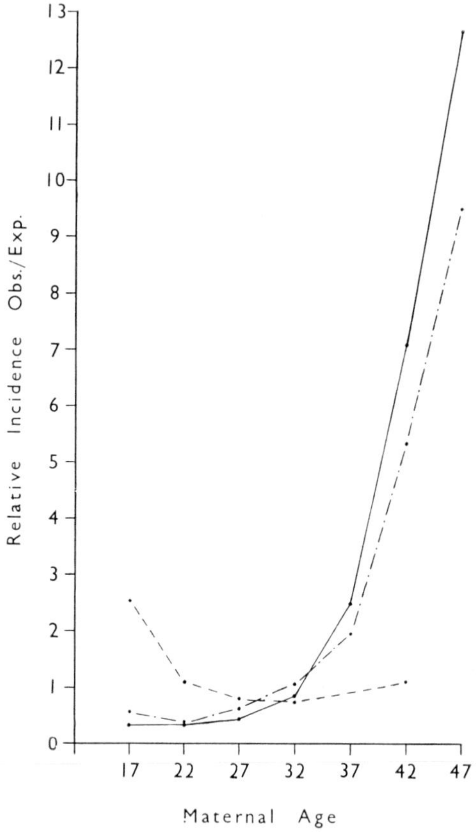

Fig. 25. The relative incidence of mongolism (data derived from Carter, 1965), Klinefelter's syndrome (data derived from Court Brown, 1964) and Turner's syndrome (data derived from Court Brown, 1964 and a personal series) at different maternal ages. Mongolism (626 cases)–; Klinefelter's syndrome (89 cases)–·–; and Turner's syndrome (68 cases)· – – – – –

five years less. The results suggest that a rise in mean maternal age of five years would about treble the incidence of trisomic conditions and that a reduction of five years would about halve this incidence. We may expect populations with different parental age structures to have quite large differences in the frequency of disorder of development due to chromosome abnormality. Parental age trends in many present-day populations are such that the frequency of chromosome abnormality as a cause of disorder of development should fall, perhaps quite markedly so.

TABLE 5. The frequency of mongolism in populations with different maternal age structure

	Actual population*		Hypothetical populations				
Maternal age	Mean maternal age 28 yr Distribution %	Mongols observed	Mean maternal age 33 yr Distribution %	Mongols expected	Mean maternal age 23 yr Distribution %	Mongols expected	Frequency in different maternal age groups %
17	4·6	5	2	3	21	30	0·04
22	26·8	29	12	17	53	76	0·04
27	31·2	45	19	27	17	24	0·04
32	21·6	92	23	91	5	20	0·11
37	11·9	154	23	273	2·7	32	0·33
42	3·6	179	17	765	1·0	45	1·25
47	0·3	41	4	453	0·2	23	3·15
		545		1,629		250	
Frequency of mongolism:		Approx. 1 in 650		Approx. 1 in 210		Approx. 1 in 1,440	

*Data derived from Penrose (1961).

B. IONIZING RADIATION

Ionizing radiation can produce gross chromosome damage with consequent structural rearrangement. Such changes are induced especially by large single doses of radiation, for example, heavy doses of X-rays (Fig. 26), or the acute radiation from an atomic accident.

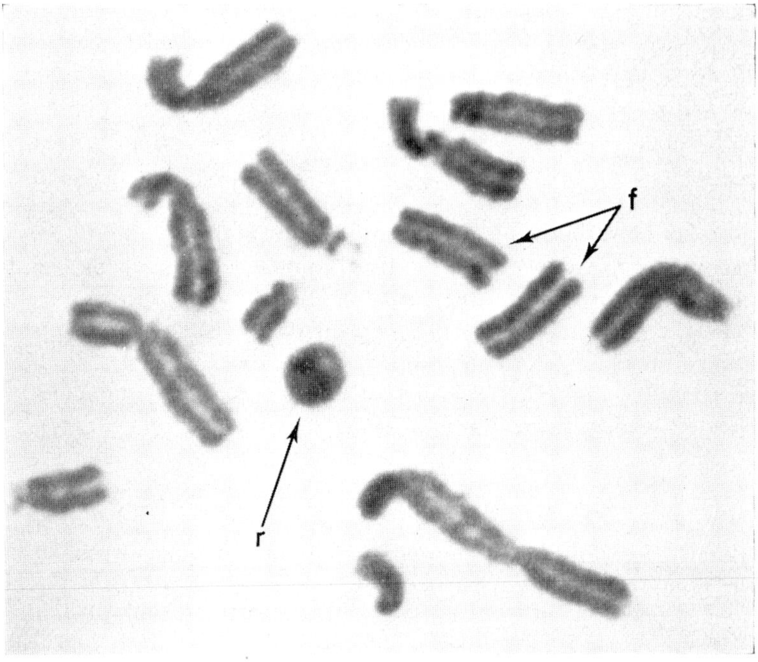

Fig. 26. X–ray damage following deep X–ray therapy for neoplasia, Lymphocyte culture. Sector of metaphase plate showing ring chromosome (r) and chromosome fragments (f) without centromeres.

Although people exposed to large doses of ionizing radiation are known to have a high incidence of leukaemia and cancer and frequently show structural chromosome abnormality in cultured leucocytes, it has never been conclusively shown that chromosome abnormality is a cause of neoplasia (see Chapter 10 for a full discussion of this subject).

It is of interest to consider the possible effect that the ever increasing use in the home, industry and medicine, of equipment producing ionizing radiation may have on the incidence of developmental defect due to structural aberration of chromosomes. This is particularly so where individuals, such as radiographers or dentists, are regularly exposed to unusually large amounts of radiation. Fortunately animal experiments on the effect of irradiation of the gonads indicate that most of the chromosome aberrations produced by ionizing radiation result in death of the effected cell or its immediate descendants and that the chance of transmission can be greatly reduced by abstention from mating for a period of weeks following exposure. The incidence of physical or mental handicap among the offspring of individuals exposed to large single doses of ionizing radiation before conception: for example the children of survivors of the atomic explosions at Hiroshima and Nagasaki, has not been found to be significantly greater than expected by chance. These groups are, however, not of great size.

C. VIRUSES AND DRUGS

Viruses may produce chromosome breaks both *in vivo* and *in vitro.* There is some evidence that the distribution of damage is not random. Chromosome breaks or irregularity of division may also be produced by a wide variety of drugs and certain drugs, like viruses, appear to produce breaks in particular sites. It is not yet clear whether chromosome damage induced by viruses or drugs is a significant factor with respect to neoplasia or developmental defect.

VI. Chromosome Analysis in Medical Practice

Accurate diagnosis is a prerequisite of good management. Chromosome analysis is clearly a particularly useful diagnostic aid in disorders of development known to be associated with a specific chromosome abnormality. Very often an appreciation of chromosome make-up, or at least of the sex chromatin state, is a crucial factor in the evaluation of anomalies of sex. When a developmental defect has been shown to be the result of a chromosome anomaly the parents can either be reassured that the risk of a handicapped child resulting from subsequent pregnancies is low or, if this is not so, the desirability of determining the fetal karyotype early enough to allow termination of pregnancy can be explained. For all that, chromosome analysis is expensive and facilities for karyotyping limited. It would not be possible, with present facilities, to study the chromosome complement of every abnormally developed child or that of every fetus carried by a woman over 35 years of age, however desirable this may be clinically.

A. DIAGNOSIS AND PROGNOSIS

When a condition known to be associated with specific chromosome abnormality is suspected on clinical grounds i.e. mongolism, Edwards' syndrome, Patau's syndrome, the Cri-du-chat syndrome, Turner's syndrome and Klinefelter's syndrome, the finding of the specific chromosome abnormality is diagnostic of the condition. Although an apparently normal chromosome complement does not necessarily refute the suspected clinical diagnosis, it does make it far less likely. Nevertheless some patients with the clinical features of Edwards' syndrome, Patau's syndrome, the Cri-du-chat syndrome or Kline-lelter's syndrome, have an apparently normal chromosome complement; these may have an undetected chromosome abnormality, e.g. mosaicism, but it is probable that in many such cases the situation is complex and the clinical features are not due to simple chromosome abnormality.

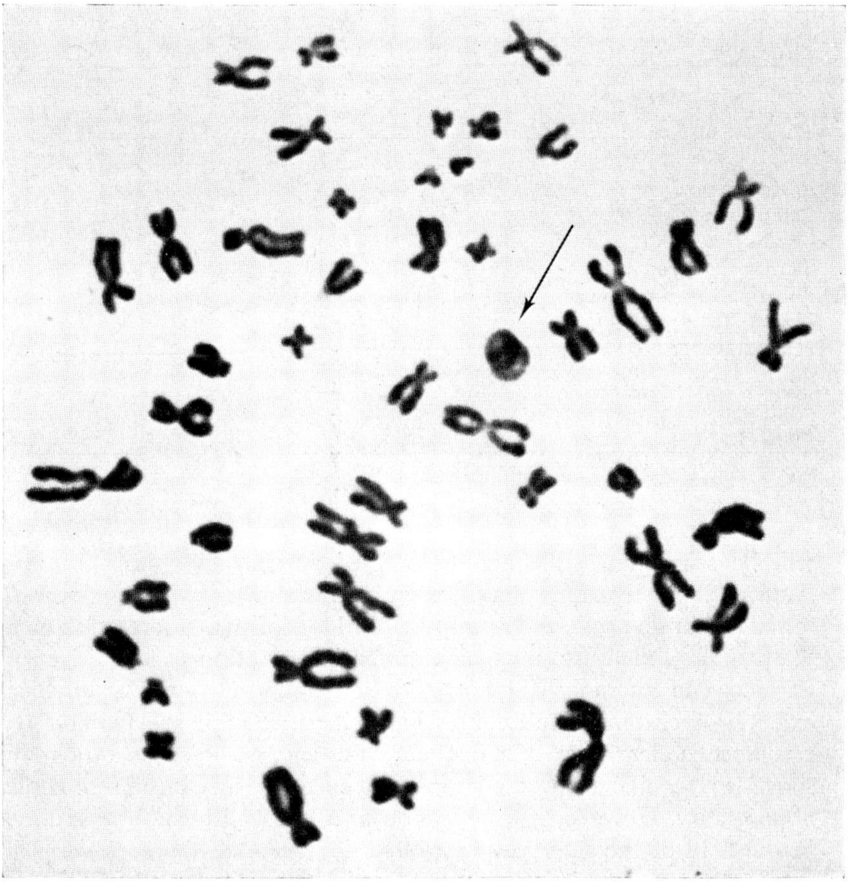

Fig. 27. Ring chromosome (arrowed) in a child with poor physical and mental development; 46,XX,1r. Deletion presumed.

A considerable number and wide variety of chromosome abnormalities have been observed in patients with non-specific mental subnormality or multiple congenital deformity. Many are associated with gene duplication, deficiency or alteration in activity and are the cause of the developmental abnormality observed; but deviation from normal chromosome morphology (Fig. 4) is also quite common in normal individuals (see also Table 1). Even if we can be reasonably sure that a particular chromosome abnormality is of pathological significance (Fig. 27) the finding of a unique karyotype in an individual with mental sub-normality or physical deformity, is rarely an aid in prognosis. The findings in 539 consecutive cases of congenital anomaly referred for chromosome analysis as an aid to diagnosis are given in Table 6. It

TABLE 6. Patients referred for chromosome analysis as an aid to diagnosis

Suspected clinical diagnosis	No. investigated	Confirmed specific chromosome abnormality
Mongolism	163	92 (56%)
Edwards' syndrome	25	7 (28%)
Patau's syndrome	19	8 (42%)
Cri-du-chat	21	3 (14%)
Turner's syndrome	76	28 (37%)
Klinefelter's syndrome	35	13 (37%)
Unspecified chromosome abnormality	200	22* (11%)
	539	173 (32%)

* Probably of pathological significance.

will be seen that while about 45% of patients exhibiting clinical features suggestive of a specific syndrome known to be of chromosomal origin do show abnormality, only 11% of patients referred as a "last ditch" measure in which a specific syndrome suggestive of chromosome anomaly did not exist had abnormal karyotypes. The expense of chromosome analysis limits it as a diagnostic facility. Its relevance is clearly greatest in those cases where clinical examination suggests a specific syndrome known to be associated with chromosome abnormality. However its application as a research technique is often in the field of the non-specific group of congenital abnormalities. In yet another context it is also useful to the genetic counsellor to know that chromosome abnormality is probably the cause of the particular clinical syndrome causing concern.

Patients with equivocal external genitalia require to be investigated in the newborn period to determine the true situation. The question to be answered is whether the child is an under-developed male or a masculinized female. Some of the latter can be so masculinized that they present as males with bilateral undescended testes and a penile urethra. Most commonly, however, they have normal female internal genitalia, masculinization occurring either in response to excessive androgen secretion by the fetal adrenals (the adrenogenital syndrome) or because of exposure to drugs with androgenic activity given to the mother in order to prevent a threatened abortion. In these cases the infant will be sex chromatin positive. If the child is sex chromatin negative then it is likely that he is an under-masculinized male. True hermaphroditism must be suspected if the external genitalia are ambiguous, one gonad is in the scrotum or inguinal canal and the patient is sex chromatin positive: the finding of a 46,XX/46,XY or similar chromosome complement would make this diagnosis virtually certain. A simple sex chromatin analysis is sufficient in most cases of doubtful sex, and may quickly exclude the adrenogenital syndrome which is potentially lethal, but a full chromosome analysis may occasionally give useful additional information.

Chromosome mosaicism may be of special prognostic significance. Normal/trisomy 21 mosaics are, as a group, better developed physically and mentally than cases of mongolism where mosaicism is not a feature. Indeed in some such cases it is controversial whether the patient should be designated a mongol at all. The degree of mental subnormality is often not proportional to the frequency of cells with the abnormal chromosome complement as observed in blood or skin specimens. Again some 46,XY/47,XXY and 45,X/46,XX mosaics may develop normal secondary sex characteristics and even be fertile. As previously mentioned normal/trisomy 21 mosaics, living in the general population, are a potential cause of familial mongolism, but unfortunately it is not possible to estimate the proportion of gametes likely to carry an extra 21 chromosome from observation of the abnormal (trisomy 21) cell type in blood and skin specimens.

B. GENETIC COUNSELLING

Most errors of division leading to chromosome abnormality are chance events, unlikely to occur more than once. Where both parents of an affected child have a normal chromosome complement, the risk of having another with an abnormal karyotype is very low, certainly less than 1 in 100 for each subsequent pregnancy, at least before the maternal age of 35 years. On the other hand, multiple congenital abnormality of unknown origin has an approximately 1 in 15 chance of repetition.

Since D/G translocation is an important cause of familial mongolism and is seen more frequently in mongols born to young mothers (p. 153), it is reasonable practice to ascertain the chromosome complement of a mongol child if the mother has the child before the age of 35 years. If translocation is present in the child then the parents are examined to determine if it is present in balanced form. If a parent has this type of balanced chromosome abnormality (i.e. he or she is a translocation carrier) then the risk of having a mongol child, or another translocation carrier is high. In practice, the risk to a female translocation carrier of having a mongol child is between 1 in 5 and 1 in 10, and about half the clinically normal children have the translocation in balanced form (Fig. 28). For unknown reasons a male translocation carrier has a much smaller risk of having a mongol child.

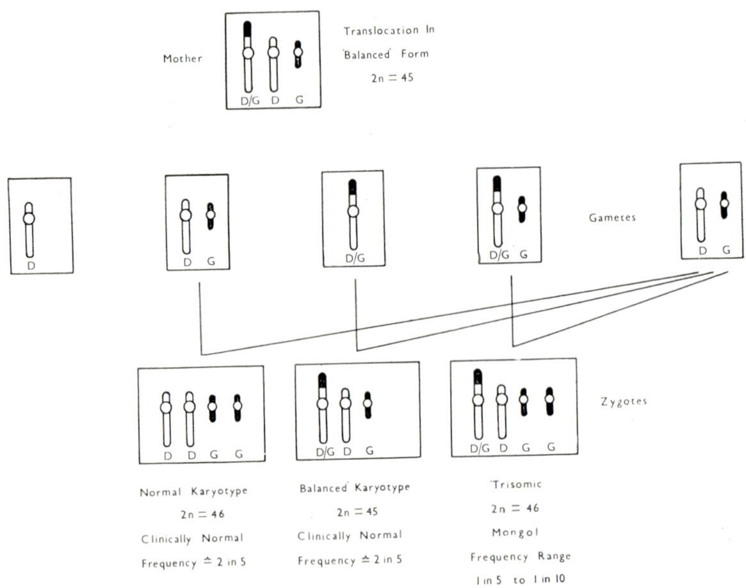

Fig. 28. D/G translocation at meiosis; only viable zygotes are considered.

In a t(21q21q) translocation carrier the risk of a mongol child is 100%. If inherited translocation as the cause of mongolism is excluded then the risk of a second mongol child is small, although still appreciably greater than that of the general population. It is probable that once inherited translocation has been excluded, and the parental karyotypes are not demonstrably mosaic, this risk is less than 1 in 100 before the maternal age of 35 years. After the age of 35 years the risk, particularly for mothers who had the first mongol child before 25 years of age, may be considerably greater than 1 in 100. About 1 in 20 mongol children born to mothers under the age of 35 years are translocation trisomics, and of these about 1 in 4 inherit the translocation from a parent with the abnormality in balanced form.

Inherited translocation duplication/deficiency also occurs in Edwards' syndrome, Patau's syndrome and in the Cri-du-chat syndrome. It is furthermore a cause of non-specific mental sub-normality and multiple congenital abnormality. A parent who has a balanced translocation and has an abnormal child not belonging to any of the groups defined above but with the translocation in unbalanced form undoubtedly has an increased risk of having a second abnormal child. It is, however, not possible on present evidence to estimate accurately the probability of this occurence. For the time being it seems only wise to acquaint the parent of the possibility of having further abnormal children and to add that the risk is probably greater than 1 in 20 and less than 1 in 3. Fortunately it is now possible (see below) to offer the facility of antenatal diagnosis with termination of pregnancy if the fetal chromosome complement is unbalanced.

C. ANTENATAL DIAGNOSIS

It is now legal in Britain to terminate pregnancy in the event of "a substantial risk that if the child were born it would suffer from such physical or mental abnormalities as to be seriously handicapped". This provision of the 1967 Abortion Act has completely altered the management of high risk families for developmental defect of chromosome origin. It is now possible to offer antenatal diagnosis and to terminate the pregnancy if the fetus has an unbalanced chromosome abnormality. Fetal cells, obtained by amniocentesis at 14–16 weeks gestation, are cultured in the laboratory and can be karyotyped 1–3 weeks later. At present hysterotomy or vacuum aspiration are the usual methods of termination adopted by the obstetrician. However, developments are taking place which make it possible that termination by drug administration will eventually be routine.

There is a risk, not yet properly evaluated but probably slight, that the technique of amniocentesis may itself cause abortion or damage the fetus and injure the mother. Another difficulty is that in perhaps 5% of cases the specimen is contaminated by maternal tissue and maternal rather than fetal cells are karyotyped. Clearly occasions where antenatal diagnosis is to be recommended are, at least for the present, limited in number. Fetal karyotyping should be confined to those situations where the risk of a child with chromosome abnormality is greater than 1 in 100:—(1) where a parent or prospective parent, has a balanced chromosome abnormality, particularly if the translocation in unbalanced form is known to be compatible with survival; (2) where a parent is a chromosomal mosaic; (3) where the mother has already had a child with chromosome abnormality and is herself older than 34 years; and (4) any pregnant woman over the age of 40 years.

The sex of the unborn child may sometimes be usefully determined as an aid to genetic counselling, e.g. where the mother is known to be a carrier of the gene for the X-linked recessive condition of pseudohypertrophic muscular dystrophy, a disease of muscle limited to boys and resulting in death in the teens, the chance that a male fetus will develop the disease is 50:50. Termination may, therefore, be recommended if the fetus is male. Fetus sex can be determined by chromosome analysis or by scoring fetal cells for sex chromatin. It is possible to score for sex chromatin state within 4 hr of obtaining amniotic fluid.

Fetal cells escape into the maternal circulation early in pregnancy and may be observed when maternal blood is cultured. However, this technique as a method of antenatal diagnosis is unreliable. The incidence of fetal cells observed is very low and difficulties of interpretation occur.

VII. Future Developments

From time to time it is said that all the major discoveries in human cytogenetics have now been made and that the field has entered a period of consolidation. There may be some substance in this view but yet new syndromes associated with chromosomes abnormality are continually being described and improved techniques introduced. Indeed, as this chapter is being written important developments involving fluorescent microscopy are being applied to cytogenetics. It seems quite probable that by the time this work is published individual chromosome pairs, or even segments of chromosomes, may be identified by the fluorescent banding pattern exhibited by mitotic metaphase chromosomes. The field of population cytogenetics has hardly been touched. The possibilities offered by cell hybridization as a research tool are only now generally being realized. Some areas of cytogenetics in which important advances may be made in the next few years are listed below.

A. Detection of Small Chromosome Abnormalities

There is reason to believe that chromosome abnormality of pathological significance is more frequent than is at present appreciated.

Small deletions, duplications and even quite gross rearrangements, e.g. paracentric inversions (see Chapter 10), can not be detected in mitotic metaphase figures by present day techniques. Such abnormalities are often recognizable in *meiotic* metaphase preparations but current techniques for the study of human meiosis are largely limited to the fertile male. It would be particularly exciting if some substance could be found that would induce meiotic-like behaviour: for example chromosome pairing, in cells usually dividing by mitosis.

B. CYTOLOGICAL MAPPING

In many organisms the recognition of particular areas of chromosomes by a characteristic banded appearance, particularly in meiotic preparations, is far advanced. Cytological mapping in man requires improved techniques for the study of chromosomes in the zygotene stage of meiosis. It may well be that the use of fluorescent stain will go some way towards providing a cytological map of particular chromosomes.

C. LOCI MAPPING

The identification of particular gene loci with particular chromosome pairs has already begun. The main lines of approach which have been usefully followed are:—(1) the study of families where functionally normal but morphologically distinguishable 'variant' chromosomes are segregating; (2) the use of abnormal: for example deleted, chromosomes; and (3) the study of the biochemical properties of hybrid cells in culture. It is probable that studies of hybrid cells in culture in cell lines containing only *one* human chromosomes may provide information perintent to the genes on that chromosome.

D. POPULATION STUDIES

Much of the drudgery of chromosome analysis may be avoided if, as is being attempted in several centres, a computer can be programmed to do the scanning and preliminary analysis. Should computer analysis of chromosome complement ever become a practical proposition whole populations could be studied. At present large surveys can only be attempted by the largest centres and, even then, they take a long time.

E. GENETIC ENGINEERING

While it is probable that organ transplantation will eventually prove to be an important tool in the control of genetic disease, for example the substitution of the liver in phenylketonuria, it is possible that the problems of immune reaction may never be satisfactorily overcome. DNA-mediated genetic transformation or transduction, although a stimulating possibility in man, has technical difficulties which at present seem overwhelming. Another approach would be to substitute, or add, not a particular locus but a whole chromosome (containing a relevant locus) to the patient's cells established in culture: for example, liver cells deficient in the enzyme responsible for the metabolism of phenylalanine (as seen in phenylketonuria). Such hybrid cell lines have already been created and kept in culture for long periods. It is possible that the reintroduction of such a cell line, with only one chromosome added or substituted to the host would induce little or no immune response. But would this hybrid cell line be maintained? Or would the extra (but essential) chromosome be lost *in vivo*?

There can be little doubt that human cytogenetics is a viable and expanding discipline with much to offer the student of biology.

Acknowledgements

I would like to thank my colleague, Dr Patricia Cooke for providing the karyotypes and diagrams. I am indebted to my clinical colleagues for permission to use photographs of their patients.

References

Carter, C. O. (1965). Personal communication.

Chicago Conference (1966). Standardization in Human Cytogenetics. Birth Defects: Original Article Series, 11:2. The National Foundation, New York.

Court Brown, W. M., Harnden, D. G., Jacobs, P. A., Maclean, N. and Mantle, D. J. (1964). Med. Res. Counc. Gt. Brit. Spec. Rep. Ser. No. 305.

Dewhurst, C. J. (1963). "Gynaecological Disorders of Infants and Children". Cassel and Co. London.

Dewhurst, C. J., Warrack, A. J. N., Blank, C. E., Bishop, A. M. and Heslop, W. B. (1965). *J. Med. Genet.* **2**, 246.

Ford, C. E. and Hamerton, J. L. (1956). *Nature, London*, **178**, 1020–1023.

Ford, C. E., Jacobs, P. A. and Lajtha, L. G. (1958). *Nature, London*, **181**, 1565–1568.

Ford, C. E., Jones, K. W., Polani, P. E., Almeida, J. C. de and Briggs, J. H. (1959). *Lancet*, **i**, 711–713.

Harnden, D. G. and Stewart, J. S. S. (1959). *Brit. Med. J.* **ii**, 1285–1287.

Hungerford, D. A., Donnelly, A. J., Nowell, P. C. and Beck, S. (1959). *Amer. J. Hum. Genet.* **11**, 215–236.

Jacobs, P. A. and Strong, J. A. (1959). *Nature, London*, **183**, 302–303.

Lejeune, J., Gautier, M. and Turpin, R. (1959). *C. R. Acad. Sci. Paris*, **248**, 1721–1729.

Penrose, L. S. (1961). In "Human Chromosome Abnormalities" (Ed. W. M. Davidson and D. Robertson Smith). pp. 116–122. Staples Press, London.

Penrose, L. S., Ellis, J. R. and Delhanty, J. D. A. (1960). *Lancet*, **ii**, 409–410.

Polani, P. E., Briggs, J. H., Ford, C. E., Clarke, C. M. and Berg, J. M. (1960). *Lancet*, **i**, 721–724.

Tjio, J. H. and Levan, A. (1956). *Hereditas*, **42**, 1–6.

Further Reading

Bartalos, M. and Baramki, T. A. (1967). "Medical Cytogenetics". The Williams and Wilkins Co., Baltimore.

Court Brown, W. M. (1967). "Human Population Cytogenetics". North-Holland Publishing Co., Amsterdam.

Ephrussi, B. and Weiss, M. C. (1969). "Hybrid Somatic Cells". Sci. Amer. **220** (No. 4), 26–35.

Miller, O. J. (1964). "The Sex Chromosomes Anomalies". *Amer. J. Obstet. Gynecol.* **90** No. (7), 1078–1139.

Mittwoch, U. (1967). "Sex Chromosomes". Academic Press, London and New York.

Penrose, L. S. and Smith, G. F. (1966). "Down's Anomaly". J. and A. Churchill Ltd., London.

Polani, P. E. (1969). "Autosomal Imbalance and its Syndromes, excluding Down's". *Brit. Med. Bull.* **25** (No. 1), 81–93.

Polani, P. E. (1969). "Abnormal Sex Chromosomes and Mental Disorder". *Nature*, **223**, 680–686.

Renwick, G. H. (1969). "Progress in Mapping Human Autosomes". *Brit. Med. Bull.* **25** (No. 1), 65–73.

Valentine, G. H. (1969). "The Chromosome Disorders – An Introduction for Clinicians". 2nd Ed. William Heinemann Medical Books Ltd., London.

12. Sex Chromatin

U. MITTWOCH

Department of Human Genetics, University College, London, UK

I. Introduction

It will have become clear from the preceding chapter that the chromosome constitution wields a decisive influence on the development of the phenotype. To begin with, a simple difference in one chromosome pair brings about the most fundamental normal dimorphism in man and other animals, namely that of sex. Secondly, a large number of chromosome abnormalities are the causes of various pathological conditions. Here again numerical and structural abnormalities of the sex chromosomes, which may give rise to abnormalities in sexual development, make up a large proportion of all abnormal chromosome constitutions.

It is one of the difficulties inherent in chromosome analysis that although the same chromosomes are present in every cell of the body, the chromosomes can be individually recognized only in cells which are in the process of cell division. At any one time, the vast majority of cells in the body are not of course dividing; they are in the resting stage, or in interphase. The chromosomes of non-dividing cells are in a dispersed state and, as a rule, individual chromosomes cannot be made out; together, they are the principal constituents of the cell nucleus. An important exception to this rule is provided by the giant used for sexing larvae prior to the development of the gonads, and Smith suggested that this method might be applied to other organisms, including mammals. This finding is of importance, not only because it was the first time that a nuclear body present in a non-dividing cell was used as a sex indicator, but also because the spruce budworm belongs to the Lepidoptera and in these insects the female is the heterogametic sex: that is to say, the female is XY and the male XX, a situation which is just the opposite to that in mammals.

(a) 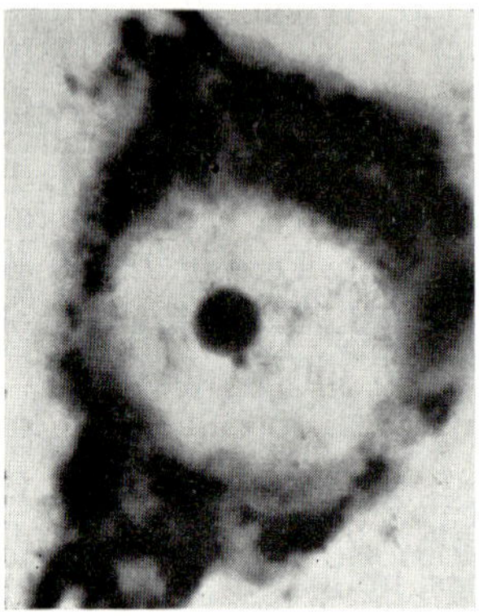 (b) 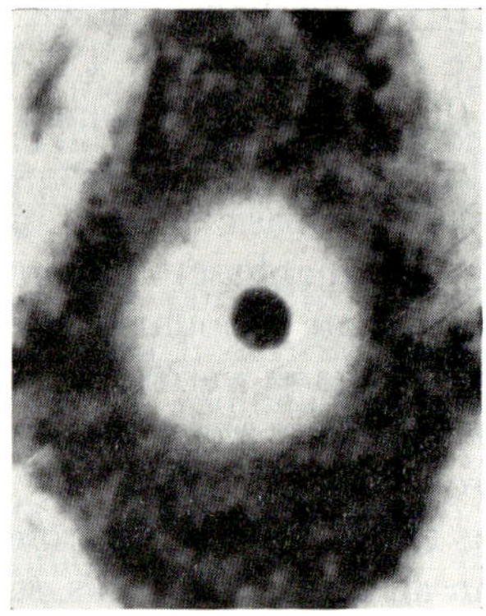

Fig. 1. Sex difference in neurones of the cat. (a) "Nucleolar satellite" adjacent to nucleolus of female; × 1 600 approx. (b) No nucleolar satellite in male; × 1 600 approx. By courtesy of Professor M. L. Barr.

In 1952, Graham and Barr reported that in the cat, sex chromatin could be seen in many different tissues. They had examined smooth, skeletal and cardiac muscle, cartilage, epidermis, endothelium, gastric epithelium, liver, pancreas, adenohypophysis, thyroid, adrenal cortex and medulla, interstitial cells of ovary and testis, kidney, uterine epithelium and prostate. They found sex chromatin in cells from all tissues from females, except in liver and pancreatic acinar cells. The sex chromatin was seen in one half to one third of cells from female cats and only very rarely in cells from males.

In the human species, sex chromatin was first seen in sympathetic ganglion cells. In living persons, sex chromatin was first demonstrated in 1953, when Moore and co-workers reported that it could be seen in preparations of skin biopsies. Included in this communication was one hermaphrodite, who thus became the first patient with an abnormal sexual development, in whom sex chromatin findings were described.

Soon afterwards, skin biopsies were examined from patients with

Turner's syndrome* by Polani and co-workers in England and by Wilkins and co-workers in the United States. The tests revealed that these patients lacked sex chromatin. By contrast, several groups of workers showed that sex chromatin was present in similar preparations from patients with Klinefelter's syndrome. While all kinds of possible reasons for these apparently aberrant results were being discussed, the techniques for preparing human chromosomes for analysis were being perfected. In 1959, the first chromosome analysis showed 47, XXY sex chromosomes in Klinefelter's syndrome and a 45, X karyotype in Turner's syndrome.

IV. Buccal Mucosa

The technique for demonstrating sex chromatin has been greatly simplified by the realization that these bodies can be seen in cells from the buccal mucosa. All that is required for this purpose is to scrape the inside of the cheek with a spatula, place the cells thus obtained on a slide and fix and stain them.

The first requirement for success with this technique is that a sufficiently large number of cells must be available. This is ensured by pressing the outside of the cheek with the fingers of one hand and scraping the inside of the cheek with a spatula held in the other hand. The cells must be fixed immediately, as they must not be allowed to dry up.

If temporary preparations are required, for immediate examination, a combined stain fixative, such as acetic orcein, may be convenient. The solution is pipetted on to the cells and a cover slip placed on top. The preparation may be examined after about half an hour. In time, the cells will become overstained and the bacteria, which are invariably present, will eventually take up the stain.

In order to make permanent preparations, buccal mucosa cells may be fixed in 95% ethanol. But in many situations a spray fixative such as "Spray-Cyte" (polyethylene glycols in isopropyl alcohol) may be preferable. Spraying fixes the cells almost instantly. After being sprayed, the slides can be packed. The preparations may be stained with one of a large number of nuclear dyes, including cresyl violet, thionin and Feulgen. Feulgen staining is often regarded as difficult, but in fact consistently good preparations can be obtained with very little practice. The fixed cells are gradually hydrated, hydrolized in 5N hydrochloric acid at room temperature for 25 min and stained in Feulgen solution for 1 h. They are then briefly rinsed, first in tap

* The reader is referred to chapter 11 for detailed descriptions of abnormalities in the human karyotype.

water, then in distilled water, after which they are dehydrated and mounted in a synthetic mounting medium, such as XAM.

A typical Barr body is a darkly stained structure of about 1μm diameter (Fig. 2). It goes without saying that in order to observe objects of this small size, it is necessary to use a high class optical system with an oil immersion objective. The majority of Barr bodies are seen at the periphery of the nucleus. As will be seen from the following section, a minority of Barr bodies occur in other situations of the nucleus, but when examining buccal mucosa preparations, it is safer to discount any bodies which are not located at the periphery. This is because a large proportion of cells of the buccal mucosa may contain non-sex specific aggregates of chromatin known as "chromocentres" and it is not always easy to distinguish non-peripherally placed Barr bodies from chromocentres.

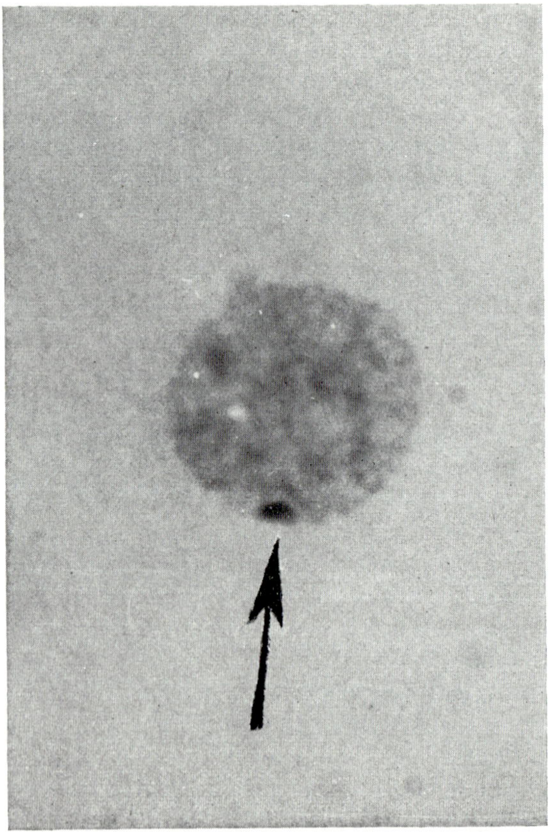

Fig. 2. Barr body in nucleus of buccal mucosa in human female. Thionin stain $\times$ 2 500.

In general, strict criteria have to be used in deciding which cells are suitable for scoring Barr bodies and which have to be discarded. Only cells in which the entire nucleus can be seen lying flat on the glass and in which there are no chromocentres should be considered when searching for Barr bodies. The proportion of such cells from normal females, showing a peripherally located Barr body, is usually relatively low, varying roughly between 20 and 35%. On the other hand, by using strict criteria, no Barr bodies at all should be found in buccal mucosa cells from normal males.

Barr bodies are found not only in cells from normal women, but also in cells from males with Klinefelter syndrome and 47,XXY chromosomes. The presence of a Y-chromosome does not seem to affect the formation of Barr bodies, which, in diploid cells, is solely dependent on the number of X-chromosomes. In buccal mucosa and all other diploid cells, the maximum number of Barr bodies is

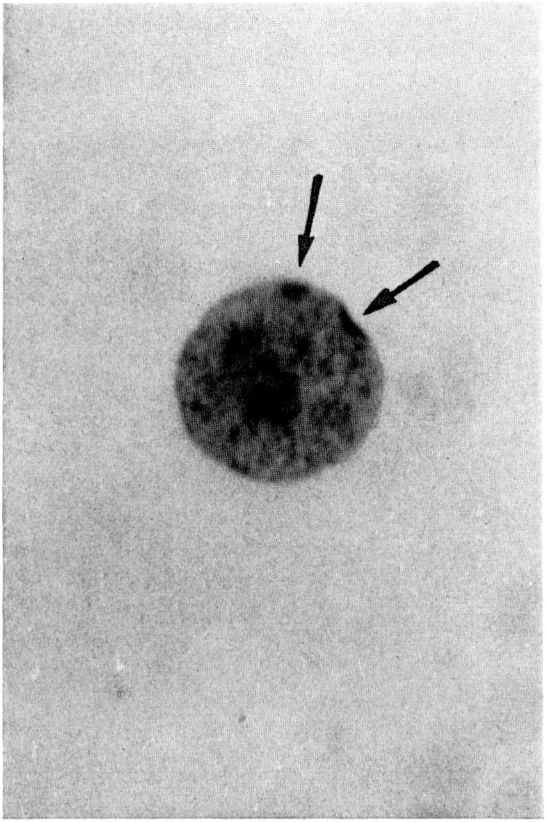

Fig. 3. Two Barr bodies in nucleus of buccal mucosa in human XXX female; × 2 500.

one less than the number of X-chromosomes per cell. This is why Barr bodies are absent in cells from normal males, as well from patients with Turner's syndrome, who are of 45,X chromosome constitution.

In cells of patients with more than two X-chromosomes, more than one Barr body may be found. Thus, in karyotypes with three X-chromosomes either 47,XXX or 48,XXXY, there will be some cells with two Barr bodies (Fig. 3) and in karyotypes with four X-chromosomes, either 48,XXXX or 49,XXXXY, some cells will contain three Barr bodies. Finally, in the rare cases with five X-chromosomes in the karyotype, 49,XXXXX, the maximum number of Barr bodies per cell is four.

In females, therefore, the buccal mucosa technique may be used to detect either a missing X-chromosome, or supernumerary X-chromosomes; while in males, it can detect the presence of supernumerary X-chromosomes. Apart from abnormal numbers of X-chromosomes, structurally abnormal X-chromosomes may be present. The presence of unusually large Barr bodies may suggest that a larger than normal X-chromosome is present, of which the presumed isochromosome (Chapters 10 and 11) of the long arm of the X-chromosome is the commonest. Similarly, deleted X-chromosomes may be associated with the presence of Barr bodies that are smaller than normal, but these may be more difficult to detect, since a distinction from non-sex specific chromatin bodies is not always possible.

It will be seen that the buccal mucosa technique is of considerable use for detecting numerical, and to a lesser extent structural, abnormalities of the X-chromosome. On the other hand, since the maximum number of Barr bodies is never present in all cells, it is likely that mosaicism would be missed by means of this technique. For instance, in a patient with Turner's syndrome and 45,X/46,XX mosaicism, Barr bodies would be present and the abnormal cell line would probably not be suspected, unless the incidence of Barr bodies happened to be unusually low. Chapter 11 gives further details about abnormalities of the sex chromosomes in the human.

The Barr body test, of course, gives no indication of the presence of Y-chromosomes. Methods of detecting Y-chromosomes in non-dividing cells will be described in Section XIV.

V. Hair Roots

A method for examining the cells in the sheaths of hair roots for the presence of Barr bodies was described by Schmid. From a single human capital hair, several thousand suitable cells can be obtained. They can be stained with acetic orcein and are therefore ready to be examined in a short time. A further advantage of the technique is that these cells are not contaminated by bacteria.

VI. Barr Bodies in Cultured Cells

Cell growing in monolayer cultures offer the best opportunities for studying the cytological details of Barr bodies. Such fibroblast–like cells flatten themselves on the cover slips, on which they actively grow and as a result their nuclei can be observed to their best advantage. Moreover, it is possible to study the behaviour of Barr bodies in different stages of the cell cycle. The technique of culturing fibroblast–like cells is as for chromosome analysis (Chapter 10) except that the addition of colchicine and hypotonic saline is omitted.

As in other tissues, the majority of Barr bodies are seen lying against the periphery of the nuclear membrane. A typical shape of such Barr bodies is triangular, the apex of the triangle pointing towards the centre of the nucleus (Fig. 4). Barr bodies may also be plano-convex or rectangular in shape. The latter shape is characteristic of those bodies which are not seen at the periphery of the nucleus (Fig. 5). These apparently centrally placed Barr bodies are often adjacent to a nucleolus.

When scoring Barr bodies, it is necessary to exclude nuclei with multiple chromocentres (Fig. 6). In different cell cultures from normal females, or other persons having two X-chromosomes, the incidence of Barr bodies in suitable nuclei may vary from about 30% to over 80%.

One factor influencing the incidence of Barr bodies is the mitotic index. In cultures which are actively dividing there tend to be fewer Barr bodies than in cultures whose cells have become confluent and where mitosis has virtually ceased. The lower frequency of Barr bodies in actively dividing cells is connected with the fact that Barr bodies are not usually formed until some time has elapsed after the completion of mitosis. They are therefore comparatively rare in cell populations containing many young cells.

Barr bodies can be seen during the S-phase, while the chromosomes are synthesizing DNA. It is possible to observe the nuclei through the microscope of an integrating microdensitometer, an instrument by means of which the density of individual cells can be measured. If the cells were stained by the Feulgen reaction under appropriate conditions, the amount of dye taken up by each nucleus will be proportional to the amount of DNA which was present, and this will be recorded by the instrument. Thus, by measuring the relative amounts of DNA in the nuclei, it is possible to classify them according

(a)
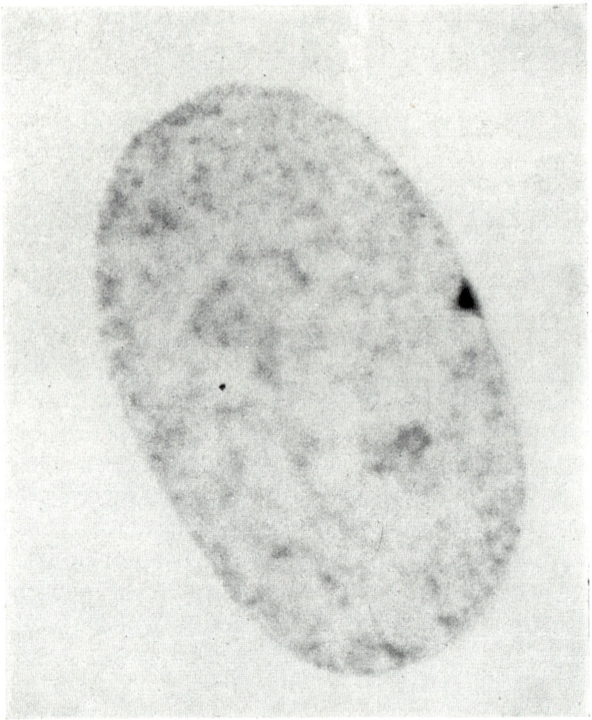

Fig. 4. Barr bodies at periphery of nuclei of human fibroblasts. Feulgen stain; × 4 000.

to whether they were in the G1, the S or the G2 phase. In this way it can be seen that the size of the Barr body increases as the cell goes through DNA synthesis (Fig. 7). The Barr body is smallest in G1, intermediate in size during S and largest during the G2 phase. These findings suggest that the X-chromosome that forms the Barr body (see Section X) synthesizes its DNA while in the condensed state.

A comparison of the incidence of Barr bodies in cells of different DNA classes in an actively growing culture often reveals that the incidence of Barr bodies rises as the cells go from G1 via S to the G2 phase. This is connected with the fact that cells in later stages of DNA synthesis are, on average, older than cells in the earlier stages.

Although Barr bodies are characteristic of interphase nuclei, structures that look like the remains of Barr bodies can sometimes be seen during mitotic prophase (Fig. 8). During the rest of mitosis, Barr bodies cannot be seen. Barr bodies have been observed to be reformed about fifteen hours after the end of mitosis. At this time, a proportion of cells will be in G1, but others will be in the S phase

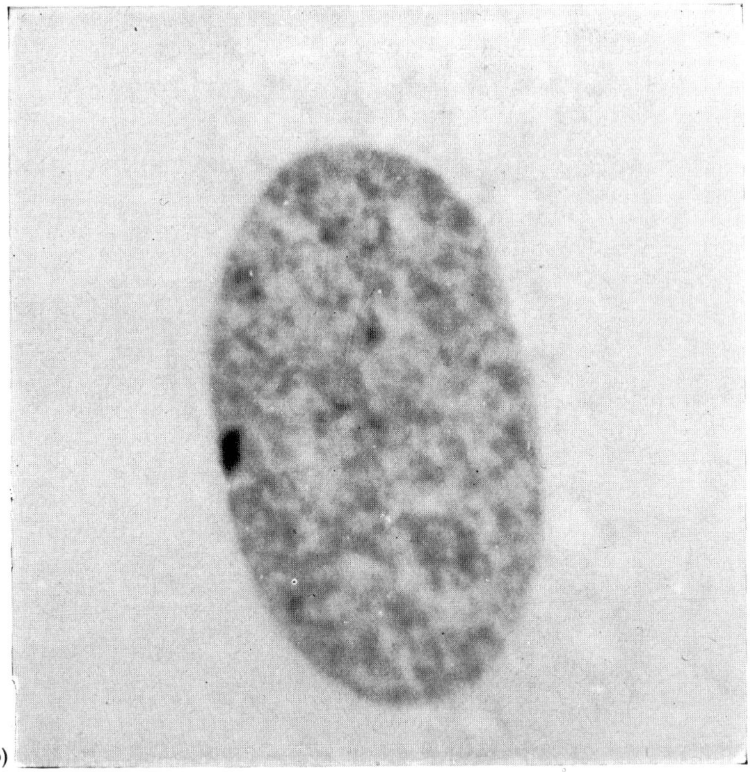

b)

or in G2. This explains why in an actively growing culture there may be a positive correlation between the incidence of Barr bodies and the amount of DNA present per nucleus.

A small proportion of cells growing in culture may become polyploid. In tetraploid cells, the maximum number of Barr bodies is two less than the number of X-chromosomes present. Since a tetraploid cell of normal female origin contains four X-chromosomes, it may contain two Barr bodies (Fig. 9). A cell of similar origin which has become octoploid may contain four Barr bodies. It follows that in evaluating multiple Barr bodies in cultured cells, the ploidy of the cells must be considered.

VII. Drumsticks

Barr bodies cannot be seen in cells of the blood. However, Davidson and Smith, in England, made a systematic search for a sex differentiator in leucocytes, which was successful in 1954, when they announced the discovery of the drumstick.

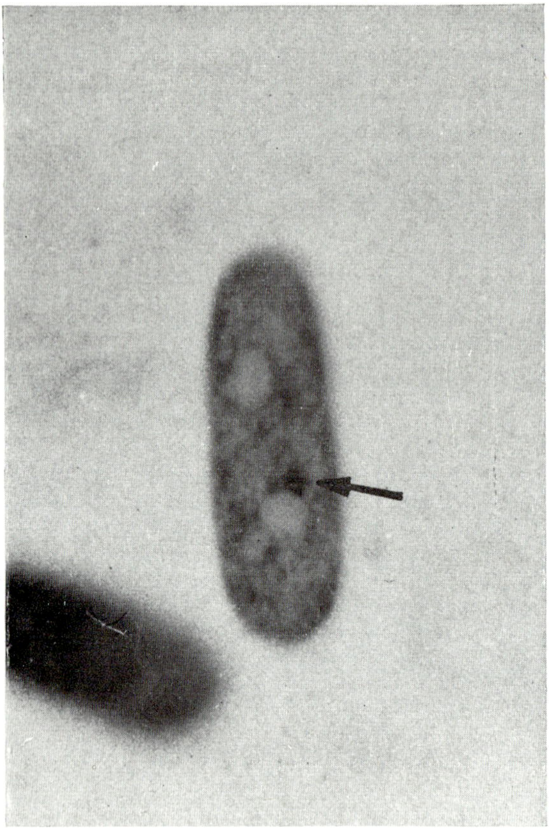

(a)

Fig. 5. Barr bodies not situated at periphery of human fibroblasts. Feulgen stain; × 2 500.

Drumsticks are nuclear appendages found in the polymorphonuclear leucocytes of females; they are absent in males. A drumstick consists of a head having a diameter of about 1·5μm, which is attached to the rest of the nucleus by means of a thin filament (Fig. 10). While the filament may be of variable length, the size of the drumstick head is an important characteristic. This is because nuclear appendages of other sizes are commonly found and these are not sex specific. The majority of these are small appendages, which may resemble drumsticks in shape, but have smaller heads. More rarely, one encounters larger nuclear appendages, which are known as minor lobes. Apart from their larger size, minor lobes also differ from drumsticks in structure. The head of a drumstick has an even outline and is densely stained all over, whereas a minor lobe is variable both in its outline and in its contours. It is undoubtedly due to the

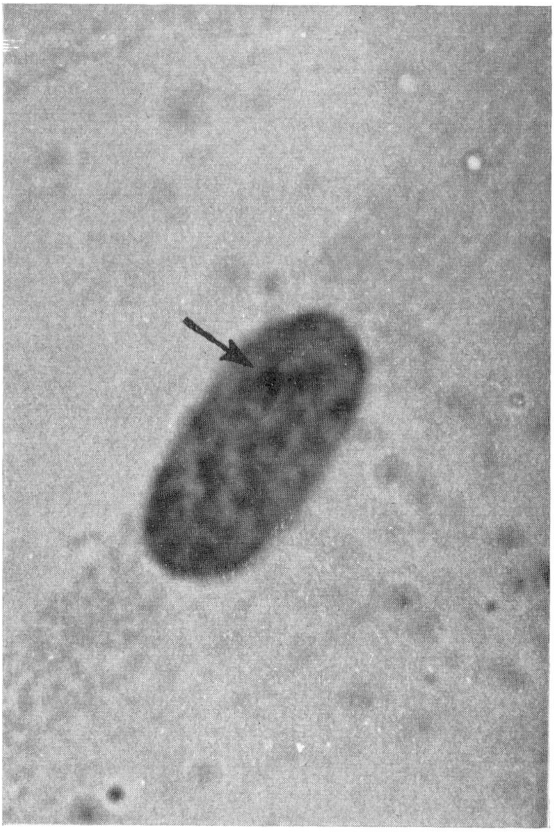

(b)

existence of these non-sex specific appendages that the drumstick was not discovered earlier.

Drumsticks without filaments are known as "sessile nodules" (Fig. 11). Although confined to the polymorphonuclear leucocytes of females or patients with two or more X-chromosomes, sessile nodules may be difficult to distinguish from other nuclear appendages and are of limited diagnostic value.

Although drumsticks can be demonstrated in ordinary blood films stained with Romanovsky dyes, the preparations must be of high technical quality. Preparations made between two diagonally opposed cover slips are preferable to those made on slides, since the cover slip method results not only in a more even distribution of different leucocytes but also in superior preservation of cytological detail.

Drumsticks are found in all types of polymorphonuclear leucocytes, but for practical purposes only those present in neutrophils are

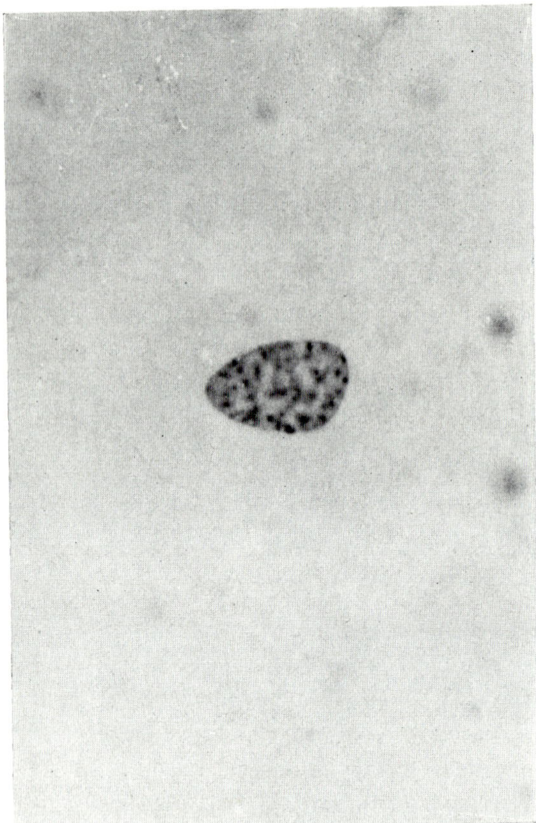

Fig. 6. Nucleus showing multiple chromocentres in human fibroblast. Unsuitable for scoring Barr bodies. Feulgen stain; × 1 800.

usually considered. This is because eosinophils and basophils are relatively rare and also because the coarse granules present in these cells tend to obscure the detailed morphology of the nucleus.

The incidence of drumsticks is much lower than that of Barr bodies. In blood films from normal women, a drumstick is present on an average in about 3% of neutrophils. There is however a certain amount of individual variation. A person's age does not seem to affect the frequency of drumsticks, but the degree of lobulation of the neutrophil nucleus is an important factor. Drumsticks are preferentially formed in cells with a high number of nuclear lobes. Consequently the incidence of drumsticks is depressed in conditions in which there is a "shift to the left" (i.e. the average number of nuclear lobes is below normal) and it is raised in conditions which show a "shift to the right" (average number of nuclear lobes above normal).

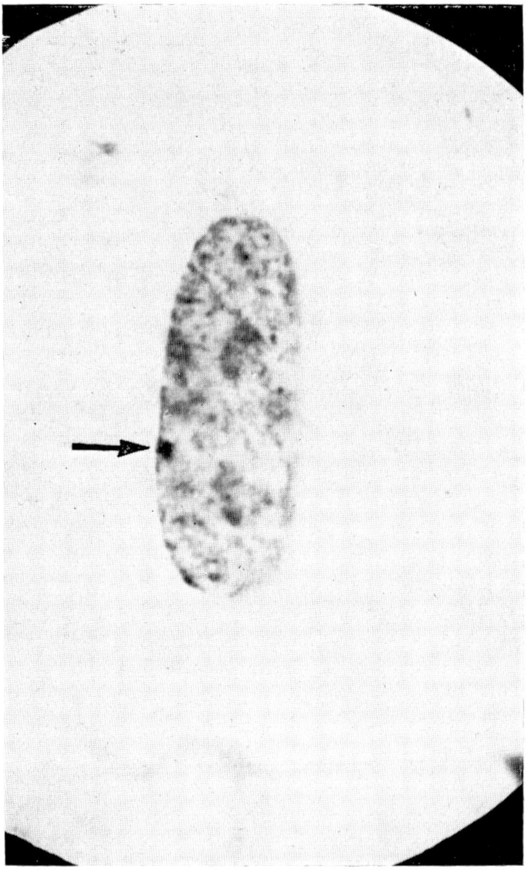

(a)

Fig. 7. Barr bodies in different stages of DNA synthesis in human fibroblasts: (a) G1 (b) S (c) G2. Feulgen stain; ×2200.

Drumsticks are absent from the neutrophils of normal males, but they are present in males with Klinefelter's syndrome. However, the incidence of drumsticks in these patients is below that in normal females. This correlates with the fact that in Klinefelter's syndrome, the mean number of nuclear lobes per neutrophil is less than normal.

In patients with three or more X-chromosomes, two or more drumsticks may be found in the same neutrophil, but such cells are exceedingly rare. This is undoubtedly connected with the low incidence of drumsticks in general. If the average incidence of drumsticks in 46,XX karyotypes is assumed to be 3%, the chance of finding two drumsticks in a cell containing three X-chromosomes might be expected to be nine in ten thousand. An incidence of this order of

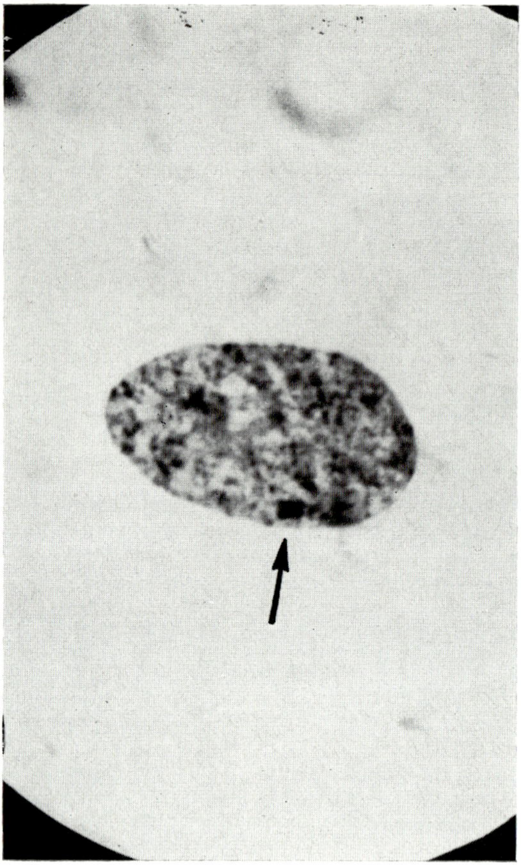

7(b)

magnitude is too low for practical consideration. We may conclude that although in theory the n – 1 rule relating numbers of X-chromosomes and sex chromatin bodies applies also to drumsticks, the incidence of drumsticks is too low for this relationship to be applicable in practice.

Drumsticks are, of course, absent in patients with the 45,X type of Turner's syndrome. However characteristic drumsticks may be present in patients having structurally abnormal X-chromosomes. In the most common type of structural abnormality, the presumed iso-chromosome of the long arm of the X-chromosome, drumsticks are present with heads that are larger than normal (Fig. 12). Moreover, these large drumsticks occur with a frequency which is well above normal; they are typically seen in 10% to 15% of neutrophils. The appearance of drumsticks, therefore, is characteristic in patients with Turner's syndrome carrying the presumed iso-chromosome of the long

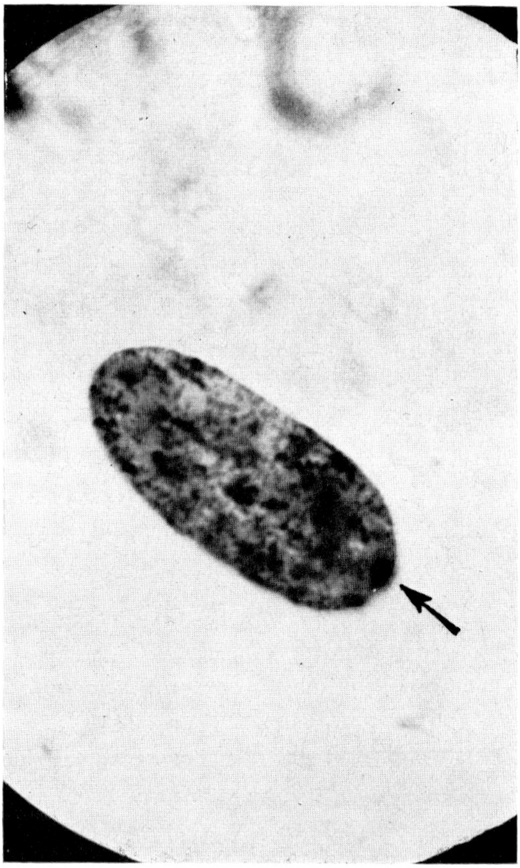

7(c)

arm of the X and, where the technique is available, it will provide an additional quick and simple test for establishing the diagnosis.

Deleted X-chromosomes are associated with drumsticks which are smaller than normal. But great care must be exercised in evaluating small drumsticks, since these are often difficult to distinguish from non-sex specific appendages.

VIII. Sex Chromatin Surveys

Systematic surveys of the sex chromatin status of large numbers of individuals have led to an understanding of the incidence of some sex chromosome abnormalities, both in the newborn population and in hospitals for the mentally retarded. The buccal mucosa technique was used almost exclusively for this purpose, though a small proportion of the work was based on the scoring of drumsticks.

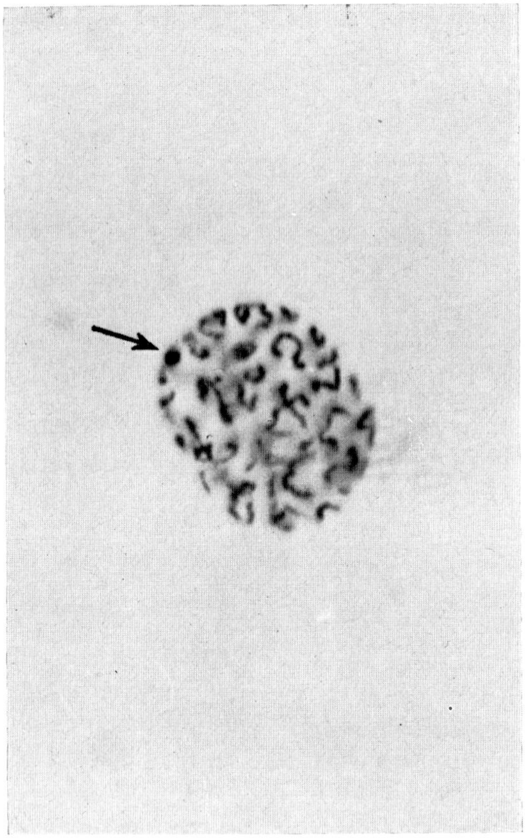

Fig. 8. Mitotic prophase in human fibroblast, showing apparent remains of Barr body. Feulgen stain; 1 800. (From Mittwoch, 1967.)

The largest available survey of sex chromatin in new born infants was carried out by Maclean and co-workers in Edinburgh. Among a total of 13 257 live male births investigated, 25 babies were found to have sex chromatin. An incidence of about 2 in 1 000 newborn males is in good agreement with results of similar surveys in Europe and North America. In Maclean's series, chromosome analysis in 21 of the abnormal males revealed that fourteen had a 47,XXY chromosome constitution, six were mosaics of the 46,XY/47,XXY type while one had 48,XXYY chromosomes. The fact that no multiple Barr bodies were found in this sample of over 13 000 children suggests that the incidence at birth of males with more than two X-chromosomes must be very low.

Surveys carried out in hospitals for the mentally retarded revealed

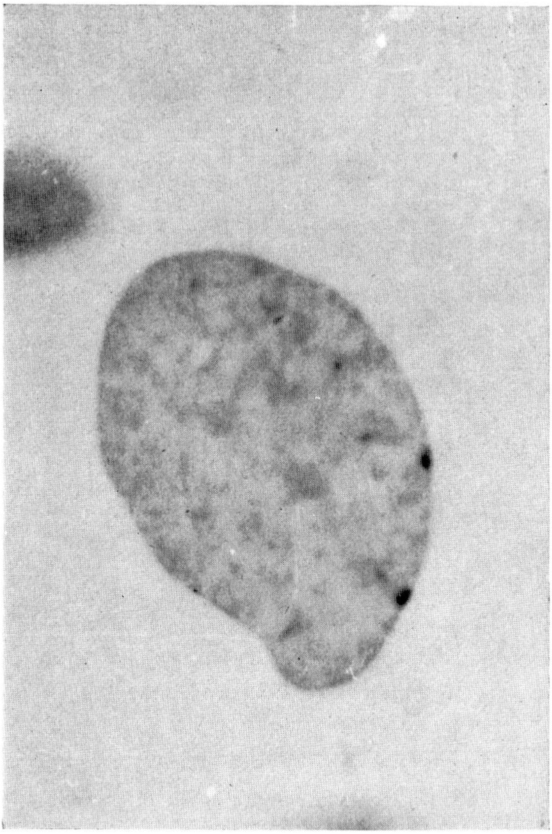

Fig. 9. Tetraploid nucleus of fibroblast from normal human female, with two Barr bodies; × 2 500.

that about 1% of the male patients had sex chromatin. This means that the incidence of sex chromatin in mentally retarded males is considerably higher than that in the newborn population and suggests that Klinefelter's syndrome predisposes to mental retardation. It is also of interest to note that among these patients the proportion of mosaics of the 46,XY/47,XXY type to non-mosaic Klinefelters has fallen to 1 to 8, compared with a ratio of 1 to 3 in the newborn population. As might be expected, the possession of a normal cell line reduces the risk of mental retardation. Maclean and co-workers also investigated to 10 000 newborn females. On the basis of this and other surveys, about 1 in 2 000 newborn females have no sex chromatin, this being the incidence at birth of the 45,X type of Turner's syndrome; while about 1 in 1 000 female infants have two Barr bodies and 47,XXX chromosomes.

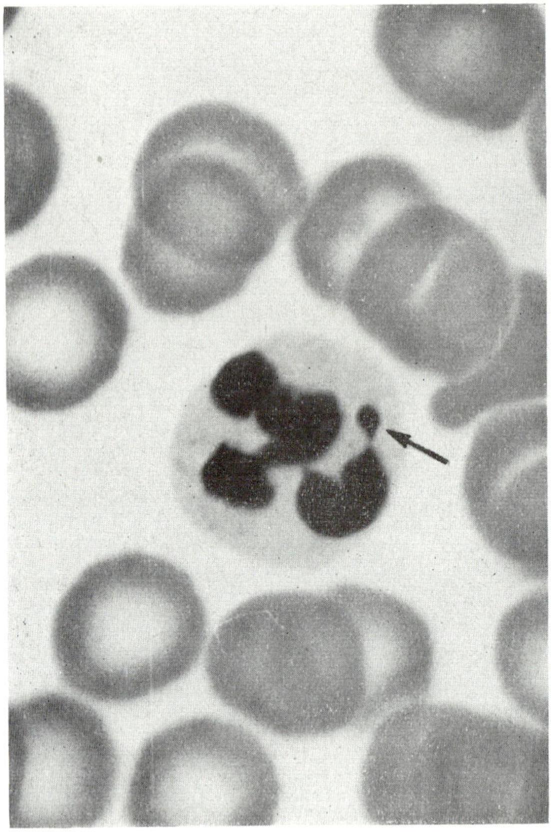

(a)

Fig. 10. Drumsticks in polymorphonuclear neutrophil leucocytes of human female; ×2 500. (a) No other appendages. (b) Cell contains minor lobe (broken arrow) and small non-sex specific appendages in addition to drumstick (arrow).

Surveys of female patients in hospitals for the mentally retarded did not reveal any appreciable increase in the incidence of Turner's syndrome as compared with the newborn population. However, the frequency of double Barr bodies among female patients is about 4·6 per 1 000, reflecting a considerable increase over the incidence at birth. Thus, the triple X condition is another chromosomal abnormality which predisposes towards mental retardation.

IX. Autoradiography and Late Replicating X-Chromosomes

The principle of autoradiography with tritiated thymidine, and its use in establishing the time at which individual chromosomes replicate their DNA has been described in Chapter 10. In cells

(b)

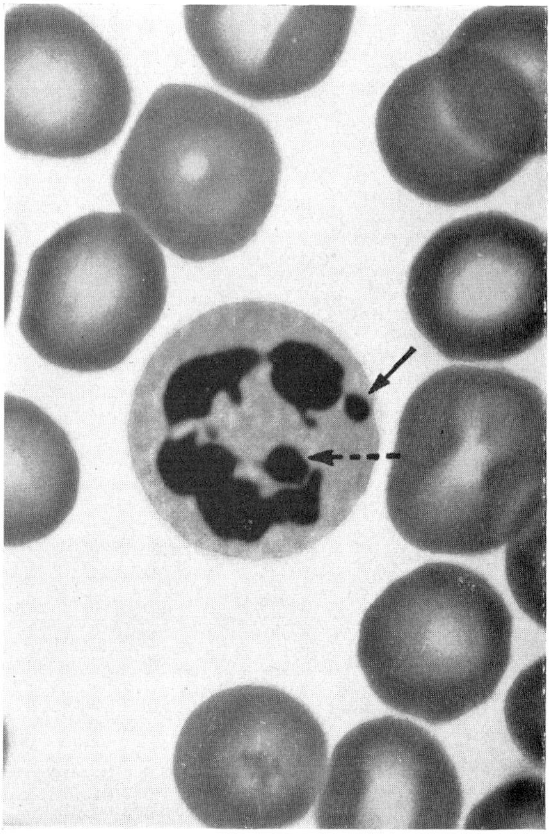

from normal females, one X-chromosome is late replicating, whereas the single X-chromosome from normal males does not replicate late. In fact, late replicating X-chromosomes follow the n—1 rule in the same way as Barr bodies do. If three X-chromosomes are present in the karyotype, many cells will contain two late replicating X-chromosomes, and so on. If a structurally abnormal X-chromosome is present, it invariably replicates late, while the normal X-chromosome replicates early, i.e. largely in synchrony with the autosomes.

X. The Origin of Barr Bodies

These findings have greatly clarified our ideas about the origin of Barr bodies. The fact that the maximum number of Barr bodies in a diploid nucleus is always one less than the number of X-chromosomes in the karyotype confirmed the idea, which was first

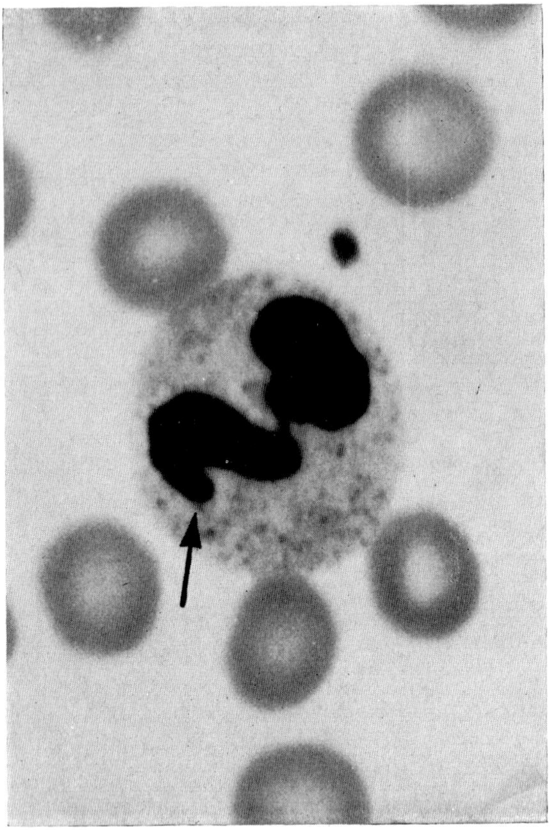

Fig. 11. Polymorphonuclear neutrophil leucocyte with sessile nodule; ×2 500.

formulated by Ohno and collaborators, that a Barr body is formed from a single X-chromosome.

A Barr body, then, represents an X-chromosome that forms a highly condensed body in the interphase nucleus, while the rest of the chromosomes are in the dispersed state and are not individually recognizable. It is remarkable that one X-chromosome always remains in the dispersed state, just like the autosomes: but if a second X-chromosome is present, it behaves differently from the first, and so do any additional X-chromosomes. In addition to forming a condensed body in interphase cell, a second X-chromosome also replicates most of its DNA later than the first X-chromosome and the autosomes. Similarly, additional X-chromosomes in pathological karyotypes replicate late, while the fact that structurally abnormal X-chromosomes, which are either larger than normal or smaller than normal, always replicate

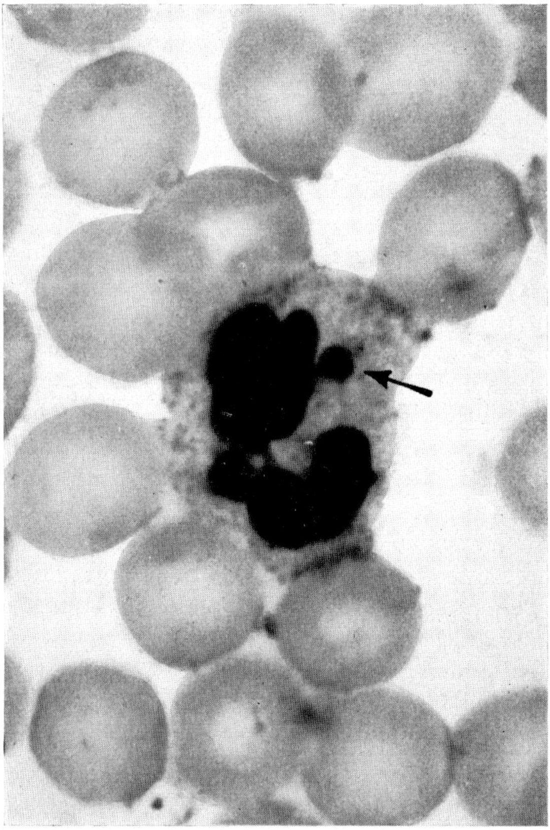

Fig. 12. Large drumstick from human female with X isochromosome; ×2 500.

late and are associated with Barr bodies which are larger or smaller than normal respectively, provides additional confirmation that a Barr body originates from a single X-chromosome, which replicates its DNA later than most other chromosomes.

In spite of the undoubted and striking correlation between Barr bodies and late replicating X-chromosomes, it should perhaps be pointed out that the two do not necessarily go together in every cell that contains two X-chromosomes. It has already been mentioned that there tend to be fewer Barr bodies in rapidly dividing cells. Apart from the age of the cells, it is likely that other factors, such as cell size and cell density also play a role, the incidence of Barr bodies increasing as the cells become smaller and more crowded. The finding that Barr bodies tend to be absent in very early embryos may be connected with these facts. Recent results on rabbit embryos suggest that the late replication pattern of the X-chromosomes is established before the formation of Barr bodies.

XI. The Nature of Drumsticks

Although the relationship between numbers of X-chromosomes and the formation of drumsticks may not seem quite so obvious as in Barr bodies, there can no longer be any doubt that a drumstick, too, consists of an X-chromosome which, being the second in the karyotype, is highly condensed and extruded from the main body of the nucleus of a polymorphonuclear leucocyte. The most decisive evidence for this interpretation comes from the large drumsticks found in patients with the presumed iso-chromosome of the long arm of the X, which is larger than the normal X-chromosome: and from the fact that small drumsticks are found in association with deleted X-chromosomes. Differences between the patterns of Barr body and drumstick formation are largely connected with the low incidence of drumsticks, which makes the scoring of multiple drumsticks in patients with three or more X-chromosomes impossible for practical purposes; while the impairment of nuclear lobulation of the leucocyte associated with various chromosomal abnormalities tends to depress the incidence of drumsticks still further.

We may conclude that the nature of the drumstick is basically similar to that of the Barr body, and that the differences in morphology and incidence are connected with the high degree of differentiation of the polymorphonuclear leucocyte, in which the drumstick occurs.

XII. The Function of Sex Chromatin—The Single Active X Hypothesis

The possibility that the sex chromatin body may represent the second X-chromosome which, being genetically superfluous is present in a condensed and inactivated form, was suggested independently by a number of authors, including Stewart, Russell and Beutler. However the single active X hypothesis in its wider form was put forward by Mary Lyon, and is now generally referred to as the "Lyon hypothesis".

The Lyon hypothesis may be divided into four parts:

(1) The first and basic postulate is that the genes on X-chromosomes that form the sex chromatin are inactive.

(2) Inactivation of X-chromsomes occurs at random, so that

paternal and maternal X-chromsomes have an equal chance of being inactivated.

(3) The process of inactivation occurs early in embryonic life.

(4) Once inactivation has occurred, the same X-chromosome will always be inactivated in the descendants of each cell.

Assuming these postulates to be true, a female mammal would consist of a patchwork of cells, each patch having either the maternal or the paternal X-chromosome in the active form. The hypothesis was originally put forward to explain the fact that female mice, which are heterozygous for coat colour genes located on the X-chromosome, are spotted.

Since the effect of the Lyon hypothesis would be that in each female cell only one X-chromosome is active, it is thought to provide a mechanism for dosage compensation in female mammals. The term "dosage compensation" means that a sex-linked gene produces equal amounts of gene products in males and females, in spite of the difference in the number of X-chromosomes (see Chapter 9).

There is evidence that dosage compensation occurs in the human species. The gene responsible for the production of glucose–6–phosphate dehydrogenase (G–6–PD) is located on the X-chromosome, yet the amount of enzyme in erythrocytes is the same in males and females, as well as in patients with abnormal numbers of X-chromosomes. Recent evidence suggests that in fibroblasts of embryos and newborn infants, the level of G–6–PD is higher in females than in males, and it is possible therefore that dosage compensation may take time to develop. Evidence in favour of postulates (1) and (4) of the Lyon hypothesis has come from tissue culture experiments. Davidson and collaborators have shown that the fibroblasts from women who were heterozygous for two different variants of G–6–PD were grown two different clones could be isolated, each containing cells of one or other variant, but not both. Clonal dimorphism has also been demonstrated for another enzyme, controlled by a gene on the X-chromosome, hypoxanthine guanine phosphoribosyl transferase (HGPT). Fibroblasts from women who are heterozygous for HGPT deficiency not only produce two types of clones, with and without enzyme activity, but this dimorphism could be demonstrated in individual cells. By contrast, in the case of a number of autosomal genes, for instance phosphoglucomutase and lactic acid dehydrogenase, no dimorphic clones could be produced from cultures of heterozygotes.

These results also confirm the second postulate of the Lyon hypothesis by demonstrating that either the paternal or the maternal X-chromosome may be inactivated, although they do not provide evidence on whether the process is truly at random.

In some special cases there is evidence against inactivation being at random. It has already been mentioned that any structurally abnormal X-chromosome always replicates late and Polani *et al.* (1970) have recently suggested that if the *Xg* blood group, which is determined by a sex-linked gene, is carried on a structurally abnormal X-chromosome it is always inactivated. This was based on the observation that patients with Turner's syndrome, due to X-chromosomes with deletions either in the long or in the short arm, lacked the *Xg* antigen, although the fathers of the patients possessed them; the implication being that whether the *Xg* gene is carried on the long or the short arm of the X-chromosome (which is at present unknown) the antigen is not expressed in a deleted X-chromosome. By contrast, the *Xg* gene appears not to be inactivated when present on normal X-chromosomes.

Other instructive evidence on the Lyon hypothesis is derived from studies of the mule. A mule is a hybrid between a mare and a male donkey. The maternal and paternal X-chromosomes of female mules can be distinguished cytologically, since the centromere in the X-chromosome of the mare is nearly in the middle of the chromosome, while in the donkey X-chromosome it is nearer one end. The bulk of the available evidence suggests that in cultures of lymphocytes and fibroblasts, the X-chromosome of the donkey replicates late in the majority of cells. These finding have been correlated with G–6–PD enzyme patterns. The gene responsible for the production of G–6–PD is sex-linked in both the horse and the donkey but the enzymes of the two species can be distinguished electrophoretically. Hamerton *et al.* (1969) found that in erythrocytes and fibroblasts of female mules there was very little donkey G–6–PD the bulk of the enzyme being derived from the horse. It should be noted that the preferentially inactivated X-chromosome in the mule is the paternal one. It seems, however, that in the reciprocal cross the X-chromosome of the donkey is also inactivated in most cells, so that the derivation of the X-chromosome—whether paternal or maternal—cannot be an important factor in deciding which one is to be inactivated. It is not yet known whether inactivation is originally non-random or whether the cells with the active X-chromosome of the horse have a better chance of proliferating. These experiments, therefore, provide good evidence in favour of the relationship between late replicating X-chromosomes and the inactivation of genes on these chromosomes, though they are inconclusive on whether this is a random event.

There are still a number of unresolved problems regarding the

Lyon hypothesis and this is hardly surprising in view of our great ignorance as to the time and place of action of any genes. One problem, raised by Grüneberg (1969), is that in female mammals which are heterozygous for a sex-linked gene affecting coat colour the distribution of the two types of pigments is not always at random, so that another mechanism controlling pattern formation must be postulated. Another apparent difficulty is that in humans, the presence of varying numbers of X-chromosomes is associated with phenotypic differences. This is generally explained by assuming either that the phenotypic differences are determined during early embryogenesis before dosage compensation is established, or that there is a part of the X-chromosome which escapes inactivation, so that karyotypes with different numbers of X-chromosomes contain different numbers of the non-inactivated parts. Probably the most likely explanation is that varying the number of any chromosome, including late replicating X-chromosomes, may alter such fundamental properties as the length of the mitotic cycle and the rate of cell proliferation and that these effects may be to a large extent independent of the activity of individual genes on these chromosomes.

We may conclude that the Lyon hypothesis has stimulated a great deal of research and from this a considerable amount of evidence has emerged that genes on one or other X-chromosome in cells of female mammals are inactive. At the same time, many aspects of the problem remain to be investigated.

XIII. Sex Chromatin in Mammals Other Than Man

After its discovery in the cat, sex chromatin has been demonstrated in a very large number of mammalian species. Nevertheless, different orders differ in the ease with which sex chromatin can be seen. In primates and in carnivores, Barr bodies have been seen in almost all cells of all the species which have been investigated. However, in rodents and in Artiodactyla, the nuclei tend to contain coarse chromocentres, which may make the identification of Barr bodies difficult or even impossible. In cattle, Barr bodies can be seen in nerve cells, buccal mucosa cells and in fetal membranes but not in most adult non-nerve cells. In the rat, too, sex chromatin cannot be satisfactorily demonstrated in most adult cells, though it has been described in Purkinje cells and spinal cord neurones. By contrast, Barr bodies are visible in many types of cells in rat embryos, as well as in fibroblasts cultured from embryonic tissues (Fig. 13). Characteristic Barr bodies can also be seen in amnion cells. In general, amnion cells are useful for establishing the chromosomal sex of embryos. In the mouse, how-

(a)

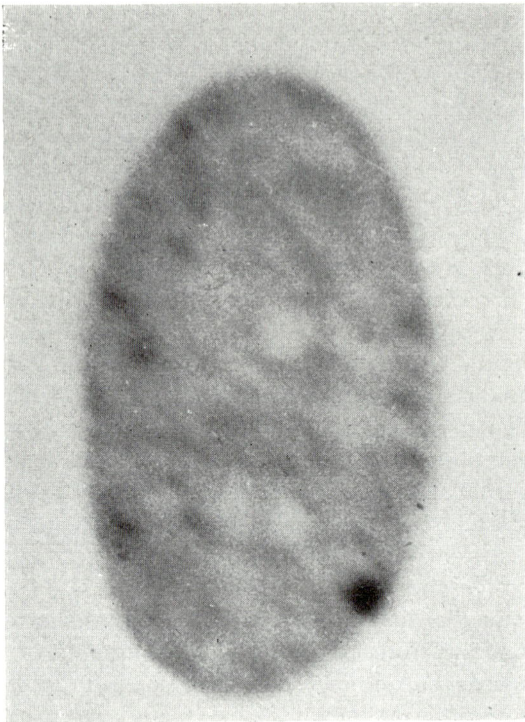

Fig. 13. Barr bodies in rat; ×5 000. (a) Fibroblast cultured from embryonic tissue. (b) Amnion epithelial cell.

ever, it is virtually impossible to demonstrate Barr bodies in any tissues. In cultured fibroblasts, there are a large proportion of cells with multiple chromocentres and in the apparently clear nuclei, the incidence of Barr bodies is too low for the technique to be of practical value. Drumsticks, too, are present in many species of mammals. They have been seen in species of primates, carnivores as well as Artiodactyla and Perissodactyla, but they cannot usually be demonstrated in rodents. In rodents, multiple drumstick-like appendages may be present in the polymorphonuclear leucocytes of both sexes.

XIV. Fluorescent Y-Bodies

For about twenty years, the demonstration of Barr bodies and drumsticks has provided simple tests for showing up the presence of two X-chromosomes in the karyotype. Recently, a new technique has been developed for identifying the presence of a Y-chromosome in non-dividing cells.

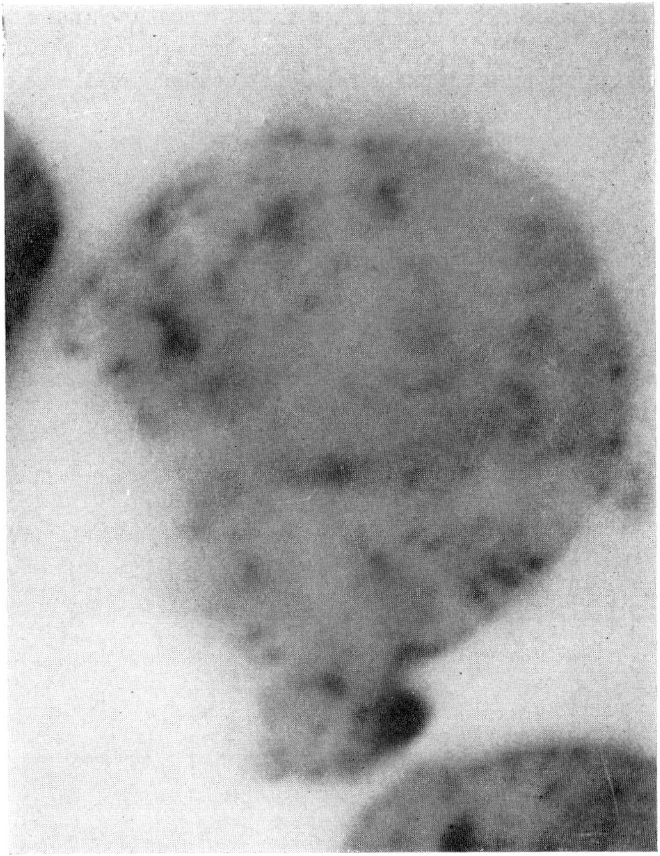

(b)

Staining is carried out using a fluorescent quinacrine derivative. When viewed under a fluorescence microscope, the Y-chromosome appears as a fluorescent body. In cells from normal males, one such body is seen in the modal number of cells, whereas in cells from 47,XYY males, the modal number of cells contains two fluorescent bodies (Fig. 14). The fluorescence of the Y-chromosome can also be seen in metaphase preparations. It is confined to a part of the long arm of the Y-chromosome. Fluorescent Y bodies can be seen in preparations of buccal mucosa as well as in lymphocytes. They can also be demonstrated in cells of amniotic fluid and this is likely to prove a valuable adjunct to antenatal sexing techniques. Fluorescent Y bodies can be seen in spermatozoa, thus providing the first means of distinguishing between X– and Y– bearing sperm. However, since these are in a fixed condition, this technique is not likely to result in parents being able to select the sex of their children.

It is now possible, by combining the fluorescent technique with the demonstration of Barr bodies and drumsticks, to identify the entire sex chromosome compliment in non-dividing cells in persons of all ages, as well as in fetuses.

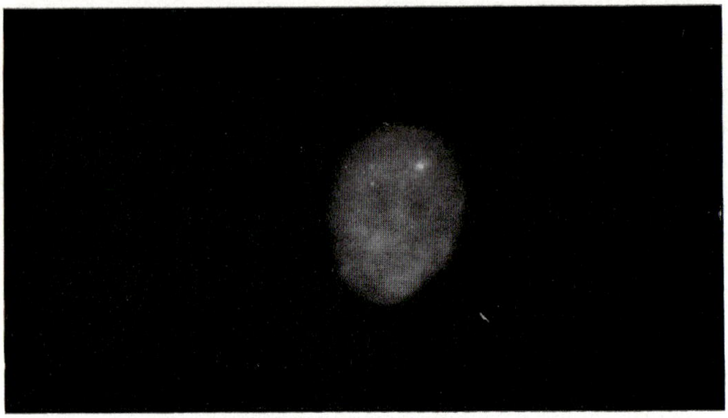

(a)

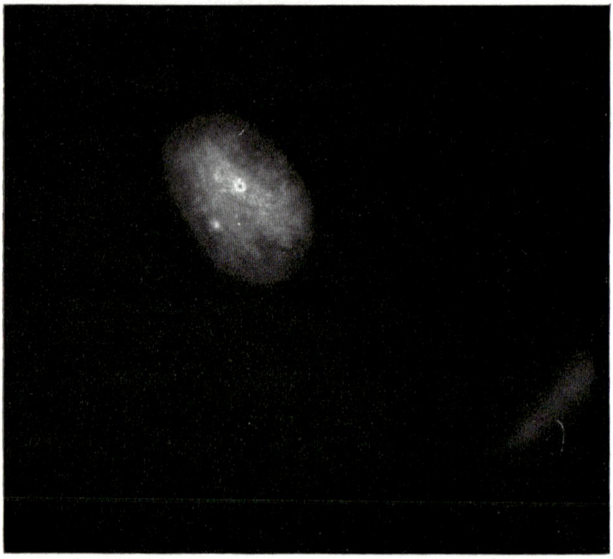

(b)

Fig. 14. Fluorescent Y bodies in cells from human buccal mucosa; ×1 800. (a) from XY male. (b) From XYY male. (Contributed by Dr. P. L. Pearson.)

References and Further Reading

Grüneberg, H. (1969). Threshold phenomena versus cell heredity in the manifestation of sex-linked genes in mammals. *J. Embryol. Exp. Morphol.* **22**, 145–179.

Hamerton, J. L., Gianelli, F., Collins, F., Hallett, J., Fryer, A., McGuire, V. M. and Short, R. V. (1969). Non-random X-inactivation in the female mule. *Nature* **222**, 1277–1278.

Hamerton, J. L. (1971). "Human Cytogenetics". Vol. 1 General. Academic Press, New York and London.

Issa, M., Blank, C. E. and Atherton, C. W. (1969). The temporal appearance of sex chromatin and of the late-replicating X-chromosome in blastocysts of the domestic rabbit. *Cytogenetics* **8**, 219–237.

Krooth, R. S. (1969). Gene action in human diploid cell strains. *In* "Comparative Mammalian Cytogenetics", (K. Benirschke, ed.), 154–179. Springer Verlag, Berlin, Heidelberg, New York.

Lyon, M. F. (1972). X-chromosome inactivation and developmental patterns in mammals. *Biol. Rev.* **47**, 1–35.

Mittwoch, U. (1967). "Sex Chromosomes". Chap. 10, Academic Press, New York and London.

Moore, K. L. (editor, 1966). "The Sex Chromatin". W. B. Saunders Co., Philadelphia and London.

Pearson, P. L., Bobrow, M. and Vosa, C. G. (1970). Technique for identifying Y-chromosomes in human interphase nuclei. *Nature* **226**, 78–80.

Polani, P. E., Angell, R., Gianelli, F., de la Chapelle, A., Race, R. R. and Sanger, R. (1970). Evidence that the *Xg* locus is inactivated in structurally abnormal X-chromosomes. *Nature* **227**, 613–616.

Race, R. R. (1972). Is the Xg blood group locus subject to inactivation? *Proc. 4th Int. Congress Hum. Benef.*

Schmid, W. (1967). Sex Chromatin in hair roots. *Cytogenetics* **6**, 342–349.

Steele, M. W. (1970). Incomplete dosage compensation for glucose-6-phosphate dehydrogenase in human embryos and newborns. *Nature* **227**, 496–498.

13. The Kinetics of Cell Production

D. BRYNMOR THOMAS
Department of Anatomy, The University of St Andrews

I. The Production of New Living Cells

A. INTRODUCTION

"Omnis cellula e cellula". (Virchow).

New living cells are invariably produced by the division of pre-existing cells. An adequate number of cell divisions is therefore necessary to ensure the growth of expanding cell populations and the maintenance of cell populations from which cells are being lost.

Somatic cells are all produced by mitotic division and all nucleated somatic cells contain the diploid number of chromosomes. In human males each somatic cell contains twenty-two pairs of autosomes, together with an X-chromosome and a Y-chromosome, while the somatic cells of human females which lack the Y-chromosome contain twenty-two pairs of autosomes and two X-chromosomes. During the formation of gametes meiotic division results in the production of haploid cells, each of which contains only half the chromosome complement of a diploid somatic cell. (See Chapter 10.)

B. MITOTIC DIVISION

The equal distribution of chromosomes between the cells formed following a mitotic division is ensured by an orderly sequence of events, which results in an equal division of nuclear material (karyokinesis), followed by division of the cell body (cytokinesis). Although cytokinesis usually results in a more or less equal distribution of material between the products of division, cytoplasmic organelles and inclusions appear to be apportioned in an arbitrary manner which depends only upon the pattern of their distribution in the parent cell. If karyokinesis is not followed by cytokinesis multinucleate cells will be produced.

The continuous sequence of events which can be observed microscopically during mitosis has been divided, for descriptive convenience, into four arbitrarily defined phases (see Mazia, 1961; Levine, 1963):—

1. *Prophase*. The commencement of prophase is difficult to recognize and therefore difficult to define precisely. It is indicated by the appearance in the nucleus of minute chromatin granules. As the chromatids, into which the chromosomes split longitudinally, coil and shorten, these granules become more numerous and aggregate to form masses; the centrioles separate and move apart, the mitotic spindle forms between the centrioles and the nucleolus disappears together with the nuclear membrane.

2. *Metaphase*. The disappearance of the nuclear membrane marks the commencement of metaphase. During metaphase the mitotic spindle enlarges and the short, thick chromosomes become arranged in its equatorial plane.

3. *Anaphase*. Anaphase commences when the chromatids separate and begin to move towards opposite poles of the mitotic spindle. Towards the end of anaphase, cytokinesis begins with the development of an annular constriction in the now elongated cell.

4. *Telophase*. The reappearance of nuclear membranes marks the commencement of telophase. During telophase the newly formed nuclei enlarge, their chromosomes disperse and cytokinesis is completed. The termination of telophase, like the commencement of prophase, is difficult to define precisely. At the termination of telophase mitosis is complete and the new cells enter interphase.

C. THE INTER-MITOTIC INTERVAL

The use of the cytophotometric analysis (see p. 208) and of autoradiographs prepared at various time intervals after the administration of radioactive DNA labels (see p. 217) has resulted in the recognition of three stages during the inter-mitotic interval (interphase) (Howard and Pelc, 1951 and 1953; German, 1964). During the first stage protein synthesis and RNA synthesis occur in the absence of DNA synthesis. Consequently the mass of newly formed cells increases while their complement of genetic material remains unchanged (G1). In cells which are destined to undergo mitotic division the commencement of DNA synthesis marks the beginning of the second stage of interphase (S). Upon the termination of DNA synthesis the cell enters the final stage of interphase (G2), which ends with the commencement of mitosis (Fig. 1).

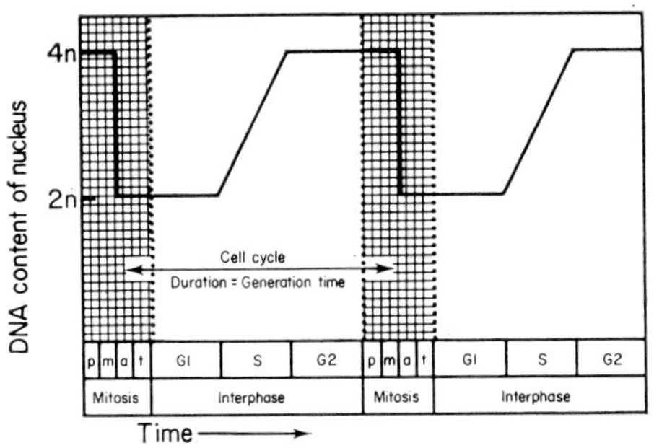

Fig. 1. At the commencement of a mitotic division the nucleus of a somatic cell contains four times as much DNA as the nucleus of a gamete (n). During mitosis the DNA content of the nucleus is halved, then during the subsequent interphase it is doubled in cells that are destined to divide again. (p—prophase; m—metaphase; a—anaphase; t—telophase.)

The synthesis of RNA and of protein continues during S and G2 at rates which do not appear to be influenced by concurrent DNA synthesis during S. A small amount of DNA synthesis during prophase appears to be of importance in cells undergoing meiotic division (Ito *et al.*, 1967) and it has been suggested that some cells which are not preparing to divide synthesize small quantities of "metabolic DA" (Pelc, 1971).

Some cells lose the ability to divide again, while others which retain this ability may divide only under appropriate circumstances (Go reserve) (see p. 205).

D. THE CELL CYCLE

One complete passage through the three stages of interphase and the four phases of mitosis is termed a cell cycle. The duration of a complete cell cycle is termed the generation time (Fig. 1). If all of the cells in an expanding population from which no cells were being lost multiplied at the same rate, the time interval required for the total number of cells in the population to double (doubling time) would be the same as the generation time. Similarly, if all of the cells divided at the same rate in a population in which the production of new cells was balanced by cell loss, so that it was undergoing continuous renewal while the total number of cells in the population remained the same, the time interval required for all of the cells to be replaced by new cells (turnover time) would be the same as the generation time. In most populations, however, the generation time is not constant for all of the dividing cells and some cells remain in interphase for prolonged periods or indefinitely. As the proportion of dividing cells decreases the doubling time or the turnover time for the population increases and ceases to bear any relationship to the generation time of those cells which continue to divide (see p. 213).

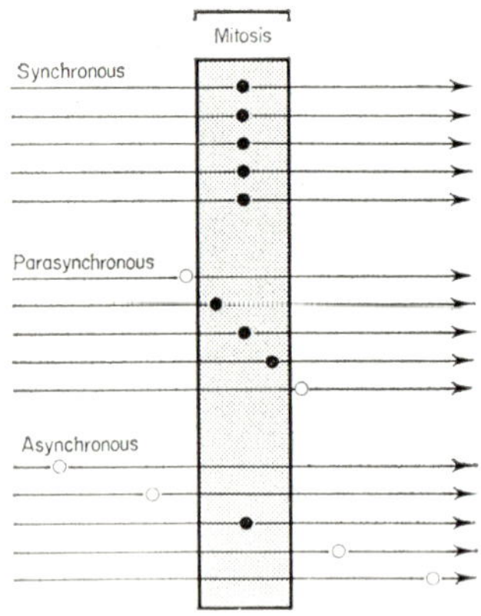

Fig. 2. All of the dividing cells in a population may enter mitosis at the same time (Synchronous), during a time interval which is of very short duration in comparison with the generation time (Parasynchronous), or over a time interval which is equal in duration to the generation time (Asynchronous).

All of the dividing cells in a population may enter mitosis at the same time (synchronous), during a time interval which is of very short duration in comparison with the generation time (parasynchronous), or over a time interval which is equal in duration to the generation time (asynchronous) (Fig. 2). In asynchronous populations the proportion of dividing cells in mitosis remains constant as cells progress through the stages of the cell cycle but in synchronous and parasynchronous populations it does *not* remain constant (Fig. 3).

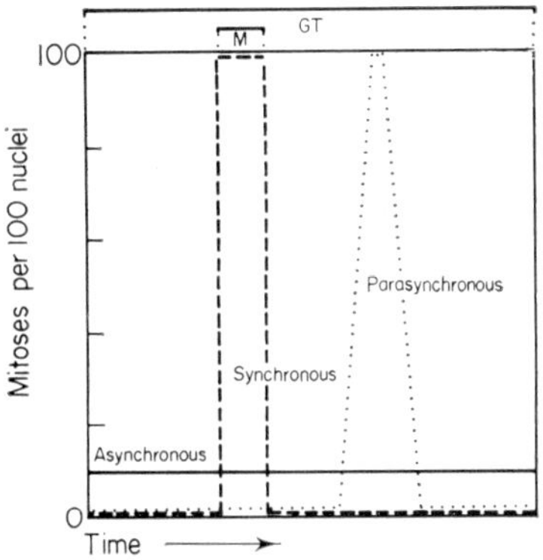

Fig. 3. In asynchronous populations the proportion of dividing cells in mitosis remains constant as cells progress through successive stages of the cell cycle but in synchronous and parasynchronous populations it does *not* remain constant. (GT–generation time; M–indicates the duration of mitosis and represents the position of mitosis in the cell cycle of the synchronous population.)

II. The Classification of Cell Populations According to Patterns of Cell Production

Populations of living cells may be classified according to the patterns of cell production which they exhibit into three groups—those in which new cells can not be produced (static), those in which cell production is matched by an increase in the total number of cells (expanding) and those in which cell production is not matched by an increase in the total number of cells (undergoing continuous renewal).

A. STATIC CELL POPULATIONS

New cells can not be produced in static cell populations which do not exhibit evidence of mitotic activity and do not increase their total DNA content unless the DNA content of individual cells is being increased. Differentiated neurons which constitute a static cell population increase in volume but not in number as the organism grows.

B. EXPANDING CELL POPULATIONS

In expanding cell populations mitotic activity is matched by an increase in size and in total DNA content. New cells are retained within such populations and so increase their size. The number of nuclei increases progressively together with the total DNA content at a rate which decreases with the growth rate of the organism. Fully differentiated cells appear to undergo mitotic division and the proportion of cells undergoing division, which may be exceedingly small in adults, can increase dramatically. Thus partial hepatectomy is followed by an enormous increase in the mitotic index of hepatic parenchymal cells (Fig. 4). The possibility has been suggested that cell populations like the parenchymal cells of the liver produce substances which inhibit their own growth. A reduction in the size of such populations would then be expected to result in increased cell production due to decreased production of inhibitory factors. The parenchymal cells of the liver, the kidneys, the exocrine glands and the endocrine glands have all been included in this group by Leblond and Walker (1956).

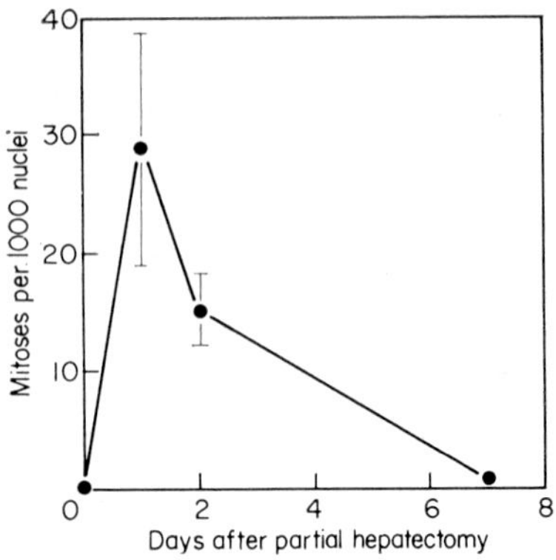

Fig. 4. Changes in the incidence of mitotic figures among hepatic parenchymal cells following partial hepatectomy. (Data of Abercrombie and Harkness (1951) for rat liver.)

C. CELL POPULATIONS UNDERGOING CONTINUOUS RENEWAL

Mitotic activity is *not* matched by an increase in the size or the total DNA content of cell populations which are undergoing continuous renewal. In these populations cell production may be balanced by cell loss. The epithelial cells which line the small intestine (p. 228) and the cells that circulate in the blood, which are derived from extra-vascular precursors (p. 229), are good examples of populations which are renewed continuously. Expansion will of course occur concurrently with renewal while such populations are growing during ontogeny, being restored following depletion or responding to an increase in the requirements of the organism. Thus the number of erythrocytes in the blood increases during development, following depletion by treatment with phenylhydrazine which damages erythrocytes or during exposure to hypoxia.

III. Methods used in the Study of Cell Production

A. CHANGES IN THE SIZE OF CELL POPULATIONS

The rate of change in the size of a cell population (S) is dependent upon the rate at which cells are being produced (P), imported (I) and lost (L) :—

$$S = P + I - L$$

If cells are neither being imported nor lost the rate of cell production is thus equal to the rate of change in the number of cells which can be calculated from direct measurements of population size. The number of free cells in a suspension can be determined by taking advantage of the methods which have been developed for the enumeration of blood cells and the number of cells in a population which can not be converted into a suspension of intact free cells can be estimated using microtome sections (p. 209) (Abercrombie, 1946; Philp and Buchanan, 1971).

If cells are being imported or lost the rate of change in the size of a population will not necessarily reflect the rate of cell production (Fig. 5). The rate at which cells are being imported or lost can however be taken into account and exploited for the study of cell production in some populations (see p. 233).

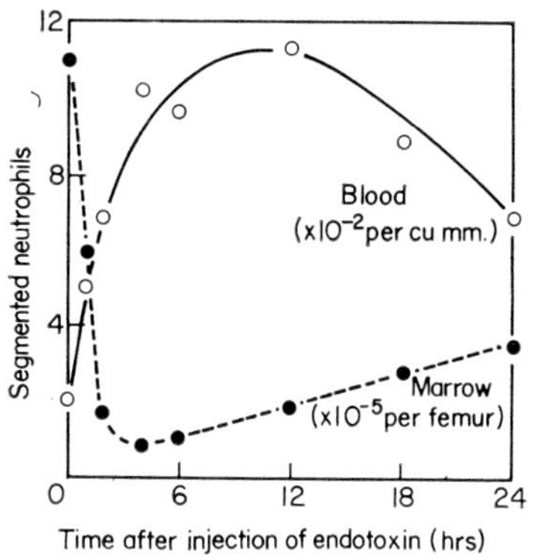

Fig. 5. The number of cells in a population may increase in the absence of cell division if the rate at which cells enter the population exceeds the rate at which cells are lost. Thus following an intravenous injection of endotoxin (0.25 ml. of TAB vaccine) the number of segmented neutrophils in the blood increases due to the transfer of cells from the bone marrow to the blood at a rate that exceeds the rate at which segmented neutrophils are lost from the blood. (Data of Thomas and Briscoe (1973) for CS1/Ash mice.)

In contrast to the tissues from which they are derived, certain tumours grow very rapidly—not because the rate of cell production is increased but because the rate of cell loss is decreased (Steel, 1968; Riddell and Thomas, 1972).

B. CHANGES IN THE DNA CONTENT OF CELL POPULATIONS

Changes in the number of cells in a population are paralleled by changes in the total DNA content of the population as the mean DNA content of individual cells does not alter appreciably if polyploid cells are not being produced and if the distribution of cells between successive phases of the cell cycle does not change. The DNA content of single cells can be determined using cytophotometry and the DNA content of the population can be determined chemically (see Enesco and Leblond, 1962; Vendrely, 1971).

Changes in the rate of increase in the total DNA content can thus be used to estimate changes in the rate of DNA production and hence changes in the rate of cell production if cells are not being lost or imported and the mean DNA content of individual cells does not alter appreciably. If the mean DNA content of individual cells throughout a period under consideration is known the rate of cell production can be calculated using measurements of the total DNA content.

Changes in the rate of DNA production in cell populations which are undergoing continuous renewal as well as in expanding cell populations can be estimated from measurements of the radioactivity of DNA extracted after the administration of labelled precursors such as tritiated thymidine or iododeoxyuridine (^{131}IUdR).

C. THE DISTRIBUTION AND INCIDENCE OF MITOTIC FIGURES

The nuclei of cells which are undergoing mitotic division can be distinguished from the nuclei of cells which are not dividing in routine

TABLE 1. Cell production in the epithelium lining the small intestine of the rat, studied using the incidence of mitotic figures. (Data of Leblond and Stevens, 1948; Lesher *et al.*, 1966.)*

Mitotic index (MI)	= 3%
Mitotic duration (MD)	= 0·7 h
Turnover time $\left(\frac{\text{MD}}{\text{MI}/100}\right)$	= 23 h
Turnover constant $\left(\frac{\text{MI}}{\text{MD}}\right)$	= 4% per h
Mitotic index after colchicine (MI_c)	= 18%
Duration of metaphase arrest (DA)	= 6 h
Turnover time $\left(\frac{\text{DA}}{\text{MI}_c/100}\right)$	= 33 h
Turnover constant $\left(\frac{\text{MI}_c}{\text{DA}}\right)$	= 3% per h
Generation time (GT)	= 10 h
Mitotic index of dividing cells $\left(\text{MId} = \frac{\text{MD}}{\text{GT}} \times 100\right)$	= 7%
Dividing cells $\left(\frac{\text{MI}}{\text{MId}} \times 100\right)$	= 43%

* The distribution of cells between successive phases of the cell cycle has not been taken into account. Together with other sources of error this may contribute to the discrepancies between the values calculated using different sets of data. In spite of these discrepancies valid generalizations can be made about the rate of turnover of the epithilial cells which line the small intestine of the rat. (See p. 229.)

histological preparations (Fig. 6) and very useful information can be obtained about cell production by plotting the distribution or estimating the incidence of mitotic figures.

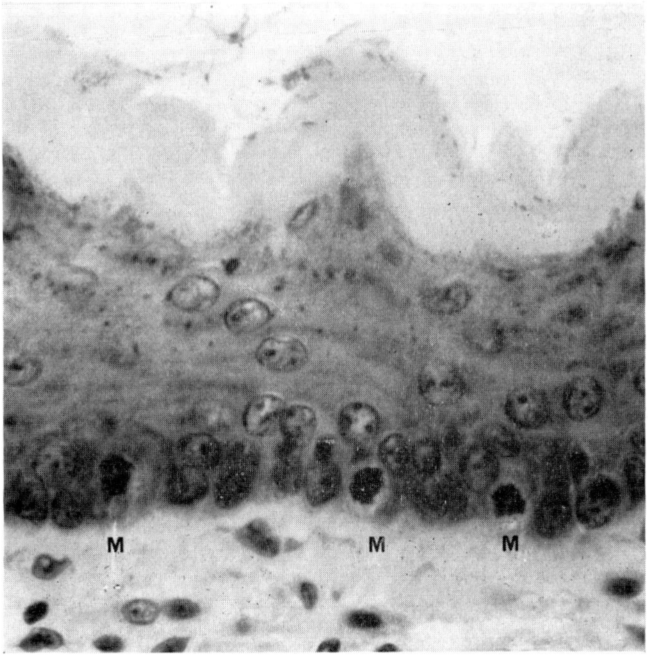

Fig. 6. Mitotic figures (M) in the stratified squamous epithelium lining the oesophagus of a mouse. (Magnification × 720.)

The incidence of mitotic figures in a cell population (mitotic index) can be determined by classifying an appropriate number of consecutively observed nuclei as mitotic or non-mitotic :—

$$\text{Mitotic index } \% = \frac{\text{Number of mitotic nuclei}}{\text{Total number of nuclei}} \times 100$$

Smears of cell suspensions or microtome sections can be employed for this purpose. If sections are employed however and the non-mitotic nuclei differ in size from the mitotic nuclei the discrepancy in size must be taken into account. This can be done by making use of the equation derived by Abercrombie (1946) which relates the average number of nuclear points—geometrical points with the same relative position in all nuclei (P)—to the number of nuclei or parts of nuclei observed in a section (A):—

$$P = A \times \frac{M}{(L + M)}$$

where M is the thickness of the section and L is the average length of the nuclei. A refinement of this relationship has been proposed by Philp and Buchanan (1971).

As the mitotic index can be determined by the use of routine methods of optical microscopy for the examination of smears or sections, without having to resort to special pre-treatment, preparation artifacts are minimized and it can be employed for the study of cell production in biopsy specimens of human tissues, or in retrospective studies of cell production in samples which have already been collected, of tissues which are not conveniently available.

The determination of the mitotic index necessitates the examination of enormous numbers of nuclei in tissues with a low incidence of mitotic figures. Besides, although the incidence of mitotic figures in a cell population provides a useful indication of the proportion of dividing cells, it depends upon the duration of mitosis as well as upon the proportion of cells undergoing asynchronous division so that variations in the mitotic index do not necessarily reflect variations in the rate of cell production. The mitotic index will be increased if the proportion of dividing cells is increased *or* if the duration of mitosis is increased relative to the generation time, provided that entry into mitosis is asynchronous. The mitotic index will not therefore reveal changes in the proportion of dividing cells if these changes are balanced by opposite changes in the relative duration of mitosis. In the absence of significant changes in the relative duration of mitosis however changes in the incidence of mitotic figures provide a useful indication of changes in the rate of cell production (Fig. 4).

If the duration of mitosis (D) is known, the mitotic index can be used to estimate the duration of the cell cycle (generation time, GT), provided that all of the cells in a closed population are dividing and that entry into mitosis is not synchronized. The distribution of cells between the series of arbitrarily defined phases of equal duration into which the cell cycle may be sub-divided must however be taken into account.

The rectangular distribution for a population in which each phase of the cell cycle contains the same number of cells (N), can be represented in a simple diagram:—

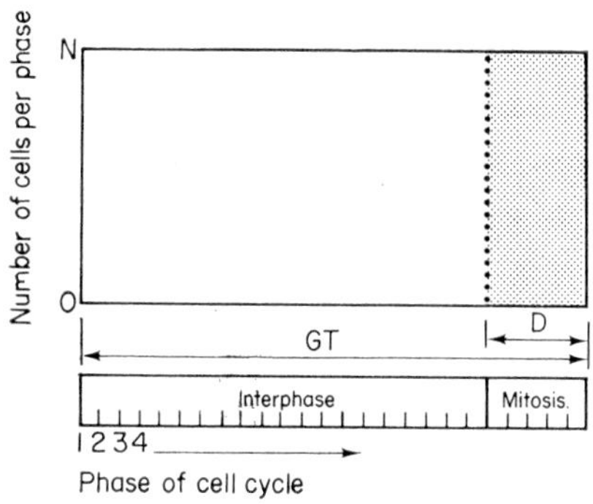

Fig. 7. The relationship between the proportion of mitotic figures, the duration of mitosis (D) and the generation time (GT), in a population of cells which is dividing asynchronously. The number of cells in successive phases of the cell cycle (N) is constant (rectangular phase distribution).

$$\text{Now, Mitotic index \%} = \frac{\text{Number of mitotic nuclei}}{\text{Total number of nuclei}} \times 100$$

$$= \frac{N \times D}{N \times GT} \times 100 = \frac{D}{GT} \times 100$$

$$\text{Whence, Generation time} = \frac{\text{Mitotic duration}}{\text{Mitotic index}/100}$$

This relationship is valid however only for a rectangular phase distribution, which will be achieved only if half of the cells produced by mitotic division are lost immediately from the population. In the absence of cell loss, at any given instant, the number of cells commencing mitosis (N, or strictly N + a) will be approximately half the number of cells completing mitosis (2N) (Fig. 8). The number of cells at successive phases of the cell cycle will decrease exponentially from 2N in the phase immediately following division to (N + a) in the phase immediately preceeding division—reflecting the exponential increase which occurs in the number of cells entering mitosis. This exponential distribution can be represented in a simple diagram and for convenience the exponential decrease may be treated as a linear decrease:—

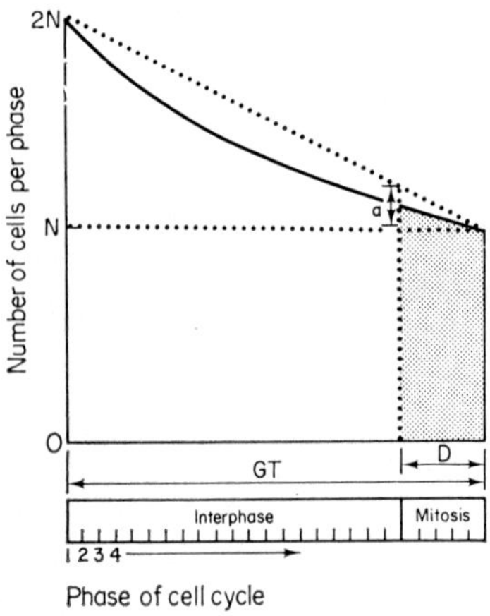

Fig. 8. The relationship between the proportion of mitotic figures, the duration of mitosis (D) and the generation time (GT), in a population of cells which is dividing asynchronously. The number of cells in successive phases of the cell cycle decreases exponentially (exponential phase distribution).

$$\text{Now, Mitotic index }\% = \frac{\text{Number of mitotic nuclei}}{\text{Total number of nuclei}} \times 100$$

$$= \frac{(N+\frac{1}{2}a)\times D}{1\frac{1}{2}N\times GT} \times 100$$

If $(\frac{1}{2}a)$ is small in comparison with N, this becomes

$$= \frac{D}{1\frac{1}{2}GT} \times 100$$

$$\text{Whence, Generation time} = 0{\cdot}667 \times \frac{\text{Mitotic duration}}{\text{Mitotic index}/100}$$

This correction factor may be calculated more accurately in a manner which allows the approximations that have been made in the foregoing simplified derivation to be avoided. It is then found to have a value of 0·693. Thus:—

Let the number of cells in a given population be N.

If all of the cells in this population are dividing, the number of cells in the population after a time interval t will be $N{\cdot}2^{t/g}$, where g is the interval between successive mitoses (generation time), provided that cells are not entering or leaving the population. If entry into mitosis is not synchronized, the number of cells in mitosis at time t will be equal to the number of cells which will be added to the population during the ensuing time interval which is equal in duration to mitosis (d)

$$= N\cdot 2^{t+d/g} - N\cdot 2^{t/g}$$

as all of the cells in mitosis at time t will complete mitosis and each give rise to two daughter cells during this period. The fraction (F) of the population in mitosis at time t

$$= \frac{N\cdot 2^{t+d/g} - N\cdot 2^{t/g}}{N.2^{t/g}}$$

$$= \frac{N\cdot 2^{t/g}.2^{d/g} - N\cdot 2^{t/g}}{N\cdot 2^{t/g}}$$

$$= 2^{d/g} - 1$$

$$\text{so } F+1 = 2^{d/g}$$

$$\log_e (F+1) = \log_e 2\cdot d/g$$

If $F \ll 1$, as it will be if the duration of mitosis is very short in comparison with the generation time, $\log_e (F + 1) \approx F$, so

$$F = \log_e 2\cdot d/g = 0\cdot 693d/g$$

and the number of mitotic figures per hundred cells (mitotic index)

$$= 0\cdot 693 \times \frac{\text{Mitotic duration}}{\text{Generation time}} \times 100$$

$$\text{Whence, Generation time} = 0\cdot 693 \times \frac{\text{Mitotic duration}}{\text{Mitotic index}/100} \quad (\textit{vide supra})$$

It has already been seen that when all of the cells in a population are dividing, the generation time is equal in duration to the doubling time, if the population is expanding without losing cells, or to the turnover time, if the population is undergoing continuous renewal while remaining the same size (p. 204). When some of the cells in a population cease to divide or remain in interphase for prolonged periods (Go), the generation time ceases to be equal in duration to the doubling time or the turnover time. The proportion of cells which are not dividing must now be taken into account in order to relate the generation time to the reduced mitotic index and the duration of mitosis but the doubling time or the turnover time can still be calculated using the mitotic index and mitotic duration. As the proportion of non-dividing cells increases the correction factor (CF) required to take account of an exponential phase distribution of the cells which are proceeding from one mitosis to the next will increase from 0·693 to approach unity and so has an increasingly insignificant effect:—

$$\text{Doubling time } \textit{or} \text{ turnover time} = \text{CF} \times \frac{\text{Mitotic duration}}{\text{Mitotic index}/100}$$

The doubling time, or the turnover time, is equal to the mean life span of all the cells in a population. If a population includes cells which are not dividing, the calculated doubling time or turnover time will therefore be longer than the generation time.

A discrepancy between the *calculated* duration of the doubling time and its *observed* duration reflects cell loss, provided that allowance is made for changes in the rate of cell production:—

$$\begin{array}{l}\text{Cell loss}\\ \text{(per 100 cells produced)}\end{array} = \left(1 - \frac{\text{Calculated doubling time}}{\text{Observed doubling time}}\right) \times 100$$

When the observed doubling time is of very long or infinite duration most or all of the cells produced in the population are lost, so that continuous renewal occurs without growth and the calculated doubling time is in fact the turnover time.

If the duration of mitosis and the generation time are known the mitotic index of the dividing cells in a population can be estimated provided that the phase distribution of the cells which are approaching mitosis is taken into account by the use of an appropriate correction factor (CF):

$$\begin{array}{l}\text{Mitotic index of}\\ \text{dividing cells (\%)}\end{array} = \text{CF} \times \frac{\text{Duration of mitosis}}{\text{Generation time}} \times 100$$

The estimated mitotic index of the dividing cells and the observed mitotic index, which is the mitotic index for the whole population, can be used to calculate the percentage of dividing cells in the population:—

$$\text{Dividing cells (\%)} = \frac{\text{Mitotic index of whole population}}{\text{Mitotic index of dividing cells}} \times 100$$

The specific birth rate in a population of dividing cells, which is the number of cells per hundred that divide in unit time, can also be calculated from the mitotic index and the duration of mitosis:—

$$\text{Specific birth rate} = \frac{\text{Mitotic index}}{\text{Mitotic duration}}$$

In view of the relationship between the mitotic index, mitotic duration and generation time it will be clear that for a rectangular phase distribution (*vide supra*):—

$$\text{Specific birth rate} = \frac{1}{\text{GT}} \times 100$$

and for an exponential phase distribution:—

$$\text{Specific birth rate} = \frac{0{\cdot}693}{\text{GT}} \times 100$$

If a population which is growing without losing cells or undergoing continuous renewal without changing in size includes non-dividing cells the specific birth rate for the population will be similarly related to the doubling time or the turnover time. In a population undergoing continuous renewal without changing in size the number of cells per hundred which divide in unit time (specific birth rate) will be equal to the number of cells per hundred which are replaced in unit time (turnover constant), provided that cells are not being imported. Likewise the number of cells which are replaced in unit time (turnover rate) will be equal to the number of cells dividing in unit time.

Since the commencement and termination of metaphase are clearly defined, unlike the commencement and termination of mitosis, the incidence of metaphases and the duration of this stage of mitosis can be determined with much greater consistency than the incidence of nuclei at all stages of visible mitosis and the total duration of visible mitosis. It is therefore advantageous to work with the incidence of metaphase figures and the duration of metaphase in preference to the incidence of mitotic figures and the duration of mitosis. Information about the duration of mitosis or of metaphase in mammalian cell populations is however often not available. The rate of cell production is therefore commonly estimated by determining the number of metaphases which accumulate during a known time interval after the completion but not the commencement of metaphase has been prevented. Advantage is taken of the fact that for each cell which enters metaphase one cell is normally added to a population.

Mitotic division can be arrested at metaphase by the administration of colchicine (Stevens Hooper, 1961). Very small doses of this plant alkaloid prevent the formation of the mitotic spindle. After the injection of colchicine into a dividing cell the effect is manifest immediately but following addition to a tissue culture or parenteral administration mitotic arrest may be detected only after a latent period. The effect of colchicine is maximal 2 h or so after the injection of 1–2 mg of the drug per kg of body weight. A smaller dose will arrest fewer metaphases while a larger dose will cause a greater proportion of metaphases to undergo disintegration. Even at the optimum time interval after the optimum dose some of the metaphases are pyknotic or degenerate and the rate of entry into prophase may be depressed. Enumeration of the number of arrested metaphases will therefore tend to underestimate the number of cells entering metaphase.

Not all tissues are equally sensitive to colchicine and different tissues vary in their ability to concentrate the drug. After the administration of colchicine labelled with a radioactive isotope the label accumulates in the intestines, the kidneys and the spleen to a greater degree than in the liver; only traces of radioactivity can be demonstrated in the blood, brain, heart and muscles.

In spite of these sources of error useful estimates of cell production rates and of the maximum duration of the turnover time in the absence of growth, or of the doubling time in the absence of cell loss, have been obtained using the metaphase index after the drug has exerted its action for a known tme interval :—

$$\text{Doubling time } \textit{or} \text{ turnover time} = \frac{\text{Duration of action of colchicine}}{\text{Metaphase index}/100}$$

If it is possible to do so, the phase distribution of the cells approaching metaphase should of course be taken into account by the use of an appropriate correction factor (*vide supra*).

The proportion of cells entering metaphase in unit time is the same as the proportion of cells entering mitosis in unit time and can be used together with the mitotic index to estimate the duration of mitosis :—

$$\text{Mitotic duration} = \frac{\text{Mitotic index}}{\text{Cells per 100 entering metaphase in unit time}}$$

The duration of metaphase, or of other sub-divisions of the cell cycle, can be similarly estimated.

In addition to providing very useful information about rates of cell production the use of colchicine has facilitated the study of the distribution of dividing cells in tissues with a low incidence of mitotic figures.

Other drugs which prevent the completion of metaphase, such as demecolcine (colcemid), vinblastine or vincristine are now often used in preference to colchicine. Ideally such drugs should satisfy the following requirements :—

(i) there should be a dose, or preferably a range of doses, over which all metaphases are arrested throughout a time interval of suitably long duration.

(ii) arrested metaphases should not degenerate and so cease to be recognizable during this time interval.

(iii) only cells in metaphase should be affected, so that the rate at which cells enter metaphase is neither increased nor decreased.

If these requirements are satisfied metaphases will accumulate at a rate which is independent of minor changes in dose and which does not decrease for several hours after administration of the drug (Fig. 9).

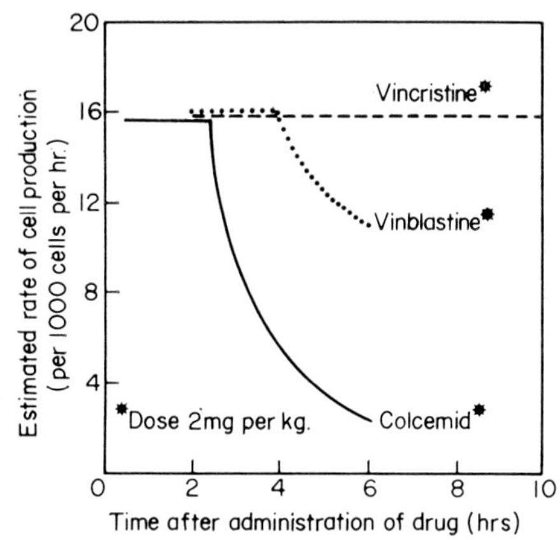

Fig. 9. The rate of cell production in tumour isografts, estimated from the rate of accumulation of metaphases following an injection of colcemid, vinblastine or vincristine. (Reprinted from Smith, Thomas and Riches (1972) with the permission of the Editor of the Journal of Anatomy.)

D. THE DISTRIBUTION AND INCIDENCE OF LABELLED CELLS

Cells which are synthesizing DNA in preparation for mitotic division can be labelled so that their distribution and incidence may be studied conveniently and used to provide extremely valuable information about cell production and migration pathways (Fig. 10). Tritiated thymidine is a specific label for cells which are synthesizing DNA. It is consumed rapidly and is therefore incorporated only into molecules of DNA which are being synthesized during the brief period for which it is available following injection. The distribution of these molecules can be revealed in autoradiographs.

For this purpose smears of cell suspensions or thin microtome sections are prepared at appropriate known time intervals after tritiated thymidine has been added to tissue cultures or injected into living animals. The smears or sections are coated in the dark with a thin layer of photographic emulsion and stored in light proof boxes.

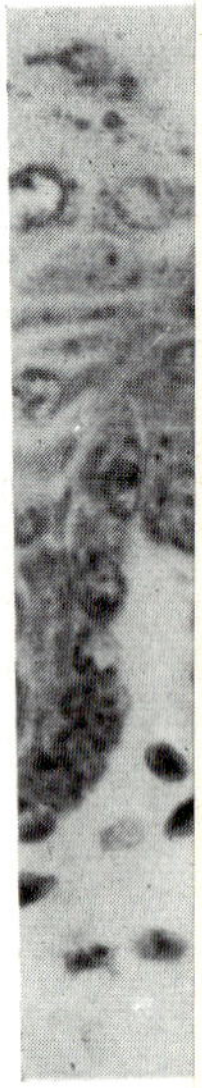

Fig. 10. Labelled cell
a mouse one hour a
× 560.)

Radioactive deca
energy β-particle
sion to strike silv
in the immediate
After the prepar
as any photogra
bromide which
distribution of si
(Fig. 11). The gr
which were use
demonstrate the
tritiated-thymidi
precision (Rogers

Before any ce
proportion of c

was available. T

ever be reduced
completed DNA
upon the relatio
thesis and the du
of cells synthesi

Similarly the mitotic index fails to reveal changes in the proportion of dividing cells which are balanced by opposite changes in the relative duration of mitosis (p. 210). Like the mitotic index and subject to similar reservations the labelling index can be used to estimate the doubling time or the turnover time of a cell population (p. 213):—

$$\text{Doubling time } \textit{or} \text{ turnover time} = \text{CF} \times \frac{\text{Duration of DNA synthesis}}{\text{Labelling index}/100}$$

The relationship between the doubling time or the turnover time and the specific birth rate has already been discussed (p. 215).

When the number of cells in successive phases of the cell cycle is not constant this must be taken into account, if it is possible to do so, by the use of an appropriate correction factor (CF). The proportion of cells which are synthesizing DNA at a given time in a population of dividing cells with an exponential phase distribution can be represented by the relationship between the shaded area and the total area in a simple diagram:—

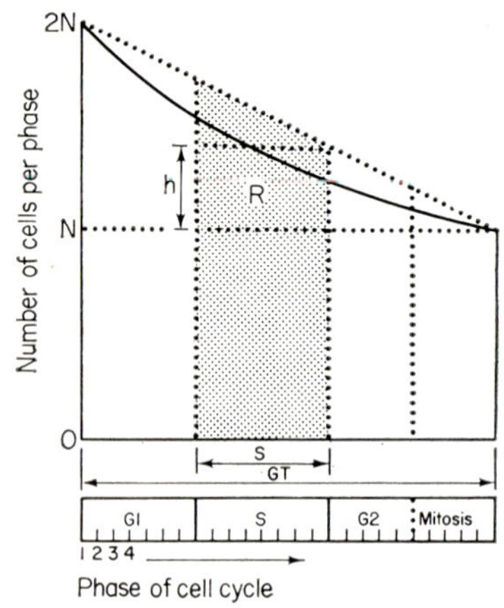

Fig. 13. The proportion of cells synthesizing DNA at a given instant in a population of cells which is dividing asynchronously. The number of cells in successive phases of the cell cycle decreases exponentially (exponential phase distribution).

The labelling index will depend upon the proportion of the generation time (GT) occupied by the stage of DNA synthesis (S) and the position in the cell cycle of the stage of DNA synthesis which will determine the height (h) and hence the area of the rectangle R.

If a population includes cells which are not dividing, the labelling index of the cells which are dividing can be estimated using the generation time and the duration of the stage of DNA synthesis, provided that the distribution of cells between successive phases of the cell cycle can be taken into account:—

$$\text{Labelling index of dividing cells (\%)} = \text{CF} \times \frac{\text{Duration of DNA synthesis}}{\text{Generation time}} \times 100$$

The proportion of cells which are dividing can now be calculated using the relationship:—

$$\text{Dividing cells (\%)} = \frac{\text{Labelling index of whole population}}{\text{Labelling index of dividing cells}} \times 100$$

The close agreement between the observed labelling index of basophilic erythroblasts and the estimated labelling index of dividing basophilic erythroblasts in the bone marrow of young rats (Roylance, 1968) implies that *all* of these cells undergo division, whereas the discrepancy between the observed labelling index of polychromatic erythroblasts (30%) and the estimated labelling index of dividing polychromatic erythroblasts (54%) implies that only 56% ($\frac{30}{54}$ × 100) of these cells are dividing (see p. 232).

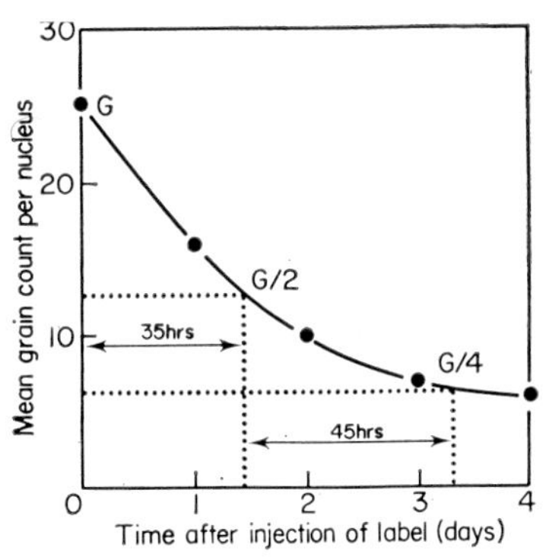

Fig. 14. The average number of grains observed over the nuclei of labelled cells at various time intervals after an injection of tritiated thymidine. (Data of Osmond and Everett (1964) for "transitional cells" in guinea-pig bone marrow.)

As labelled cells divide and replace unlabelled cells, which will be lost continuously from the post-mitotic compartment of a population which is undergoing continuous renewal, the proportion of labelled cells in the post-mitotic compartment will increase; concurrently the number of grains per dividing cell will decrease as the label is shared between increasing numbers of cells (Fig. 14).

The duration of the time interval required for the mean graint count of dividing cells to be reduced to half its initial value, is equal to the duration of Mitosis + G2 + S in a population which is not importing labelled cells. In contrast, the time interval required for it to be reduced from one half to one quarter of its initial value is equal in duration to the cell cycle (generation time, Mitosis + G2 + S + G1).

Extrapolation of the line fitted to the proportion of labelled cells at various time intervals after the administration of tritiated thymidine (Fig. 15) or any other suitable label (Fig. 16) allows the turnover time in the post-mitatic compartment of a population which is undergoing continuous renewal to be estimated.

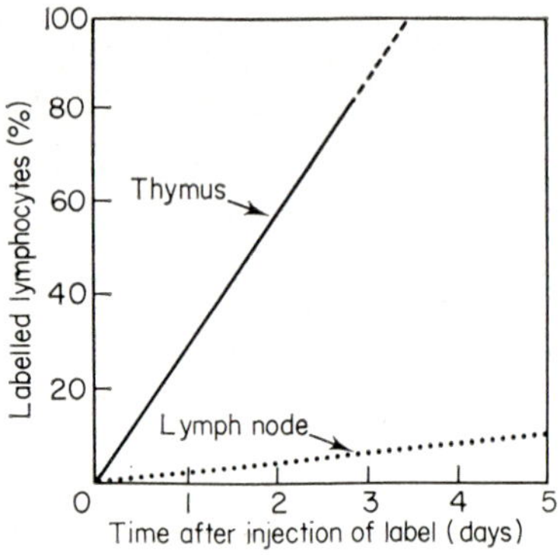

Fig. 15. The proportion of labelled lymphocytes in the thymus and in a subcutaneous lymph node at various time intervals after an injection of tritiated thymidine. Whereas the lymphocytes in the thymus appear to be renewed every three days or so, those in the subcutaneous lymph node appear to be renewed every fifty days or so. (Data of Metcalf (1966) for C57BL mice.)

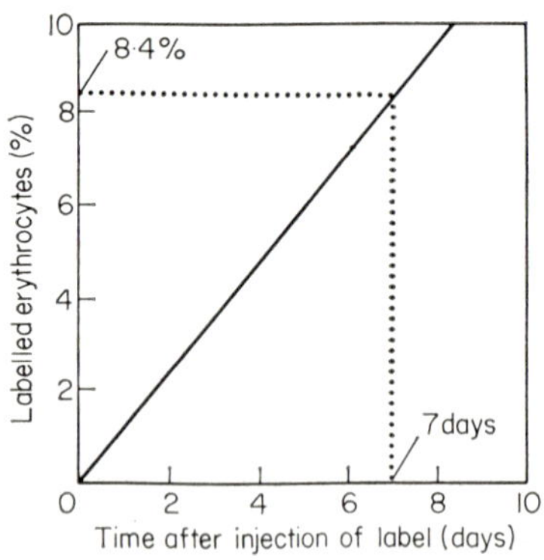

Fig. 16. The proportion of labelled erythrocytes in the blood at various time intervals after an injection of ^{59}Fe. As 1·2% (8·4/7) of the erythrocytes in guinea-pig blood appear to be renewed each day, complete renewal will occur every eighty days or so. (Straight line fitted to the data of Everett and Yoffey (1959) for guinea-pigs.)

For a given proportion of dividing cells in a population the mitotic index will reflect the duration of mitosis and the labelling index following a single injection of tritiated thymidine will reflect the duration of the stage of DNA synthesis. As the stage of DNA synthesis is usually of very much longer duration than mitosis, the labelling index is usually very much higher than the mitotic index. The distribution and incidence of dividing cells can therefore be established far more conveniently by studying labelled cells than by studying mitotic figures, unless colchicine or some other drug has been used to prevent the completion of metaphase (p. 215). In addition the migration pathways of new cells and their progeny can be traced readily in autoradiographs prepared at various time intervals after the label has been provided.

The proportion of labelled cells in a population which is undergoing continuous renewal can be increased until all of the cells are labelled, by prolonging the period for which the label is available (Fig. 12). This can be done either by giving repeated injections or by administering continuous infusions or by ensuring regular ingestion (Fig. 17).

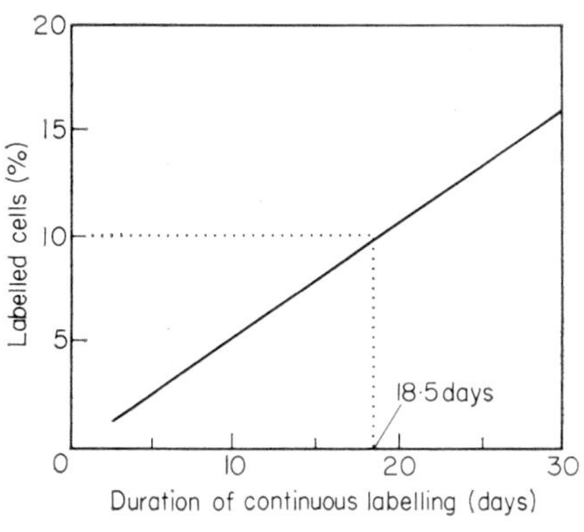

Fig. 17. The accumulation of labelled alveolar cells in the sub-mandibular gland during continuous administration of tritiated thymidine. If labelled cells continued to accumulate at this rate all of the alveolar cells would be labelled after 185 days. (Line fitted to the data of Cameron (1971) for mice.)

As the duration of availability increases, the proportion of cells which have been able to incorporate tritiated thymidine throughout the entire duration of the stage of DNA synthesis increases. Consequently the average number of grains per nucleus increases to approach a plateau value (Fig. 18).

Cells which have been labelled by ^{14}C–thymidine can be distinguished in autoradiographs from cells which have been labelled by tritiated (^{3}H–) thymidine, because the β particles from ^{14}C atoms travel a mean distance of about 50 μm and so produce widely dispersed grains, whereas those from ^{3}H atoms travel a mean distance of only about 1 μm and so produce sharply localized grains. Cells labelled by ^{3}H only may therefore be distinguished both from cells labelled by ^{14}C or by $^{14}C + {}^{3}H$ but cells labelled by ^{14}C cannot be distinguished from cells labelled by $^{14}C + {}^{3}H$. If an injection of ^{3}H–thymidine is followed by an injection of ^{14}C–thymidine after a known time interval, which is not longer in duration than G2, the relationship between the number of cells labelled by ^{3}H alone and the number of cells labelled by ^{14}C (^{14}C *or* $^{14}C + {}^{3}H$) can be used to estimate the

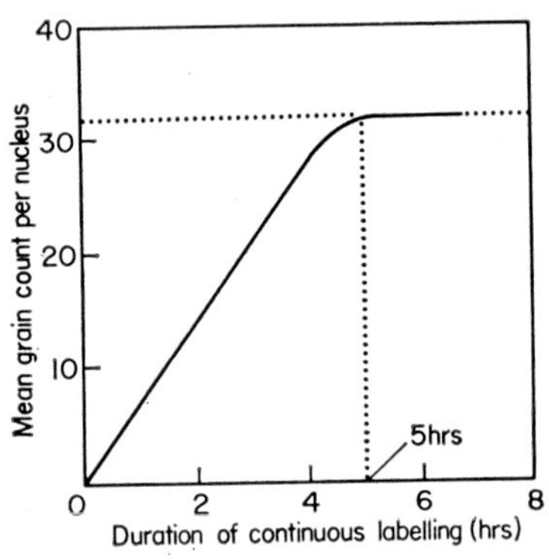

Fig. 18. The relationship between the average number of grains per nucleus, seen in autoradiographs, and the duration of continuous infusion of tritiated thymidine. (Smooth curve fitted to the data of Maloney, Patt and Weber (1962) for canine granulocyte precursors (myelocytes).)

duration of the stage of DNA synthesis. If the number of cells progressing through the stage of DNA synthesis in unit time does not alter appreciably, the number of cells labelled by ^{14}C (N_{14C}) will be proportional to the duration of DNA synthesis (t_s), while the number of cells labelled by ^{3}H only (N_{3H}), which have moved into G2, will be proportional to the known time interval between the two injections (ta), so :—

$$\frac{N_{3H}}{N_{14C}} = \frac{ta}{ts}$$

This technique of *double labelling* can also be used to estimate the duration of mitosis, G1 and G2 under favourable circumstances. (See Wimbler and Quastler, 1962.)

TABLE 2. Cell production in the epithelium lining the small intestine of the rat, studied using the incidence of labelled cells one hour after an injection of tritiated thymidine. (Data of Messier and Leblond, 1960; Lesher *et al.*, 1966.)*

Labelling index (LI)	= 15%
Duration of DNA synthesis (DS)	= 7 hr
Turnover time $\left(\frac{DS}{LI/100}\right)$	= 47 hr
Turnover constant $\left(\frac{LI}{DS}\right)$	= 2% per hr
Generation time (GT)	= 10 hr
Labelling index of dividing cells $\left(LI_d = \frac{DS}{GT} \times 100\right)$	= 70%
Dividing cells $\left(\frac{LI}{LI_d} \times 100\right)$	= 21%

* See footnote to Table 1.

E. THE INCIDENCE OF LABELLED MITOTIC FIGURES

Immediately after the administration of tritiated thymidine autoradiographs prepared from any tissue in which mitotic divisions are occurring will contain labelled interphase nuclei but *no* labelled mitotic figures. Later labelled mitotic figures will appear. The appearance of labelled mitotic figures indicates the entry into mitosis of cells which were almost at the end of the stage of DNA synthesis (S stage of interphase) when the label was provided. The interval between the administration of tritiated thymidine and the appearance of labelled mitotic figures reflects the occurrence of an interval between the completion of DNA synthesis and the commencement of mitosis. This interval (gap) which has been termed G2 is the third and last stage of interphase. The interval (gap) between the completion of mitosis and the commencement of DNA synthesis, which is the first stage of interphase has been termed G1 (see p. 203).

As more cells which were synthesizing DNA while the label was available for incorporation enter mitosis, the proportion of labelled mitotic figures increases until all of the mitotic figures are labelled. Then cells which were in G1 while the label was available begin to enter mitosis and the proportion of labelled mitotic figures begins to decrease. Eventually the labelled cells will divide again and when they do labelled mitoses will reappear (Figs. 19 and 20). Variations in the duration of the interval between successive mitoses, the failure of a proportion of cells to go on dividing and dilution of the label at successive mitoses will however reduce the height and increase the width of the second peak on the curve fitted to the percentage of labelled mitotic

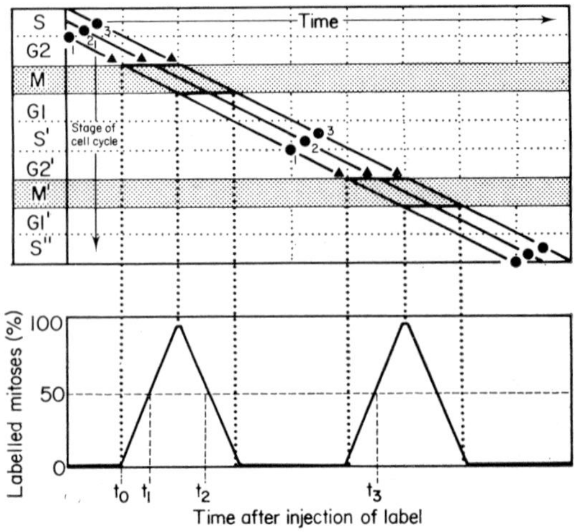

Fig. 19. Schematic representation of the changes which occur in the incidence of labelled mitotic figures as cells which have been labelled following the administration of a single injection of tritiated thymidine progress through successive cell cycles. The time intervals t_0, t_1, t_2 and t_3 are used to estimate the duration of the cell cycle (generation time) and the stages into which it is subdivided. For convenience the stages of the cell cycle have been represented as being of equal duration. (Compare with figures 20 and 21.)

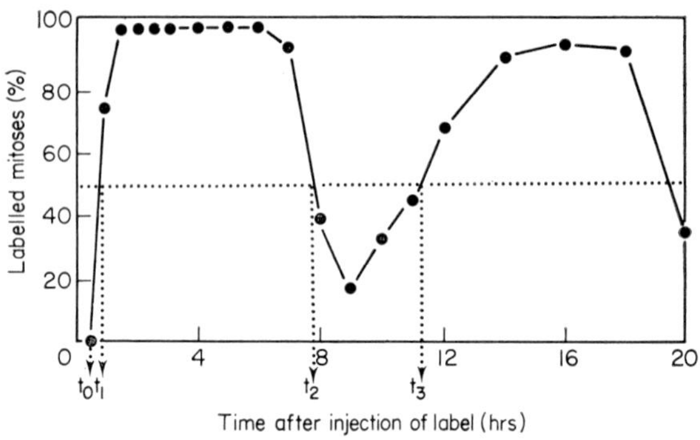

Fig. 20. The percentage of labelled mitotic figures in the epithelium lining the small intestine (duodenum) at various time intervals after a single injection of tritiated thymidine. The time intervals t_0, t_1, t_2, and t_3, are used to estimate the duration of the cell cycle (generation time) and the stages into which it is subdivided. (Data of Lesher *et al.*, (1966) for rats.)

figures at various time intervals after the administration of tritiated thymidine. The third peak which might be expected to occur eventually will be even lower and wider than the second peak and may fail to be detected (Fig. 21).

The duration of the cell cycle and of its four component stages can be estimated by utilizing measurements of four time intervals on the

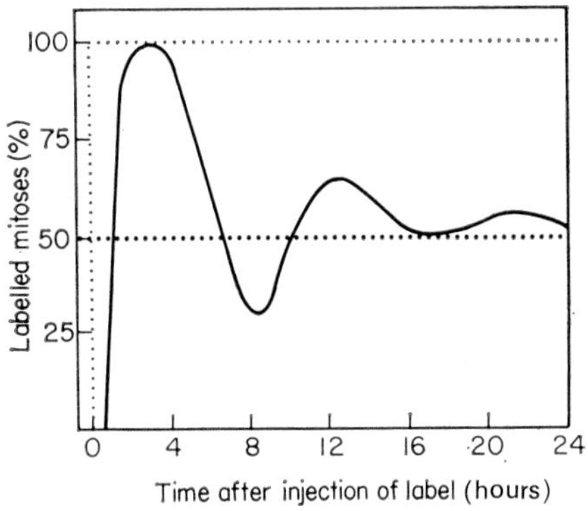

Fig. 21. The incidence of labelled mitotic figures among proerythroblasts and basophilic erythroblasts (see p. 232) at various time intervals after a single injection of tritiated thymidine. (Smooth curve fitted to the data of Roylance (1968) for young rats.)

labelled mitosis curve. The median time interval between administration of the label and the appearance of labelled mitoses (t_o) is estimated by extrapolating the linear increase in the percentage of labelled mitoses back to its intersection with the time axis. The other three intervals (t_1, t_2, and t_3) are measured from the time of administration of the label to each of the first three occasions when the curve crosses the 50% line (Figs. 19 and 20). The duration of G2, M (mitosis), S, G1 and GT (generation time) are calculated from the values of t_0, t_1, t_2 and t_3 using the following equations:—

$$G2 = t_0 \quad \dots\dots \quad (1)$$
$$M = 2(t_1 - t_0) \quad \dots\dots \quad (2)$$
$$S = t_2 - t_1 \quad \dots\dots \quad (3)$$
$$G1 = t_3 - t_2 - 2t_1 + t_0 \quad \dots\dots \quad (4)$$
$$GT = t_3 - t_1 \quad \dots\dots \quad (5)$$

Equations 1 and 3 are obtained directly. Equations 2 and 4 are obtained by substitution in equations 6 and 7 :—

$$G2 + \tfrac{1}{2}M = t_1 \quad \dots\dots \quad (6)$$
$$G1 + G2 + M = t_3 - t_2 \quad \dots\dots \quad (7)$$

These procedures avoid the use of the time interval between administration of the label and the appearance of 100% labelled mitoses, which may be impossible to measure accurately, as the 100% value may be approached very gradually (Lesher *et al.*, 1966). If t_0 can not be determined accurately, the duration of the generation time may be estimated together with the duration of S, G2 + $\frac{1}{2}$M and G1 + $\frac{1}{2}$M using t_1, t_2 and t_3. A computer programme may be employed to analyse the relationship between the percentage of labelled mitoses and the time interval after administration of tritiated thymidine (Roylance, 1968; Tarbutt, 1969; Mendelsohn and Takahashi, 1971; Steel, 1972).

TABLE 3. The duration in hours of the generation time (GT) and of its four components (G1, S, G2 and M) in epithelial cells of the small intestine (duodenal crypt cells). Estimated from the incidence of labelled mitotic figures at various time intervals after the administration of tritiated thymidine. (From Lesher *et al.*, 1966).

Species	G1	S	G2	M	GT
Rats (54 days old)	2·2	7·0	0·5	0·7	10·4
Mice (54 days old)	1·3	7·5	0·5	1·0	10·3

IV. Cell Production in Populations Undergoing Continuous Renewal

A. The Spatial Relationship between Precursors and Progeny is Preserved (Renewal of the Epithelial Cells Lining the Small Intestine.)

The simple epithelium which covers the villi and lines the crypts of the small intestine consists predominantly of columnar chief cells together with scattered mucus secreting goblet cells, a few Paneth cells and occasional argentaffin cells. Although the size of this cell population is not increasing progressively in adults it includes very numerous mitotic figures. New cells are therefore being produced at a rate which must be balanced by the rate of cell loss. The turnover time of the epithelial cells lining the small intestine of the rat, the mouse and the cat, estimated from the mitotic index and the metaphase index following colchicine arrest ranges from 0·6 to 2·8 days (Leblond and Walker, 1956). Turnover times estimated using the results of autoradiographic studies also lie within this range (Messier and Leblond, 1960).

Appreciable numbers of mitotic figures do not occur over the intestinal villi however; they are virtually confined to the crypts of Lieberkühn. Nearly a century ago this distribution of mitotic figures lead Bizzozero to suggest that newly formed cells move from the crypts to contribute to the epithelial covering of the villi. Evidence in support of this hypothesis was not provided for nearly sixty years until Friedman (1945) observed that goblet cells which were swollen after being exposed to X-rays could be traced from the crypts to the villi. He implied that all of the cell types produced in the crypts move to the villi in a similar fashion. Subsequently a massive migration of newly formed cells from the crypts to the villi was demonstrated by following the distribution of cells containing DNA, which had incorporated radioactive isotopes, in autoradiographs prepared at various time intervals after administration of the isotopic labels (Fig. 22).

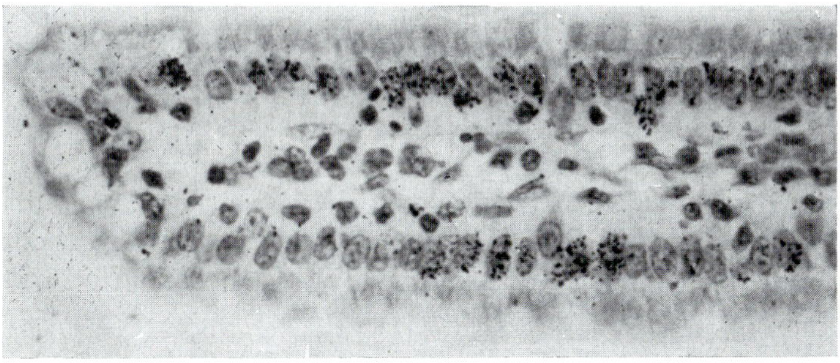

Fig. 22. Labelled cells are distributed along the length of this villus in the small intestine of a mouse thirty-six hours after an injection of tritiated thymidine (25μCi). One hour after such an injection labelled cells are restricted to the crypts of Lieberkühn. (Magnification $\times$ 500.)

Epithelial cells move from the crypts to the tips of the villi and are then lost into the lumen of the intestine (Messier and Leblond, 1960). Thus the cells of the crypts continuously export new cells to the intestinal villi without themselves becoming depleted. About one third of the cells which constitute the epithelium lining the small intestine of the rat divide every ten hours. This cell population is therefore renewed every thirty hours or so. (Tables 1–3.)

When necessary this very rapid rate of cell renewal can be accelerated. For instance following exposure to continuous gamma irradiation (50r per day for 12 days) the generation time of the dividing intestinal epithelial cells is reduced from 10·4 h to 8·1 h in rats and from 10·3 h to 8·4 h in mice. As a result the rate of cell production is increased and sufficient cells are provided to replace those damaged by irradiation, so that the integrity of the epithelium covering the intestinal villi is maintained (Lesher *et al.*, 1966).

B. THE SPATIAL RELATIONSHIP BETWEEN PRECURSORS AND PROGENY IS LOST (ERYTHROPOIESIS)

1. *Erythrocyte production*

The cells which circulate in the blood include several populations in which continuous renewal is occurring. Unlike the epithelial cells of the intestinal villi which preserve their spatial relationship with their precursors in the crypts of Lieberkühn, these cells are not spatially related to their precursors nor are their precursors spatially related to one another. The maturation sequences which result in the production of blood cells are not therefore displayed in the convenient fashion in which the maturation sequence of epithelial cells in the small intestine is displayed along the length of a villus. Maturation sequences are inferred by arranging the various blood cells and their precursors to form spectra of morphological continuity and can only be established when the validity of the inferred transformations can be demonstrated. The study of erythrocyte production can be used conveniently to illustrate the methods which are utilized to analyse the kinetics of proliferation and the patterns of differentiation in haematopoietic cell populations.

Subject to reservations which are discussed elsewhere (Riches *et al.*, 1973), the total number of erythrocytes in the blood can be estimated from the erythrocyte content of unit volume of blood and the blood volume. If the life span of the erythrocytes or the rate at which erythrocytes are replaced is known the fraction of the total population replaced every day can be calculated and the number of erythrocytes replaced each day can be estimated. Thus in guinea pigs weighing 400 grams the blood contains $5 \cdot 02 \times 10^6$ erythrocytes per μl, the blood volume is 24 ml and the lifespan of the erythrocytes is about 80 days (Fig. 16) (Yoffey, 1966), so :—

Total number of erythrocytes	$= 1 \cdot 2 \times 10^{11}$
Fraction of erythrocyte population replaced daily	$= 1/80$
Number of erythrocytes replaced daily	$= 1 \cdot 5 \times 10^9$

The rate of erythrocyte production per unit body weight in the guinea pig ($3{\cdot}8 \times 10^8$ cells per day per 100 gm body weight) is rather greater than that in man ($3 \cdot 0 \times 10^8$) but slower than that in the rat ($6{\cdot}0 \times 10^8$) and in the mouse ($1{\cdot}5 \times 10^9$). In these mammals the rate of erythrocyte production per unit body weight decreases as the body weight increases.

Even the very high rate of erythrocyte production in the mouse can be increased dramatically in response to an increase in the requirements of the organism or acute depletion. Thus following an injection of phenylhydrazine, which causes destruction of erythrocytes, a sevenfold increase in the rate of erythrocyte production occurs within five days (Fig. 23).

2. *Erythrocyte precursors which have been identified.* (*Erythroblasts*)

These enormous numbers of new erythrocytes are derived from extra-vascular precursors. Under normal circumstances during postnatal life erythrocyte production in man is confined to the marrow which occupies the medullary cavities of the bones. In some mammals erythrocytes are also produced in other situations—such as the spleen of the mouse—but extra-medullary sites of erythrocyte production are not normally used after birth in man as the space provided by the medullary cavities is more than adequate to accommodate the precursor cells required to maintain erythrocyte production. If however the medullary cavities cannot accommodate a sufficiently large number of precursor cells erythrocytes are produced in extra-medullary sites and in particular in the spleen. This happens in, for instance, myelofibrosis, a disorder in which the volume of the medullary cavities is reduced by the accumulation of fibrous tissue. During ontogeny erythrocytes are first produced in the wall of the yolk sac; later the liver is the main site of erythrocyte production.

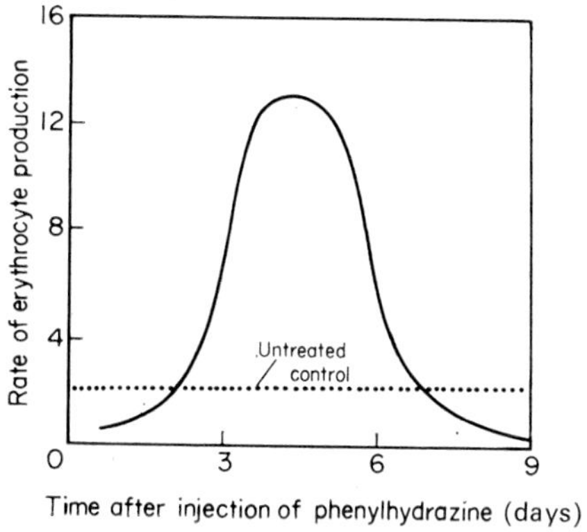

Fig. 23. Changes in the rate of erythrocyte production (new erythrocytes per cu. mm. per day $\times 10^{-5}$) following the administration of a single injection of phenylhydrazine (4 mgm). (Smooth curve fitted to unpublished data of Stevens and Thomas for CS1/Ash mice.)

Although erythrocytes and other blood cells may be produced in considerable numbers in these various extra-medullary sites the properties of the haematopoietic microenvironments which they provide are different from those of the haematopoietic microenvironment provided by the medullary cavities. Thus in comparison with that of the medullary cavities the microenvironment provided by the spleen is more conducive to erythrocyte production than to granulocyte production (Thomas, 1973).

The immediate precursors of erythrocytes are nucleated cells termed erythroblasts. These cells have distinctive morphological characteristics and can be recognized quite easily in stained smears prepared from suspensions of haematopoietic cells (Hayhoe and Flemans, 1969). Erythroblasts can be sub-divided conveniently into four groups which are defined according to morphological criteria and correspond to successive maturation stages. The cells in these groups are termed proerythroblasts, basophilic erythroblasts, polychromatic erythroblasts and orthochromatic erythroblasts. During the course of erythroblast maturation the proportion of dividing cells decreases progressively. Thus all of the proerythroblasts and basophilic erythroblasts are capable of dividing while only half of the polychromatic erythroblasts and none of the orthochromatic erythroblasts are capable of doing so (p. 221) (Roylance, 1968). The cytoplasmic basophilia of the earliest stages in this maturation sequence (proerythroblasts and basophilic erythroblasts) reflects the presence of abundant RNA which is distributed among free ribosomes (Fig. 24). As maturation proceeds the RNA content of the cytoplasm decreases while its haemoglobin content increases, in consequence the affinity of the cytoplasm for basic dyes decreases and its affinity for acidic dyes increases (Harris, 1963). The staining properties of the cytoplasm of orthochromatic erythroblasts are very similar to those of erythrocytes and a complete spectrum of morphological continuity links the erythrocyte with the proerythroblast.

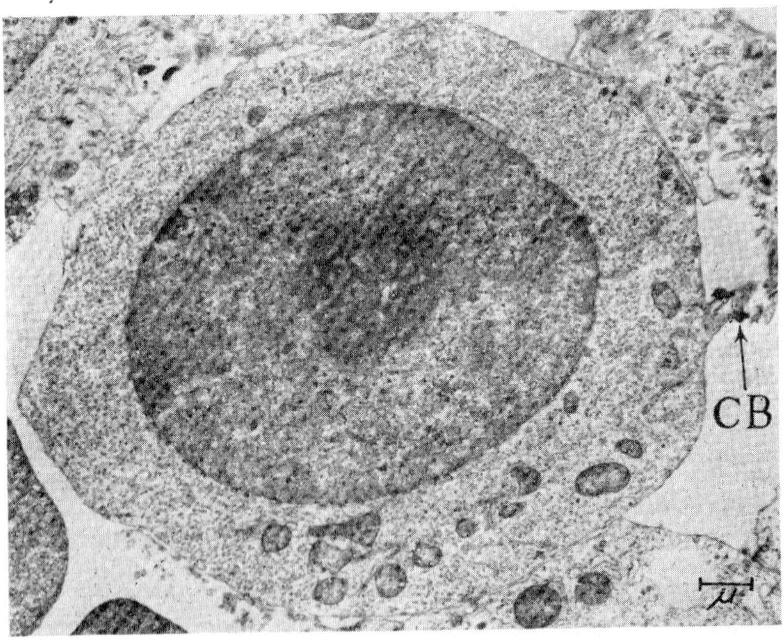

Fig. 24. The rim of cytoplasm surrounding the relatively large nucleus of this blood cell precursor contains numerous free ribosomes, which confer upon it an affinity for basic dyes. The remnant of the cytoplasmic bridge which connected this cell to its sister cell at the end of teleophase is indicated (CB). (Reprinted from Thomas (1966) with the permission of the Editor of the Journal of Anatomy.) (Magnification ×7 000.)

It has become clear however that erythroblasts do not constitute a self-maintaining population but require a continuous input from another population of cells, the morphological characteristics of which are still uncertain (Alpen and Cranmore, 1959; Blackett *et al.*, 1964; Erslev, 1959; Gurney *et al.*, 1962).

Under normal circumstances the erythroblasts in adult mammals constitute a *steady state population* which is undergoing continuous renewal without increasing in size (S = 0). The rate of cell production in the erythroblast compartment (P) is therefore equal to the difference between the rate at which cells are being lost (L) and the rate at which cells are being imported (I). (See p. 206.) The cells which leave the erythroblast compartment all enter the erythrocyte compartment so the rate at which cells are lost from the erythroblast compartment (L) is equal to the rate at which cells enter the erythrocyte compartment which can be estimated relatively easily (see p. 230). The rate at which cells enter the erythroblast compartment (I) can also be estimated but an account of the method for doing so is beyond the scope of the present discussion (see Tarbutt, 1967; Tarbutt and Blackett, 1968). In the rat $2{\cdot}4 \times 10^7$ cells per hour per 100 gm body weight (L) leave the erythroblast compartment to enter the erythrocyte compartment, while $2{\cdot}4 \times 10^5$ cells per hour per 100 gm body weight (I) enter the erythroblast compartment from the unidentified stem cell population. The rate of cell production in the erythroblast compartment (P = L − I) is therefore $2{\cdot}4 \times 10^7$ cells per hour per 100 gm body weight (I is insignificant in comparison with L in this particular instance).

For every cell which enters the erythroblast compartment a hundred cells or so leave. If all of the cells which leave the erythroblast compartment are derived ultimately from the stem cell compartment this amplification factor can be used to estimate the average number of mitotic divisions which occur during the course of erythroblast maturation (n):—

$$2^n = \frac{L}{I} = \frac{2{\cdot}4 \times 10^7}{2{\cdot}4 \times 10^5}$$

$$\text{whence } n = 7 \text{ (approximately).}$$

The first three divisions in this sequence take place during the twenty-nine hours which is spent at the proerythroblast stage, the next two during the twenty-three hours spent at the basophilic erythroblast stage and the remainder during the eighteen hours spent at the polychromatic erythroblast stage. After completing the polychromatic stage of maturation erythroblasts spend four hours at the orthochromatic stage without undergoing further mitotic divisions and then lose their nuclei giving rise to new erythrocytes—reticulocytes— which are retained in the bone marrow for seventeen hours before entering the blood.

The transformation of a stem cell into a reticulocyte thus takes three days. During the course of this transformation mitotic division occurs initially every ten hours or so. Erythroblast maturation is thus punctuated by nearly seven mitoses and as a result a hundred or so reticulocytes are produced for every stem cell which enters the erythroblast compartment.

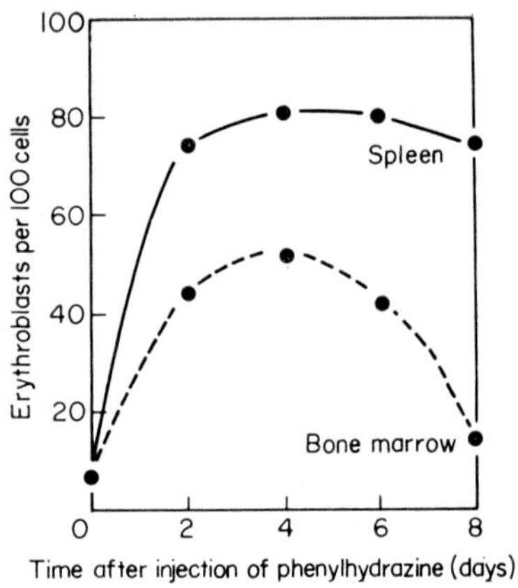

Fig. 25. Changes in the relative size of the erythroblast compartment following the administration of a single injection of phenylhydrazine (4 mgm). (Unpublished data for CS1/Ash mice.)

A dramatic increase in the rate of erythrocyte production following treatment with phenylhydrazine, has already been described in mice (p. 230). A similar increase in the rate of erythrocyte production in rats treated with phenylhydrazine has been attributed by Tarbutt (1969) to an increase in the number of cells entering the erythroblast compartment of the bone marrow (from $2{\cdot}4 \times 10^5$ to $7{\cdot}3 \times 10^5$ cells per hour per 100 gm body weight) to which any increase in the input to the splenic erythroblast compartment must be added. In the mouse a very considerable increase in the input to the splenic erythroblast compartment is suggested by a ten-fold increase in the relative size of this compartment which accompanies enlargement of the spleen under circumstances in which the transit time of cells passing through it would certainly not be expected to increase (Fig. 25).

In contrast the unidentified precursors are unable to maintain their output to the erythroblast compartment in rats exposed to continuous

γ-irradiation (43 rads per day for ten weeks). The input to the erythroblast compartment falls from $2 \cdot 4 \times 10^5$ to between $4 \cdot 5 \times 10^4$ and $9 \cdot 0 \times 10^4$ cells per hour per 100 gm body weight (Tarbutt, 1969). The deficiency is compensated for, however, by an increase in the number of mitotic divisions which occur during the course of erythroblast maturation so that the normal rate of erythrocyte production ($2 \cdot 4 \times 10^7$ cells per hour per 100 gm body weight) is maintained.

These considerations highlight the importance of obtaining information about the population of unidentified cells which provides a continuous and variable input to the erythroblast compartment.

3. *Erythrocyte precursors which have not been identified. (Haematopoietic stem cells)*

Although the morphological characteristics of the cells which provide a continuous input to the erythroblast compartment have not been established information about the properties of these cells is accumulating (Lajtha, 1970; Thomas, 1973).

Clearly the cell population which is the ultimate source of all the erythrocytes produced by an organism must be capable of maintaining an adequate output and therefore of exceeding a certain minimum size throughout the life of the organism. Animals do not appear to have enormous populations of haematopoietic stem cells which are progressively depleted by the transfer of cells to the maturation compartments. On the contrary the size of the stem cell population appears to be maintained by the continuous production of new cells which replace those that are being lost. Following acute depletion the size of the haematopoietic stem cell population may be restored extremely rapidly (Fig. 26).

Like the erythroblast compartment the maturation compartments for the other varieties of blood cells and blood platelets depend upon an input from the stem cell population. Each of these maturation compartments appears to derive cells from a single multipotent stem cell population rather than from several unipotent stem cell populations, each of which supplies cells to a single maturation compartment (Thomas, 1973). If this is so a haematopoietic stem cell must be capable of transforming in a manner which is determined by the environmental influences to which it is exposed. The stem cell population from which lymphocytes are derived may however be distinct from that which is the source of erythrocytes, granulocytes and platelets (Hayhoe *et al.*, 1964).

Several assay techniques which are now being exploited extensively allow changes in the size of haematopoietic stem cell populations to be studied quantitatively (*vide infra*).

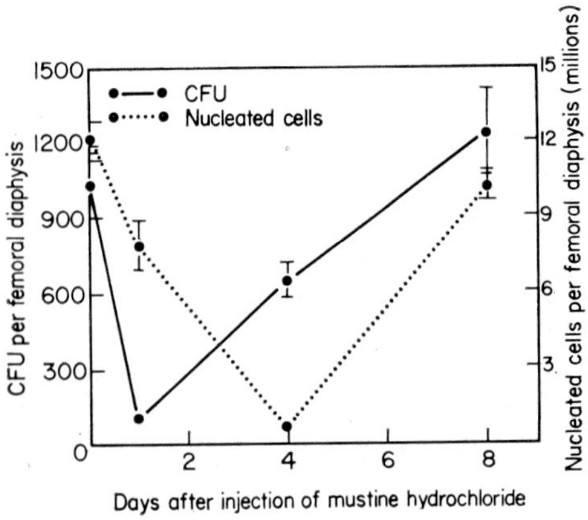

Fig. 26. Between the first day and the fourth day after an injection of mustine hydrochloride a rapid increase in the stem cell content of murine bone marrow is suggested by the increase in the number of colony forming units (CFU) which it contains (see p. 237). During the ensuing four days the nucleated cell population of the bone marrow is restored. (Reprinted from Sharp and Thomas (1971) with the permission of the Editor of Acta Haematologica.)

4. *Methods of assessing changes in the size of haematopoietic stem cell populations*

Suspensions of haematopoietic cells which are maintained *in vitro* under suitable conditions for a week or so can form visible colonies which consist predominantly of granulocytes. Changes in the number of such colonies formed under standard conditions can be used to assess changes in the stem cell content of haematopoietic cell suspensions (Bradley and Metcalf, 1966), although it is important to recognize the possibility that some of the colonies may not be derived from undifferentiated stem cells but from cells which have already entered the granulocyte maturation compartment.

Changes in the size of haematopoietic stem cell populations can be assessed *in vivo* by measuring the response to erythropoietin after erythropoietin production has been suppressed and erythrocyte production has been inhibited by hypertransfusion or starvation (Gurney and Wackman, 1961; Gurney *et al.*, 1962; Porteous *et al.*, 1963). Erythropoietin is a hormone which regulates erythrocyte production. Alternatively changes in the proliferative capacity either of cells derived from a shielded region such as the femur *or* of injected cells can be measured in irradiated animals and used to assess changes in the stem cell content of haematopoietic cell populations.

The proliferative capacity of haematopoietic cell populations which have been transplated to lethally irradiated recipients provides a most valuable index of their stem cell content and can be measured in several different ways (Thomas, 1973).

5. *The proliferative capacity of transplanted haematopoietic cells*

In animals which have been exposed to a lethal dose of whole body X-irradiation the bone marrow is rendered virtually acellular and the lymphoid tissues severely hypotrophic (Fig. 27). If the dose administered does not cause a significant degree of irreparable damage to the central nervous system or to the epithelium lining the gastro-intestinal tract death will be due to haematopoietic insufficiency. Haematopoietic insufficiency can be prevented by injecting suspensions of living bone marrow, spleen or foetal liver cells. Cells derived from such suspensions repopulate the haematopoietic and lymphoid tissues of the lethally irradiated recipient and thus restore cell production in these tissues (Fig. 27c) (Micklem and Loutit, 1966).

For a given genetic relationship between the recipient and the donor, the pattern of recovery in the tissues which are undergoing repopulation depends upon the number of cells which have been injected. Differences in the patterns of recovery which follow the administration of different cell suspensions in comparable doses have therefore been taken to reflect differences in the proportion of cells capable of effecting repopulation which they contain. Such cells which must of course be capable of sustained proliferation are by definition haematopoietic stem cells (Thomas, 1973).

Initially changes in the number of haematopoietic cells required to protect animals against the lethal effects of whole body X-irradiation were used to assess changes in their stem cell content. More recently mortality studies have been superseded by measurements of the ability of transplanted cells to restore erythrocyte production (Blackett *et al.*, 1964), to restore granulocyte production (Hellman and Grate, 1967), to effect an increase in the weight of the spleen (Fig. 28), or to effect an increase in the cellularity of the bone marrow (Stevens and Thomas, 1970).

These methods cannot of course be used to determine the absolute number of stem cells in haematopoietic cell suspensions but only to assess changes in the proportion of stem cells which they contain, relative to the proportion of stem cells in a standard suspension of haematopoietic cells—such as bone marrow cells from an untreated donor. The ability of haematopoietic cells to form macroscopic nodules in the spleen of an irradiated animal is therefore of considerable importance, as it can be exploited to determine the absolute number of cells capable of forming such nodules as well as to measure changes in the relative size of this cell population (Fig. 29).

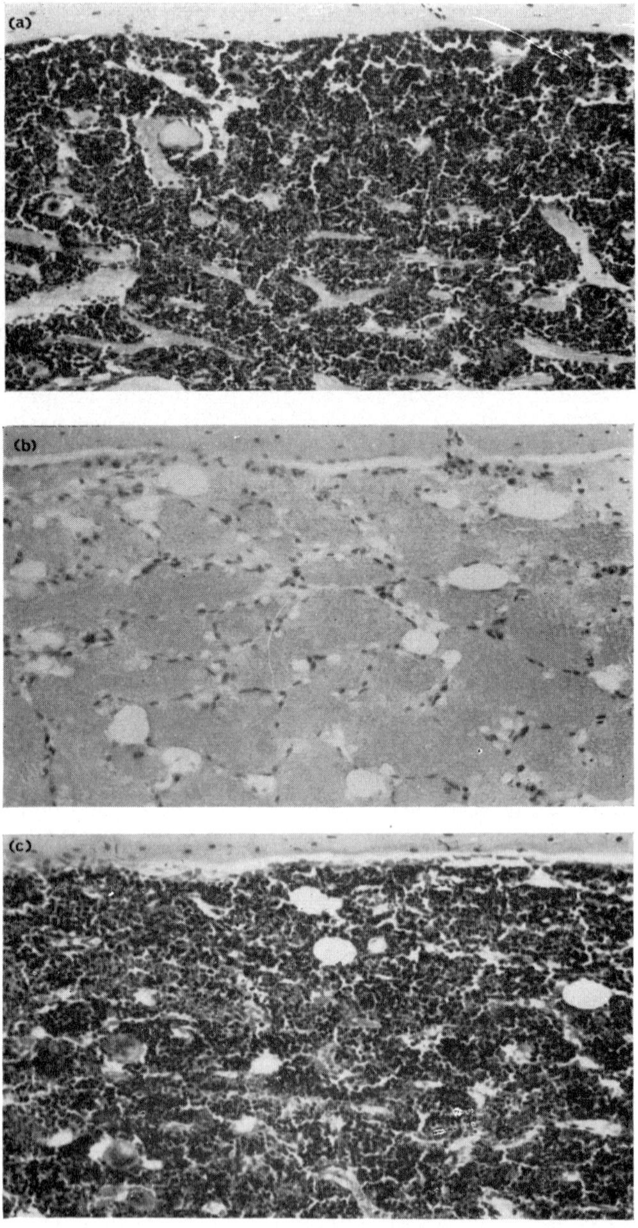

Fig. 27. The effects of a lethal dose of whole body x-irradiation (820 rads) on the femoral bone marrow of the mouse and the modification of these effects by the intravenous injection of living bone marrow cells (5×10^6). (a) Untreated control; (b) Ten days after x-irradiation; (c) Ten days after x-irradiation, followed within an hour or so by an injection of bone marrow cells. (Magnification $\times 140$.)

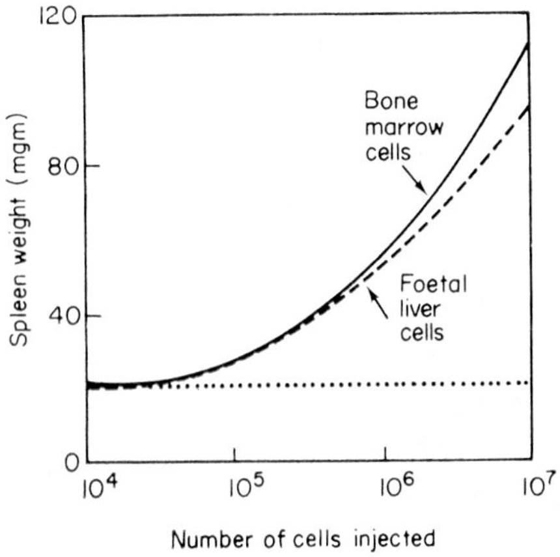

Fig. 28. The relationship between the weight of the spleen and the number of haematopoietic cells injected into lethally irradiated recipients ten days previously. (Smooth curves fitted to unpublished data for CS1/Ash mice.)

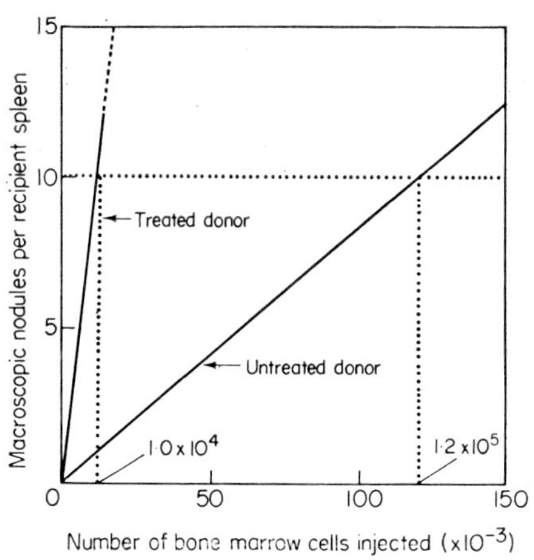

Fig. 29. The relationship between the number of macroscopic nodules (colonies) formed on the spleen and the number of bone marrow cells injected into lethally irradiated recipients ten days previously. Treated donors received a single intra-venous injection of mustine hydrochloride (100μg.) four days before donating bone marrow cells. (Straight lines fitted to the unpublished data of Sharp and Thomas for CS1/Ash mice.)

The formation of macroscopic nodules in the spleen following the intravenous injection of haematopoietic cell suspensions into heavily irradiated mice was described by Till and McCulloch (1961). These nodules appear to be clones of cells each derived from a single cell in

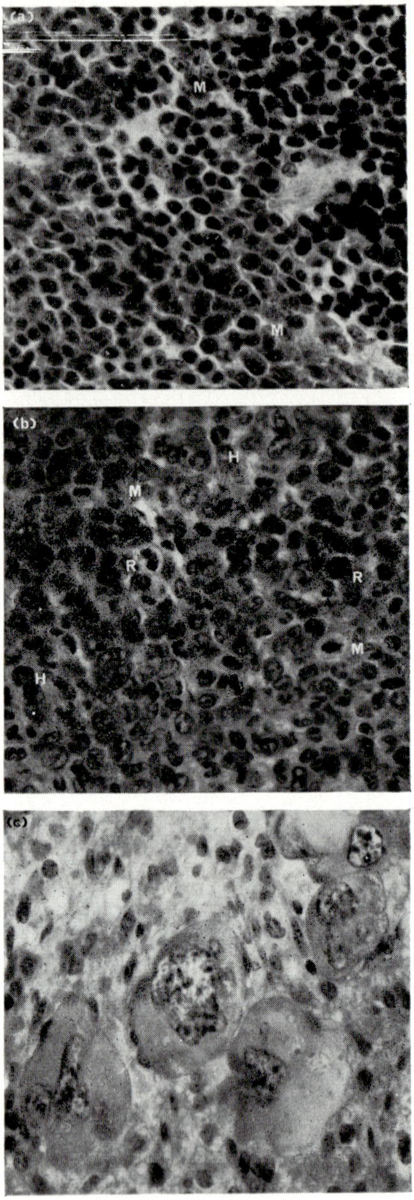

Fig. 30. The histological appearances exhibited by spleen colonies. a. Colony composed predominantly of erythroblasts. b. Colony composed predominantly of granulocytes in which the scattered ring (R) and horseshoe (H) nuclei of metamyelocytes are a distinctive feature. c. Colony of megakaryocytes. (M-mitotic figures. In each instance the cell designated is to the left of the label. Magnification $\times 380$.)

most instances (Chen and Schooley, 1968) and the number of nodules observed on the surface of the spleen is related to the number of haematopoietic cells which have been injected. Although only a fraction of the injected cells which are potentially capable of forming macroscopic nodules settle in the spleen this fraction can be estimated (Siminovitch *et al.*, 1963) and taken into account subject to certain reservations (Kretchmar and Conover, 1969).

Histological studies have revealed the existence of nodules composed of erythroblasts, granulocytes or megakaryocytes and of mixed colonies (Fig. 30). Colonies which contain appreciable numbers of erythroblasts accumulate iron in preparation for haemoglobin synthesis and can be demonstrated autoradiographically (Thomas *et al.*, 1970; Thomas, 1972). Like the number of surface colonies, the number of zones of iron concentration observed in autoradiograms is related to the number of haematopoietic cells which have been injected (Marshall *et al.*, 1971). Cells isolated from a single colony can give rise to colonies composed of erythroblasts, granulocytes and megakaryocytes following re-transplantation to irradiated secondary recipients (Wu *et al.*, 1967). Thus the single cell which gave rise to the original colony has produced progeny of each variety.

6. *The morphological characterization of the haematopoietic stem cell*

Recently measured changes in the proliferative capacity of transplanted bone marrow cells have been related to measured changes in the distribution of cells between the various morphological categories. A pronounced increase in the proportion of stem cells in the bone marrow four days after the treatment of mice with a high dose of mustine hydrochloride is suggested by comparison of the number of bone marrow cells from treated and untreated donors required to establish a given number of macroscopic nodules in the spleen, to effect a given increase in the weight of the spleen or to ensure a given increase in the cellularity of the bone marrow (Fig. 29). This increase in the proportion of stem cells is balanced however by a decrease in the number of bone marrow cells so that the absolute number of stem cells is not altered appreciably (Fig. 31A) (Sharp, Thomas and Briscoe, 1971). Cells of the only morphological category which satisfy these criteria are illustrated in Figs. 31B and 32. Similar cells have previously been described in human foetal bone marrow (Yoffey and Thomas, 1961) and have been subjected to extensive morphological study in guinea pig bone marrow (Yoffey *et al.*, 1965).

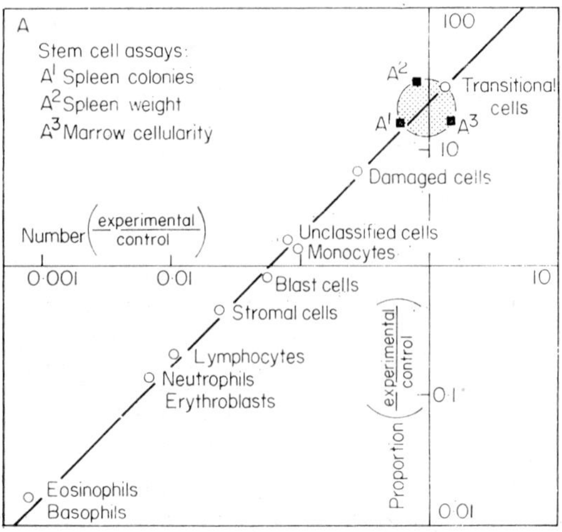

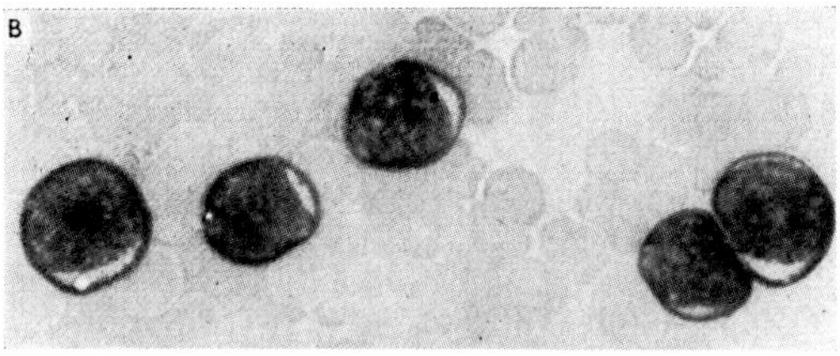

Fig. 31. Cells of only one morphological category behave in the manner predicted for haematopoietic stem cells by stem cell assays (p. 237) in animals treated with mustine hydrochloride (A). These cells were termed "transitional cells", when their role was uncertain because their morphological characteristics are intermediate between those of small lymphocytes and the earliest blood cell precursors that are generally recognized (blast cells) (B × 1 000). (Reprinted from Sharp, Thomas and Briscoe (1971) with the permission of the Editor of the Journal of Anatomy.)

V. The Regulation of Cell Production

The homeostatic mechanisms which regulate the rate of cell production have not yet been fully elucidated for any population of cells. It is clear however that in certain populations the rate at which new cells are produced can be adjusted to ensure prompt restoration following depletion or to match the changing requirements of the organism efficiency (Figs. 4, 23 and 33). As a result pronounced changes can occur in the size of cell populations and in the rate at which cells are made available for export.

Factors which induce cells to undergo mitotic division presumably contribute to the regulation of cell production. Such factors, secreted under the control of negative feedback mechanisms, may help to ensure that cell production is matched to the requirements of the organism. Similar factors may be released from damaged cells and serve to accelerate the restoration of cell populations which have been depleted by injury (Abercrombie, 1957). Thus the regeneration of damaged skin may be accelerated by the release of a factor which has been extracted from thin slices of skin and shown to be capable of inducing epidermal cells in Go or G1 to commence DNA synthesis (Hell, 1970).

Tissue specific mitotic inhibitors may also contribute to the regulation of cell production (Iversen, 1961) and it has been suggested that the equilibrium of certain cell populations may be dependent upon

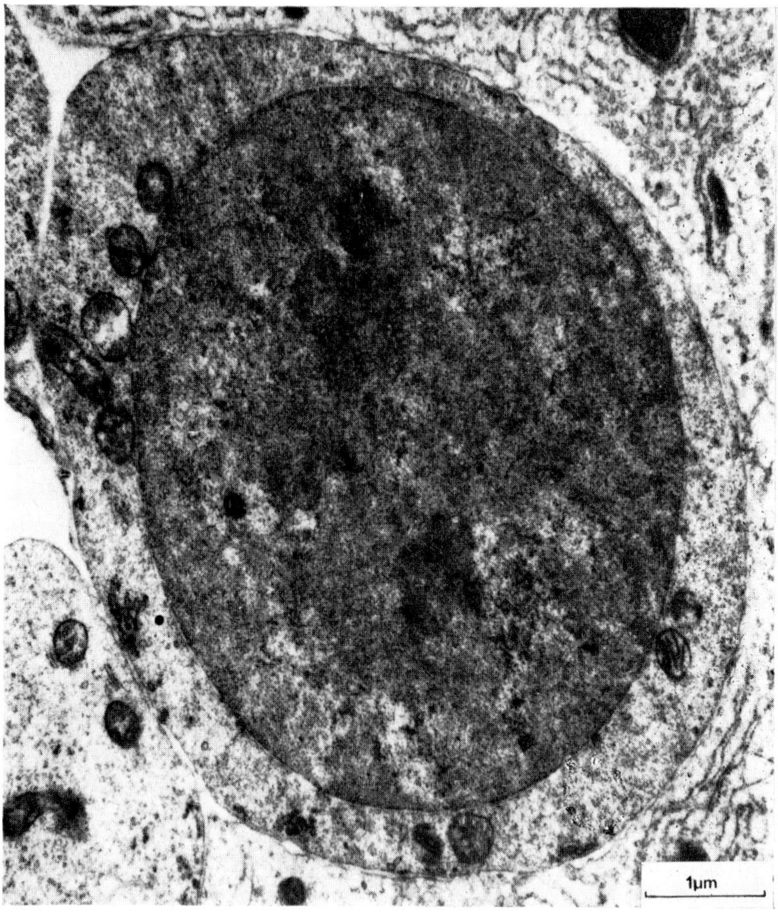

Fig. 32. "Transitional cell" in murine foetal liver on the fifteenth day of gestation. (Magnification ×16 000.) (Reprinted from Thomas (1973) with the permission of the Editors.)

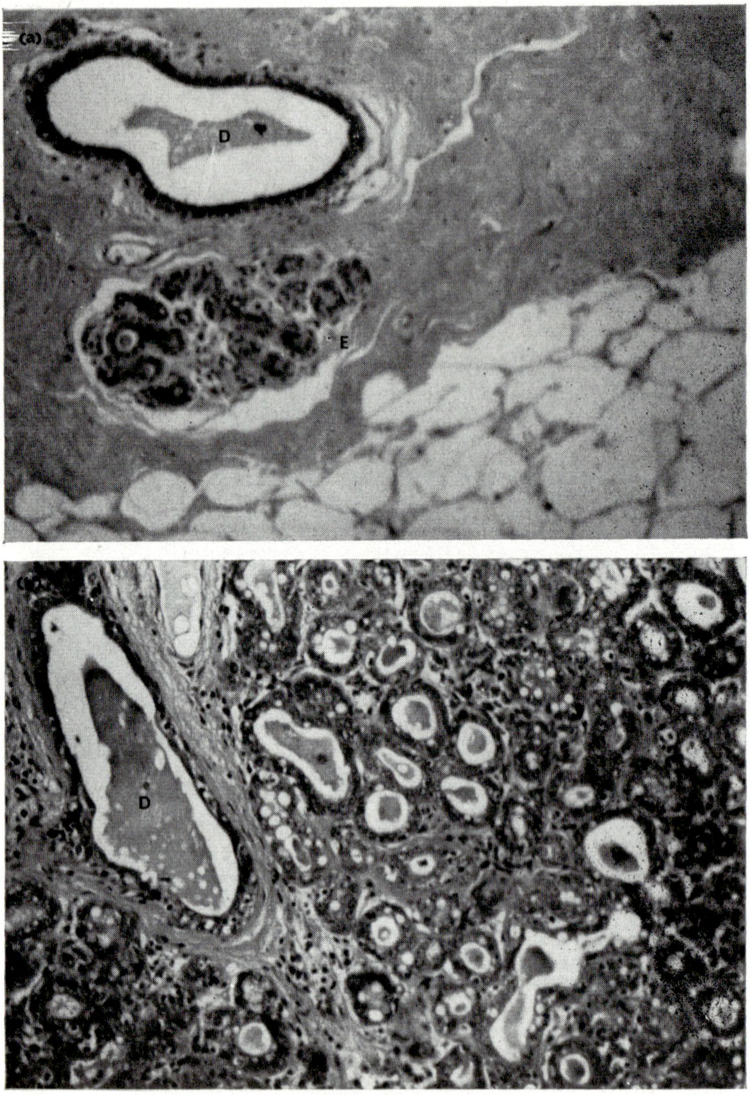

Fig. 33. The small collections of epithelial cells (E) seen in the breast of a non-pregnant female (a) become very much more numerous during pregnancy in preparation for lactation (b). (D—duct. Magnification ×140.)

such inhibitors (*chalones*), which prevent the cells from reverting to a "stable programme of recurrent mitosis" (Bullough, 1965). The possibility that some cell populations may regulate their own size by producing factors which inhibit mitosis has already been considered (p. 205). When the number of cells in such a population is decreased production of intrinsic mitotic inhibitors is also presumed to decrease and so allow mitotic activity to increase. Then as mitotic activity restores the population to its original size production of these inhibitors is presumed to increase and so to depress mitotic activity.

Recommended Reading

Baserga, R. (1971). "The cell cycle and cancer". New York, Marcel Dekker, Inc.

References

Abercrombie, M. (1946). Estimation of nuclear population from microtome sections. *Anat. Rec.*, **94**, 239–247.

Abercrombie, M. (1957). Localized formation of new tissue in an adult mammal. *Symp. Soc. exp. Biol.*, **11**, 235–254.

Abercrombie, M. and Harkness, R. D. (1951). The growth of cell populations and the properties in tissue culture of regenerating liver of the rat. *Proc. Roy. Soc. B.*, **138**, 544–561.

Alpen, E. L. and Cranmore, D. (1959). Observations on the regulation of erythropoiesis and on cellular dynamics by ^{59}Fe autoradiography. *In*: "The Kinetics of Cellular Proliferation." p. 290. (Stohlman, F. Jr., ed.) Grune and Stratton, Inc., New York.

Blackett, N. M., Roylance, P. J. and Adams, K. (1964). Studies of the capacity of bone-marrow cells to restore erythropoiesis in heavily irradiated rats. *Brit. J. Haemat.*, **10**, 453–467.

Bradley, T. R. and Metcalf, D. (1966). The growth of mouse bone marrow cells in vitro. *Aust. J. Exp. Biol. Med. Sci.*, **44**, 287–300.

Bullough, W. S. (1965). Mitotic and functional homeostasis: a speculative review. *Cancer Res.*, **25**, 1683–1727.

Cameron, I. L. (1971). Cell proliferation and renewal in the mammalian body. *In* "Cellular and Molecular Renewal in the Mammalian Body", (Cameron, I. L. and Thrasher, J. D., eds), pp. 45–85. New York and London: Academic Press.

Chen, M. G. and Schooley, J. C. (1968). A study on the clonal nature of spleen colonies using chromosome markers. *Transplantation*, **6**, 121–126.

Enesco, M. and Leblond, C. P. (1962). Increase in cell number as a factor in the growth of the organs and tissues of the young male rat. *J. Embryol. exp. Morph.*, **10**, 530–62.

Erslev, A. J. (1959). The effect of anaemic anoxia on the cellular development of nucleated red cells. *Blood*, **14**, 386.

Everett, N. B. and Yoffey, J. M. (1959). Life of guinea pig circulating erythrocyte and its relation to erythrocyte population of bone marrow. *Proc. Soc. exp. Biol. Med. N.Y.*, **101**, 318–319.

Friedman, N. B. (1945). Cellular dynamics in intestinal mucosa: The effect of irradiation on epithelial maturation and migration. *J. exp. Med.*, **81**, 553–558.

German, J. (1964). The pattern of DNA synthesis in the chromosomes of human blood cells. *J. Cell Biol.*, **20**, 37–55.

Gurney, C. W., Degowin, R., Hofstra, D. and Byron, J. (1962). Applications of erythropoietin to biological investigation. *In*: "Erythropoiesis", p. 151. (Jacobson, L. O. and Doyle, M. eds)., Grune and Stratton, Inc., New York.

Gurney, C. W., Lajtha, L. G. and Oliver, R. (1962). A method for investigation of stem-cell kinetics. *Brit. J. Haemat.* **8**, 461–466.

Gurney, C. W. and Wackman, N. (1961). Impairment of erythropoiesis by irradiation. *Nature, Lond.* **190**, 1017–1018.

Harris, J. W. (1963). The red cell, production, metabolism, destruction: normal and abnormal. Harvard University Press, Cambridge, Mass.

Hayhoe, F. G. J. and Flemans, R. J. (1969). "An atlas of haematological cytology". Wolfe Medical Books, London.

Hayhoe, F. C. J., Quaglino, D. and Doll, R. (1964). "The cytology and cytochemistry of acute leukaemias". Her Majesty's Stationery Office, London.

Hell, Elizabeth (1970). A stimulant to DNA synthesis in guinea-pig ear epidermis. *Br. J. Derm.* **83**, 632–636.

Hellman, S. and Grate, Helen E. (1967). Production of granulocytic progeny by transplanted bone marrow in irradiated mice. *Blood*, **30**, 103–111.

Howard, A. and Pelc, S. R. (1951). Synthesis of deoxyribose nucleic acid and nuclear incorporations of ^{35}S as shown by autoradiographs. In *Isotopes in biochemistry*. CIBA Foundation Conference, (G. E. W. Wolstenholme, ed.), Churchill, London.

Howard, A. and Pelc, S. R. (1953). Synthesis of deoxyribonucleic acid in normal and irradiated cells and its relation to chromosome breakage. *Heredity*, Supplement to Volume 6, 261–273.

Ito, M., Hotta, Y. and Stern, H. (1967). Studies of meiosis in vitro. II. Effect of inhibiting DNA synthesis during meiotic prophase on chromosome structure and behaviour. *Devel. Biol.* **6**, 54–77.

Iversen, O. H. (1961). The regulation of cell numbers in epidermis. A cybernetic point of view. *Acta path. microbiol. scand. Suppl.* **148**, 91–96.

Kretchmar, A. L. and Conover, W. R. (1969). Colony-forming cells in the spleen. *Transplantation*, **8**, 576–581.

Lajtha, L. G. (1970). Stem cell kinetics. *In*: "Regulation of Hematopoiesis", (Gordon, A. D. ed.) (2 volumes), Appleton-Century-Crofts, New York.

Leblond, C. P. and Stevens, C. E. (1948). The constant renewal of the intestinal epithelium in the albino rat. *Anat. Rec.* **100**, 357–377.

Leblond, C. P. and Walker, B. E. (1956). Renewal of cell populations. *Physiol. Rev.* **36**, 255–276.

Lesher, S., Lamerton, L. F., Sacher, G. A., Fry, R. J. M., Steel, G. Gordon, and Roylance, P. J. (1966). Effect of continuous gamma irradiation on the generation cycle of the duodenal crypt cells of the mouse and rat. *Radiat. Res.* **29**, 57–70.

Levine, L. (ed.) (1963). "The cell in mitosis". Academic Press, New York and London.

Maloney, M. A., Patt, H. M. and Weber, C. L. (1962). Estimation of deoxy ribonucleic acid synthetic period of myelocytes in dog bone marrow. *Nature, Lond.* **193**, 134.

Marshall, Moyra, Riches, A. C. and Thomas, D. Brynmor (1971). The development and frequency of zones of iron concentration in the spleen of the mouse radiation chimera. *J. Anat.* **109**, 350–351.

Mazia, D. (1961). Mitosis and the physiology of cell division. *In*: "The Cell", (Brachet, J. and Mirsky, A. E. eds.) Volume III, pp. 77–412. Academic Press, New York and London.

Mendelsohn, M. L. and Takahashi, M. (1971). A critical evaluation of the fraction of labelled mitoses method as applied to the analysis of tumour and other cell cycles. *In*: "The Cell Cycle and Cancer" (Barserga, R. ed.), 58–95, Marcel Dekker, Inc., New York.

Messier, B. and Leblond, C. P. (1960). Cell proliferation and migration as revealed by radioautography after injection of thymidine —^{3}H into male rats and mice. *Am. J. Anat.* **106**, 247–285.

Metcalf, D. (1966). The native and regulation of lymphopoiesis in the normal and neoplastic thymus. *In*: "The Thymus: Experimental and Clinical Studies". CIBA Foundation Symposium, pp. 242–263. (Wolstenholme, G. E. W. and Porter, R., eds.) Churchill, London.

Micklem, H. S. and Loutit, J. F. (1966). "Tissue Grafting and Radiation." Academic Press, New York and London.

Osmond, D. G. and Everett, N. B. (1964). Radioautographic studies of bone marrow lymphocytes in vivo and in diffusion chamber cultures. *Blood*, **23**, 1–17.

Pelc, S. R. (1971). Influence of metabolic DNA on determinations of the cell cycle. *Cell Tissue Kinet.*, **4**, 577–583.

Philp, J. R. and Buchanan, T. J. (1971). Quantitative measurement on finite tissue sections. *J. Anat.* **108**, 89–97.

Porteous, D. D., Alexanian, R. and Lajtha, L. G. (1963). The fasted mouse in the study of bone-marrow stem-cell kinetics. *Int. J. Radiat. Biol.* **7**, 95–1C0.

Riches, A. C., Sharp, J. G., Thomas, D. Brynmor and Vaughan Smith, S. (1973). Blood volume determination in the mouse. *J. Physiol.* **228**, 279–284.

Riddell, A. J. and Thomas, D. Brynmor (1972). The kinetics of cellular accumulation in a murine thymic lymphoma. *J. Anat.*, **113**, 285–286.

Rogers, A. E. (1967). "Techniques of Autoradiography". Elsevier Publishing Company, Amsterdam, London, New York.

Roylance, P. J. (1968). An evaluation of erythropoiesis in the young rat. *Cell and Tissue Kinetics*, **1**, 299–308.

Sharp, J. G. and Thomas, D. Brynmor (1971). The effects of mustine hydrochloride on the colony forming units of murine bone marrow. *Acta haemat.* **46**, 271–274.

Sharp, J. G., Thomas, D. Brynmor and Briscoe, Cecilie V. (1971). An approach to the morphological characterisation of the haematopoietic stem cell. *J. Anat.* **108**, 597–598.

Siminovitch, L., McCulloch, E. A. and Till, J. E. (1963). The distribution of colony-forming cells among spleen colonies. *J. cell. comp. Physiol.* **62**, 327–336.

Smith, S. R., Thomas, D. Brynmor and Riches, A. C. (1972). Metaphase accumulation in tumour isografts following the administration of colcemid, vinblastine or vincristine. *J. Anat.* **111**, 480.

Steel, G. G. (1968). Cell loss from experimental tumours. *Cell Tissue Kinet.* **1**, 193–207.

Steel, G. G. (1972). The cell cycle in tumours: an examination of data gained by the technique of labelled mitoses. *Cell Tissue Kinet.* **5**, 87–100.

Stevens, R. F. and Thomas, D. Brynmor (1970). Bone marrow regeneration as a measure of the proliferation of transplanted haematopoietic cells. *J. Anat.* **107**, 394.

Stevens Hooper, C. E. (1961). Use of colchicine for the measurement of mitotic rate in the intestinal epithelium. *Am. J. Anat.* **108**, 231–244.

Tarbutt, R. G. (1967). A study of erythropoiesis in the rat. *Expl. Cell Res.* **48**, 473–483.

Tarbutt, R. G. (1969). Cell population kinetics of the erythroid system in the rat. The response to protracted anaemia and to continuous X-irradiation. *Br. J. Haemat.* **16**, 9–24.

Tarbutt, R. G. and Blackett, N. M. (1968). Cell population kinetics of the recognizable erythroid cells in the rat. *Cell Tissue Kinet.* **1**, 65–80.

Thomas, D. Brynmor (1966). Hepatic haematopoiesis in the mouse foetus. *J. Anat.* **100**, 932–934.

Thomas, D. Brynmor (1972). The relationship between zones of iron concentration and macroscopic surface colonies in the spleen of the mouse radiation chimera. *J. Anat.* **114**, 95–100.

Thomas, D. Brynmor (1973). The radiation chimera as an experimental model for the study of haematopoietic stem cell populations. In Haemopoetic Stem Cells. Ciba Foundation Symposium, (Wolstenholme, G. E. W., O'Connor, M., eds), Associated Scientific Publications, pp. 71–103.

Thomas, D. Brynmor and Briscoe, C. V. (1973). The effects of intravenous injections of endotoxin on the distribution of cells between murine blood and bone marrow. *J. Anat.*, **114**, 407–419.

Thomas, D. Brynmor, Steers, D. J. F. and Riches, A. C. (1970). A macroautoradiographic method for the demonstration of iron-concentrating spleen colonies. *Transplantation* **10**, 274–275.

Till, J. E. and McCulloch, E. A. (1961). A direct measurement of the radiation sensitivity of normal mouse bone marrow cells. *Radiation Res.* **14**, 213–222.

Vendrely, C. (1971). Cytophotometry and histochemistry of the cell cycle. *In*: "The Cell Cycle and Cancer" (Baserga, R. ed.), pp. 227–268. Marcel Dekker, Inc., New York.

Wimber, D. E. and Quastler, H. (1962). A C^{14}— and H^{3}— thymidine double labelling technique in the study of cell proliferation in tradescantia root tips. *Expl. Cell. Res.* **30**, 8–22.

Wu, A. M., Till, J. E., Siminovitch, L. and McCulloch, E. A. (1967). A cytological study of the capacity for differentiation of normal hemopoietic colony-forming cells. *J. Cell. Physiol.* **69**, 177–184.

Yoffey, J. M. (1966). Bone marrow reactions. Edward Arnold Limited, London.

Yoffey, J. M., Hudson, G. and Osmond, D. G. (1965). The lymphocyte in guinea-pig bone marrow. *J. Anat.* **99**, 841–860.

Yoffey, J. M. and Thomas, D. Brynmor (1961). The development of bone-marrow in the human foetus. *J. Anat.* **95**, 613.

14. Cell Differentiation

R. BELLAIRS

Department of Anatomy, University College London, UK

As an embryo develops it grows and it differentiates. This Chapter and the next will deal with differentiation and growth in embryonic tissues.

I. Ideas on Differentiation

What do we mean by the term, differentiation? Broadly, we may define it as the increasing complexity that occurs in a developing system The cells in each part of the body are modified for their various roles. They become different from one another, and individually each changes so that it differs from its former self.

Until recently, the term "differentiation" was used by embryologists solely to describe morphological changes that took place in cells and tissues. For instance, when ectoderm cells become changed into neural cells they become larger and more elongated, so that the entire tissue becomes thickened. If we examine these cells with the electron microscope, other and more subtle changes can be seen, such as an increase in the amount of endoplasmic reticulum (Fig. 1). These changes are signs enabling us to say that the ectodermal cells are differentiating into neural tissue.

Recently, developmental biologists have begun to look at differentiation in a new light, and to emphasise that morphological changes are merely evidence of previous biochemical events. Thus, the morphological changes that take place in the ectodermal cells as they differentiate are due not only to the shifting around of the components of these cells, but also to the synthesis of new and specialized proteins. Certain proteins are present in all cells, whether embryonic or adult. But other proteins are specific to certain organs in which alone they are found. As we shall see, we have direct evidence in several developing tissues that changes in protein synthesis precede morphological differentiation. One example which will be discussed (p. 299) occurs in the differentiation of the lens where the production of specialized proteins, the crystallins, takes place before any morphological differentiation of the lens cells.

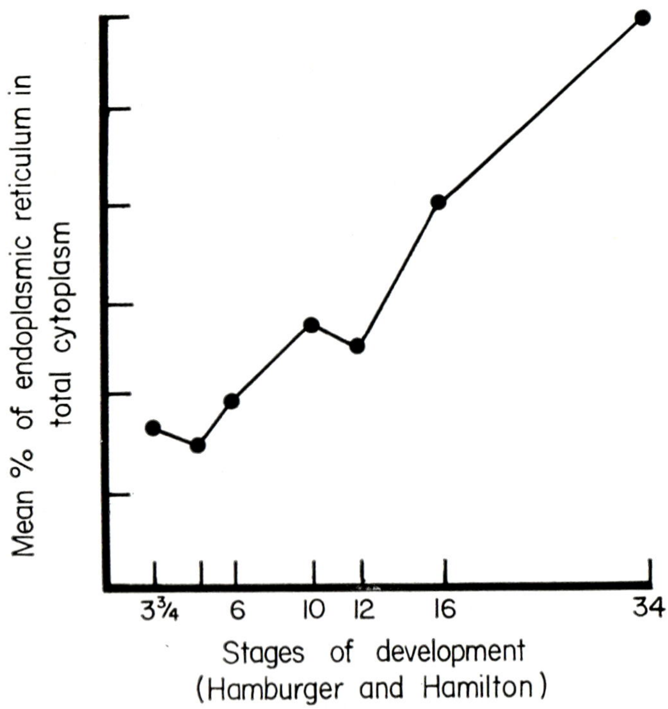

Fig. 1. Graph to show how the percentage of endoplasmic reticulum in the cytoplasm of chick neurons increases during development. The data was obtained from the analysis of electron micrographs. The youngest embryos were at the primitive streak stage and the oldest had reached 18 days of incubation. (After Bellairs, 1959.)

It is convenient to consider differentiation separately from growth, but we should not forget that the two processes are connected. In some tissues differentiation is heralded by a burst of cell division (pp. 256 and 278), and in others, the very shape of the developing organ is influenced by the fact that proliferation is exceptionally high in some regions or unusually low in others (p. 278).

II. Classifying Differentiation

Biologists have in the past recognized two types of embryonic development, mosaic and regulative. In embryos exhibiting mosaic development, the fertilized egg is considered to be a mosaic of all the future parts of the body. The development of each part of the mosaic egg is fixed, so that it is capable of forming only appropriate organs and no others. The programme for this type of development is so inflexible that destruction of any region at an early stage of development should result in the absence of some tissue or organ. In modern terms, we might say that the genes are "switched on and off" in a predetermined way at appropriate times and that there is no possibility of altering the programme. Nowadays it is considered doubtful if mosaic development ever occurs in so extreme and rigid a manner. But in many invertebrates, such as molluscs, a certain inflexibility does set in at an early stage of development.

By contrast, the characteristic of regulative development is that it allows for adjustment and change, at least during the early stages of embryonic life. For example, if a mouse embryo at the 32-cell stage is cut into two equal parts, each part does not form merely half an embryo but it differentiates into one complete though small individual. Conversely, if two early mouse embryos are fused together, they do not remain as two recognizable individuals who are permanently attached to each other, but rearrange themselves to form one single though double-sized embryo. In doing so, some at least of the cells have changed their course of development. This changing and adjusting of the cells so that they co-operate with the other cells and give rise to a normal tissue or individual is given the name of regulation.

Unlike the genes of mosaic embryos, those of regulative organisms do not come into play according to a prefixed programme, but in response to the needs of the embryo. Regulation can therefore be regarded as the process whereby the genes respond to the changes that are taking place in other parts of the embryo.

Regulative development is characteristic of the early stages of all vertebrates. But the time comes when regulation of all the parts to form a complete individual is no longer possible, though reorganization of a small region of the body can still take place. Once an embryo (or part of an embryo) has reached this stage, we say it has become determined. Thus, although the mouse embryo may be capable of regulation at the 32 cell stage and at the 64 cell stage, it has lost this ability by the time a neural tube and somites have formed. If it is then cut in two, the separate halves never develop into normal single embryos.

Determination is not however a state that is suddenly achieved, like hatching or birth. During the early stages regulation is at its maximum and, in vertebrates, there is no determination. As development proceeds the powers of adjustment (i.e. regulation) are reduced and the tissues become more determined, until the ability to regulate is lost. Thus, at first it is possible for the entire embryonic axis to become rearranged: cells originally destined to be, say, somites are able to change and become heart. After this stage the ability to regulate becomes restricted and more localized. Somite cells may no longer be able to form heart but are still perhaps capable of differentiating into either myotome or sclerotome. As differentiation proceeds, even this localized type of regulation becomes impossible. With increasing differentiation of the tissues there is an increase in determination. Expressed another way, we can say that with increasing determination the genes become less able to respond to changes in other parts of the embryo.

Nowadays, the rigid distinction between mosaic and regulative development has been largely discarded, and it is usual to regard mosaic embryos as those that have lost their regulative ability at an especially early stage. Our knowledge of the degree of regulation of which an embryo is capable is derived from experiments. By devising new and better experiments it is sometimes possible to demonstrate that regulation can occur in species that were formerly thought to be entirely mosaic in character. For instance, it used to be considered that the ascidians (tunicates) were entirely mosaic until it was shown that regulation was possible during the first stages of cleavage.

III. Stability of Differentiation

A. THE DIFFERENTIATED CELL

We may now ask: Can determination ever be absolute? Once determination has set in, is the power to regulate lost for ever? The answer to both questions appears to be No, at least from the morphological point of view. Many cells of late embryos and even adults are able to display some regulative ability if they are removed from the body and grown in tissue culture (see Chapter 1). If pieces of well-differentiated chick heart are grown *in vitro,* many of the cells take on a new morphological appearance and two distinct cell types become visible, the so-called fibroblasts and macrophages. If cells, which have undergone such modifications, are now grafted back into the body they can often regain their normal appearance. This process, by which the cells can change their morphology with the environment, is sometimes known as modulation.

Another example of reversibility of determination occurs during the processes of regeneration and tumour formation. In both situations the morphology and physiology of the cells changes. For instance, in certain amphibian embryos, such as the axolotl, it is possible to obtain regeneration of the limbs. If a limb is amputated many of the cells in the stump begin to lose their distinctive characters and take on the appearance of the relatively undifferentiated mesenchyme cells of the early embryo. In sections through this region—i.e. through the so-called blastema—it is not possible to say which cells are derived from different tissues such as muscles and blood vessels. It is sometimes said that these cells have become de-differentiated. Most cell biologists nowadays are perhaps a little chary of this term, for it implies a return to the physiological state of the early embryo: an inference for which there is no evidence. Nevertheless, the fact that this term is sometimes used demonstrates that the blastema cells have undergone morphological changes which would not have been possible if they had been completely determined.

B. THE UNDIFFERENTIATED CELL

We have seen that the problem of differentiation is bound up with that of increasing determination. A differentiated cell is one that is specialized for a particular task, and it may or may not be completely determined. The converse, however, is not true. An undifferentiated cell is not necessarily an unspecialized one. The prime example is the fertilized ovum, which is, in some ways, the least differentiated of all cells though a highly specialized one, containing as it does an unusual type of *m*RNA, specialized cell organelles such as yolk, and frequently a modified cell surface (discussed on p. 291 *et seq.*).

IV. Control of Differentiation

Differentiation is essential for a multicellular organism. Without it an animal would develop merely into a massive ball of cells. What then controls differentiation? How does it happen that the organs and tissues come to lie in the correct relationship to one another? Even within a tissue the cells must be orientated appropriately. This question of how control of development occurs has now become one of the central themes of developmental biology. What processes are involved, and how do they start, how are they integrated and how is it that they stop at the appropriate time?

There are of course many control mechanisms in an embryo, and these operate on nuclear, cytoplasmic, cellular, tissue and organ levels. Probably none act independently, for the overall picture of development suggests that there is interaction and co-operation between the processes going on within a cell, of one cell with another and of one tissue with another. Regulation has often been considered as an activity that occurs only when the embryo needs to replace damaged or dead tissues. It seems more probable, however, that continual adjustments are made during development to the protein synthesis that goes on in each cell and hence to each step in differentiation. Only in this way is an orderly integrated embryo likely to form.

A. CONTROL WITHIN THE CELL

During the last twenty years interest has centred on the genetic control of development. It is now well known that the genes are composed of DNA and it is generally believed that with certain exceptions (see below) each cell in the embryo contains the same set of chromosomes as all the others, and so has the same set of genes. This means that each cell is not only capable of manufacturing those proteins needed by all living cells, but is also potentially able to synthezise many specific proteins it does not, and will not, need. For example, respiratory enzymes are needed by all cells, such as those of the heart, gut and kidney, but cardiac myosin is only needed by the cells of the heart. How then does a cell select which genes to use and which to suppress? This is one of the basic problems of molecular biology and is dealt with in detail in Chapter 33.

A number of theories have been put forward:

The first is that although all cells in the early embryo have the same amount and type of DNA this does not remain so, as development proceeds. Generally, however, there is little evidence to suggest that the amount of DNA varies except in a few special cases. In the nematode worm, *Ascaris,* there is a loss of DNA in the somatic cells, and in certain tissues (such as the mammalian liver) polyploid cells are found, which possess an exceptionally high amount of DNA. But in most animals the amount of DNA remains the same in all cells of the body throughout life, except during the later stages of the mitotic cycle when it becomes temporarily doubled, and during meiosis when it becomes halved.

The structure of the DNA also seems to remain the same in different cells of the developing embryo. Some of the most compelling evidence that this is so has been provided by experiments in which the nuclei have been removed from *Xenopus* eggs and have been replaced with nuclei from the gut of a *Xenopus* tadpole. The older nuclei appear to have lost none of their potentiality for directing development, for the egg has differentiated normally in many of these experiments.

The second mechanism by which the genes may be controlled is thought to be by the process known as variable gene activity. This supposes that part of the DNA may not be available for transcription at certain times. The concept of Jacob and Monod (see Chapter 34) is based on this idea, but it is important to remember that this theory was initially devised for bacteria, and that the embryos of higher vertebrates are more complex organisms. Even their DNA differs, for in the bacteria it lies freely in the cytoplasm, whereas in higher organisms it is enclosed in a nucleus. Moreover, the DNA of bacteria is not organized into several chromosomes, nor does it complex with histones. There can be few biologists who imagine that the Jacob–Monod model can be applied without modification to explain the events in a complex multicellular organism.

Unfortunately, we still do not know whether the control of gene activity is brought about in multicellular organisms by suppression or activation. That is: Are all genes normally active unless they are specifically suppressed, or are they normally inactive unless specifically activated? On the one hand, there is evidence that histones and some other nuclear proteins, both of which form part of the chromosomes, may play an important part in repressing gene activity. On the other hand, Britten and Davidson (1969) have recently produced arguments in favour of the alternative hypothesis. They have proposed that the activators are special RNA molecules confined to the nucleus (see Chapter 33).

A well known modification of the chromosomes may also make the DNA "unavailable" for transcription. This occurs in the "inactive X" condition which is a characteristic of all female mammals (discussed in Chapter 12). One of the X chromosomes becomes contracted into a densely staining clump and it has been suggested that this coincides with an inactivation of that chromosome (Lyon, 1968).

The third way in which cells control the production of proteins is thought to be during translation rather than by acting directly upon the genes themselves. This mechanism occurs during the later stages of oogenesis, when the messenger RNA becomes "blocked" or "masked" (see p. 295) and only unblocked or unmasked upon fertilization (Davidson, 1968). It may also occur at other stages.

So far we have looked upon the nucleus, and in particular the genes, as if they dominate the cell. But the nucleus is itself subject to control by the cytoplasm. This is shown by experiments (similar to those mentioned above) in which nuclei have been taken from one type of cell and injected into another. For instance, the nuclei from the oocytes of *Xenopus,* have been replaced by those from cleavage or gastrula cells or even from adult brain neurons. Whatever the origin of these nuclei, however, they appear to take on different characteristics when implanted into another cell, so that they come to resemble the type of nucleus characteristic of that cell. It appears therefore that the cytoplasm influences the implanted nucleus. Gurdon and Woodland (1968) who have reviewed the subject have suggested that in normal cells also, the cytoplasm influences the nucleus so that it comes to function in a particular way. They have suggested that during mitosis, when the nuclear membrane has broken down, the chromosomes come into contact with cytoplasmic proteins in some way, which results in the reprogramming of the gene activity. This idea is an attractive one, for it suggests that there is a feedback mechanism that occurs with great regularity. It is an idea that ties in well with the concept that differentiation is often preceded by a burst of mitotic activity (see pp. 251 and 278 and p. 14 Chapter 1). Further evidence for the cytoplasmic control of gene activity is provided by experiments in which two cells of different histological type are fused together to form a single cell with two nuclei (a heterokaryon). For example, if a cell which is no longer capable of synthesizing nucleic acids (such as a nucleated erythrocyte) is fused with one that is still active, the active cell stimulates the synthesis of nucleic acids and even new proteins, by its partner's nucleus (see also Chapter 10).

B. CONTROL WITHIN A TISSUE: THE INTERACTION OF CELLS

The process of co-ordinated development within a tissue is often known as "pattern formation". This term was introduced to describe the fact that the tissues become arranged in a regular manner. For instance, in the early cleavage stages of the mouse embryo, the cells become arranged in a ball and not in a linear way. Later, at the blastocyst stage, a cavity, the blastocyst cavity, forms, and a group of cells, the inner cell mass becomes distinguishable to one side of it. Later still, after gastrulation the somites form in an orderly relationship to one another, in pairs, down either side of the neural tube.

The way in which the orderliness is brought about and maintained during development is still one of the major problems of biology. A number of theories have been put forward.

The first was the concept of mosaic development but, as we have already seen (p. 251) this is applicable only to certain invertebrates. In a regulative system, therefore, some control must be exerted over the individual cells. In the past, attention has often been concentrated on the idea of a certain *dominant region* which has been supposed to override and control all parts. The prime example is the so-called amphibian organizer, discussed below (p. 267). In recent years, however, attention has begun to shift to the idea that the control lies within the individual cells themselves, and that these *interact* and co-operate with one another to produce an orderly individual. An example of this type of scheme is Wolpert's "positional information" (p. 274). But before we can discuss these ideas we must become familiar with certain aspects of early development of amphibian, bird and mammalian embryos, for these are the three classes of vertebrates whose development has received the most study. Indeed most of our concepts are based on work carried out on amphibian and chick embryos and it is only within the last few years that any appreciable amount of work has been performed with mammalian embryos.

V. Early Stages of Development

In all vertebrates, the single–celled egg rapidly becomes converted after fertilization into an embryonic axis, consisting of neural tube, somites, lateral plate mesoderm and a simple gut. In the chick embryo this change has occurred by about 48 h after fertilization; in the mouse by about 7 days. In each species there has been a dramatic re-organization of the living material, and this has taken place in two main phases—cleavage and gastrulation. Cleavage involves a great increase in the number of cells: in the mouse embryo several hundred cells are present at the end of cleavage. Gastrulation consists of a shifting and rearrangement of these cells. Detailed descriptions of the events involved will be found in most textbooks of embryology.

The main morphological difference between the amphibians, birds and mammals is in the amount of yolk within their eggs. Amphibians possess a moderate amount of yolk, birds a large amount and placental mammals virtually none at all. This difference is usually attributed to evolutionary changes related to the environment that each group has come to occupy (see discussion by Bellairs, 1971).

In the amphibian egg there is a large amount of yolk at one side, the vegetal pole, and very little at the other, the animal pole (Fig. 2a). Conversely, there is a high concentration of cytoplasm at the animal pole, and practically none at the vegetal pole. There are thus two gradients in the distribution of yolk and cytoplasm, and these gradients are inversely proportional to one another. Cell division appears to take place with more facility through the regions with a high cytoplasmic content: i.e. at the animal pole rather than at the vegetal pole. There is no growth in size of the egg during cleavage so that with continual cell division the initial egg becomes divided into smaller cells. Since the division proceeds more rapidly at the animal pole than at the vegetal a gradient in size of the cells soon becomes apparent, the cells at the animal pole being the smallest, those at the vegetal pole the largest (see Fig. 2, c and d).

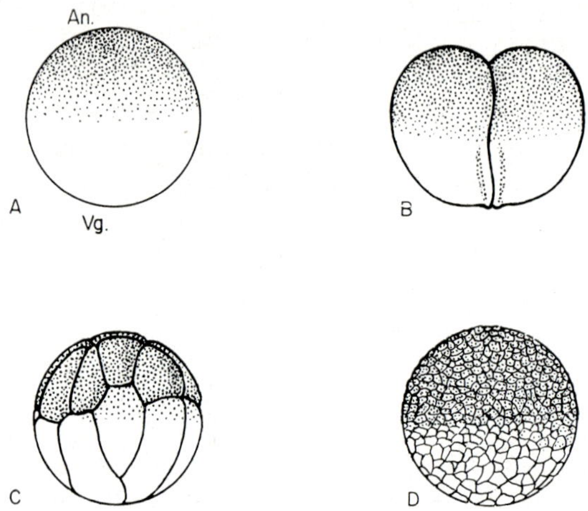

Fig. 2. Cleavage of an amphibian egg. A: At the animal pole (An.) there is a high concentration of pigmented cytoplasm (shown as stipple). At the vegetal pole (Vg.) there is a high concentration of non-pigmented yolk. B., C. and D. show the division into many cells, those at the animal pole being smaller than the ones at the vegetal pole.

We know relatively little about the mechanisms involved in amphibian cleavage, compared with those that play a role in the later stages of development. It does appear however that although the distribution of the yolk is related to the pattern of cleavage, it is not an essential factor in controlling this type of cleavage. If an amphibian egg is punctured so that large amounts of the yolk are lost then the rate and pattern of cleavage remain much the same (Waddington, 1952a). It is also known that the division of the cytoplasm in cleavage can take place even if the nucleus has been removed (Waddington, 1956). Experiments of this type therefore support the idea that much of the control of development at this stage lies in the cytoplasm rather than in the nucleus.

Because the very earliest stages of cleavage of the chick embryo occur within the body of the hen, our knowledge of them is poor. During oogenesis, the cytoplasm has streamed to one side of the oocyte. The cleavage furrows however scarcely extend into the yolk in the bird embryo and the cytoplasm becomes a flattened disc sitting at the surface of the yolk. By the time the egg is laid, two regions can be distinguished. These are best seen in the blastoderm if it is dissected from the yolk and examined by transmitted light. They are the area pellucida, an inner relatively clear area, and the area opaca, the outer, less transparent region (Fig. 3). The difference in appearance is due to the fact that many large intracellular yolk droplets are contained by the cells of the area opaca, whilst the yolk droplets in the cells of the area pellucida are smaller and fewer. The area pellucida gives rise to the embryo proper whilst the area opaca develops into the yolk sac. At one side of the blastoderm a further region can be distinguished, the "embryonic shield", and this marks the future posterior end. It is here that the primitive streak (p. 260) begins to form.

Although, like birds, all mammals have internal fertilization our information about the events involved in the process is not so limited, for it is possible to fertilize and maintain mammalian eggs in tissue

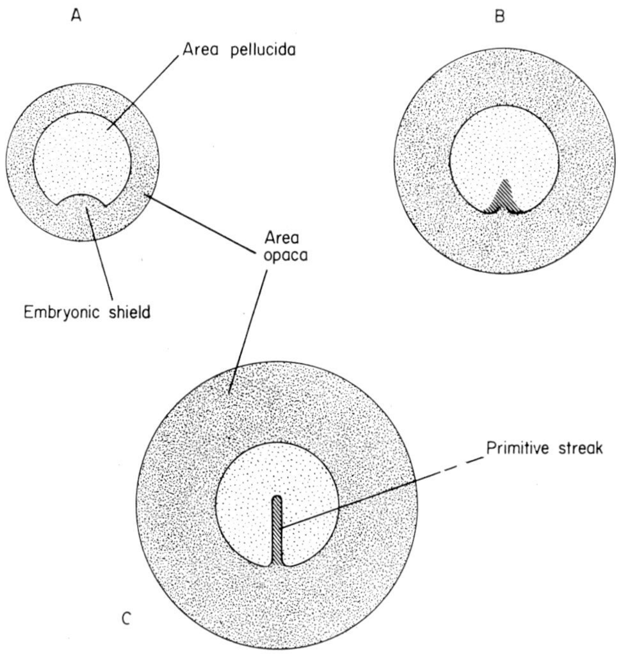

Fig. 3. Early stages in the development of the chick embryo. The blastoderm has been dissected from the surface of the large extra-cellular bag of yolk and is viewed by transmitted light. A: Pre-primitive streak stage. B: early primitive streak stage. C: Full grown primitive streak.

culture. Indeed, development will take place under artificial conditions right up to the time of implantation. Most of the investigations have been performed on mouse embryos. Because the egg is almost yolkless, cleavage divides it into a ball of cells which are all of nearly similar size. These are known as the blastomeres and the embryo is called a morula. A cavity, the blastocyst cavity, develops within the embryo and becomes filled with fluid. When it has appeared, the embryo becomes known as a blastocyst. If a section is taken through it, the inner cell mass can be seen at one end (Fig. 4). It is from this region that the embryo will form. The wall of the blastocyst is the trophoblast which forms much of the placenta.

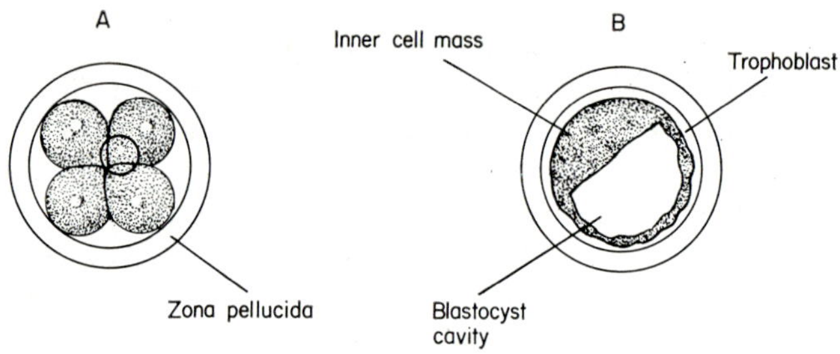

Fig. 4. Early stages in the development of a mouse embryo. A: 4-celled stage with polar body. B: Blastocyst stage.

We know little of the factors that determine when an embryo shall switch from cleavage to gastrulation. It has been established, in some species at least, that there is a surge of protein synthesis which coincides with the onset of gastrulation. Experiments on echinoderms and amphibians have shown that this increase is under the control of events that occurred during cleavage. Thus, if cleaving amphibian eggs are treated with actinomycin D (which inhibits the production of RNA) they continue to divide, but they fail to undergo normal gastrulation (Brachet *et al.*, 1964). This suggests that the RNA necessary for the protein synthesis that takes place in gastrulation is synthesized during cleavage.

The main characteristic of gastrulation is that cells become shifted around so that by the end of gastrulation they lie in the appropriate positions ready to start development into the organs of the body. The morphological result of gastrulation in all vertebrate species is that the endoderm and mesoderm come to lie internal to the ectoderm. In amphibians the endoderm and mesoderm invaginate through the blastopore, in amniotes through the primitive streak. The shifting of the cells is brought about by massive streaming migrations. The two main types of movements are the process of invagination (the tucking inwards of the cells beneath the outer layers), and epiboly (the stretching of the cells on the surface of the embryo).

Our knowledge about the routes taken by the migrating cells comes from experiments in which marks of various kinds have been applied to the surfaces of early embryos. The sort of marks used in the past were particles of carbon jabbed into the cells, or relatively harmless stains (vital dyes) such as Nile blue sulphate or neutral red. Nowadays, however, a more sophisticated technique involves the use of radioactive tracers. A piece of tissue is removed from a normal untreated embryo and replaced by a graft taken from the identical region of another embryo which has been treated with tritiated thymidine. The nuclei of the cells in the graft have incorporated the tritiated thymidine into their DNA and have thus become radioactive. After the graft has been implanted, the host embryo is allowed to develop for a further period and is then examined by autoradiography. The tissues derived from the graft can then be identified by the density of the silver grains lying over them.

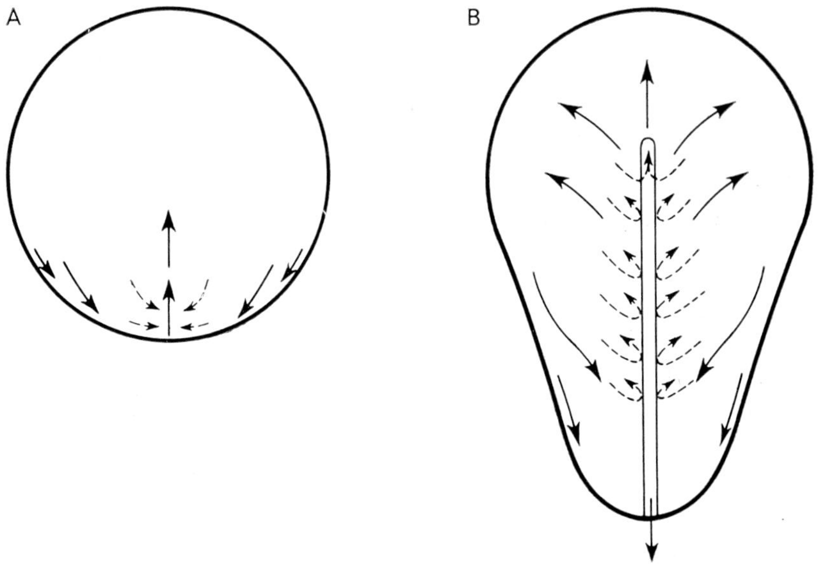

Fig. 5. Cell movements during gastrulation in the chick embryo. A: Short primitive streak stage. B: Full length primitive streak stage. The patterns of these movements were analysed by marking regions of the embryo with vital dyes (e.g. Nile blue sulphate). The unbroken arrows indicate the movements of the ectoderm. The broken arrows show the movements of the mesoderm. (After Pasteels, 1937.)

As gastrulation takes place the marks become shifted into new positions. It is thus possible to plot out the paths taken by the moving cells (Fig. 5). When this type of experiment is repeated for all regions of the early embryo, it becomes possible to recognize a pattern of movements taking place which is consistent for any species.

Further information obtained from these marking experiments is about the fate of the migrating cells. Thus we can say that a specific group of cells will move toward the blastopore, will invaginate and subsequently give rise to notochord. Another group of cells however will remain on the surface and be fated to develop into epidermis or perhaps neural cells. In other words we can deduce from our experiments the likely fate of any patch of cells present in the late blastula. We can draw maps of the future development of these regions (Fig. 6). We call these "presumptive maps", the presumption being that development will take place normally. We must not forget however that these are regulative eggs (see p. 251) and are thus capable of modifying their development.

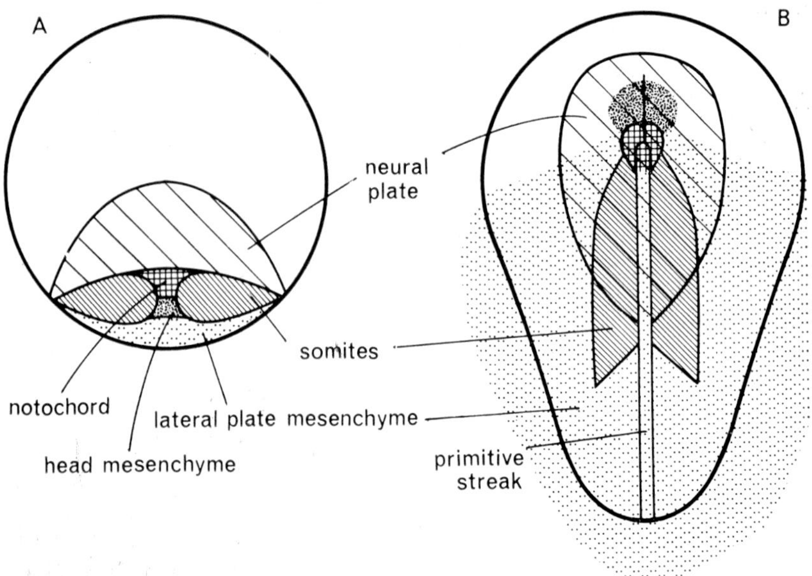

Fig. 6. Presumptive maps of the chick blastoderm. A: Short primitive streak stage. B: Head process stage. (After Pasteels, 1937.)

One of the main ways of investigating the differentiation of embryos is that of experimental embryology. This usually involves interfering with the normal development of the embryo by such techniques as extirpation of parts of it, or the transplantation of cells from one region into another. The importance of the presumptive maps is that they enable us to assess how far our experiments have influenced the development of the tissues. In this way it is sometimes possible to make deductions about the influences that one tissue may have upon another during development.

Once gastrulation has occurred, the tissues are ready to begin differentiating into simple organs. They do so in an orderly sequence which starts at the head end and gradually moves towards the tail end. In amniotes, the most anterior end of the primitive streak is known as the node (Hensen's node, primitive node, primitive knot) and the presumptive notochord is located in the mesoderm of this region. There is a short forward migration of some of these cells to form the head process, the most anterior part of the notochord. In the chick embryo however, the rest of the node has been shown to move back along the primitive streak, laying a trail of notochord as it goes. Thus the entire notochord is probably formed solely from the cells which initially occupied the node region.

The posterior migration of the node is not an isolated event, for it is accompanied by similar though lesser movements of cells in the regions on either side of it (see Fig. 9b). There is thus a streaming of cells towards the posterior end of the area pellucida, with the most pronounced movement occurring in the midline. Accompanying this mass migration, and perhaps even the result of it, is a general elongation and stretching of the area pellucida. At the same time the neural plate, the somites and the lateral plate mesoderm appear.

As we shall see, this backward migration of the node probably plays an important part in controlling differentiation of the embryo at this stage. It is of especial interest since there is nothing that strictly corresponds with it in amphibians; although like the chick embryo, the amphibian undergoes a stretching and lengthening of the embryo at this time.

It is generally assumed that the ancestors of modern mammals laid yolky eggs similar to those of birds and modern reptiles. We have seen that the type of cleavage undergone by modern mammals is very unlike that of birds. The mode of gastrulation however remains similar, for a flat blastoderm forms, just as if a large volume of yolk was still present, and a primitive streak differentiates. Marking experiments show that some at least of the morphogenetic movements which take place through the primitive streak are similar to those occurring in the chick (Fig. 7). So little work has however been carried out on young mammalian embryos that we do not yet know whether regression movements take place, though there is some indirect evidence that they do (Deuchar, 1969).

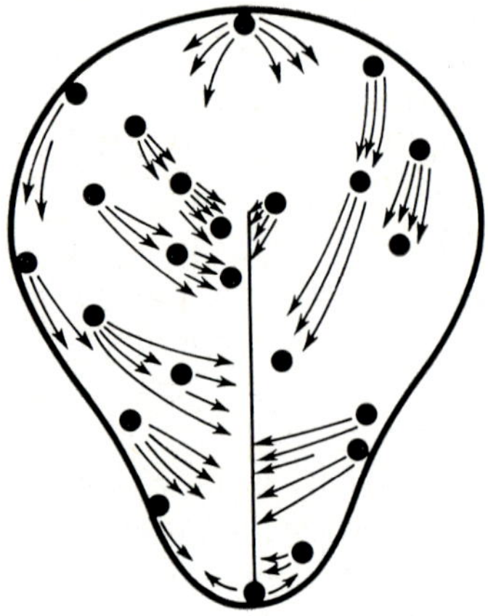

Fig. 7. Cell movements during gastrulation in the rabbit. The black circles indicate regions marked with neutral red. The arrows show the directions in which the marks were displaced. (From Daniel and Olson, 1966.)

The result of gastrulation is similar in the mammals to that of amphibians and birds. That is, an embryonic axis consisting of neural plate, notochord, somites and lateral plate mesoderm becomes laid down.

VI. Controlling Processes

Having briefly reviewed the basic events of early development we are left with a number of problems. First, we need to know about the processes which control these cell movements and the role that these movements themselves play in controlling development. We must then return to discuss the two major ways of looking at development that were mentioned on p. 257, the concept of control by a dominant region and the alternative hypothesis of multiple interaction of cells.

A. Cell Movements

The mechanisms of cell movement

Most of our knowledge about the mechanisms of cell movement is derived from the study of cells in tissue culture. Two main techniques have been used. In the first, small pieces of tissue such as chick heart are grown in hanging drop culture, and the cells that migrate from the explant are filmed by time-lapse cinematography. By examining the film it is possible to make deductions about the behaviour of individual cells, and of populations of cells in culture. The other method is to disaggregate the tissue into individual cells and to study their behaviour, either alone or in combination with other types of cells.

Several kinds of cell behaviour have been discovered which may also be characteristic of cells migrating within the living embryo. The first of these is *contact guidance*. This principle was originally put forward by Weiss (1934) who found that the fibroblasts move along grooves or lines of stress in the culture medium. A number of situations are known in embryos where one type of cell migrates over another. For example, in the early amphibian gastrula, the invaginating notochord migrates beneath the enveloping ectoderm. In the young chick embryo the migrating patterns of the presumptive heart mesoderm are thought to be guided by the morphological characteristics of the underlying endodermal cells. The latter are spindle-shaped and appear to "direct" the migrating mesodermal cells to their appropriate region (DeHaan, 1965).

Contact guidance probably also operates at the extreme periphery of the area opaca. The blastoderm lies at the surface of the yolk, and is maintained there by means of its attachment to the inner surface of the vitelline membrane, the transparent elastic sac which encloses the yolk of the hen's egg. The blastoderm is attached to the vitelline membrane only by the outermost edge of the area opaca. The vitelline membrane is composed of two main layers which can be distinguished both biochemically and by light and electron microscopy. It is significant therefore that the edge cells are capable of migrating only over the inner and not over the outer surface of the vitelline membrane (New, 1959). They thus appear to have a high degree of specificity for this inner surface and this is borne out by the fact that their powers of migration over other substrates are inferior.

Other situations in which contact guidance may play a role are in the migration of individual cells through the body of the embryo, such as the cells of the neural crest, and the migration of spinal nerves from the central nervous system to the organs which they innervate. Nerves are frequently associated with blood vessels and it seems possible that the outgrowth of the nerves may sometimes be guided by the layout of the blood vessels (Hughes, 1968).

It has been found that most cells if removed from the body and placed in suitable conditions are capable of migrating. One of the conditions that is necessary is that there should be a suitable surface (usually called the substrate) on which they can move. Cells suspended in a fluid medium are not capable of directed movements. In the living embryo, cells are probably always presented with a substrate.

Most cells migrating in tissue culture are under some tension. This appears to be necessary for contact guidance; it is interesting therefore that the chick blastoderm is also under tension and if this tension is released the ability of the edge cells to migrate is greatly reduced.

The second major principle of tissue culture is *contact inhibition* and this also may have some relevance within the living embryo. This concept was put forward by Abercrombie and Heaysman (1954) to explain the fact that certain cells in culture will stop moving (at least temporarily) when they make contact with another cell. If the cell becomes surrounded by other cells it ceases to move at all. It is not clear what the term "contact" means in electron microscopical or biochemical terms. It is merely a term which was introduced to describe contact that is visible by light microscopy. Usually this means that there is contact of the so-called ruffled membranes. A ruffled membrane is a thin, fan–like membrane which appears at the leading edge of many migrating cells. It receives its name from the fact that waves of ruffling and folding of these membranes occurs. Ruffled membranes have been discovered primarily by means of time–lapse cinematography of tissue cultures, and this has also revealed that ruffled membrane activity becomes greatly reduced or even ceases with contact inhibition.

Contact inhibition probably also occurs within the living embryo. It seems to occur where epithelial sheets meet during the closure of wounds. Ruffled membranes can certainly be seen in films of the edge of the chick blastoderm as they migrate beneath the surface of the vitelline membrane (Bellairs *et al.*, 1969).

Abercrombie (1966) has suggested that most cells in the body are subject to contact inhibition, and that the ones that are not are the ones that migrate within the body and infiltrate other tissues. He suggests that contact inhibition results in the trapping of moving cells so that they fill up empty spaces within the body. Thus, he postulates that a cell will migrate if it is not completely surrounded by other cells, but that once it does become surrounded it will cease to move. This is one of the most stimulating ideas in developmental biology, but it should be emphasized, that firm evidence of contact inhibition in any situation *in vivo* has not yet been produced.

Indeed, there are many examples of cell movement in the embryo where it seems unlikely that contact inhibition can be at work. During the invagination of mesoderm cells through the primitive streak itself each cell is closely surrounded by similar cells. It would appear to be the ideal situation for contact inhibition—and yet the cells keep moving. Similarly, the primordial germ cells and the neural crest cells also migrate through the body before they reach their destination. Certain organs are composed of cells from two sources such as the liver, which has an endodermal and a mesodermal component. It seems unlikely that these cells can be subjected to contact inhibition.

A third way by which cells may be guided during development is by *chemotaxis*. This supposes that some cells migrate along an increasing (or alternatively, decreasing) gradient in the chemical concentration of some component. The sort of situation where this has been envisaged is in the migration of the primordial germ cells towards the developing gonad. The concept of chemotaxis has perhaps found less favour with cell biologists than the ideas of contact guidance or contact inhibition. It may be noted however that there is little positive evidence that any of the three play the sort of role within the embryo that they seem to play among cells in the artificial conditions in tissue culture.

The control of differentiation by the cell movements

The movements of the cells during gastrulation are essential in bringing the developing tissues into the correct positions in relation to one another. In at least one other situation (viz. the early development of the somites in the chick) the movements themselves seem to act as controllers of differentiation, for without them the somites do not develop properly (discussed on p. 271).

Examples can however be found in which differentiation is not controlled by the cell movements. In the chick, the heart is formed from two primordia which move together and fuse in the midline. If the two halves are prevented from fusing, two separate, small, hearts develop. That is, each part has differentiated on its own. Thus, the movements of these cells to form the organ are not essential for the differentiation of the tissues.

B. THE IDEA OF A DOMINANT CONTROLLING REGION

The fact that the eye cup could control the development of the lens was discovered as early as 1901, but it was not until about twenty years later that Hans Spemann, working in Freiburg, made his profound discoveries about the dorsal lip of the blastopore. In this experiment he took a piece of tissue from the dorsal lip of the blastopore of an early amphibian gastrula and grafted it into the flank region of another embryo (Fig. 8a, b). The result was that an additional embryonic axis formed in the host embryo. Thus, the host embryo now possessed its normal axis and an additional one in the flank region.

When he repeated the experiment he took his graft from a species which was lightly pigmented and implanted it into an embryo of a highly pigmented species. It was then possible to distinguish those tissues which had developed from the graft from those which had formed from the host. It was apparent that both graft and host tissues had co-operated in producing this embryonic axis (Fig. 8c). Grafts taken from other regions of the donor did not lead to the same result.

Spemann concluded from these experiments that the graft had influenced the host tissues to develop in this way and he deduced that the dorsal lip of the blastopore was a dominant controlling region. He called it "the organizer"—or primary organizer to distinguish it from other controlling regions found at later stages. The term "inducer" was also introduced at about the same time. To some extent it is interchangeable with the term "organizer", though a little more specific. It is generally used to refer to one tissue which acts upon another. The eye cup is an inducer because it acts upon (induces) the ectoderm with which it comes into contact to form a lens. The terms acting and reacting tissue are also used, the eye cup being an acting tissue, the ectoderm being a reacting tissue.

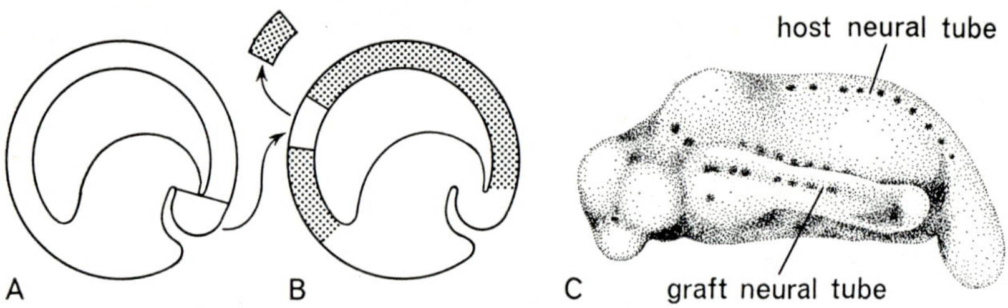

Fig. 8. The classical "organiser" experiment. A. and B. A graft from the dorsal lip of an amphibian embryo was implanted into the flank of another embryo. C. A secondary embryonic axis formed. (After Spemann, 1938.)

The history of the studies that followed Spemann's discovery is an instructive one for those interested in the effects of chance and fashion on scientific research. It is understandable that most of the investigators of the period assumed that the organizer region (the presumptive chorda mesoderm) carried out its domineering task by means of a chemical of some sort which it emitted. (Confusingly, this supposed chemical was also called the "organizer" by many, though not by Spemann himself.) Search for this substance thus became a preoccupation in many biological laboratories all over the world. In particular, attention was concentrated on the problem of trying to find out the nature of the supposed chemical substance emanating from the presumptive chorda mesoderm during neural induction. It was soon discovered that other tissues could also bring about neural induction if implanted beneath flank ectoderm. Extracts of many tissues were therefore investigated in an attempt to identify the responsible component. But, disappointingly, it proved an impossible task for not only did almost every diffusible chemical investigated have this effect but even inert substances such as fragments of glass produced a similar result. An account of this early work was published by Spemann in 1938.

Eventually, most investigators became discouraged and abandoned their search. Work continued in a few laboratories, however, notably that of Toivonen and Saxén in Helsinki. Much of the post-war research is discussed in a book by Saxén and Toivonen (1962). Perhaps the main difference between the ideas of the pre-war workers and those of Saxén and Toivonen is that the earlier theories supposed that induction was brought about by the diffusion of a single chemical substance, but Saxén and Toivonen concluded that several inducing substances were present.

One of the problems have always been to understand how the neural tissue comes to show regional differences along its length. Some workers supposed there to be three single substances emanating from different parts of the chorda mesoderm, each responsible for inducing a specific region: i.e. a head organizer, a trunk organizer, and a tail organizer. Saxén and Toivonen however suggest that there are probably two substances and that these are each produced all along the archenteron roof, but that they differ in their distribution. The regionalization of the neural tissue is thus determined by the relative proportions of the two substances.

With the excitement of Spemann's discovery, scientists were stimulated to try to find out if the supposed organizer was a characteristic only of amphibians, or whether something comparable was present throughout the animal kingdom. In 1932, Waddington showed that similar results could be obtained in chick embryos. If Hensen's node (the primitive knot) was cut out of a chick at the primitive streak stage and grafted into the area pellucida of another embryo of the same stage, a secondary embryonic axis was often induced there. Similarly, it was found that if the corresponding region of the primitive streak of a rabbit embryo was grafted into the area pellucida of a chick embryo, it too could induce a secondary embryonic axis. For a summary of many of the experiments on amniote embryos, see Waddington (1952b).

As a result it appeared that the anterior end of the primitive streak was the equivalent of the amphibian organizer. This seemed at the time to be a particularly satisfactory idea, since each is composed largely of presumptive notochord, and both the blastopore and the primitive streak are regions through which mesoderm invaginates.

This concept of the node as the organizer still holds sway in certain quarters. Spratt, in particular, has carried out an impressive range of

But it is known from many experiments that regression movements can occur in the area pellucida even when the node region is not present. Thus the area pellucida cannot be entirely under the control of the node. Indeed, there is evidence that the converse is true. Abercrombie (1950) demonstrated that the node itself was controlled by the area pellucida. He extirpated the primitive streak and replaced it in the area pellucida with reversed orientation. It might be expected from this experiment that if the node were truly the dominating region, the embryo would develop with reversed orientation: its head forming in the new position occupied by the node and its posterior end developing in the place originally occupied by the node. Contrary to expectation, however, the embryo developed as if no operation had been performed at all. The original node had apparently been suppressed in its new position in the area pellucida and a new node was presumed to have formed.

We may conclude that the node owes its properties largely to its position in the area pellucida and that the regression movements, and consequently differentiation of the entire axis, are ultimately under the control of the area pellucida itself. Furthermore, we can terminate this section on the concept of the dominant controlling region by the idea that, although such regions do appear to exist, they are probably always themselves the result of a cell to cell interaction, and that they owe their properties to their position in the early embryo.

C. THE IDEA OF CONTROL BY CELL POPULATION

Intertwined with the idea that development is controlled by a dominant "organizer" region, has been another concept. This supposes that there is some sort of overall pattern in the embryo brought about by a generalized interaction between all its cells. This pattern thus affects every cell, and each cell affects every other.

Embryonic Gradients

Perhaps the best known example is that of the *embryonic* gradients. It is not difficult to define a gradient in a generalized way. One can say, for instance, that it is the distribution of a substance that is present in large amounts in one region, the high point of the gradient, and that the concentration gets less and less as we get further and further in all directions from that point. The most obvious example is probably the yolk gradient of amphibian eggs. Here the high point of the yolk gradient lies at the vegetal pole, and the low point at the animal pole (Fig. 2). There is also a gradient in the distribution of the cytoplasm which is the obverse of that of the yolk. Here the high point is at the animal pole and the low point at the vegetal pole. The yolk gradient is considered to play an important role in establishing the dorso–ventral polarity of the amphibian embryo. This was shown by experiments in which the amphibian eggs were turned over into abnormal positions so that the yolk shifted. The surface of the egg, the so-called cortex, also moved, and it was found that a new blastopore developed in the region where yolk came into contact with the gray crescent (Pasteels, 1964). Another gradient that has some meaning is that of *r*RNA. In amphibian embryos (Fig. 10) there is a high point at the animal pole and a low point at the vegetal pole (Brachet, 1967). This gradient is, however, probably a reflection of the distribution of the cytoplasmic gradient.

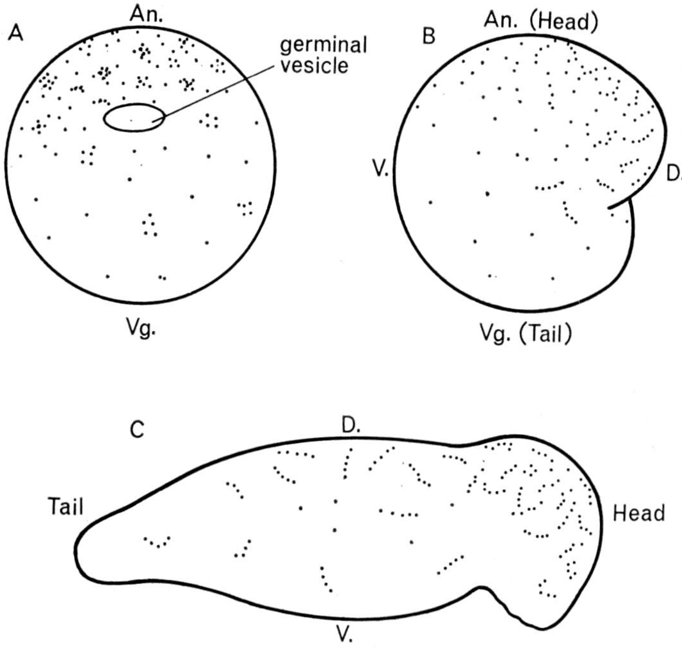

Fig. 10. Gradients of *r*RNA. A: In the unfertilized amphibian egg the ribosomes are more numerous toward the animal pole (An) than toward the vegetal pole (Vg.). Some ribosomes are clumped together to form polysomes. B: In the gastrula the ribosomes are more numerous near the dorsal lip (D). C: In the neurala the ribsomes are most highly concentrated in the head. Most are in the form of polysomes. (After Brachet, 1967.)

Other gradients which have been demonstrated are usually related to the ability of the different regions of the embryo to respond to obnoxious substances. For example, if the chick blastoderm is treated with various poisons, there is usually a graded response, the most sensitive point being Hensen's node. There is no evidence, however, that such discoveries bear any relationship to the normal physiology of the developing embryo.

The concept of gradients has often been invoked to explain both normal development and the response of the embryo to unusual situations. For example Saxén and Toivonen have suggested that there is a gradient in concentration along the archenteron roof of each of the two inductive substances, one having a high point at the head and, the other at the tail end (see p. 269). Unfortunately although many theories put forward are frequently ingenious, and therefore highly appealing, they are often impossible either to prove or to dismiss. One reason is that the concept has been used in a vague and imprecise way. If it is not possible to specify the exact nature of the chemical, then it becomes impossible to test the hypothesis directly.

Grobstein's ideas on cell populations

A second example of the idea that control of development is distributed among the population of cells is that of Grobstein on the sizes of cell populations. This author has studied the differentiation of tissues explanted from a variety of organs in tissue culture. In particular he used explants of developing kidney and pancreas. He found that development proceeded best when the explant was above a minimum size. Below this size differentiation did not take place. As a result he postulated that a minimum mass of cells was necessary if differentiation was to occur in the tissue (Grobstein, 1966). The situation is a complex one, however since if the mass of undifferentiated pancreas cells is increased, this does not lead to precocious differentiation (Wessels, 1968).

"Positional Information"

The term "positional information" was coined by Wolpert (1969) in an important theoretical paper. Previously, a number of authors had independently suggested that each part of the embryo was somehow "aware" of what was happening in other parts of the body. If this were not so, then cells would be unable to respond in an orderly manner, and the developing embryo would be a disorderly jumble of cells and tissues (as is to some extent the situation in the type of tumour called a teratoma). From this arose the idea that the individual cell must in some way be "aware" of its own position in the tissue, and in the body. Similarly, cells in tissue culture must each be able to assess their position in the explant.

Such an assessment implies that each cell is continually receiving information from other cells. The sort of information it receives will depend on the neighbouring cells. For example, a mesoderm cell which lies in the presumptive somite region is in contact with similar presumptive somite cells, and the whole group of cells is overlain by ectoderm and underlain by endoderm. Bonner (1965) has suggested that each cell in the developing embryo is constantly carrying out tests to find out such things as whether it lies close to another cell, or whether it is in contact with another cell, or whether it lies against a basement membrane. Wolpert has elaborated this idea and has proposed that each cell may have its position specified with respect to one or more points in the developing system. This idea implies that each cell continually takes a bearing on its position. In this way it always "knows" its position; it has, in Wolpert's terminology, "positional information".

The idea of a continual test has led to a further supposition: that there is a rhythm in development so that testing can take place at regular time intervals. Goodwin and Cohen (1969) have suggested that within every differentiating system there is a periodically recurring event which synchronizes the cells and so enables them to measure time. They say: "It is as though every cell has access to and can read, a clock and a map."

In the same paper Goodwin and Cohen (1969) postulated that, periodically in each cell, a brief signal, involving a number of biochemical reactions, is transmitted to neighbouring cells. These would perhaps be able to respond or perhaps ignore the signal, depending on their own condition at the time of receiving it. These authors liken the cells in such a system to a set of coupled non-linear oscillators, and point out that such a system must be controlled by the fastest oscillator among them. They suggest that the interactions between the cells in an embryonic tissue are sufficiently strong to result in the establishment of a single frequency of the periodic event in all the cells of the tissue. Another idea is that the regular rhythm of mitosis provides the timing mechanism for the embryo (see p. 256).

Ideas of this sort, though still unproved, are stimulating ones providing new ways of looking at embryonic differentiation. They enable us to see that the path of differentiation a cell will follow is the result of three factors: that is, of its own genetic constitution, of the development it has already undergone, and of its position within the embryo and within the tissue.

The concept of regulation has already been discussed (p. 251). The idea was originally put forward to explain the response of cells or tissues to injury or to experimental interference. Viewed in modern terms we may now consider regulation as the response of the cells to the continual feedback of information from neighbouring tissues.

Indeed regulation may now be seen to include all those minute shifts of protein synthesis and of cell behaviour that apparently take place in a normal differentiating embryo.

Even the idea of embryonic induction may also be brought under the umbrella of this hypothesis. The implication here is that a dominant controlling region, such as the dorsal lip or Hensen's node, becomes dominant because it is situated in a particular position in the embryo. Because of all that has happened in the embryo during cleavage and gastrulation, the cells of the dorsal lip are able to interpret and respond to their new position in a particular way, by inducing the neighbouring tissues. Indeed, it is possible that all the results of experimental embryology which have been taken as evidence of embryonic inductions, could now be reconsidered as evidence merely of positional information! This idea is an appealing one since the concept of the organizer and of other dominant regions has perhaps carried somewhat mysterious connotations in the past. Not all developmental biologists would agree, however, that the organizer concept should be dropped, and they argue that there is as yet no more positive evidence in favour of Wolpert's theory than there is in support of the organizer concept. Nevertheless, although we should not forget these criticisms, these new ideas are important because they have led to a reconsideration of the mechanisms by which development is controlled. Control of differentiation may now be considered to be a more broadly based phenomenon than was formerly suspected.

VII. The Ways in Which Cells May Communicate with One Another

We have seen that throughout the entire embryo cells appear to be continually communicating with, and responding to, each other. Unfortunately, we know little about the ways in which information is passing from one cell to another. The search for chemical inducers has been disappointing (p. 268) and the evidence for chemotactic stimuli is slight (p. 267). In the early stages of development, before the endocrine organs form, there can be little control of development by hormone secretions, except possibly in mammalian embryos where certain hormones may pass from mother to offspring across the placenta. There is, however, evidence that the cells in the young embryo may interact with each other in at least two different ways. The first is by the passage of ions from cell to cell. The second is by the rearrangement of the molecules at the surface of the cell as a result of changes in the neighbouring cells.

A. IONIC EXCHANGE

It has now been demonstrated in a variety of types of cell, both adult and embryonic, that there is sometimes a passage of ions from one type of cell to another at the surfaces where the cells are in contact. This has been shown by experiments in which microelectrodes have been inserted into individual cells (discussed in Chapter 22). Sheridan (1966) inserted micro-electrodes into a neural plate cell and a notochord cell in a young chick embryo. Although the two cells were widely separated from each other by other cells, it was found that current flowed from one to the other, thus indicating that there was a passage if ions through the tissue.

It appears that in the early embryo most of the cells in the organism are linked in this way. As differentiation proceeds, however, the numbers of pathways seems to become reduced. We can interpret this by supposing that in the earliest stages all cells must be in communication and the organism is therefore capable of widespread regulation. By the time the organs have become partially differentiated, however, such widespread communication is not only less necessary, but also less desirable since greater stability in behaviour and differentiation is now required. Thus the possibilities for communication become more restricted.

Attempts have been made to correlate certain morphological regions at the surface of the cell with the passage of ions from cell to cell. Most authorities now favour the idea that it takes place at the region of so-called gap junctions (Bennett and Trinkaus, 1970). But so far we have little clear evidence. The problem is discussed further in Chapter 3.

B. THE CELL SURFACE

There is some evidence that the cell cortex, (i.e. the outermost region of the cell, including the plasma membrane) plays an important role in the early differentiation of the amphibian egg (Curtis, 1967). The cell surface is also important in that it is the first part of a cell to be affected by any change in the environment. It seems likely that it is the first part of the cell to respond, possibly by a chemical reaction, or even by a realignment of existing molecules. Unfortunately, we know very little about such changes though we can imagine that they play a crucial role in the chain of events which culminates in the switching of protein synthesis in differentiation.

VIII. The Role of Mitosis in Differentiation

A stimulating idea that has been put forward recently is that mitosis plays an important role in controlling differentiation. Before discussing the evidence, let us first consider the part played by mitosis in the general growth, or increase in mass, that is so essential a part of development.

Mitosis takes place at a rapid rate during cleavage and gastrulation. In the chick embryo at the primitive streak stage, mitotic figures can be seen in about 5–7% of the cells at any one time, though the rate varies a little from tissue to tissue. After this early period the mitotic rate falls throughout the embryo. This low rate is not necessarily maintained for the rest of the embryonic period, for localized bursts of mitosis play an important part in the gross structuring of certain organs. The heart of the chick embryo owes much of its shape to regions of increased miotic rate (regions of localized growth).

Less obvious, however, is the possibility that the process of mitosis affects the differentiation of the cells. One of the first exponents of this view was Balinsky (1955) who reported that during embryonic induction there was an increase in the mitotic rate of acting and reacting tissues. In particular, he found that the mitotic rate was raised in tissues when they are stimulated to form a limb, but that there was also a high rate of division in the regions capable of inducing limb development. Similarly, there appears to be an increase in the mitotic rate in the splanchnic mesoderm just before the foregut develops in the chick (Bellairs, 1957). The splanchnic mesoderm is thought to induce the foregut.

Recently, more elaborate studies have indicated that in a number of tissues there is a drop in the mitotic rate with the onset of differentiation. In myoblasts the rate of cell division falls abruptly when they start to manufacture myosin. Similarly, the rate drops in erythrocytes when they begin to synthesise haemoglobin. Findings of this type have led to two ideas: (1) That differentiation cannot take place during periods of rapid cell division and that this is because the two processes are incompatible. (2) That each tissue must pass through a period of rapid cell division before it can begin to differentiate.

It is not easy to test either of these hypotheses in the entire organism, since most tissues contain both dividing and nondividing cells and they frequently consist of several different types of cell. For these reasons, much of the work has been performed on cells growing in tissue culture.

We have already seen that there are a number of tissues whose differentiation can be measured by their production of special molecules. They include chondrocytes, whose early steps in differentiation are indicated by the appearance of chondroitin sulphate. Specialized molecules are similarly produced by the pancreas (amylase), by cardiac muscle (myosin), and by the lens (crystallins). Early experiments suggested that the production of specialized molecules could occur only in non-dividing cells. Improved techniques of tissue culture, however, have resulted in better differentiation of tissues *in vitro* and it is now apparent that under suitable conditions mitosis and differentiation are not necessarily incompatible phenomena: and so the first hypothesis has become less tenable.

The second hypothesis, that each tissue must pass through a phase of rapid cell division before it can begin to differentiate, has received more support. The virgin mammary gland will respond to treatment by prolactin and hydrocortisone not only morphologically, but also by producing casein. These responses only occur if the epithelium has previously undergone a burst of mitosis. Furthermore, if the tissue is treated with mitotic inhibitors such as colchicine the production of casein also fails to take place. Similar results have been obtained with other tissues such as the pancreas.

This type of experiment has led to the suggestion that cells become reprogrammed during the final mitotic division before differentiation (Holtzer, 1968). If this is so, then the final division is in some way a special one. We know little of the factors that may influence a cell to "decide" which division is to be the last before differentiation. But it seems likely that the environment may play a decisive role. In cells growing in tissue culture the composition of the culture medium may affect differentiation (see also Chapter 1). Similarly, the density of the other types of cells in the culture dish exert an effect on differentiation. Within the embryo, also, there is a continual change in the environment of each cell. As we have seen (p. 275) each cell is somehow able to assess these changes.

Recently, these ideas on the controlling role of mitosis in differentiation have been extended further. Most organs are composed of several distinct types of cell and these are usually arranged in a regular pattern. The retina of many vertebrates contains several different types of receptor and these are distributed in a repeating pattern. It appears that an orderly sequence of mitosis may play a role in the laying down of different cell types within the retina.

Morris (1970) who has studied the pattern in the chick retina has concluded that it may arise as the result of an orderly sequence of mitoses. The pattern she has described is a mosaic which is composed of several sub-patterns. One of these consists of rows of receptors which intersect at angles of 60° (Fig. 11). Each of these rows is composed of repeating sequences of 1 rod and 2 other types of receptor. Another sub-pattern is indicated by the broken lines: broken line 1 is a row of rods, broken line 2 is a row of single cones, broken line 3 is a row of double cones. A third sub-pattern which occurs through-

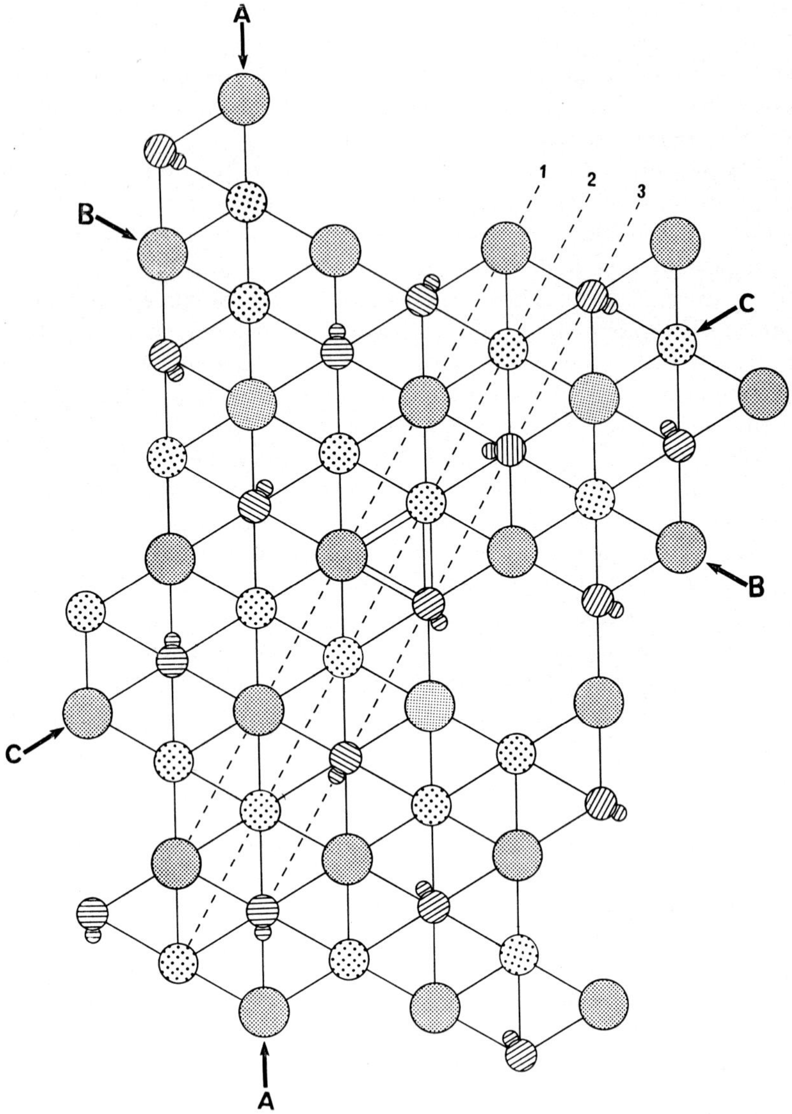

Fig. 11. Diagram of the pattern of receptors in the peripheral retina of the chick. Rows of receptors (e.g. AA, BB, and CC) intersect at angles of 60°. Each row is composed of a repeating sequence of one rod and two other types of receptor. Another pattern indicated by the broken lines. Broken line 1 is a row of rods. Broken line 2 is a row single cones. Broken line 3 is a row of double cones. A third regular pattern which visible throughout the lattice is a triangle of receptors consisting of one rod, one sing cone and one double done. (After Morris, 1970.)

out the mosaic is indicated with double lines and consists of 1 rod, one single cone, and one double cone. Morris regards this triangle as the basic developmental unit from which the overall pattern is constructed. She suggests that this unit is probably derived from a stem cell which underwent an orderly sequence of mitoses. At the first division the stem cell might give rise to a rod, at the second to a single cone and at the third division it might convert itself into a double cone.

IX. Concluding Remarks

During the last decade, ideas about the way that differentiation is controlled have undergone modification. Formerly it was thought that the first stages in the development of the organs were controlled by a dominant region, the organizer, and that it carried out its task by secreting chemicals which diffused into other cells. This idea is nowadays considered by many developmental biologists to be too simple a one. It has largely been replaced by the concept that the control of differentiation is directed by an interaction of cells throughout the embryo as a whole. These interactions may take place in a variety of ways. For example, the cells can affect each other as they migrate over one another; and they are influenced by the density with which they are packed together in a particular tissue. It is now supposed that each cell is able to assess the events that are taking place in the tissues that surround it. Probably it does so by analysing the changes occurring at its own cell surface. One recent theory carries this idea further and suggests that each cell "knows" its own position within the tissue and even within the embryo as a whole. It "knows" this because of the information it continually receives from the other cells and it responds to this information by differentiating in a way which is appropriate for its position, and for the stage of development of the embryo. In this way, differentiation may be controlled and integrated throughout the organism.

References

Abercrombie, M. (1950). *Phil. Trans. Roy. Soc. London Ser. B.* **234**, 317–318.
Abercrombie, M. (1966). *Nat. Cancer Inst. Monogr.* **No. 26**, 249–277.
Abercrombie, M. and Heaysman, J. E. M. (1954). *Exp. Cell Res.* **6**, 293–306.
Balinsky, B. I. (1955). "The Problem of Development". Inaugural Lecture. Witwatersrand University Press.
Bellairs, R. (1957). *J. Embryol. Exp. Morphol.* **5**, 340–350.
Bellairs, R. (1959). *J. Embryol. Exp. Morphol.* **7**, 94–115.
Bellairs, R. (1963). *J. Embryol. Exp. Morphol.* **11**, 697–714.
Bellairs, R. (1971). "Developmental Processes in Higher Vertebrates" Logos Press, London.

Bellairs, R., Boyde, A., and Heaysman, J. E. M., (1969). *Arch. Entwicklungsmech. Organismen.* **163**, 113–121.

Bennett, M. V. L. and Trinkaus, J. P. (1970). *J. Cell Biol.* **44**, 592–610.

Bonner, J. T. (1965). "Size and Cycle". Princeton University Press.

Brachet, J. (1967). *Compr. Biochem.* **28**, 23–54.

Brachet, J., Denks, H. and DeVitry, F. (1964). *Develop. Biol.* **9**, 398–434.

Britten, R. J. and Davidson, E. H. (1969). *Science*, **165**, 349–357.

Curtis, A. S. G. (1967). "The Cell surface: Its Molecular Role in Morphogenesis." Logos Press, London.

Daniel, J, C., Jr. and Olson, J. D. (1966). *Anat. Rec.* **156**, 123–128.

Davidson. E. H. (1968). "Gene Activity in Early Development". Academic Press, New York and London.

DeHaan, R. L., (1965). In "Organogenesis" (R. L. DeHaan and H. Ursprung. eds), pp. 377–420. Holt, Rinehart and Winston, New York.

Deuchar, E. (1969). *Acta Embryol Morphol. Exp.*, 157–167.

Goodwin, B. C. and Cohen, M. H. (1969). *J. Theor. Biol.* **25**, 49–108.

Grobstein, C. (1966). *Am. Zool.* **6**, 89–95.

Gurdon, J. B. and Woodland, H. R. (1968). *Biol. Rev. Cambridge Phil. Soc.* **43**, 233–267.

Hamburger, V. and Hamilton, H. L. (1951). A series of normal stages in the development of the chick embryo. *J. Morph.* **88**, 49–92.

Holtzer, H., (1968). In "Epithelial-Mesenchymal Interactions" (R. Fleishmajer and R. E. Billingham, eds), pp. 152–164. Williams and Wilkins, Baltimore.

Hughes, A. F. W. (1968). "Aspects of Neural Ontogeny". Logos Press, London.

Lyon, M. F. (1968). *Annu. Rev. Genet.* **2**, 21–52.

Morris, V. B. (1970). *J. Comp. Neurol.* **140**, 359–397.

New, D. A. T. (1959). *J. Embryol. Exp. Morphol.* **7**, 146–164.

Pasteels, J. (1937). *Arch. Biol.*, **48**, 381–488.

Pasteels, J. (1964). *Advan. Morphog.* **3**, 363–388.

Saxén, L. Toivonen, S. (1962). "Primary Embryonic Induction". Logos Press, London.

Sheridan, J. D. (1966). *J. Cell Biol.* **37**, C1–C5.

Spemann, H. (1938). "Embryonic Development and Induction". Yale University Press, New Haven, Connecticut.

Waddington, C. H. (1952a). *J. Exp. Biol.* **29**, 484–489.

Waddington, C. H. (1952b). "The Epigenetics of Birds". Cambridge University Press.

Waddington, C. H. (1956). "Principles of Embryology". Allen and Unwin, London.

Weiss, P. (1934). *J. Exp. Zool.* **68**, 393–448.

Wessels, N. K. (1968). In "Epithelial-Mesenchymal Interactions" (R. Fleischmajer and R. E. Billingham, eds) pp. 132–151. Williams and Williams, Baltimore.

Wolpert, L. (1969). *J. Theor. Biol.* **25**, 1–48.

Wolpert, L. (1971). *Current Topics Dev. Biol.* **6**, 183–224.

Note added in proof.

The concept of contact guidance has recently been re-investigated by Dunn (1973).

Dunn, G. A. (1973). In "Locomotion of Tissue Cells". Ciba Foundation Symposium **14**, 211–232.

15. Cell Growth and Morphogenesis

R. BELLAIRS
Department of Anatomy, University College London, England

In the previous chapter certain principles of development were outlined, and we saw that differentiation may be controlled in a variety of ways. In the present chapter we shall first consider the process of growth and will then outline the development of a few selected types of cells and tissues which illustrate special aspects of development.

I. Growth

We have already seen that growth and differentiation are closely related. In general, the growth rate is highest in an embryo during the earliest, post-cleavage stages, and it gradually declines throughout embryonic life and childhood. Similarly, differentiation is at its height during the early part of development, though in most organs it is usually completed long before birth. The enlargement and elaboration of the organs that occurs after differentiation may take place in several different ways (see below).

A. Definition

It is difficult to define the term "growth" with precision since it is commonly used to describe a range of very different situations. In this chapter we are concerned principally with embryonic and post-natal growth, and also with the growth of tissues under experimental conditions. The definition ought however also to embrace those growth processes involved in wound healing, in regeneration, in the growth of tumours, and in the general replacement and turnover of tissues in the adult body (Chapter 13). Compensatory hypertrophy, such as the compensatory growth of one kidney to make up for the loss of the other, must be included.

In all these growth processes there is an increase in the mass of tissue and this increase is brought about by cell division. At the same time a small number of cells die and are lost. Growth is therefore usually defined as being "the *net* increase in the mass of the system", which takes into account the fact that some tissue has been lost by catabolism.

Even such a broad definition as this however does not cover all the situations where the term "growth" is used. For instance, we speak of the epidermis as continually "growing" in the adult, and yet in this tissue there is normally no appreciable increase in mass. This is because the dead cells are sloughed off at the free surface at about the same rate as new cells are produced in the basal layer. This special type of growth is more appropriately described as "turnover" (Chapter 13) and is especially interesting since the rates of sloughing and of mitosis appear to be linked. We understand little about the nature of the controls involved however. (For a discussion of turnover in cells and tissues, see Goss, 1970.)

Another situation in which growth may occur without an increase in mass has been described in starving amphibian larvae. These animals may continue to increase in length (linear growth) whilst nevertheless losing weight. In this situation "growth" is probably due to the rearrangement of the available material, and certain organs may increase in mass whilst others decrease.

An organ or tissue may also "grow" by the accretion of cells or material formed elsewhere in the body. For instance, the deposition of material in the reticulo–endothelial system may lead to a measurable increase in mass. This type of "growth" is however not included in this discussion, which will be restricted to processes directly dependent on cell division.

B. TYPES OF GROWTH

There appear to be three possible ways in which a tissue or organ may grow by cell division. They are: by the renewal of cell populations, by the expansion of cell populations, and by the maintenance of static cell populations (discussion by Goss, 1964). The kinetics of cell production is discussed in Chapter 13.

In *renewing cell populations* (Fig. 1) there is a steady loss of cells, and these are continually replaced by division in a special region of cells, the stem cells. Examples include epidermis, spermatocytes and haematopoietic cells. In *expanding cell populations* (Fig. 1) there is a generalized growth by cell division throughout the tissue. This situation is found in many organs of embryos and juveniles, e.g. in the liver and spleen. In most organs the rate falls to a low level or ceases when the appropriate size has been reached, though it may be raised again under certain conditions such as in regeneration in the liver or in the formation of tumours. Individual cells may also divide in a tissue of this type to replace dead cells. Probably all the cells in an expanding population are capable of division and this may be correlated with the absence of stem cells, for all the cells are differentiated.

In *static cell populations* (Fig. 1) the number of cells remains unchanged, but some or all of the cells grow in size and may also change in shape. An example is the cells of the nervous system, where mitosis ceases after the initial stages of development, and changes in mass and shape take place largely through alterations to individual neurons.

In this section, attention will be concentrated on expanding cell populations. We will concern ourselves with the factors affecting the size and proportions of the developing individual and its organs.

C. THE OPTIMUM SIZE

In a normal animal, the size of each organ is proportional to that of the whole body. Furthermore, the ultimate size is one that is characteristic for the species, although some variation occurs from one individual to another. It appears therefore that there is an optimal size for each species or race. The limit in size is related to the physiology and morphology of that species and of that individual, and there frequently appear to be factors which place an upper limit on the size which can be achieved. It is often said that body surfaces may be a limiting factor in determining sizes of certain animals. For instance, it is thought impossible for insects to achieve the sizes of the larger land vertebrates because of the physical limitations of their respiratory mechanisms. It appears therefore that the size of any animal is largely determined by the evolutionary background of the species, as well as by the particular genetic constitution of that individual. Superimposed on these is the effect of the general environmental conditions experienced by the growing organism.

Much has been learnt by studying the effects of domestication. In some domestic animals great differences in size are apparent between different breeds of the same species. This is especially marked in horses

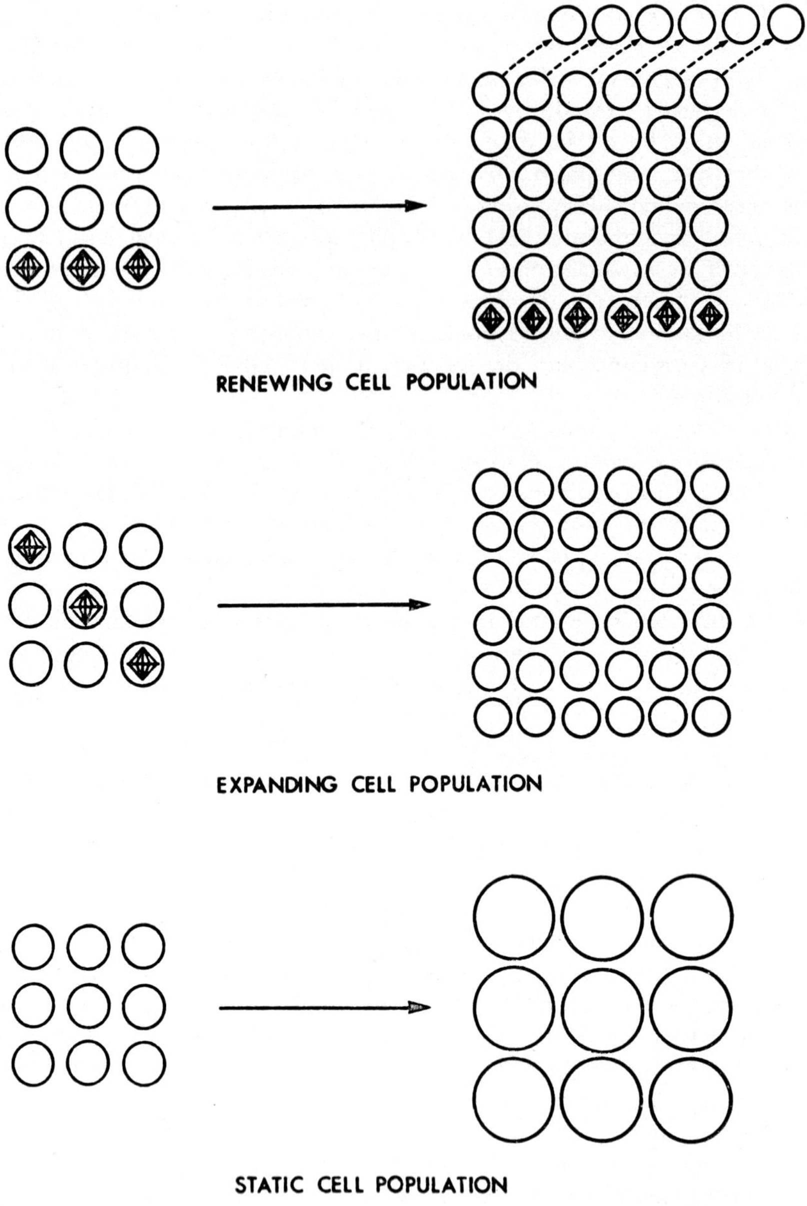

Fig. 1. Diagram to illustrate three different types of growth. In renewing cell populations there is a steady loss of cells which is made good by the division of a set of stem cells. In expanding cell populations there is cell division throughout the tissue. In static cell populations the individual cells grow in size (and may also change shape) but there is no cell division except in the earliest stages of development. (From Goss, 1964.)

and dogs. There is some evidence that larger dogs tend to have a greater development of the intestinal surface than smaller ones (Bertalanffy, 1960). The process of growth has also been compared in the embryos of wild ducks (Mallards) and their domesticated relations (Pekins). The Pekins are larger than the Mallards, but it appears that their pattern of growth has not changed with domestication: growth merely goes on for a longer period (Prince *et al.*, 1968).

Growth may be compared in different species by adjusting the separate growth curves to a single plot. When pre-natal growth was analysed in this way in fifteen species of birds and mammals it was found to follow the same pattern (Laird, 1966). The main difference was in the time scale, those species that were born at earlier developmental stages were born earlier on the normalized growth curve (Fig. 2).

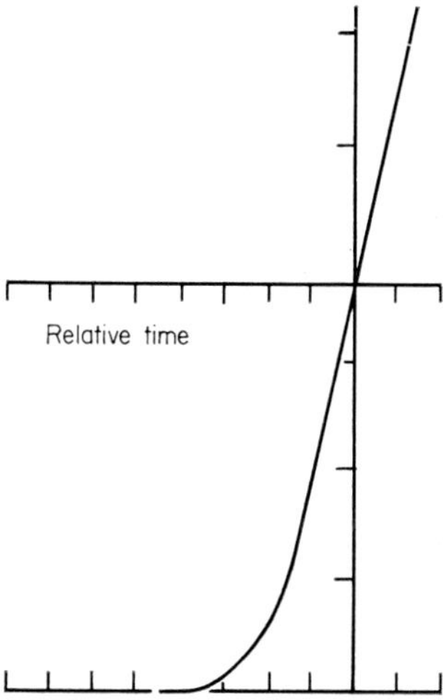

Fig. 2. A graph in which growth measurements of 15 species of birds and mammals have been superimposed after adjustment of the units of time and weight on the two axes. The growth curve follows much the same course in all the species.

The growth of cells and pieces of tissue has frequently been studied in tissue culture. Waymouth (1960) has reviewed the advantages and disadvantages of this approach and has pointed out that it allows the various processes to be both experimentally and conceptually separated. The changes in the cells and in the populations of cells can be seen and followed more readily *in vitro* than *in vivo*. The most important factors in the environment have been found to be adequate nutrition and a suitable temperature: without these, extensive proliferation will not take place.

D. CONTROL OF GROWTH

Many theories exist which attempt to explain how growth is controlled (reviewed by Goss, 1964). The earlier ones were usually based on the idea that growth was the direct response of an organ or tissue to functional overload. The more recent ones suppose that growth is controlled by humoral agents that pass from one cell to another.

One of the major difficulties is that attempts are frequently made to find a universal law that will explain growth control in all systems. It is often assumed that any theory put forward to suggest how embryonic growth is controlled must also be applicable to other situations, such as the control of growth in regeneration, or during organ hypertrophy or in response to experimental interference. This assumption seems naïve, since special conditions may operate in each circumstance. In cells growing in tissue culture growth is often limited by inadequate nutrition, but this factor is seldom critical in a living embryo.

In his book, Goss (1964) recognises three main types of growth theory. Basic to them all is the idea of a feedback mechanism, and this is imposed on the genetic background of the cells concerned. They are:—

(1) Control by tissue mass.
(2) Control by functional demand.
(3) Control by growth regulating factors.

Evidence in support of each theory has been obtained from a variety of experiments. No one theory has however been shown to be applicable in all situations.

(1) *Control by tissue mass*

This theory supposes that an organ stops growing when it has formed a certain, critical mass of tissue and, that if part of this organ (or tissue) is extirpated further growth will occur until the mass has been restored. It supposes that the feedback mechanism is a humoral (chemical) one and that it is not the direct result of a functional overload. It is therefore interesting to study the effects of removal of one of a pair of organs (such as the lachrymal glands) which act independently of one another. Each gland independently secretes into either the right or the left eye. Thus if the left gland is removed, the right one will be unable to take over its function, since it can only secrete tears into the right eye. Nevertheless, if the theory is correct, the right gland should be stimulated to proliferate. On the whole, however, the evidence suggests that there is no increased mitotic rate in the right lachrymal gland if the left is removed. As Goss (1964) points out the results are not clear-cut even when the discussion is restricted to experiments in adult vertebrates. We possess little information about the situation in embryos.

(2) *Control by functional demand*

This theory supposes that so long as an organ is being used it will continue to grow, at least until it reaches a certain size. Thus, if part of an organ (or one of a pair of organs) is extirpated the work load on the remainder is increased, and this stimulates growth of the organ. If, for example, the kidney is extirpated, excretion must become carried out entirely by the remaining kidney which undergoes hypertrophy.

Useful though this concept is in explaining the results of many experiments on adult organs, it has less value for explaining growth control in embryos. In many organs, such as the lungs and gonads, the size that will be achieved by birth is apparently determined when little if any functional activity is taking place. It seems unlikely therefore that functional demand plays a significant role in the early growth of these organs.

(3) *Growth regulating factors*

These theories suppose that specific chemical substances are passed from one part of the body to another. Some theories are based on the idea that the substances inhibit growth and others on the concept that they promote it.

One of the best known in embryos is the epidermal growth factor (E.G.F.) of Cohen (1965). This substance obtained from the submaxillary gland of mice, has been found to have a stimulatory effect on chick epidermis growing in tissue culture. It not only brings about an increase in mitosis but also promotes keratinization. Untreated control cultures show very little proliferation or keratinization unless dermis is present. The E.G.F. appears to act by stimulating the synthesis of proteins at the level of the ribosomes.

Hormones are perhaps the best analysed growth-regulating substances affecting post-natal tissues. Some hormones also play an active role in controlling the growth of organs in the late embryo. The pituitary growth hormone has been shown to be necessary in chicks and mammals since hypophysectomized embryos tend to be smaller than normal ones. In the absence of the pituitary the thyroids of chick embryos remain small and produce very little colloid. The testes of hypophysectomized chicks and rabbits are small and poor in interstitial cells. Similarly, there is evidence that the thyroid and gonads may be actively secreting hormones during this period (discussion by Bellairs, 1971).

It is often difficult to pinpoint the stages at which an endocrine organ begins to secrete in an embryo, largely because of the small quantities of material involved. It is clear, however, that secretion cannot begin until an organ has at least achieved *some* basic histological differentiation.

Other examples of growth regulating substances are the "chalones" of Bullough (1962) which are thought to inhibit mitosis. The evidence is based mainly on the fact that a substance extracted from the mouse epidermis will inhibit mitosis in mouse epidermis. Another example is the nerve growth factor of Levi-Montalcini. Similarly, humoral growth-regulating factors have been invoked to explain the results of experimental interference in the liver.

(4) *Control of Shape*

Closely related to the size of the developing organs and tissues is the control of the shape. This appears to be largely dependent on the shape and spatial relationships of the individual cells and of the intercellular materials. For example, the characteristic lentoid shape in the lens is due to a gradual increase in length of the individual fibres, the longest being in the fattest part of the lens, the shortest in the thinnest region (see p. 299).

The epigenetic factors that cause cells to change their shapes during differentiation are less well understood. There is evidence that pressures and tensions play a role in shaping the tissues of the eye (p. 302). Tension is also important in promoting normal development in the earliest stages. The young chick embryo is maintained in a state of tension beneath the vitelline membrane. If this tension is relaxed the embryo fails to grow at the normal rate, differentiation is poor and the shapes of the organs may be distorted. Tension effects are themselves related to the specific adhesions between cells. Thus the shapes of the cells and their communication with one another are intimately related.

As we have seen, function may affect the growth of an organ; it may also affect the shape. The pharyngeal arteries of amniotes are initially present on both left and right sides of the embryo. In birds, the systemic arch develops from the 4th arch on the right side, and the corresponding artery on the left side disappears. If the right arch is ligatured, however, the blood passes along the left arch which then adapts to form the systemic arch. Similarly, in the differentiation of the extra-embryonic blood vessels of the yolk sac, a series of possible channels is initially laid down, but the final shape and structure of the vessels depends on the details of the flow of the blood.

It has sometimes been suggested that the shape of the organs may be affected by areas of localized growth (Sissman, 1966). Conversely, localized regions of cell death appear to be associated with shaping the development of the limb bud in chick embryos. Initially, the limb buds are paddle-shaped structures but, as they elongate, a necrotic patch develops in the posterior margin of the wing. Saunders (1966) estimated that as many as 1 000–2 000 cells may die in an 8 h period in this region. Similarly, necrotic patches develop between the fingers and toes of most vertebrate embryos, so that the digits become separated from one another. It is significant that necrosis is slight in the webbed region of the duck's feet.

II. Some Special Aspects of Development

A. THE OOCYTE AND FERTILIZATION

It is helpful to start our discussion by considering the problems cell. By this we mean that although it has not yet developed into embryonic tissues and organs, it is nevertheless highly adapted for its own particular mode of life. The development of the oocyte occurs solely within the ovary. Each oocyte is enveloped within a capsule of maternal cells, the follicular or granulosa cells, and everything that enters the oocyte must do so through these cells. Thus a study of the relationship with the follicle cells is basic to our understanding of the development of the oocyte.

It is helpful to start our discussion by considering the problems faced by the oocyte. They are:

(a) It must halve the number of chromosomes from the diploid to the haploid number, in preparation for fertilization.
(b) It must be ready at fertilization to switch to a new physiological phase.
(c) In most animal species it must store yolk within itself as food for the developing embryo. (The oocytes of placental mammals are an exception.) It must become covered by various membranes to protect it from the environment after it has been ovulated, and if it is from an oviparous (egg laying) species, also after it has been laid.
(d) Whilst preparing in these ways for the events that occur after ovulation, it also needs to maintain itself as a living cell.

The ways in which these problems are tackled by the oocyte will now be considered.

Reduction division

The reduction of the number of chromosomes begins to take place in early oogenesis but is not completed until the time of ovulation, or even fertilization. The delay in the process may last for many years in such species, as man. This is because the earliest stages of oogenesis occur very early in the development of the mother and the later ones do not take place until after she has reached sexual maturity. This fact suggests that a block to the completion of meiosis takes place during the late stages of embryonic life.

The meiotic process is completed in most vertebrates soon after ovulation, though in some species such as the dog and the fox, the first and second polar bodies are not thrown off until after fertilization.

During the later phase of oogenesis, the nucleus of the oocyte swells and becomes known as the germinal vesicle. When the nuclear membrane breaks down at the first meiotic division, the contents of the germinal vesicle are shed into the cytoplasm. Various authorities have considered that the nuclear sap contributed some important component to the cytoplasm at this time which was essential if fertilization was to take place. But recent work throws doubt on this idea (see p. 296).

Protein synthesis and preparation for fertilization

(1) *Nucleic acid synthesis in oogenesis.* RNA is synthesized at two sites within the oocyte nucleus, the chromosomes and the nucleoli. DNA is probably replicated only on the chromosomes, but there is evidence that after it has been formed some of this DNA shifts to the nucleoli. The nucleolar DNA thus provides additional foci for RNA synthesis and this therefore enables a more rapid synthesis to occur. This process, whereby the chromosomes are supplemented, is known as "gene amplification".

(a) Synthesis on the chromosomes.

In the oocytes of many species, the chromosomes are particularly large at the diplotene phase: when the chromosomes are starting to split during the first meiotic division. These large chromosomes have a characteristic appearance which is reminiscent of the brushes which were formerly used to clean oil lamps. They were thus given the name "lampbrush chromosomes" by a previous generation of cytologists. Their appearance is due to the fact that loops of material project from the main axis of the chromosome. These loops are formed of strands of chromatin which have become extended outwards.

The loops become coated with additional material which, since it can be digested with ribonuclease, is considered to be composed of

RNA. Thus there is synthesis of RNA and it takes place along all the lampbrush loops. This RNA is DNA-like in nucleotide composition, at least in *Triturus* where it has been investigated, and so it may be messenger RNA. This conclusion is based on experiments where RNA has been extracted from lampbrush chromosomes, dissected by hand from the nucleus of *Triturus* (Edstrom and Gall, 1963). Davidson and Mirsky (1965) who analysed the RNA in the nuclei of *Xenopus* oocytes at this stage, found that 97% was ribosomal. As we shall see, however, some of this RNA is synthesized on the nucleoli.

Lampbrush chromosomes are visible at the time of greatest synthetic activity in the oocyte. The main difference between them and the chromosomes in somatic cells is thus probably a quantitative one, but it is possible that there are also small qualitative differences. Lampbrush chromosomes are visible in the oocytes of many species including the human. We know little about the synthesis of RNA in the oocytes of those species that apparently do not have lampbrush chromosomes (e.g. the worm, *Gordius*). It seems possible that species of this type may also undergo a massive synthesis of RNA, but that this takes place over a longer period. If this is so, it might explain why the chromosomes are not so dramatically enlarged.

(b) Synthesis in the nucleoli.

Somatic cells seldom possess more than about three nucleoli. Oocytes, however, often possess many. It has been estimated that more than 1 000 nucleoli are present in the occyte nucleus of the newt, *Triturus* (Gall, 1954). These nucleoli contain DNA which has itself been synthesized on the chromosomes. The evidence is based mainly on autoradiographic experiments which have shown that the label accumulates in the nucleoli at least sixty times faster than it does in the lampbrush chromosomes. (For a discussion of the evidence, see Davidson, 1968.)

Little is understood about the mechanism by which the DNA is constantly replicated during gene amplification. The significance of gene amplification, however, is that it speeds up the process of RNA synthesis. In *Xenopus* some of *r*DNA is amplified about 1 000-fold (Dawid, Brown and Reeder, 1970). It has been calculated that if the synthesis of *r*RNA took place at the same rate in the oocyte as it does in the somatic cells, it would take $1{\cdot}7 \times 10^5$ days to produce the appropriate amount of ribosomes. In reality, the presence of so many identical copies of the genes means that the process is completed in 3–6 months (discussed by Davidson, 1968).

Some of the *r*RNA produced in the nucleoli of oocytes passes through the nuclear pores into the cytoplasm. This has been visualized as a densely staining material in both light and electron microscopic sections of the oocytes of a number of species, e.g. the rat (Szollosi and Ris, 1961).

(2) *Protein synthesis in oogenesis.* Protein synthesis is most pronounced during the early stages of oogenesis, at the time when nucleic acid synthesis is also maximal. In the later stages of oogenesis the synthesis of both nucleic acids and of proteins drops, and rises again only at fertilization. The new protein that is formed in the early oocyte is probably used in the formation and replication of cell membranes and of other membranes within the cell, such as those of the mitochondria and the endoplasmic reticulum.

Nucleic acid synthesis and fertilization

The morphological events of fertilization will not be discussed here. For a general account of the significant stages in vertebrate fertilization, see Bellairs (1971). The present discussion will be restricted to a consideration of the process of "activation". This term may be taken to imply a series of cortical and cytoplasmic changes in the ovum that are initiated by the sperm as soon as it touches the cell membrane. The stimulus provided by the sperm is apparently a non-specific one since the same changes can be brought about experimentally in some species by a variety of artificial stimuli. This type of activation without the intervention of sperm is known as parthenogenesis, and occurs spontaneously in certain species (e.g. in certain generations of aphides), and it has even been shown to take place among the vertebrates. The most highly authenticated examples in the amniotes are in a strain of inbred turkeys, *Beltsville small whites*, which have an incidence of nearly 40% parthenogenesis. Some of these survive to adulthood, and, since they are diploid, they apparently have doubled their chromosomes at some stage during early development. Spontaneous parthenogenesis has also been found in two wild populations of lizards.

Spontaneous parthenogenesis has never been demonstrated conclusively in mammals, though there is now satisfactory evidence that it can be induced experimentally. Tarkowski *et al.* (1970), for example, have brought about parthenogenesis in mice by mild electric shock, and Graham (1970) has produced parthenogenetic development in the same species by treating the unfertilized eggs *in vitro* with hyaluronidase.

The significance of parthenogenesis (whether spontaneous or experimental) is that it demonstrates that activation is not always dependent on sperm entry. Moreover, it shows that some development can occur without the presence of the sperm nucleus. The extent to which development proceeds varies. Generally, if the activated egg fails to double its chromosomes, and so retains the haploid number, then it probably fails to proceed beyond cleavage. This is because, in most species studied, a new phase of protein synthesis begins during gastrulation, which differs from that during cleavage. (In mammalian and avian species the new phase begins before the end of cleavage.) The evidence is as follows.

During the later stages of oogenesis the oocyte is impermeable to radioactive protein precursors even though protein synthesis is taking place. Immediately after fertilization however, the precursors are readily taken up, and a sudden rise in the rate of protein synthesis occurs. It is generally considered that the protein synthesizing mechanism of the oocyte is blocked, and that activation at fertilization results in the removal of this block.

One of the reasons for the blockage is that the messenger RNA in the late oocyte has been "masked", so that it is temporarily unable to fulfil its normal function. The masking is probably due to the combination of the RNA with a protein, since unmasking can be brought about experimentally by treatment with trypsin.

The masked messenger RNA is formed before fertilization, and has thus been manufactured entirely under the influence of the maternal genome. After its unmasking at fertilization, it is used during the first stages of development (throughout cleavage, at least in those echinoderms and amphibians that have been studied). Thus, the protein synthesis at this time is under the control of maternally–derived *m*RNA.

Similarly, the ribosomes that take part in protein synthesis at this period are entirely of maternal origin, since they were formed during oogenesis. New ribosomal RNA, made under the influence of the embryo's own genome, does not appear until later. In amphibians no new *r*RNA is manufactured until gastrulation, although in mammals and birds it appears to be formed at an earlier stage. In all species studied however a new phase of protein synthesis begins when the embryo's own genes take over. It is not surprising, therefore that the proteins synthesized by the rabbit blastocyst are not identical electrophoretically with those synthesized during cleavage.

Let us return to consider the processes occurring at activation. We start by deciding on what events are involved in activation. It will then be possible to discuss the way in which these processes are controlled.

First, an egg can be activated only if it is already physiologically "mature". That is, it must have reached a certain stage of development. In a frog, *Rana pipiens,* a physiologically mature egg is one that has developed as far as 2nd meiotic metaphase. Secondly, activation is accompanied by certain cytological changes, and in *Rana pipiens* they include cortical changes, elevation of the vitelline membrane and a redistribution of cytoplasmic components. These changes are thus a useful indicator of whether activation has occurred or not. Indeed, they were used in this way in a recent experimental investigation by Smith and Ecker (1969) who were studying the role of the germinal vesicle in the maturation of the oocyte of *Rana pipiens.*

It had frequently been considered that the breakdown of the germinal vesicle was an essential prerequisite for activation. This idea was based on the fact that in echinoderms, at least, non-nucleated eggs could not be activated by the sperm, and nucleated eggs could only be activated after the germinal vesicle had burst. Less direct evidence pointed to a similar state of affairs in amphibians. The experiments of Smith and Ecker however show that the bursting of the germinal vesicle is not necessary for activation in *Rana pipiens.* These authors enucleated the eggs by poking out the nuclei with a glass needle. They were able to show that these eggs could nevertheless become activated again if given an artificial stimulus. For example, protein synthesis was unaffected by whether the germinal vesicle was present or not. The essential stimulus was shown to be a hormonal one: *in vivo* the eggs come under the influence of pituitary hormones, and the effect could be mimicked *in vitro* by treating the eggs with progesterone.

The enucleated eggs were however unable to undergo genuine cleavage. This did not appear to be because they were lacking in nuclei, since they failed to cleave even after injection of a somatic nucleus. If, however, germinal vesicle material was injected into eggs from which germinal vesicles had previously been removed, these eggs became capable of cleaving. These authors therefore concluded that, in this species at least, the mixing of the germinal vesicle material with the cytoplasm is not essential for activation, though it appears to be required for cleavage to take place.

The laying down of the yolk reserves

One of the reasons why we regard the ovum as a highly specialized cell is that in most species it contains yolk. What is yolk? It is not easy to give a precise definition to cover all types of yolk. Probably the best we can say is that yolk is composed chiefly of lipids and proteins laid down in the oocyte and subsequently used as foodstuffs by the embryo. The composition and proportions of the lipids and proteins vary greatly from species to species. The best analysed is the yolk of the domestic hen (Bellairs, 1964). There is also a great variation in the quantity of yolk present in different species. For instance the ovum of the domestic fowl is about 3·5 cm in diameter and most of the bulk is due to the yolk. The ovum of the mouse is about 80–90μ in diameter, and is nearly yolkless. The amount of yolk is so great in a large yolked egg that the cytoplasm becomes restricted to a small disc at one side (p. 259). Between these two extremes are eggs containing moderate amounts of yolk.

How is the yolk formed? In the domestic hen, the lipids and proteins are manufactured in the liver and are then transported in the blood stream to the ovary. There they must pass through the capsule of follicle cells into the oocyte. During the final stages of oogenesis, yolk deposition takes place at an impressively high rate. Indeed, most of the raw materials for the yolk enter the oocyte during the last ten days before the egg is laid, and these materials must all pass through the follicle cells. It is for this reason that the relationship between follicle and oocyte has excited so much attention.

Despite the fact that much research has been carried out on the problem we still understand little of the processes involved, although we know that it is ultimately under hormonal control. Hens under stress not only go "off lay" but they also cease to deposit yolk in their oocytes. We also know that during the early stages of oogenesis in the unstressed hens yolk materials do not accumulate in the oocyte. It has frequently been suggested that a selective barrier exists either in the follicle cells or at the surface of the oocyte.

During the later stages this selective barrier disappears, for a wide variety of foreign proteins can pass from the blood stream into the oocyte. This implies that important changes are taking place at the surface of the oocyte and, indeed, we possess a clue as to their nature. During the early stages of oogenesis, the oocyte cell membrane is a continuous structure of the type sometimes called "unit membrane" (see Chapter 5). As the oocyte grows the cell membrane becomes extended into projections and this presumably increases its absorptive surface. This presumption is supported by the fact that there are many more so-called pinocytic vesicles present in this region, as seen by electron microscopy. With further growth, the inner part of the vitelline membrane begins to be laid down around the oocyte, and so takes over from the cell membrane the function of mechanically supporting the oocyte. Finally, the oocyte cell membrane breaks down into a number of vesicles; and it is probably during this stage that the selectivity of the cell membrane is lost for this corresponds with the time of great yolk deposition. Various authors have also noticed that during the final yolk deposition phase, the follicle cells not only become flattened but that the intercellular spaces between the cells increase. It has therefore been suggested that this enables the yolk components to pass between the cells and aids the speed of flow.

The mechanisms by which the raw materials are transformed into yolk after they have entered the oocyte are also not well understood. In the hen's oocyte, vesicles appear within the cytoplasm and these increase in size. As they do so the yolk becomes laid down within them (for a further discussion, see Bellairs, 1964). This mechanism does not hold for all species, however. In amphibians there is a close association between yolk and mitochondria, and it appears that yolk is laid down within the substance of the mitochondria. These differences in the formation of the yolk are parallelled by differences in the morphology of avian and amphibian yolk, the former consisting of an emulsion of spherical droplets floating in a watery fluid, the latter of platelets with a crystalline structure.

The protective membranes

Much of this discussion has been concerned with the high degree of specialization exhibited by the female germ cell as a cell. Perhaps the most exceptional aspect however is that in all animals the ovum becomes covered by protective membranes. These membranes are non-cellular and generally secreted by some part of the female genital tract.

The membranes vary greatly in both number and structure throughout the animal kingdom (see Raven, 1961), but the type of coverings that are provided for the egg are appropriate for the environment. For example, the eggs of teleost fishes are ejected directly into water and fertilization is external. They are usually covered by a capsule which is furnished with a small hole, the micropyle, through which the sperm enters. The eggs of reptiles and birds are laid in a terrestrial environment and so must be equipped to cope with the problem of dessication. Appropriately, they are covered by more elaborate membranes than are the eggs of aquatic vertebrates. The hen's ovum is surrounded by a thin transparent membrane, the vitelline membrane, the inner part of which is secreted around the oocyte as it lies in the follicle. The outer part of the vitelline membrane is laid down immediately after ovulation in the cranial end of the oviduct. Fertilization also takes place in this upper part of the oviduct. As the egg passes down the oviduct the egg white, the shell membranes and the shell are secreted on to it.

The egg membranes of mammals have apparently been modified

by viviparity. The shell proper and shell membranes have been lost and the only true membrane that remains is the zona pellucida (Fig. 4, Chapter 14). This is a granular structure consisting largely of mucopolysaccharides. Various roles have been suggested for the zona pellucida (discussion by Bellairs, 1971). One possibility is that it may provide an immunological barrier to prevent the egg and early embryo from being attacked immunologically by the mother.

A thick mucin layer (65–130 μm) is present around the rabbit egg and this surrounds the zona pellucida. Many mammalian ova are also surrounded by the cumulus in the upper part of the oviduct. This is composed of follicular cells which are shed from the ovary at the time of ovulation, and is not a true membrane. It may be noted that all these coverings must be penetrated by the sperm.

The mammalian embryo must escape from the zona pellucida before it can implant. There appear to be two ways in which this may happen (McLaren, 1970). The first is the result of changes in the environment; the uterus is under hormonal control and, in the rat at least, variation from the normal hormonal conditions affects shedding, probably because the uterus produces a lysin as a result of stimulation by oestrogen. The other method is by rhythmical contractions and expansions of the blastocyst which have been followed by time–lapse cinématography. These movements can apparently lead to "hatching" of the embryo.

The relative importance of the two methods varies with the species but is also affected by the local conditions. For instance, in tissue culture, mouse embryos escape by their own exertions. Within the uterus even unfertilized mouse eggs can become freed from the zona pellucida although they are immobile, and this is probably because of lysis.

B. THE LENS

The vertebrate lens has proved to be one of the most useful organs in which to study differentiation. This is largely because it lies on the surface of the embryo and so is easy of access for experimental investigations. It also has the advantage of being a clearly defined structure, not only morphologically but also biochemically (see p. 302). It is also important in being an organ with a good regenerative capacity, even in higher vertebrates.

Morphological events in normal lens development

The vertebrate lens is formed in the early embryo from the ectoderm of the head. In the higher vertebrates this becomes thickened to form a placode and it then undergoes invagination to form a hollow spherical structure which sinks beneath the surface and is subsequently known as the lens vesicle (Fig. 3). The central cavity soon becomes pushed to one side and eventually disappears. The inner wall of the lens vesicle becomes greatly thickened, and this is because the cells become lengthened. The thick inner wall thus becomes pressed against the thin outer wall (Fig. 3E).

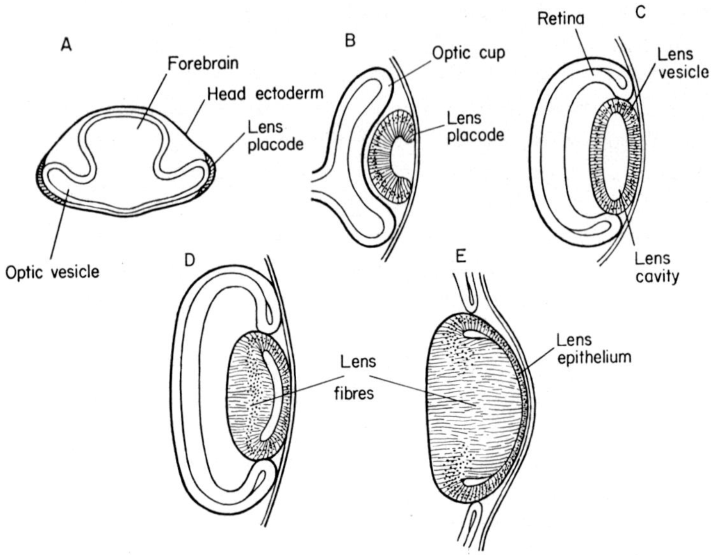

Fig. 3. Morphological stages in the development of the lens.

Initially, the cells of the inner region are quite distinct, and their nuclei lie in a conspicuous band passing through the centre of the lens. Eventually, however, these cells differentiate into lens fibres. The cells in the outer layer become the lens epithelium.

Lens induction

The lens is induced from the overlying ectoderm by the optic cup. The evidence is based on the following experiments, mainly on amphibian embryos. First, the ectoderm which normally gives rise to the lens will not do so if the optic cup is prevented from coming into contact with it. It will however develop a lens if another optic cup is placed beneath it. Secondly, if the optic cup is removed from its normal position in the head and transplanted beneath ectoderm elsewhere in the body a lens will often develop from that ectoderm. In the third experiment, the epidermis, which lies on the side of the head and would normally give rise to lens, is removed. It is then replaced by epidermis taken from another region of the body. If the optic cup comes into contact with the new epidermis a lens may develop. Much of the early work carried out on *Rana fusca* is summarized in Spemann's classical book (1938).

The situation is however more complex than the early experiments suggested. It soon became apparent that there was considerable variation from one species to another. For instance, the lens was able to form in its normal position in *Rana esculenta* even though the optic cup had been removed. Furthermore, although in some species (e.g. *Rana palustris* and *Rana sylvatica*) a lens could be induced from epidermis in any region of the body, in other species (such as *Rana esculenta*) the ability was restricted to the region of the head that normally formed a lens.

Another complication is that the presumptive lens ectoderm has some contact with the foregut endoderm and with the head mesenchyme, before its induction by the optic cup. In certain species, these contacts appear to be important in preparing the ectoderm for induction (i.e. making it competent). Certainly, in most amphibian species the presumptive lens ectoderm is incapable of responding to the inductive activity of the lens until it has passed through these early stages.

We know little about the chemical nature of lens induction. In certain amphibians it is affected by temperature. Ten Cate (1953) showed that it took place more quickly at low temperatures when head mesenchyme was the inductor than when "normal" optic cup was inducing.

It has been suggested that in the chick there is a passage of RNA from the optic cup to the lens at the time of induction. The evidence is based on histochemical studies which have shown a decrease in basophilia in the optic cup and a concomitant increase in basophilia in the developing lens placode (McKeehan, 1956). Interesting though these results are it seems unlikely that they do more than indicate changes associated with protein synthesis taking place in the cells. They cannot be taken as evidence of diffusion.

Mitosis in the lens and its relation to differentiation

As the lens of the embryo differentiates it also increases in size. It cannot do this by direct mitosis throughout its substance, since, as we have already seen, the lens cells lose their nuclei as they become transformed into fibres. Recent experiments in which mammalian and chick embryos have been labelled with [H^3]–thymidine have shown that the new cells are supplied entirely from the lens epithelium. As they form, they move from the lens epithelium into the fibre region.

After this they undergo no more division but become transformed into lens fibres.

This transformation is accompanied by the production of several specific proteins, the crystallins, which are not formed in any other organ. The production of these proteins can be taken as evidence that differentiation has begun. The fact that they appear in cells that have lost their nuclei indicates that they are specified by *m*RNA produced whilst the nuclei were still active. This point is especially important, since the distribution of the different crystallins within the chick lens changes as it grows in size. Thus α-crystallin appears at first in some of the centrally located fibres of the chick lens at about 3 days of incubation, but during the next four and a half days it spreads to all the lens cells (Ikeda and Zwaan, 1967).

It appears therefore that in some way the cells are able to respond to their position so that the type of crystallin they possess is appropriate to the role they will subsequently play in focussing the light which enters the eye. If the cells of the lens epithelium are maintained in suitable tissue culture conditions they elongate and undergo some cytological changes appropriate for developing lens fibres. They are not however capable of complete differentiation unless they are returned to the eye (Philpott and Coulombre, 1965).

Relationship of the lens to the optic cup

The relationship between the lens cells and the other structures within the eye has also been demonstrated in two other ways. The first shows an influence of the eye on the lens. If the lens of a young chick embryo (5 days incubation) is experimentally reversed so that its epithelium now faces the neural retina, the lens cells become reorganized. Those cells which have already begun to elongate do not continue to do so; the cells of the lens epithelium however now begin to change into lens fibres (Coulombre and Coulombre 1963). Thus it appears that the influences which inform the cell of its initial position must be continually being reinforced by further information from the neighbouring tissues.

The second set of experiments show the influence of the lens on the eye cup. It appears that the growth in size of the developing lens is closely related to that of the surrounding tissues. Coulombre and Herrmann (1963) experimentally altered the growth rate of the outer eye coat (cornea and sclera) of the chick embryo. This was done by introducing a small glass drainage tube through the eye wall at 4 days of incubation, so that the vitreous fluid escaped through it. With the reduction in internal pressure the rate of increase in size generally became abnormal, and this was shown to be due to a decreased rate of cell division. Yet the result did not affect the increase in size of the lens, nor its change in shape. On the contrary, although the outer coat did not appear to control lens growth and shape, the lens appeared to exert some control over the outer coat of the eye. If the lens was absent the vitreous material failed to accumulate, and, absence of vitreous material lead to abnormalities of the outer eye coat.

Regeneration of the lens

Perhaps the inter-relationships between the cells of the lens and of its neighbours is most clearly demonstrated in the way that the lens regenerates in vertebrates. Most of the work has been carried out on the larvae and adults of amphibians (for a review see Reyer, 1954). The main interest lies in the fact that once the entire lens has been removed from an eye there is no remnant of similar tissue to form a focal point for regeneration. Thus the new lens forms from a totally different tissue: that is, from one that has already differentiated for some other "purpose".

One of the earliest workers to study lens regeneration in amphibians was Gustav Wolff (1895), who confirmed an earlier observation of Colucci (1891) and showed that the new lens was formed in *Triturus* from the margin of the dorsal side of the iris. This process has since been known as Wolffian lens regeneration. (For references and a review of the earlier work see Reyer, 1954.) Wolffian regeneration has also been found to take place in other amphibians (e.g. various species of *Amblystoma, Hynobius, Pleurodeles* and *Salamandra*) as well as in the chick.

Lens regeneration of a slightly different type has been reported in some mammals that have undergone surgical removal of the lens. It is surprising that so little attention has been given to these findings, in view of their potential importance for patients subjected to cataract operations. For instance, it is possible that even Wolffian lens regeneration could be stimulated by surgical interference with the dorsal side of the retina at the time of operation. This has certainly proved to be the situation in rabbits.

The initial observations of lens regeneration were made as early as 1899 by Randolph (reviewed, by Stewart, 1960), but since then a variety of experiments have been carried out. The main fact established is that, if the normal contents of the lens capsule are removed, a new lens will regenerate but that it will be optically useless. If some cytolysing fetal eye-lid tissue is placed on the lens capsule the lens that regenerates is optically effective. The reason for this striking result is not clear.

Most of the analyses of Wolffian lens regeneration have been carried out on species of *Triturus*. The first deduction that we can make applies to many other species. That is, that since Wolffian regeneration does not occur in a normal eye then the normal lens prevents the formation of an additional lens from the dorsal part of the iris.

The evidence is based on the results of many types of experiment. For example, if the lens is removed and immediately replaced no regeneration follows. If the dorsal side of the iris is injured, but the normal lens remains *in situ*, no regeneration takes place. Similarly, if the lens remains in the eye but is tilted into a new position no regeneration occurs. The experimental manner in which the lens is removed does not appear to affect the result. Thus, lens regeneration took place just as readily in amphibia when the lens was extirpated by way of the vitreous chamber (either dorsally or through the mouth) as when it was taken through the front of the eye.

Several theories have been put forward to explain the role of the normal lens in preventing regeneration. They are:

1. The normal lens affects the iris by pressing on it.
2. The normal lens and suspensory ligament act as a barrier, cutting off the iris from the posterior part of the eye. This may be a mechanical or a chemical barrier or both.
3. The normal lens secretes a substance which inhibits the differentiation of another lens.

Most of the evidence derived from experiments on amphibians appears to support some form of the third hypothesis. It may well be that several factors are involved.

The experiments that have been carried out to test these hypotheses fall into two groups. The first is composed of those in which the effect of abnormal lens tissue has been investigated. For example, if the lens is only partially removed from the eye, the remnants may differentiate into normal-looking, though small lenses, and these seem to have the power to suppress most, if not all, the regeneration of a new lens. If the lens from an *Amblystoma* larva is transplanted into a lens-less *Triturus* eye, regeneration is inhibited and will only take place if the *Amblystoma* lens is removed again. Most important of all, if aqueous humour is taken from eyes containing normal lenses and inoculated into eyes from which lenses have been removed, lens regeneration is inhibited. In control experiments, injection of saline fails to inhibit lens regeneration. All these experiments lend credence to the idea that a chemical agent diffuses from the normal lens.

The second group of experiments has been aimed at testing the idea that the normal lens affects the eye by its shape and by the pressure it exerts on the dorsal part of the iris. Lens-shaped objects have been implanted into lens-less eyes but have usually failed to prevent regeneration, though they have sometimes retarded the process. Occasionally, however, lens regeneration takes place in eyes where a foreign body—such as glass or silver—has been implanted between the normal lens and the dorsal iris. If foreign material such as agar, silica gel, or pieces of brain or liver were implanted into a lens-less eye, they usually succeeded in inhibiting regeneration but only if they were in contact with the dorsal side of the iris.

Results of this type therefore have often been taken as evidence that the lens exerts a mechanical effect on the iris. Reyer (1954) in discussing the problem concludes that not only the size and physical condition of the lens are important, but also the distance of the lens from the reacting iris tissue.

After the extirpation of the simple lens, the new lens always regenerated from the margin of the dorsal side of the iris. The ability to regenerate is however not limited to this region but can occur under suitable experimental conditions from other parts of the iris. For instance, lenses can form from the nasal or temporal parts of the iris if the eye chamber is divided into compartments, or if a ventral piece of iris is inserted in place of the mid-dorsal region. Lenses may also form from head ectoderm if it heals over the eye.

Recent work has been concerned more with the molecular and cytological aspects of Wolffian regeneration in adult newts. There is first a depigmentation of the iris epithelium, the melanin being removed by macrophages. The nucleoli become more numerous and increase in size. This step is correlated with an increase in amount of RNA, first in the nucleus and then in the cytoplasm, as well as with an increase in protein synthesis. DNA is also synthesized at this time and mitosis takes place.

When the lens starts to form, a well defined lens vesicle is present. The inner cells of this vesicle become columnar and differentiate into lens fibres. At about this time the crystallins become detectable. Experiments using actinomycin D indicates that the RNA which specifies for the lens proteins is formed in the earlier stages (discussion by Reyer, 1965).

These events underline for us the similarites between the normal and regenerating lenses. The main difference is that they are formed from cells with very different histories before their differentiation into lenses. Moreover, the environment of the developing lens cells is not the same. In the normal condition the main influence is the developing optic cup of the embryo. In the Wolffian regenerates, the influence comes from the eye as a whole.

C. THE FORMATION OF MUSCLE

It has long been known that soon after the somites form they become modified into three regions. These are the myotome, which forms much of the muscular system, the sclerotome, which gives rise to the axial skeleton, and the dermatome which differentiates into the dermis of the skin of the dorsal body wall. The latter may even form the dermis of the entire body, but there is little evidence on this point.

The lateral plate mesoderm lies on either side of the somites. In the chick, at least, some of the muscles of the limbs and of the body are formed from the somatic layer of the lateral plate mesoderm, that is, from the part which lies beneath the ectoderm. This has been shown by experiments in which the cells have been marked, either with carbon particles or with [H^3]-thymidine. Seno (1961) concluded that in the chick the somites formed the entire musculature of the abdominal wall, but that the somatic mesoderm formed the pectoralis muscle.

Before the muscles begin to form, the cells of the myotome and splanchnic mesoderm must often undergo extensive cell migration. Deuchar (1958) showed that when carbon marks were placed on the occipital somites of chick embryos they were subsequently found in the mandibular arch and, later, in the tongue. This was of especial interest in that it confirmed the anatomical deduction that since the tongue is innervated by nerve XII its muscles must have originated in the occipital somites. Similar results have been obtained by Hazelton (1970) using tritium-labelled tissues.

By the time the future muscle cells have reached their final destination they have already undergone some biochemical differentiation. This occurs before any histological or cytological differentiation is visible. The muscle proteins, actin and myosin, are already present in chick embryos of 30–35 pairs of somites, that is at a stage prior to the migration of the myotomes.

The morphological stages in muscle formation occur slightly later. Essentially, they consist of the myoblasts (the cells derived from the myotomes) becoming spindle shaped. The myoblasts then fuse together to form myotubes, and the latter ultimately give rise to muscle fibres. Two main problems have been studied: the first is the formation of the myotubes.

The formation of myotubes

Until recently a controversy existed about the way in which the myotubes formed, one theory being that individual myoblasts fused together and the other, that they formed by a progressive budding from a single myoblast. The problem has been solved in several ways. Konigsberg (1965) grew myoblasts in tissue culture and was able to show by cinematography that the fusion concept was correct. This has been confirmed in an ingenious way by others. It is possible to take mouse blastocysts and disaggregate their cells. If the cells from two or more individual blastocysts are now mixed together *in vitro* they will reaggregate and co-operate with one another to form a new individual which can be re-implanted into a suitable mouse mother and allowed to come to term. These baby mice are thus composed of cells from two different individuals and, if genetic mutants with recognizable cellular characteristics have been selected as the donors, it is possible to identify the two types of cells. Using this technique, Mintz and Baker (1967) took a mutant strain whose cellular characteristic was an abnormal form of isocitrate dehydrogenase: They found that many of the individual muscle fibres in these mice of dual origin contained both the normal and the abnormal enzyme (Fig. 4). Thus they concluded that the muscle fibres formed by a fusion of the myoblasts.

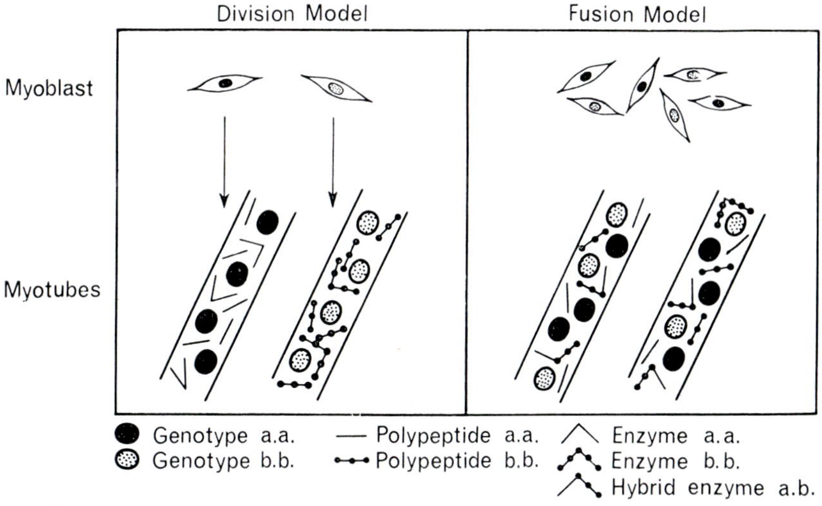

Fig. 4. Formation of muscle syncytia. The diagram shows two hypothetical ways in which syncytia might form: by *division* of a single initial myoblast (left) or by *fusion* of several myoblasts (right). These theories were tested by obtaining mouse morulae from two genetically different strains, the adults of which differed from each other in possessing specific enzymes in their skeletal muscle. Many of the skeletal muscle fibres of the hybrid individuals were found to contain hybrid enzymes, thus demonstrating the correctness of the fusion model. (After Mintz and Baker, 1967.)

The "stabilization" of muscle cells

The second problem is concerned with the influence which causes the muscles to form from the myoblasts. Until a few years ago, it appeared as if the sequence of differentiation from myotome, to myoblast, to myotube to muscle fibre, was the result of an inductive influence. Many experiments were performed which seemed to show that the notochord and/or the floor of the neural tube, were responsible for inducing the muscle. Recently, however, a number of authors have demonstrated that differentiation into muscle cells can even take place when somite cells are growing alone in culture. The effect seems to be that cells which already have a tendency to form muscle become "stabilized" to do so, if the environment is suitable. In the embryo proper it may be that this stabilization occurs during the period of cell migration.

The implications of such ideas are far reaching. We have already seen (p. 274) that in development each cell must "know its place" if it is to develop appropriately. It now begins to seem that much evidence for secondary inductions—and perhaps even for primary induction—could be re-interpreted in a different way. Thus, perhaps even the lens of the normal eye forms not by direct induction, but by a response to a specific environment, the environment that embraces and surrounds the optic cup.

D. MIGRATORY CELLS

We have already seen that muscle cells migrate during their differentiation, and do not begin to form until they have reached their final destination. The cells of the neural crest and those of the primordial germ cells also migrate through the tissues (for discussion, see Bellairs, 1971).

Various theories have been put forward to explain how these cells migrate in the correct direction so that they eventually arrive at their true destination. The first possibility is that there is a chemotactic control, the cells continually moving to a more favourable environment. The alternative hypothesis is that they are guided there by the substrate.

The Neural Crest

There is little evidence of chemotactic control of the neural crest migration, though it may be a factor in the movement of the primordial germ cells (see below). Weston (1963), who carried out experiments in which the neural crest of the chick embryo was labelled with [H^3]–thymidine, suggested that they move by means of contact guidance. He found that the cells migrated out on each side of the neural tube in two distinct streams. One moved dorso–laterally into the ectoderm, whilst the other passed ventrally between the neural tube and the developing myotome.

The Primordial Germ Cells

The primordial germ cells are unusual in that their mode of migration differs among the classes of vertebrates. In most if not all they arise from the wall of the yolk sac and move from there to the gonad where they give rise to the gonocytes. In some vertebrates they migrate by amoeboid activity through the tissues (e.g. the mouse). In others, like the chick, they are carried passively in the blood stream, though they probably indulge in some amoeboid activity at the time of leaving and entering the blood vessels. We should however be careful not to assume that amoeboid activity occurs in all species of mammals and passive transport in all species of birds. Certainly in the reptiles, different methods appear to be used by quite closely related species (see Bellairs, 1971).

We know nothing of the factors that inform the neural crest cells and the primordial germ cells that they have arrived at their destination. Although it is possible to imagine that contact guidance may help the amoeboid primordial germ cells to reach the appropriate place, this is unlikely to be the answer for those cells carried passively in the blood.

We also know little about the factors which stimulate these cells to undergo differentiation when they have reached their destination. Primordial germ cells taken from the gut of 9 day mouse embryos and transplanted to the anterior chamber of the eye of adult mice failed to undergo further differentiation. It seems possible that if suitable conditions could be found, the primordial germ cells might be persuaded to differentiate even in an abnormal environment.

One problem that does not seem to have been investigated is whether the migration of the cells plays some role in their differentiation. If one were designing an embryo, it is unlikely that one would incorporate a process of extensive migration, since it would be simpler if the germ cells were formed *in situ* in the gonad. It would perhaps be unwise to look for a "purpose" in these migrations, but it is worth considering that some important influence is brought to bear during this event. This might be a positive influence that helps to promote their future development. Alternatively, it is possible to think of the cells as being held back from differentiation during the motile period. Thus, the migration might act as a timing device which enables some cells to remain relatively undifferentiated until the embryo has reached the appropriate stage of development. The primordial germ cells give rise to gonocytes, and it would be inappropriate if they began to differentiate before the stroma of the gonad was sufficiently advanced to provide the correct environment.

III. Concluding Remarks

Much of the previous chapter was concerned with the ways in which differentiation is controlled. The first part of this one deals with the control of growth.

Differentiation and growth are the two main aspects of development and, although they are frequently discussed separately, it is important to remember that they are closely related. This relationship has been demonstrated in this chapter by examining the development of a few selected organs. For instance, the morphology of the lens is dependent not only on the biochemical differentiation of the individual cells but also on the changes in shape that are related to this process. Furthermore, the lens cells lose the power to undergo mitosis shortly before they differentiate into lens fibres.

In the past, embryologists have tended to regard development largely as the interactions that occur between neighbouring tissues. Thus, in the pre-war and immediately post-war periods, the phenomenon of embryonic induction was given preference. Recently, the emphasis has shifted to the changes that occur at the molecular level within the cell, and in particular the question of how the genes control development. It has been my aim in these two chapters to point out that we are now ready for the next phase in our study of development—an understanding of how the cells interact throughout the embryo as a whole.

References

Bellairs, R. (1964). *Advan. Morphogen.* **4**, 217–272.

Bellairs, R. (1971). "Developmental Processes in Higher Vertebrates." Logos Press, London.

Bertalanffy, L. von. (1960). In "Fundamental Aspects of Normal and Malignant Growth" (W. W. Nowinski, ed.) pp. 127–259. Elsevier, London and New York.

Bullough, W. S. (1962). *Biol. Rev. Cambridge Phil. Soc.* **37**, 307–342.

Cohen, S. (1965). *Develop. Biol.* **12**, 394–407.

Coulombre, J. L. and Coulombre, A. J. (1963). *Science*, **142**, 1489–1490.

Coulombre, A. J. and Herrmann, H. (1963). *Exp. Eye Res.* **4**, 302–311.

Davidson, E. H. (1968). "Gene Activity in Early Development". Academic Press, New York and London.

Davidson, E. H. and Mirsky, A. E. (1965). In: "Genetic Control of Differentiation". 18th Brookhaven Symp. pp. 77–98. Brookhaven National Laboratory, New York.

Deuchar, E. M. (1958). *J. Embryol. Exp. Morphol.* **6**, 527–529.
Edstrom, J. F. and Gall, J. C. (1963). *J. Cell Biol.* **19**, 279–284.
Gall, J. G. (1954). *J. Morphol.* **94**, 283–351.
Goss, R. J. (1964). "Adaptive Growth". Logos, London.
Goss, R. J. (1970). *Advan. Cell Biol.* **1**, 233–296.
Graham, C. F. (1970). *Nature, London* **226**, 165–167.
Hazelton, R. D. (1970). *J. Embryol. Exp. Morphol.* **24**, 455–466.
Ikeda, A. and Zwaan, J. (1967). The changing cellular localization of α-crystallin in the lens of the chicken embryo, studied by immunofluorescence. *Dev. Biol.* **15**, 348–367.
Konigsberg, I. R. (1965). In: "Organogenesis" (R. L. DeHaan and H. Ursprung, eds.) pp. 337–358. Holt, Rinehart and Winston, New York.
Laird, A. K. (1966). *Growth*, **30**, 263–275.
McKeehan, M. S. (1956). *Amer. J. Anat.* **99**, 131–155.
McLaren, A. (1970). *J. Embryol. Exp. Morph.* **23**, 1–19.
Miller, L. and Gurdon, J. B. (1970). *Nature, London* **227**, 1108–1110.
Mintz, B. and Baker, W. W. (1967). *Proc. Nat. Acad. Sci. US.* **58**, 592–598.
Philpott, G. W. and Coulombre, A. J. (1965). *Exp. Cell. Res.* **38**, 635–644.
Prince, H. H., Siegel, P. B. and Cornwell, G. W. (1968). *Growth*, **32**, 225–233.
Raven, C. P. (1961). "Oogenesis". Pergamon Press, London.
Reyer, R. (1954). *Quart. Rev. Biol.* **29**, 1–46.
Reyer, R. (1965). In: "Eye Structure", Vol. II. (J. W. Rohen, ed.) pp. 495–513. Schattauer-Verlag, Stuttgart.
Saunders, J. W., Jr. (1966). *Science* **154**, 604–612.
Seno, T. (1961). *Acta Anat.* **45**, 60–82.
Sissman, N. J. (1966). *Nature, London* **210**, 504–507.
Smith, L. D. and Ecker, R. E. (1969). *Develop. Biol.* **19**, 281–309.
Spemann, H. (1938). "Embryonic Development and Induction". Yale Univercity Press, New Haven, Connecticut.
Stewart, D. S. (1960). *Trans. Opthalmol. Soc. U.K.* **80**, 357–372.
Szollosi, D. and Ris, H. (1961). *J. Biophys. Biochem. Cytol.* **10**, 275–283.
Tarkowski, A. K., Witkowska, A. and Nowicka, J. (1970). *Nature, London,* **226**, 162–165.
ten Cate, G. (1953). "The intrinsic Embryonic Development". North Holland, Amsterdam.
Waymouth, C. (1960). In: "Fundamental Aspects of Normal and Malignant Growth" (W. W. Nowinski, ed.) pp. 546–587. Elsevier, London and New York.
Weston, J. A. (1963). *Develop. Biol.* **6**, 279–410.

Supplementary Reference
David, I. G., Brown, D. D. and Reeder, R. H. (1970). *J. Mol. Biol.* **51**, 341–360.

16. Ageing

D. BELLAMY

Department of Zoology, University College, Cardiff, Wales

1. Definitions

The study of the biology of ageing is described by the term "Gerontology" and it is the aim of gerontologists to explain why it is that organisms have characteristic life spans.

To the layman ageing simply means growing older. To many biologists, the term is used to describe any time-dependent change with respect to life history. It is necessary, however, to place some restriction on the definition in order to arrive at a category of changes which is useful in formulating problems open to experiment. For example, ageing may be distinguished from the time-dependent changes of development: developmental events aid survival until the individual is a reproductively competent adult, whereas ageing processes lead to failure to adapt to the environment and result in death. Points of difficulty arise because many developmental events are the obvious precursors of ageing phenomena and some changes that begin before or shortly after birth continue throughout life (Fig. 1). Despite this, many workers would restrict the term "ageing" and eliminate from consideration any changes which do not render the individual more likely to die in a given time as he grows older. A good example of a post-reproductive decline in function that does *not* show up as an increase in mortality, is the regression of the female reproductive system. This particular difficulty of definition has been recognized and another category, senescence, introduced to include only those events that contribute to the decreased resistance to death. However, it is generally accepted that of the many ageing phenomena, very few could be proved not to influence mortality. Also, the force of mortality depends very much on the environment. For example, it is of no obvious disadvantage for civilized man not to be able to run a five-minute mile at the age of fifty. Nevertheless, the well-documented decline in human athletic performance from the second or third decade could be included as an early ageing process in a less sophisticated society, where predators had the ability to run faster than human prey.

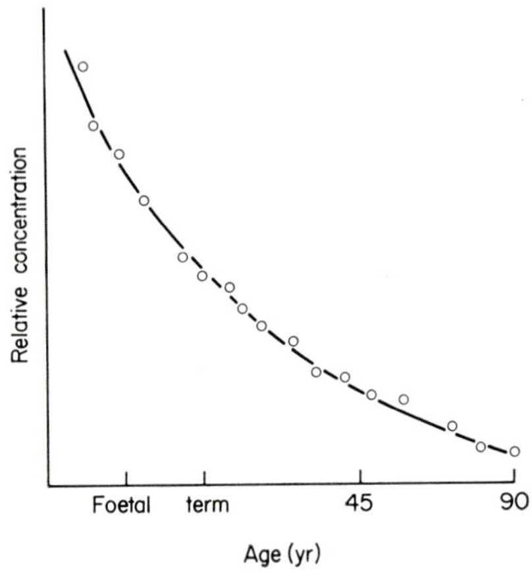

Fig. 1. Galactosamine content of human costal cartilage.

Perhaps our difficulties arise because it is implicit in most current definitions that ageing phenomena have unique causes. It is by no means clear that this is so and the view may be advanced that differentiation and ageing result from the operation of the same basic cellular processes. Differences between early and late events are that early events lead to perfection of function and late events result in the deterioration of function.

For the purpose of the present discussion, ageing is defined as "a decrease in adaptation as a consequence of loss of tissue and functional reserves". The definition implies that a major feature of ageing is a decline in the ability of the homeostatic systems of the body to cope with fluctuations in the external world. On this basis, ageing begins long before there is a marked increase in mortality. In the human, a steady decline in physiological performance is first noticeable during the third decade of life; the corresponding stage in the laboratory rat (life span 800 days) occurs at about 200 days; in the fruit fly, *Drosophilia* (life span 27 days), ageing of some systems begins at about

10 days. Thus, it appears that in this context, ageing phenomena are noticeable for well over half the life span.

II. Life Tables

When a population of animals is maintained under controlled laboratory conditions all members have the same environmental history, yet they do not die at the same instant in time. That is to say, when the number of individuals remaining alive at set intervals out of a given number recorded at birth is plotted against time, instead of a rectangular graph, a curve is obtained (Fig. 2). There is a wide spread in the age at death and this is taken to mean that there is variability in resistance to death in the population. The actual shape of the curve, called a life-table, gives no information about the rate of ageing in individuals. It is merely an accurate age-frequency distribution at the moment of death: the shape of the curve is not fundamentally related to the rate or kind of ageing of a given population.

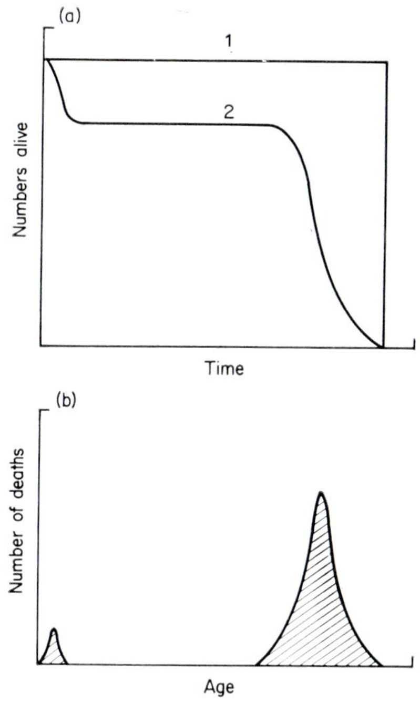

Fig. 2. Mortality curves and frequency of deaths. (a) Graph 1 shows a hypothetical result for a uniform population in a constant environment. Graph 2 shows an actual result for an inbred population maintained under laboratory conditions. (b) Analysis of the data in Graph 2 above as frequency of deaths in different age groups.

Ruling out accidents, the spread in the timing of death indicates that either the population is not uniform in genotype, or that if it is genetically homogeneous, events have occurred at random since birth which have resulted in different rates of ageing. Life tables are useful in making comparisons between the ageing process in two or more populations which differ with respect to environment, genetics, or the experience of random internal events which may influence the rate of ageing. They show the times of decreasing "fitness" in terms of death rate and it is possible by comparing life tables to detect the differential effects of a changed environment upon the life expectancy at different ages, implying that the fitness of individuals varies with age.

Life tables are simply constructed by counting the numbers of individuals of different ages in a population. This tends to give a static picture, whereas the population is made up of at least five distinct components. An input of individuals occurs through births and immigration : an output occurs through deaths (natural and by predation) and emigration. Also, with the passage of time, the size of the total population exhibits change.

Mortality in natural populations is simple to measure when reproduction and migrations can be ignored and the population examined repeatedly as it decreases in size. However, this is only possible with animals with an annual life cycle, a short reproduction period and a small range of movement. More often than not, the death rate in a population is difficult to measure and one must resort to estimating the accumulation of animal remains that are resistant to decay, such as bones in vertebrates and the cuticle of insects.

It is also difficult to assess the age of many wild type species. Evidence of age may be obtained from annual events which leave a permanent alteration in the hard parts, such as rings or bands of different consistency in bone, teeth and shells. On the other hand, physiological periodicity leaves its mark in some organs, particularly the ovary, where the number of empty ovarioles in the ovary of insects or the number of corpora lutea in the ovary of mammals may be used to assess age. Few of these methods are sufficiently reliable to be used on individuals and the sparse physiological and biochemical data that are available for non-captive animals are mean values referred to mean size as the only practical temporal quantity.

Although it is often difficult to assess the sex of many animals without dissection, it appears that a differential mortality rate is a sex-linked feature of many animals. In humans, the male death rate is higher than that of the female over the entire life span. At birth, the ratio of males to females is 106 : 100 and there is evidence that the higher death rate of males is a feature of intrauterine life with estimated primary sex ratios of between 110–160 males to every 100 females. In some birds, reptiles and some insects, the female has the shorter life span. However, a consistent feature appears to be that the shorter-lived sex is the one that is heterogametic. The higher death rate in heterogametic individuals is possibly due to the action of recessive alleles on one chromosome for which there are no counteracting alleles in the other.

Because of the difficulties in obtaining reliable data from natural non-human populations, most life tables have been constructed using laboratory species or domesticated populations. It is well-known that these tables frequently show a life expectancy much higher than for the same species in the wild. For example, the mean life expectancy of wild mice which are not subjected to predation is only a few months, whereas in the laboratory the same strain lives for between two to three years. It may be that part of this increase in life span in captivity is due to the experimenter selecting genotypes for good adaptation to laboratory conditions, but a more likely explanation is that the major causes of death in wild populations are physical and biotic insults of a fluctuating environment. It is the natural life span in the wild that must be under selection pressure. Consequently, in asking which processes render the individual more likely to die in a given time interval as it grows older, we are likely to obtain different answers from laboratory strains compared with wild types.

III. The Nature of Ageing

Ageing of machines occurs because frictional forces result in the wearing-out of moving parts and also because there is an inevitable deterioration of key components through chemical changes defined variously as rusting and perishing. With regard to the former category, there is no evidence that components of living tissue age in proportion to usage. Indeed, it is a characteristic of life that the working capacity of organs increases with the demands made upon them. The two phenomena of work hypertrophy and compensatory hypertrophy do, however, decline with age. In relation to the machine analogy, it does appear that degenerative changes due to chemical interactions between molecules and their environment akin to the phenomena of "perishing" are responsible for the age-deterioration of connective tissue. However, on the whole the machine analogy is of little use in Gerontology.

The fact that organisms age and die implies that imperfections appear at the chemical level which cannot be corrected either during

parental age is the determining factor (Table 1). For example, in the species *Philodina citrina*, reproductive maturity is reached in 5 days with a mean life span of about 23 days. Eggs taken at 11 days and 16 days produced clones which had lowered mean life spans, and which both died out after 4 and 3 generations respectively. This implies that the reproductive system of the parents deteriorates in old age. Both egg size and egg shape show a greater variability with increasing parental age and it has been argued that this is the basis for a corresponding variability in fecundity of offspring: offspring developing from large eggs have a very high fecundity and those coming from small eggs have a low fecundity.

TABLE 1. Life spans of clones of a rotifer (Philodina citrina) derived from parents of different ages

Age of parent	Life span: 5 days (Mean)	5 days (%)*	11 days (Mean)	11 days (%)*	16 days (Mean)	16 days (%)*
Generation						
0	24·2	—	24·2	—	24·2	—
1	22·8	94·2	20·8	86·0	22·0	90·9
2	23·5	97·1	18·3	75·6	20·2	83·5
3	19·2	79·3	22·5	93·0	Died out	
4	24·3	100·4	22·6	93·4		
5	24·5	101·2	Died out			
6	24·3	100·4				
7	27·8	114·9				

*3 Clones were produced using eggs from the same parental strain (mean life span 24·2 days) which were taken at different phases of the life cycle. Life span expressed as a percentage of that of parental strains.

The shortening of life span leading to extinction of offspring from old parents is not the result of a genetic change in the offspring because individuals developing from eggs taken early in the life of a clone doomed to subsequent extinction—because it was derived from a senile parent—show the life span and pattern of development of early eggs produced by the original stock.

These phenomena observed in rotifers stress two common findings in other animals: the influence of environment on longevity and the increased phenotypic variation with age.

Extracellular components

The only phenomenon at the organismic level that has given rise to a plausible theory of ageing concerns the accumulation of extracellular structures composed of the proteins collagen and elastin (see also Chapter 23). These proteins, once formed, have a very low rate of turnover and they appear to undergo a series of alterations akin to the perishing of rubber, where the individual fibres shrink and harden. Chemical changes of this nature are responsible for wrinkling of the skin and loss of elasticity of the tendons, which are two of the most telling characteristics of the ageing mammal.

Most investigations into the ageing of extracellular materials have been carried out using easily accessible connective tissue such as skin, tendon, arterial walls, ligaments and lungs. All of these tissues contain cellular components but usually these are of minor importance in terms of bulk and have been ignored in most generalized studies.

At the highest level of organization, there is much evidence for degenerative changes in connective tissue. In old skin, for example, haemorrhages frequently occur (senile purpura) which are due to the failure of the matrix of collagen fibres surrounding the capillaries to cushion them against mechanical damage. From the mechanical behaviour of intact skin, it has been proposed that ageing proceeds in two phases: initially, there is a stiffening of the structure leading to a reduced extensibility of fibrous components. In the human, this process predominates up to the age of 40. Later, the structure becomes looser.

These changes at the organ level have their origins in the extracellular molecules. The extracellular phase of connective tissue consists of fibrous components composed mainly of two proteins, collagen and elastin, with a less well-defined matrix which supports and lubricates the fibres. Age changes in the organization of the fibrous components may be seen under the light microscope. The structure and arrangement of the fibres varies somewhat between different tissues, but the age changes are similar. In general, there is an increase in material that takes up the elastin stain. There is some doubt as to the chemical nature of the new material that accumulates with age, as its properties seem to be intermediate between those of collagen and elastin.

Elastin is characterized chemically by its insolubility in water and by the fact that it is only slightly hydrolyzed by pepsin or trypsin; it consists mainly of amino acids with non-polar side chains. It swells only slightly in water. Collagen, on the other hand, is definitely hydrolyzed by pepsin, albeit slowly, but not by trypsin. Like elastin, it is composed of about 30% glycine and 12% proline. Collagen, but not elastin, contains a large proportion of hydroxyproline (13%). Sequence analysis has shown that the amino acid composition is not random along the peptide chains. But the protein has a much more regular arrangement of its component residues than elastin. When heated in water, it is converted into gelatine.

Evidence for chemical changes in ageing of fibrous components was first provided by studies of the human cardiovascular system, when it was found that the elastin of old aortae contained higher concentrations of basic and acidic amino acids. It was later found that old elastin could be separated into two fractions, one of which was identical in amino acid composition with young elastin. The fraction with the different amino acid composition is thought to be a contaminating glycoprotein. The second chemical change in elastin is due to the accumulation of calcium in the elastin molecules bound by co-ordinate bonds. This accumulation amounts to almost a hundred-fold increase in human aorta over seven decades.

As mentioned earlier, collagen is a highly orientated molecule, the individual fibres having a well-defined striated appearance under the electron microscope. The predominant component (tropocollagen) consists of three peptide chains intertwined in a helical arrangement. The proportions of the three components of tropocollagen alters with age—two of the sub-units increasing at the expense of the third.

A basic assumption in this field is that the collagen ages by the accumulation of cross-links between peptide chains, although the nature of the links has not been defined with certainty. Cross links certainly appear in collagen after it has been synthesized and lead to a decrease in solubility. However, some workers would describe these as a process of maturation of collagen. What is perhaps more important is that collagen steadily accumulates in most organs throughout life (Fig. 4).

Do these changes in connective tissue components lead to loss of organ efficiency? Failure in the homeostatic systems of the body may be visualized at three levels: receptors tend to show a higher stimulus threshold; the central nervous system handles information at a slower rate; effectors become less efficient in making the appropriate response.

Deterioration in the collagen-containing stroma of key organs such as lungs, heart and kidneys has been proposed as a key event in the ageing process, where degenerative changes interfere with various aspects of homeostasis by reducing the mechanical efficiency. The continuous accumulation of collagen is also thought to impede the supply of nutrients to the cells. Although the evidence on molecular ageing of collagen and related compounds has firm chemical and histological backing, there is no clear-cut experiment showing that these changes actually result in death. However, it is reasonable to assume that accumulation of collagen and its loss of elasticity

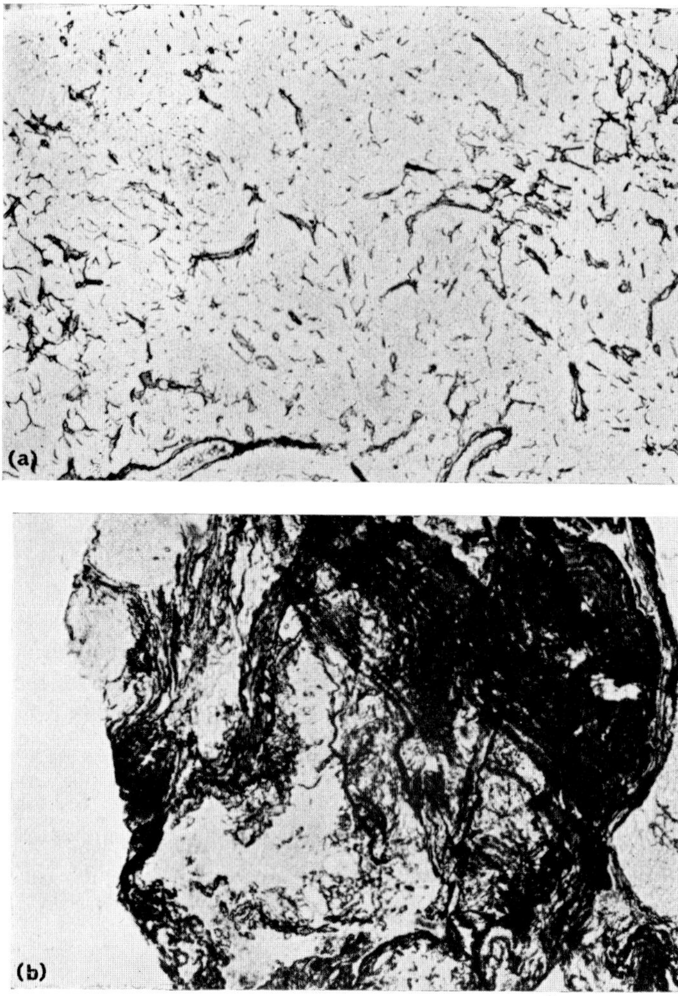

Fig. 4. Collagen accumulation in mouse thymus. Sections through mouse thymus were stained for collagen by silver impregnation—dark material. (a) $4\frac{1}{2}$ months old (b) 32 months old.

contribute to the mechanical failure of some effectors, notably skeletal muscle.

Homeostasis and ageing in mammals

If the composition of the blood is taken as an indicator of the overall state of homeostasis, there appears to be no impairment of functional capacity with age. For example, the acid-base balance of the blood in resting humans varies little between the ages of 25 to 85 and there are no significant changes in the carbon dioxide tension, total carbon dioxide content and bicarbonate concentration. In agreement with this, the mean pH only changes from 7·400 to 7·368 between the second and eighth decades of life. Organic constituents of blood show a similar stability. The general constancy of the ionic composition of the body is also borne out by studies on individual tissues and subcellular fractions. No significant age-changes are observed in total water, fat, potassium and the sodium content of liver, whilst the intracellular water, nitrogen, potassium and phosphorus in muscle decrease only by between 5 and 8%. There is no change in muscle lactic acid and creatine and analysis of human aortae taken over an age range from 2 to 69 years reveals no significant alterations in either total nitrogen or sulphur; aortic elastin and creatine also remain relatively constant. A constancy of composition is also found to be a feature of the ribonucleic acid in ventricular muscle of the rat from 100 to 1 200 days of age. In keeping with this, the mean DNA per nucleus and the mean volume of nuclei in rat liver do not change over an age range of 12–27 months.

The failure to detect age-changes indicative of a large-scale deterioration in tissue function impressed the early workers using histological methods. Only moderate signs of ageing are found in the histochemistry of liver and there are no obvious large-scale functional deficiencies at this level in the brain. Kidney tissue is marked by having the largest age alterations in structure, but this appears to be a consequence of a large decrease in the relative cell mass without a marked change in composition of the remaining cells.

Despite the evidence in favour of an overall stability of tissue structure, age-changes in composition are frequently observed. This applies to a small number of blood constituents. Often, age-changes are not uniform between tissues: although there is no variation in the riboflavin content of human brain, heart and skeletal muscle over seven decades, aortic tissue is characterized by a 60% fall in the riboflavin content over the same period. Also, in contrast to ventricular tissue the folia tissue of the cerebellum loses 30% of its ribonucleic acid throughout the life-span.

Some of the largest alterations in tissue composition have been found in the lipid fraction. Cholesterol in the blood of healthy women increases twofold in concentration from the age of 20–60. There are also general substantial increases in the concentration of elastin, collagen, mucopolysaccharides and calcium in many tissues. However, these changes are not a universal phenomenon.

Taking the available evidence on the gross composition of tissues, it appears that on the whole, age changes are not very marked in those intracellular components, such as water, nitrogen and inorganic ions, that are concerned with fundamental cellular organization. On the other hand, there are marked changes in the chemistry of the extracellular compartment, notably in the ground substance. These changes are probably a reflection of the general tendency for there to be an age-dependent shift in the balance of cell populations. For example, the eosinophil count in rats decreases progressively with age, there being a 50% drop between the ages of 300 and 800 days. Despite this loss, there is little change in other blood cells. In ageing cattle, it has been observed that a decrease in blood lymphocytes is accompanied by a rise in the proportion of blood neutrophils and a fall in the total leucocyte count.

TABLE 2. Effect of age on the cell population of mouse thymus

	Percentage of cells*	
Age	5 months	32 months
Cell type		
Lymphocytes	92·7	73·2
Epithelial cells	5·8	7·8
Macrophages	0·6	9·6
Fibroblasts	0·2	3·4
Plasma cells	0·2	1·1

*The overall cell density was about 50% less in the thymus of old mice.

Changes in cell population are most marked at the histological level. An almost universal feature is that of cell loss (age involution). Organ involution is a well established characteristic of the thymus. Here, the process which results in a progressive general decrease in cellular mass is accompanied by shifts in the ratio of lymphocytes to macrophages and fibroblasts (Table 2, Fig. 4). Age involution is also a feature of skeletal muscle and some rats show a 30% loss of thigh muscle between the first and second year of life. Taking total body potassium as a measure of cell mass, the number of cells per unit body weight shows a steady decline in humans from the late teens and, apart from tissue involution, there seems little change in functional organization of the cells. This points to the fact that the ageing process is manifest for the most part at a physiological level.

Although physiological parameters that depend upon a complete interaction of several organs, such as ion balance and blood pressure, hardly change with age, there is an increasing tendency for departures from these norms to take place. Thus, while the systolic blood pressure of CBA mice under controlled conditions remains within the range 126 ± 12 mm Hg throughout the life span of $2\frac{1}{2}$ years, the normal distribution curve for rat blood pressure has been observed to become biphasic with age. When these rats pass the age of 600 days, about 30% of males and 20% of females become hypertensive with marked renal disease. The latter condition is probably the main cause of the hypertension.

A decline in the ability to cope with increased functional demands due to endogenous processes can be clearly seen in the reproductive capacity of rodents. After 14 months, the litter-size of hamsters is greatly reduced and the length of gestation increases. Similarly, data for the mouse suggest that parental age influences the length of gestation, the litter-size and the perinatal mortality rate. Resorption during the late phase of gestation is thought to be a contributory factor in the reduction in litter-size in senescent females.

A failure of physiological adaptation is most clearly seen under laboratory conditions in response to specific stresses. Old C57 B2 female mice have a high mortality rate following exposure to a temperature of 6–7°C which is not seen in young, mature females. The failure of integrative function which is responsible for the death of older animals appears early in the experimental period. Old mice that survive the first 2 days of cold exposure have a good chance of surviving for the next 12 days in the cold. Their chances are also improved when they are slowly adapted to the lower temperature, but the adaptive process is less effective than in the young.

A number of the individual physiological responses of rodents to cold such as the increased oxygen uptake, rise in blood sugar, weight loss and changes in blood cell counts also become less marked with age. Studies on humans also suggest that there is a similar deterioration in mechanisms for minimizing heat loss.

The response of the human body to heat stress has also been examined for possible age effects and in general, there appears to be an increased degree of discomfort at high temperatures with the physiological responses in the old becoming more variable. One relevant feature is that men over the age of 40 have a greatly increased vasodilation in response to heat. Since in a hot environment heat loss is almost entirely by evaporation, the greater skin blood flow in old subjects would not be expected to have a marked effect on their heat regulation. Age differences in rectal and skin temperatures are not significant in heat-exposed subjects, therefore the view may be taken that the faster skin circulation of older men is unnecessary for heat regulation and places a needless burden on the general circulation. Responses to temperature are intimately linked with the basal metabolic rate. There is a significant fall in basal heat production and carbon dioxide elimination as humans age. There is also a larger day-to-day variation in several aspects of respiratory function which may be of significance with regard to the efficiency of adaptation to extreme temperatures.

Starvation, another common stress encountered by animals in the wild, may be examined conveniently in the laboratory. However, there have been few investigations of starvation in relation to age. With regard to rats, no significant age differences have been observed in either the rate of weight loss in long-term starvation or the survival-time after food deprivation.

There is a marked loss of adaptation to experimental hypoxia in old animals. With respect to traumatic injury, the data for humans do not support the concept that older individuals have less chance of survival. On the other hand, this conclusion relates to data only on traumatic abdominal injuries. Regeneration of the liver of rats after partial hepatectomy occurs to an equal extent in the young and old but old animals take more time to form the requisite number of new cells. Following unilateral nephrectomy, senescent rats have the same capacity for compensatory growth as young ones. However, as for liver regeneration, the kidney of old animals responds slowly.

There is no evidence that new cells produced in response to unilateral nephrectomy by old rats are biochemically different from those in young rats. Despite this finding, the chronic hypertrophy of kidneys in rats given a pathogenic diet is associated in old animals with a decreased capacity of mitochondrial oxidative phosphorylation.

Responses to drugs, anaesthetics and toxic agents have also been used to measure age-changes in adaptation. Increased age is often associated with a loss of drug responsiveness. For example, patients with arthritis are more responsive to steroid therapy in the third decade than in later years. This difference is not related to the duration of the disease. Hexabarbital anaesthesia is prolonged in old rats compared with young ones. Alongside this age-change there is an increased mortality attributable to the drug. The endocrine status is thought to be an important factor underlying this rise in toxicity with age, but neurophysiological and neuropathological factors may well be involved. In this connexion, the toxic action of uranium salts is manifest locally by specific degenerative changes in the kidney tubule which are more marked in old animals. These histological changes in the epithelial cells go in parallel with the loss of functional capacity of the kidney.

Age-dependent failures of adaptive renal responses have been the subject of much detailed investigation. Studies on *p*-aminohippurate clearance indicate that renal plasma flow is diminished in old subjects, although there is no impairment of the ability of the kidney to extract *p*-aminohippurate from the tubules. In contrast, the capacity to reabsorb glucose from the glomerular filtrate shows a linear decline with age (Table 3), which amounts to about 10% from the third to the ninth decade. Changes in glomerular filtration parallel the changes in reabsorptive capacity (Table 3) which may be of significance in relation to the age-dependent depression of glucose tolerance in humans and the failure to induce a reduction in the alkali reserve after treatment of old dogs with anaesthetics. To summarize: there are many physiological functions that deteriorate with age, any one of which may be equated with a loss of vigour in a changing environment.

TABLE 3. Age changes in human kidney function

	Glomerular filtration rate		Tubular reabsorption of glucose	
	Relative mean	Coefficient of variation (%)	Relative mean	Coefficient of variation (%)
Decade				
3rd	1·00	2·86	1·00	5·97
4th	0·93	2·55	0·95	4·33
5th	0·87	5·75	0·92	5·85
6th	0·76	4·96	0·78	6·88
7th	0·71	4·41	0·72	5·18

There is also a considerable body of statistical evidence attesting that, on the average, the ability to exercise and carry out physical work declines with age. This may be related to changes in several factors such as oxygen uptake, pulmonary ventilation, diffusing capacity of lungs, work pulse and blood lactate. Between the ages of 30 and 70, there is a two-fold increase in the time for recovery from exercise.

An important aspect of many adaptive responses is the ability to make rapid neurophysiological adjustments. Thus, there is a consistent decline in the capacity to see at low light intensities and an increased limitation upon the extent of dark adaptation. There are also age-dependent losses in the ability to move rapidly in response to a range of environmental changes.

In long-term responses involving locomotory activities, the impairment in ventilation and gas exchange appears to underlie the age deterioration. However, in short-term reflex delays in the aged, it is suggested that the increased reaction time is due to decreased excitatory influences from higher levels and resides in the nervous system. Apart from the well-established age-involution of muscular tissue and the appearance of "age-pigment" in neurones of the central nervous system there are few indications of the cause of the failure of nervous co-ordination in old individuals. The excitation threshold for electrical stimulation of nerves is higher in old animals than in young ones. Also, there is a decrease in the activity of choline acetylase and choline-esterase, the key enzymes in transmission of the nerve impulse. It may well be that the observed decline in the number of neurones and motor end-plates with age also contributes to the loss of flexibility of nervous responses. However, the most likely cause of the decline in vigour in relation to long-term environmental changes lies in the organization of the endocrine system.

Hormones and ageing

There are indications from parabiosis in rats that systemic factors are important in controlling the ageing process. Also, some of the age-related changes in tissues that are target organs for hormones seem to be reversed when old tissue is transplanted into younger animals. However, evidence for intrinsic tissue-specific ageing processes is very convincing. For example, old ovaries and skin of rodents grafted into young individuals are not rejuvenated. Also, experiments involving the extirpation of endocrine glands and the injection of hormones indicate that hormones by no means fully determine the characteristics of ageing. Endocrine glands are often involved in the manifestation of some signs of old age, but hormones can only modify the rate of ageing in certain tissues.

If it is anticipated that a hormone deficiency is responsible for an ageing phenomenon, the obvious test is to treat the animal with the hormone in an attempt to restore the youthful characters. Studies of this nature with sex steroids indicate that age differences in biochemistry, physiology and behaviour are not related solely to hormone deficiencies.

In view of the possible connexion between the cessation of growth and the onset of ageing, growth hormone treatment has been tried in an effort to retard ageing. However, although the maximum body weight attained during growth in rodents is increased by exogenous growth hormone, the termination of growth is not due to a lack of this hormone in the plasma and life span of rats is not increased by growth hormone.

With regard to other hormones of the pituitary gland, it has been pointed out that several features of old age are similar to those encountered in hypothyroidism. However, there is no evidence for a general decline in the concentration of thyroid hormone with age and most symptoms of old age cannot be alleviated by thyroxine.

In connexion with the neurohypophysis, the characteristic age-related changes in the water and electrolyte content of cells are similar to those in diabetes insipidus and point to an impairment of neurohypophysial function. Support for this idea comes from experiments in which the pattern of salt and water distribution in old rats was partially restored to that characteristic of earlier stages of development by treatment with neurohypophysial hormones. Further, the combined treatment of old rats with posterior pituitary powder and cortisol results in marked prolongation of life. In the old rats used in the above experiments, the neurohypophysis was depleted of neuro-secretory material and failed to respond adequately to an osmotic stimulus. An impairment of pituitary function is also inferred from other experiments, but the histological evidence is often contradictory. From similar work, it is thought that changes in water metabolism in old rats are due mainly to an intrinsic regression in renal function and are not mediated solely through an alteration in the neurohypophysial system.

Homeostasis and mortality

With regard to the failure of homeostasis during ageing, it has been suggested that, in general, there is a diminution of neural influences upon effector organs whilst the sensitivity to hormones increases. As already pointed out, there is much evidence in support of the failure of nervous co-ordination. The evidence for increased hormonal sensitivity comes from experiments on the action of exogenous thyroxine. The data for other hormones are, in the main, in favour of a decreased sensitivity, although the degree of hormonal stimulation may depend upon the previous experience of the animal. In many glands, there is a fall in the rate of hormone turnover.

There can be little doubt that marked age deterioration is a feature of many organs, particularly those with an endocrine function, and the cardiovascular system. Despite this, the individual cells of kidney, muscle and brain do not appear to undergo a drastic loss of functional capacity: it is probably true to say that for the former two organs, the deficiency in old age lies in the loss of cells. Despite changes in cell mass, the ability to maintain day-to-day resting stability in an unchanging environment is largely unimpaired. Lack of adaptability in a disturbed environment is not due to the failure of any single component. The endocrine system probably fails to respond on several counts, bound up with the decreased capacity to secrete sufficient hormone and the failure of the hormone to evoke the correct degree of response in the target organ.

The major problem in the field of gerontology is to relate age changes in function to the increasing probability of death. Where a particular reflex action or rapid neuromuscular response is of importance to an organism in its search for food and shelter, the reasons for the increased chances of mortality may be more or less defined. For example, the sensitivity of rat aorta to catecholamines is reduced in old animals. This is associated with a coarsening of the muscle fibres of the tunica media and an increased thickness and disarrangement of the elastic lamella. At a purely chemical level, the old tissue appears to be more efficient in that it contains a higher concentration of protein and DNA. However, it is concluded that the structural changes alone are sufficient to account for the observed decrease in hormonal sensitivity through a limitation on the mechanical response. The gradual deterioration of effector organs containing muscle fibres and elastic tissue is probably an important factor in the poor adaptive responses of old organisms, where the adaptation requires a rapid neurophysiological response at a much higher intensity than normal. Nevertheless, it must be stressed that animals still die over a narrow age-range when maintained under constant laboratory conditions where no sudden large demands are made upon the body.

It must be admitted that we are largely ignorant of the immediate cause of death in old age. This problem is distinct from the underlying molecular events that produce age-changes at a higher level of organization. From this standpoint, death is a consequence of events of a more complex order and is open to study without reference to the long-term progressive changes in function that lead up to it. It is true that the slow linear trend by which some physiological characters change with age is theoretically sufficient to account for the exponential increase in mortality observed at the population level (see page 314). But death is an individual process and involves events which take place over a small fraction of the life span, possibly amounting to only a few hours. Death, in this context, is the result of an internal fluctuation which cannot be contained through homeostatic regulation; a small shift in metabolism which could be counteracted in youth becomes amplified to the point of preventing a vital function. Thus, death in old organisms may result from minor changes in the environment.

Few experiments have been designed specifically to test the possibility that an increased degree of oscillation about the norm occurs in old animals. With regard to future work, it is desirable to concentrate on the fluctuations in homeostatic control. In the past, too much emphasis has been placed on the maintenance of mean values which are regarded as constant norms. Data for many individual animals are required to establish these statistical norms and the short-term fluctuations within any one individual are largely ignored. In order to understand the increased tendency for homeostasis to fail during ageing, we must first determine just how stable the "constant" internal environment is at all ages, and investigate the reasons for changes in norms and "overshoot phenomena" which in the old organism may ultimately result in death.

Variability and age

It is frequently observed that, as a population of animals ages, the variability of physiological and biochemical parameters increases. There are three possible sources of variability. First, an increased error of measurement. Published data usually give no indication of experimental error and this can only be ruled out as a source of increased variability where there is no change in the mean value with age. The second source of variation is the drift in norms within individuals as expressed by increased genetic and phenotypic variation. In this sense, there are certainly shifts in population norms with age. Thirdly, there is a possibility of increased random variation within individuals due to a greater overshoot and undershoot among the various homeostatic systems at both a physiological and a biochemical level.

The only way to distinguish between an age-dependent variability in norms and a variability in individual homeostasis is to carry out longitudinal studies, but as yet there are not sufficient data from such studies which can be analysed in this way. There are many indications of increased population variability with age and it is unlikely that they are all related simply to measurement errors.

One of the areas of possible functional age-variation is that of blood homeostasis, since the composition of the blood would be expected to mirror any general deterioration in regulatory systems. There is certainly a great variability in the blood constituents amongst humans, but an exhaustive analysis of 21 norms in 478 individuals and 15 in 284 subjects has led to the conclusion that the observed variability is independent of age. This does not rule out age-linked loss of accurate homeostasis because homeostatic oscillations could be masked in natural populations by virtue of great intrinsic genetic and phenotypic variation.

Within inbred laboratory stock, variability does increase with age. One of the first indications of this came from the work on ageing of rotifers, where it was observed that homozygous populations derived from a single egg gradually become more variable with regard to reproductive function. It is difficult to avoid the conclusion that this is due in large part to some kind of random drift in individuals. In-bred rodent populations also show increased variability at both the physiological and biochemical level. However, from the point of view of mechanisms, variability in structure offers the opportunity of distinguishing events at the level of individual cells. Thus, within the age group 65–69 years, there is an increased coefficient of variation for the numbers of mitochondria per human hepatic cell, and the circumference and total size of these organelles (Table 4). Further, the inability to maintain the norm for mitochondrial structure is manifest in the injurious effects of thyroxine which appear with increasing frequency in the cells of senile rats.

TABLE 4. Age changes in the variability of human liver mitochondria

	Coefficient of variation (% of mean)		
	Number/cell	Circumference	Total size
Age			
Under 49	18·8	28·4	23·7
50–59	15·5	27·0	25·4
60–64	18·4	30·3	27·7
65–69	32·1	41·5	41·1
Over 70	19·4	40·6	34·8

Increased variability within a population as a concomitant of age is important because it may increase the chances of death in individuals showing extreme variations. This is particularly so if the variation is associated with a decline in the mean resistance to death. A useful model system has been examined where the resistance to death was measured by the volume of blood that had to be taken from a rat in order to kill it. A steady decline in the resistance to death by experimental haemorrhage of only 12% per decade, when coupled with an increased variability in resistance, resulted in an exponential rise in the chances of death occurring at a particular blood loss (Fig. 5).

B. CELLULAR AGEING

Working on the assumption that the ultimate control of the direction and force of development resides in the cells, it is logical to investigate cell function for the underlying causes of ageing. For the purposes of discussion, there are basically two types of cells in the body. Cells in some adult mammalian tissues may be classed as mitotic cells and are capable of multiplication and replacement. However, highly specialized tissues, such as the central nervous system, together with cardiac and striated muscle, are composed of cells which have lost the power to divide. These post-mitotic cells make up the bulk of living tissue.

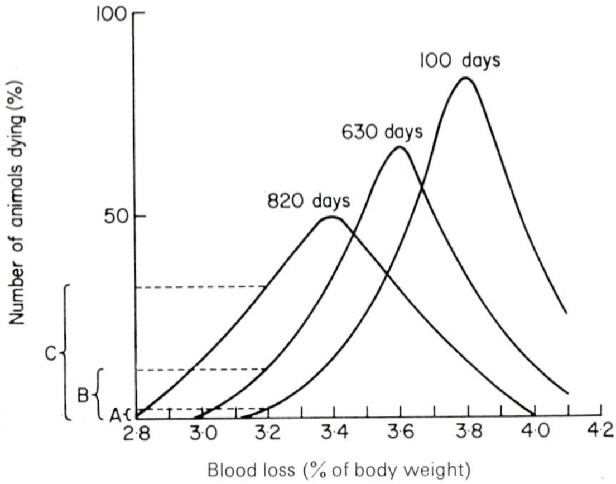

Fig. 5. Experimental haemorrhage as a cause of death in rats. Rats of different ages were bled to the point of death. As they increased in age, the mean amount of blood that had to be removed to cause death decreased and the variability in individual response increased. At a blood loss of 3·2%, there was a non-linear rise in the chances of death occurring with increasing age measured by the heights of A, B and C.

Mitotic cells

For many years, it was held that dividing cells were potentially immortal. Recently, however, experimental work has clearly shown that populations of normal vertebrate somatic cells cultured *in vitro* cannot be maintained indefinitely. For example, cultures of human diploid embryonic fibroblasts undergo approximately fifty doublings then die out, often showing well-defined histological changes with marked chromosome abnormalities. Similar conclusions on the finite life of cells come from experiments in which normal cell populations of skin, bone marrow and mammary tissue have been serially transplanted as their hosts age. There is as yet no clear indication of the mechanism involved in the eventual dying out of these cell populations. Diploid fibroblasts from adults grown *in vitro* show less then half the doubling potential of embryonic cells, suggesting that the cells change progressively *in vivo* from birth. There is evidence that the life span *in vitro* may be extended by modifying the environment, particularly by increasing the concentration of steroid hormones.

Possible explanations for the gradual deterioration of the population involve a progressive loss of some vital self-duplicating cellular entity —death occurring when the concentration of this entity falls below a critical level. The loss of such an entity could occur either through the cumulative effects of random losses or because the entity was itself dividing at a rate somewhat slower than that of the cells. On the other hand, the cells may be progressively changing by the operation of processes akin to differentiation. Workers in this field have suggested that the evidence for a finite lifetime of diploid cells *in vitro* may be the cellular manifestation of ageing *in vivo*.

Normal cell populations which duplicate *in vitro* have been described as cell "strains" to distinguish them from cell "lines" which may arise during the culture of normal cells. Cell lines are capable of indefinite proliferation *in vitro,* but have many abnormal properties. In particular, they may behave like cancer cells on inoculation into living organisms. The existence of cell lines suggests that mechanisms exist for perfect control of cell metabolism. Indeed, one would expect this from the fact that a continuous series of perfect cell division lineages has been maintained through evolutionary time to the germ cells of all present-day organisms. It is also a strong possibility that cell proliferation in other tissues, such as the intestinal epithelium, is a perfect process.

As an alternative to perfect duplication and self-maintenance, it is possible that mechanisms exist to counteract imperfections which inevitably occur. The elimination of imperfections may take place through the presence of error-correcting codes where the genome in any cell is maintained in a large number of replicate forms and errors are periodically eliminated by comparison of all replicates with each other; DNA sequences that do not conform to the predominant pattern are corrected. Cell lines may appear in tissue culture because of the unmasking of a self-correcting code which is normally only expressed in the germ cells.

The fact that there are cells in many organisms which are capable of division raises the possibility that a process of selection could occur within the body to favour the multiplication of cell variants which, during the lifetime of the individual, might prove harmful to the organism. This idea forms the basis for the "clonal selection" theory of ageing. Thus, in cells which generate antibody-forming lymphocytes, random mutation or the loss of mechanisms governing discrimination could allow a lymphocyte population to arise that would not be able to distinguish its own cells as being self. The outcome would be an internal antigen–antibody reaction resulting in the progressive loss of cells from a wide range of organs. This phenomenon, termed "auto-immunity", has been proposed as a mechanism that leads to the progressive loss of post-mitotic cells and evidence has been obtained for autoimmune reactions developing in old individuals.

Ageing in Protozoa. Although cultures of micro-organisms cannot be regarded as analogous to cell cultures, they have been examined from time to time in an attempt to throw light on cellular ageing. Early workers predicted that because protozoans were pure germ plasma, they must inevitably be immortal. However, it was soon found that ciliates underwent a series of temporal changes akin to sexual maturation, maturity and senescence: they aged and died. In the laboratory, a particular cell clone will eventually die out through the fulfilment of this life cycle. New clones may arise by the processes of *conjugation* and *autogamy* which entail, respectively, the introduction of individuals of a different clone and a *kind* of internal fertilization.

Conjugation is a process peculiar to micro-organisms. It is the union of two individuals of the same species for the purposes of exchanging nuclear material. In *Paramecium caudatum*, two organisms come into contact at their oral surfaces. The meganucleus in each individual breaks down and the micronucleus divides by meiosis to give two haploid nuclei. Each new micronucleus divides and of the four, three disappear. The remaining nucleus divides by mitosis and one of the products of this division is exchanged and fuses with the stationary nucleus in the other conjugant. The conjugants separate and the nucleus divides three times. Four of the eight nuclei enlarge to form meganuclei. Each animal divides twice, giving four organisms each with one mega- and one micronucleus (Fig. 6).

Autogamy is a process whereby there is nuclear reorganization without pairing. It occurs at frequent intervals, often alternating with mitosis. The nuclear events are similar to those in conjugation, except that nuclei are not exchanged. Two haploid nuclei formed in the same organism at the fourth division of the micronucleus fuse together. Animals that have passed through autogamy have two sets of identical genes because the fusing nuclei were formed from a haploid nucleus by mitosis. Autogamy constitutes the most extreme example of inbreeding. From a genetic point of view, the advantages gained are obscure since one would not expect any further genetic changes when the organism had passed once through autogamy, although possible re-alignment of genes cannot be ruled out.

The nuclear apparatus is replaced during the both processes. In conjugation, male nuclei are reciprocally exchanged between organisms and a fertilized nucleus is formed in each male. In autogamy, haploid male and female micronuclei from the same individual unite to form a fertilization nucleus. Both kinds of reorganized cells become the starting points for new clones. Progeny of autogamy (exautogamonts) produced early or late in the senescent period develop in several possible ways; to give fully viable clones, which begin new life cycles; to give clones which are either non-viable or divide at markedly reduced rates and soon die out; to give clones with intermediate characteristics. Despite this variability of response, it is clear that the probability of obtaining a new viable clone from a senescent clone is inversely proportional to the age of the parental cells. In extreme old age, new clones either cannot be formed at all, or if formed always die off.

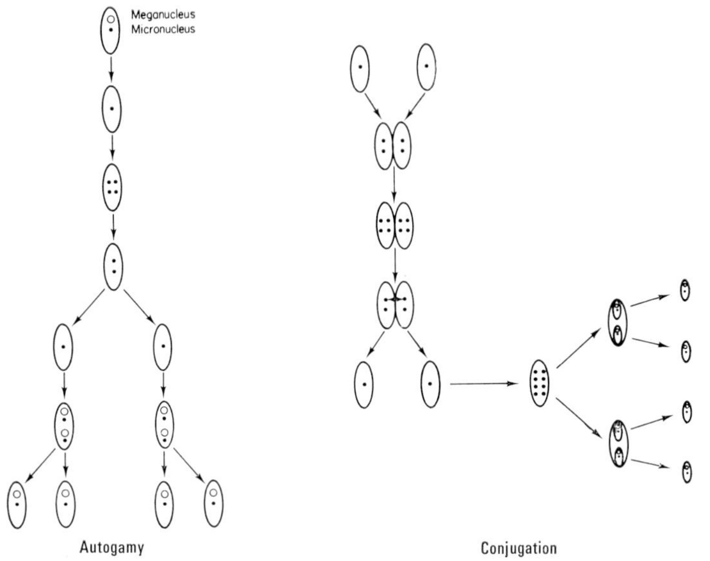

Fig. 6. Conjugation and autogamy in *Paramecium*. For explanation see text.

The earliest evidence for senescence in *Paramecium aurelia* is the appearance of abnormal numbers of micronuclei. In young cells, the norm is 2 : in old cells, the range is 1 to 4. Old cells also have abnormalities in the micronuclear spindles and chromosomes. Other cell components, such as cilia and the gullet, also experience characteristic age changes. By breeding experiments, it has been found that old clones accumulate many detrimental and even lethal mutations which are expressed as both dominant and recessive characters. These genetic changes in the larger of the two nuclei appear rapidly when conjugation is not allowed to occur : lethal changes are noted after only 80 fissions and by about 200 fissions all of the progeny are non-viable.

Under circumstances where a culture has the opportunity of periodic nuclear reorganization, it may be maintained for over 1 000 fissions and old age is avoided.

There is much evidence pointing to the meganucleus as the central organelle which controls the activities of the micronuclei—via the intervening cytoplasm. During late senescence, various structural abnormalities appear in the meganucleus which are not inconsistent with a central hypothesis that senescence involves progressive changes in the function of macronuclear genes. There is also ample data indicating that ageing in ciliates is the result of differential gene expression. The two questions central to the phenomenon of ageing are: What is the mechanism by which cells belonging to a common mitotic lineage and cultivated in a constant environment come to develop an orderly sequence of diverse phenotypes? How does it happen that nuclear reorganization halts the progress of ageing, transforming a given cell with but a limited future into a totipotent individual? The answers to these questions are also pertinent to ageing in higher organisms which resemble protozoans in the progressive manifestation of two inevitable signs of ageing—phenotypic instability and loss of reproductive function.

Despite the obvious age-deterioration in the nucleus, it is by no means certain that the key to ageing phenomena lies primarily in this organelle. There is no reason why we should not turn our attention to the cytoplasm or even the cellular environment for causes of ageing. It may also be wise not to overlook the fact that highly differentiated cells can be derepressed and revert to an unspecialized form so that the ageing process may begin all over again. Neglected material in this context may be found in the invertebrates (de-differentiation of planarians), vertebrates (conversion of iris epithelial cells to lens cells, Chapter 15) and plants (conversion of leaf epithelial cells into meristematic tissue). Many of these events appear to be initiated by virtue of the cells finding themselves in a new environment.

Post-mitotic cells

Because post-mitotic cells cannot be replaced if there is a failure in function, changes in non-dividing cells with the certainty of accumulated faults are likely to be an important cause of ageing. There are three views as to the way in which post-mitotic cells could deteriorate with the passage of time.

Accumulation of inert metabolites. One of the most commonly observed age-deteriorations accepted widely by histologists is the progressive accumulation of lipofuscin granules in certain cells of the body. This "age pigment", as it has been termed, appears in the light microscope as brown granules in nerve cells, seminal vesicles, adrenal cortex, interstitial cells of the gonads and is particularly prominent in cardiac muscle. The accumulation of lipofuscin in the myocardial cells results in a large displacement of muscle volume, and one would anticipate that this would result in a loss of efficiency of the contractile elements. However, there is no correlation with cardiac disease or heart failure.

Age pigment apparently occurs in increasing amounts in a variety of nondividing cells in long-lived animals, but it is not prominent in the cells of short-lived mammals such as rats and mice.

Age pigment has been isolated and attempts made to characterize it in chemical terms. It has a blue-green fluorescence and is partially soluble in aqueous and lipid solvents. Chemical analysis indicates that it is predominantly composed of lipid and protein. The lipid factor is similar to the constituent lipid of the tissue of origin. Some of the components appear to be oxidized products of unsaturated fatty acid residues which are probably responsible for the colour, the fluorescence and the general resistance to enzymic degradation. The protein component is relatively rich in glycine and valine.

Another accumulation, termed "an amyloid deposit", has been observed in old tissues from a range of animal types and its appearance seems to be part of the normal ageing process. There is no firm evidence as to its role in the ageing process, although one group of workers has suggested its appearance indicates the development of auto-immune reactions. In this connexion several human pathological disorders are associated with accumulation of amyloid and this also is put down to linked disturbances of an autoimmune nature.

A recent refinement of the "accumulated waste" theory holds that there are interactions between macromolecules and metabolites or highly reactive intermediates of metabolism (free radicles) which result in the linking together of proteins into large inert aggregations. It is postulated that many of these links are formed by non-enzymic mechanisms, particularly through oxidative reactions. This theory awaits the firm chemical identification of cross-linked polymers and the establishment of a rise in concentration with age.

Somatic mutations. One of the most widely held theories on ageing holds that spontaneous chemical changes occur in the DNA of somatic cells which are analogous to mutations in germ cells, in that they give rise to proteins with abnormal amino-acid sequences. Somatic mutations would be expected to alter the function of those organs composed largely of post-mitotic cells as the tissues gradually become occupied by malfunctioning cells. Alternatively, mutations could result in cell death with organ function deteriorating through the loss of its parts. Major problems surround the identification of mutations. Chromosome abnormalities appear in histological sections of cells in regenerating liver and the frequency of the aberrations increases with age. These aberrations have been equated with mutations. However, there are serious difficulties in accepting the theory of somatic mutations. The frequency of aberrations does not always fit observed differences in life spans. Also, the life-shortening action of radiation, which has been taken to involve somatic mutations, may be explained in other ways. For example, gamma irradiation of Drosophila adults at specific times between 1–20 days of age results in death at a constant time post-exposure. Age at irradiation influences survival time only when the flies are irradiated 30–90 days after emergence. These results are taken to mean that the life-shortening effects of irradiation result from a radiation-induced sickness which, at least for 1–20 day old flies, is unrelated to ageing.

It is established that ionizing radiations may cause specific mutations in the skin pigment cells of mice at the same rate as in spermatogonia. On the other hand, irradiation at doses which would be expected to result in somatic mutations, does not affect the rate of ageing. For humans, the evidence is somewhat indirect and comes mainly from the report of the Atomic Bomb Casualty Commission of the USA. This report covers more than two decades of longitudinal studies on the survivors of the atomic bomb explosions at Hiroshima and Nagasaki. The relevant evidence arises from a comparison of a group who were within two kilometres of the explosion of one of the bombs and had survived acute radiation symptoms, indicating that they had been subjected to a very high dose of radiation, with a matched non-exposed group. There is no evidence of any inter-group differences in cardiovascular function, sensory acuity and the cosmetic aspects of ageing. Immunological tests give no indication that the irradiated group behaved in any respect as if they were older than their chronological age. Thus, there is an absence of evidence to support the idea that human ageing is accelerated by a high dose of radiation and this is also the case with regard to reproductive capacity, ageing of collagen and timing of the incidence of cancer in irradiated mice.

In summary, whilst it is highly probable that somatic mutations occur with age, the available evidence is inadequate to establish that this process is responsible for a general deterioration in function. Even in protozoa, where both dominant and lethal mutations are produced on a fairly large scale in very old individuals, there is some considerable doubt as to whether this is a cause or an effect of ageing.

Incorrect translation. Failing a primary change in DNA through mutation, an abnormal amino acid sequence could be inserted into a protein due to a loss of competence of the cytoplasmic apparatus to correctly translate the genetic message into the polypeptide specified by the nucleus. Such an error would concern the specificity of information handling enzymes and lead to the synthesis of malfunctioning enzymes by virtue of the substitution of abnormal amino acids. Following treatment of micro-organisms and invertebrates with amino acid analogues which give rise to abnormal proteins through amino acid displacement, the organisms show premature senescence and a reduced life span. These results suggest the probable outcome of disturbing the process of translation but, as yet, there is no direct evidence for amino acid substitution occurring on a large scale in ageing organisms. The outcome of either mutation or faulty translation would most likely be a steady decline in biological activity of tissue proteins. As will be made clear in the next section, this is not a general feature of ageing at the molecular level.

Unselected development. Once a mitotic cell has differentiated into a post-mitotic cell, it has entered upon an irreversible developmental pathway and there is a certain degree of probability that it will result in cell death. An extreme form of this theory states that there is an ageing pathway in all post-mitotic cells which is controlled by a special group of "ageing genes" and that in this sense, all specialized cells are dying cells. The origins of this idea stem from a consideration of the proliferation of epidermal cells which, in the process of dying, are converted into the important protective covering of the body.

A "programme theory" of ageing is the only alternative to the mutation theory as an explanation for a change in the direction of metabolism which would alter the relative activities of various genes to produce progressive alterations in the concentrations of intracellular enzymes. This shift in enzyme pattern need not involve special ageing genes and would be no different in principle from those events responsible for differentiation. That is to say, a pattern of genetic information develops in post-mitotic cells with time, which is influenced by events in the extra-nuclear phase. New protein patterns would differ from those of development, in that the changes would be quantitative rather than qualitative. Ageing, by this mechanism, differs from development because the process leads to a loss of efficiency and there has been no selection against random disproportionate alterations in enzyme activity which, on the time scale of ageing, may cause inefficiency at the organ level and possibly death of the cell.

Up to a certain point in time, it may be postulated that these changes are potentially reversible. In other words, there has been no change in the chemistry of DNA, only a progressive opening and closing of information channels leading to other parts of the cell. From this viewpoint, the regeneration of protozoa after autogamy may well be the key to an understanding of the nature of ageing in that "regeneration" could occur by a process which was, in effect, the chemical isolation of DNA from the cytoplasm and its re-establishment in a "youthful" environment, so allowing the specific developmental programme to unfold once more. Evidence that ageing is linked with development comes mainly from experiments showing that slowing down development increases life span.

There is no shortage of evidence that changes in enzyme pattern occur in post-mitotic tissues, although only a few of these changes are obviously connected with a failure in tissue function. On the whole, the age-dependent variations are not very marked. Few experiments have been designed to follow age changes in enzyme activity throughout the life span. Data from laboratory rodents suggests that the period of rapid change occurs during the first few months of life and by the end of the first year of life, the activities have reached a stable level.

Although few comprehensive comparative studies have been accomplished, it would appear that a high proportion of cells in the body maintain enzyme concentrations in old age that are close to those found in the mature adult. As is the case of physiological and morphological processes, it appears that biochemical processes involving enzyme synthesis exhibit ageing mostly in terms of a less effective response to a given stimulus.

Conclusions as to the stability of enzyme concentration patterns rest, for the most part, in the lack of statistical evidence for change. It has been found many times that there is a large difference in mean values in a two-point experiment, but that the deviation from the mean is a large fraction of this difference. Some of the differences in means may become significant if more individual measurements were made (n is usually not more than 10 and often less than 10). Another factor is that by sampling at more than 2 points in time, consistent trends might be established. Thirdly, by analysing for many enzymes simultaneously instead of the usual one or two, patterns might be established.

As an indication of the biochemical constancy at the cellular level, the basal oxygen uptake of humans, expressed per unit of intracellular water, does not show a regression with age. From this it is concluded that there is no impairment in the respiratory metabolism in old cells. The basal whole-body oxygen consumption of adult rats actually increases with age between 10% and 13%. This discrepancy between rats and human subjects has not been resolved, but it is likely that the decrease in muscle mass of rats is counteracted by a higher rate of respiration of cells in other tissues. On the other hand, no age changes have been detected in aerobic metabolism of tissue preparations from rat liver. Similar experiments with kidney indicate only a small age-decrement in aerobic metabolism, possibly due to a fall in the number of mitochondria per cell. The maintenance of aerobic metabolism at the cellular level is borne out by results obtained for guinea-pig tissues. Further, the specific dynamic action of a standard protein meal expressed as excess oxygen uptake is essentially the same in old (mean age 77 years) and young (mean age 24 years) men, although older subjects show a slower response.

Little is known about anaerobic metabolism. Indirect evidence is suggestive of a decrease in the availability of anaerobically produced energy in the ageing brain. Some glycolytic enzymes in rat kidney have been observed to decrease with age by between 15% and 20%. However, these changes are not associated with an impairment of glycolysis measured *in vitro*.

Ageing rat liver and brain are remarkable for their metabolic stability. Liver does not undergo age involution and there is no alteration in protein turnover. It also appears that the enzyme pattern of liver and brain is largely independent of age. Despite this, there is a decrease in nitrogen content of liver, attributable mainly to losses of mitochondria and other membrane structures and a decrease in the RNA content per cell. In humans, there is a rise in the volume ratio of nucleus to cytoplasm. The actual quantity of RNA per nucleus is increased and there is an alteration in the base composition. These latter changes may be connected with the marked increase in gross RNA turnover which is observed in nuclei of a number of tissues in old rats. Despite an alteration in RNA metabolism, cytophotometric measurements of Feulgen-stained nuclei in neurones of the cerebellum and liver cells indicate that there is no loss of DNA from individual cells during ageing. However, a comparison of this work with a histochemical study of the ageing rat submandibular gland, which revealed a decrease in the activity of several enzymes and a fall in RNA concentration, stresses the lack of an ageing pattern common to all tissues.

Most work on the biochemistry of ageing has been carried out using the laboratory rat and it could be argued that the biochemical features of ageing in the laboratory rat are a consequence of inbreeding. This objection has been overcome with regard to some of the phenomena described above, in that they are also found in the ageing wild rat, although there are quantitative differences indicative of a slower rate of ageing in the wild strain when both strains are housed in the laboratory.

Taken together, the biochemical work suggests that each tissue is characterized by a particular pattern of ageing, but to support this idea it is desirable that comprehensive studies be made on each tissue. So far, only one such analysis has been carried out on human blood vessels which indicates that there is a considerable variation in both the extent and pattern of change in enzyme activity. Some enzymes increase in activity (maximum, 3·5 fold rise), others hardly change and some at first increase and then decrease in activity. They may be divided into four groups according to the time when the activity reaches a maximum, but there are no obvious similarities in function of enzymes within any group. Enzymes in group 1 show a steady decline in activity from the first decade amounting to about −5% per decade. Those in group 2 have a peak activity in the second decade (mean 50% increase) then show a decline, which between the third and eighth decade proceeds on average at a rate similar to that for enzymes of group 1. Enzymes in group 3 reach their maximum activity in the third decade (mean, twofold increase) then decline at a rate similar to those in groups 1 and 2. Group 4 contains enzymes that reach their peak between the third and ninth decades. The mean rate of increase to the peak is about the same for all groups.

Taking all enzymes together, there is no evidence to support the idea that there is a general deterioration in metabolism with age. By the ninth decade about 50% of the enzymes retain more than 90% of the activity characteristic of the first decade; only one enzyme retains less than 50% of its activity in the first decade.

Another study, although not so many enzymes were measured, has been carried out on mouse prostate. In this organ, there is no detectable change in total DNA content as the animals age, indicating that cell loss did not occur on a large scale. Enzyme activities, expressed per unit of DNA, fell into two basic patterns: starting at age 9 months, some enzymes showed a steady decline; others showed a rise, followed by a fall. At the age of 30 months, only 5 enzymes out of 11 showed an activity per unit DNA which was less than at 9 months.

Regarding evidence for the operation of programmed ageing at the level of cellular enzymes, it is probable that genotypic or phenotypic influences would, given time, eventually lead to ratios of key enzymes in some individuals which were incompatible with life at the cellular level. This might account for the random death of cells. The expression of this phenomenon at the organismic level might be expected to take the form of, first an increase in variability followed by a decline in variability in the relevant enzymes at the time of peak mortality. This situation is presented diagrammatically in Fig. 7. The shift in mean plus the lowered variability would be the result of selection against those phenotypes with greatly disturbed enzyme patterns, leaving a small fraction of the population to survive into extreme old age with an enzyme pattern closer to that of youth.

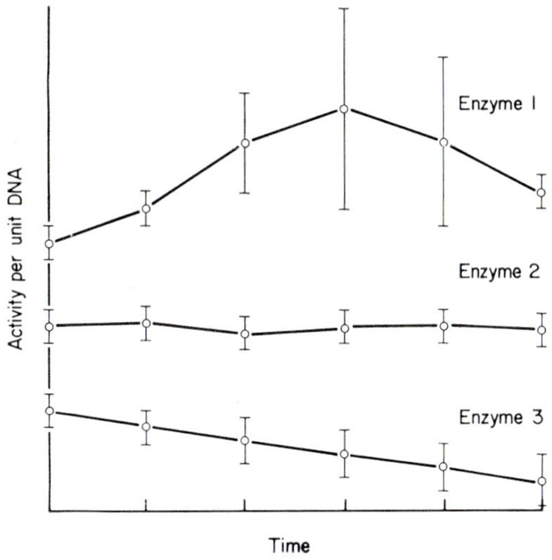

Fig. 7. Hypothetical changes in activities of three types of enzyme with age. 1. Population mean increases with increasing age and shows increased phenotypic variation up to the time of peak mortality. Thereafter, population mean reduced with low variability. 2. No change in the mean concentration and variability. 3. Population mean decreases and shows increasing variability.

It is not an essential feature of this model that all cells in a given tissue change in the same way at the same time. There are variations in the environment of cells dependent upon local differences in the rate of blood flow and the presence of adjacent extracellular structures and other cell types. Thus, basic chemical differences probably exist for cells to take up an individual ageing pattern. Indeed, there is clear histological evidence for age-dependent random variations in cells. For example, hepatic parenchymal cells, scattered either individually or in small groups, show variations in the morphology of cell organelles and functional differences are apparent through histochemical tests. Changes in enzyme pattern, once set in motion, may lead to an early death of the cells and may prove the explanation for the apparent random nature of cell death in post-mitotic tissues.

IV. Evolution and Ageing

In many organisms, death may be regarded as a programmed event, in that it is an essential part of the life cycle. For example, annual plants develop in such a way that food is transferred from parent to seeds during the terminal phase of the reproductive cycle, resulting in the rapid death of the parent plant. In many animals, both invertebrate and vertebrate, similar developmental processes have evolved. In the salmon, for example, during the up-river spawning migration, the irreversible utilization of post-mitotic tissues, mainly skeletal muscle, to provide amino-acids for the formation of eggs and sperm, culminates in rapid death after the gametes are shed and it is difficult to escape the conclusion that the reproductive process is the major cause of death. In these organisms, death is an obvious outcome of the natural selection of an efficient reproductive process. On the other hand, for most organisms, ageing and death appear to be merely the outcome of the extension of a sequence of chemical events which was selected to establish an efficient mechanism for maturation and reproduction. Malfunctions that occur beyond the reproductive phase cannot be corrected because events occurring during this period, which are advantageous in prolonging life, cannot be perpetuated in the gametes through the forces of natural selection. From this aspect, cell loss begins in some organs before the animal is adult. Indeed, it is an essential feature of differentiation that entire organ systems be dismantled to make way for more advantageous structures. During post-natal development of vertebrates, post-mitotic cells begin to disappear from key organs well before the onset of sexual maturity, and the rate of loss continues steadily throughout life. The existence of this phenomenon may be explained, in that organisms can tolerate a certain degree of imperfection in function in keeping with the maintenance of reproductive efficiency. The random appearance of imperfections seriously interferes with the life process only when the organism is allowed to live beyond the natural life span selected through evolution. This applies to animals in captivity and modern man.

V. Control of Ageing

A. Possibilities for Control

Animals tend to have a life span characteristic of the species, implying that ageing is under genetic control. It is this genetic element in the control of longevity which accounts for the thousand-fold variation in the life span between species.

With regard to human ageing, studies on senescent twins have shown that the differences in life span are smaller for monovular than for binovular twins, thus supporting the idea of genetic control of human longevity. The fact that there is a tendency for children to have a life span similar to their parents may be obscured because of the impact of an adverse environment. For example, although having four grandparents surviving to eighty years of age gives four years greater than average life expectancy in the same environment, it has been estimated that life in a city reduced average life expectancy by five years compared with rural living.

During the last hundred years, man has been moving towards a uniform optimal environment where better living conditions involving improved sanitation, housing, medical care and food have contributed towards the lengthening of the average life span. Gains in life-expectancy have occurred almost entirely in the younger age groups. In effect, a person has today a greater chance of reaching old age than in previous times, but if he has already reached old age he cannot expect a much greater chance of living beyond the biblical three score years and ten.

Although it is conceivable that if humans mated selectively according to ancestral longevity, a generally longer life span would evolve, it is unlikely that future improvements in social welfare and medicine will result in a dramatic increase in mean life span. Indeed, if a cure for cancer were found, it has been estimated that the present life expectancy would increase only by between one and two years. At the moment, most deaths occur close to the biological life span and differences in longevity are increasingly due to genetics and congenital influences.

A significant slowing down of the process of human ageing, together with a prolongation of life, are likely to come only through interference with the body's chemistry. At the moment, practical possibilities for the pharmacological control of ageing fall into two categories corresponding to the feeling that the deterioration of the body may result from various inert extra- and intra-cellular accumulations, on the one hand, and from a too rapid rate of metabolism on the other.

Attempts to stimulate the removal of various inert substances such as collagen and lipofuscin have been made on experimental animals and this type of research is in its early stages. Also, taking the view that cross-linking oxidative reactions are responsible for cellular ageing, anti-oxidants are being administered to rodents in an effort to retard ageing. So far, none of these experiments has produced a basis for the practical control of human ageing.

The most successful laboratory experiments in prolonging life involve treatment designed to reduce the rate of metabolism. This may be achieved in cold blooded animals by maintaining a low environmental temperature and in mammals by under-feeding or continuous treatment with catabolic hormones. The starting point for these experiments is that the rate of ageing is bound up with the rate of chemical change and that the latter may be slowed down, either through reducing the availability of metabolities or by administering metabolic inhibitors.

In these experiments, as in others of the same kind, we are handicapped because of the lack of fundamental knowledge of the chemical basis of ageing phenomena. Indeed, if there was a sudden empirical discovery of an anti-ageing compound, we would not know where to begin in order to discover its mode of action.

So far, no discoveries have been made to indicate that ageing is *not* amenable to chemical manipulation and it would be wise to assume that a form of treatment will materialize in the future that will affect some, if not all, of the undesirable aspects of human ageing.

B. DESIRABILITY OF CONTROL

Medical planning to formulate specific therapies for controlling human ageing must take account of the various points at which the ageing process may be attacked. So far, human gerontology has been mainly concerned with the alleviation of the ills that overtake us in old age. Geriatric rehabilitation has been achieved largely through artificial aids for the sense organs, improved welfare services and the design of living conditions which are in keeping with changed inclinations and reduced capabilities. Since it is with the encouragement of society that more and more people will experience the distressing aspects of old age, there can be little argument against developing an effective palliative geriatric service. It also follows that society should make better use of its older citizens.

With regard to the deliberate control of ageing, research has not raised hopes for a rejuvenatory potion and, at best, it seems that man may only look towards a slowing down of certain age changes.

Most relevant laboratory work has been orientated towards increasing life span. At its worst, the application of this approach to humans could result in a long drawn-out childhood and corresponding extensions of middle age and old age. Resultant demographic effects would entail a catastrophic social upheaval.

A more realistic form of control which has the support of many gerontologists is the selective prolongation of the period of life which is the most socially productive. For the majority of the population, this period appears to coincide with the peak in physical fitness between the ages of 20–24 years. However, most assessments of fitness show that physical capabilities describe a broad curve and it is not until the age of about 50 that the capacity to engage in physical work has declined to what it was at the age of 16. Similar curves have been established for scientific productivity and achievement, although there is evidence that physical capabilities develop sooner and decline more rapidly than psychological and intellectual capacities. Thus, a top-level achievement in sports declines by the age of 45 to a level not reached until 70 in the sciences. On the other hand, social skills mature later in life and are often maintained at a high level throughout the remainder of the life span.

The question arises, assuming that we cannot increase our peak efficiency, how should the curve be modified? Should the peak be converted to a plateau? Other difficult questions relate to how many years should be added at peak efficiency. Would we be satisfied with a uniform bonus of, say, five years at the age of 25, followed by a normal rate of ageing, or would it be better to allow individuals freedom to mark time for a limited period at an age of their own choice?

At a deeper level, the significance of the control of ageing in relation to human productivity may be considered in relation to the optimal population size. In terms of increased productivity, the average number of productive years per individual in western society has been increased by over 20 in the last 50 years, mainly through improvements in medicine. Without control of the ageing process, it is likely that future productivity will be sustained only through the maintenance of a large population. In this respect, a prolongation of adult vigour could pave the way to high productivity, coupled with a small population size.

Research has not made it clear whether a prolongation of active life has to be followed by a normal rate of ageing when treatment ceased. Thus, the gain in productivity in youth might be offset by a longer period of non-productive old age. If life span was not prolonged in proportion to the gain in productive years, we would have to face a more rapid rate of deterioration once our bonus period was exhausted.

To use productivity as the main argument for controlling ageing is to over-simplify the situation. Surveys of the mean relative productivity of workers usually show that output remains stable well into the mid-fifties, with only a slight falling-off up to retirement age. Also, there is a great variability between individuals. These results probably reflect the fact that modern working conditions are designed to allow an output that is well within the limits of individual performance. In those jobs that do demand near maximum output, there is a tendency for workers to shift to a less demanding job in later life. Agitation to lower the individual work load is likely to continue in the future. Set against these trends, it is difficult to use productivity as a yardstick to measure the advantages of allowing everyone to prolong their most productive years.

The other argument that has been advanced to highlight the desirability of controlling human ageing rests on the feeling that a new opportunity to determine one's pattern of ageing would add an extra dimension to man's psychological life. An important qualification is that immortality is not within man's grasp and freedom to add a specified number of years to one's life, and no more, is likely to turn out to be no freedom at all. Also, it is difficult to escape the conclusion that, should an anti-ageing pill be developed, society should have strict control of its use according to some desirable demographic plan.

In conclusion, although gerontology is only just beginning to unravel the biological aspects of ageing, it is gaining momentum in the search for an antidote for human ageing, despite the fact that there seem to be no overwhelming advantages to be gained. This does not mean that we must revert to a policy of "let well alone". Working simply on the principle that there is always some environment that is optimum for the age and state of a particular organism, society has a mandate to devote more money and effort to adjusting man's physical, biological and social conditions so that his extra years won over the last century may be accommodated "successfully". There can be little doubt that pharmacological control will be applied increasingly to this area in the future and experience gained in this field should pave the way to a better understanding of the implications of controlling man's biological age.

Further Reading

Allison, R. S. (1962). "The Senile Brain". Edward Arnold, London.
Birren, J. E. (1964). "The Psychology of Ageing". Prentice Hall, New Jersey.
Comfort, A. (1965). "The Process of Ageing". Weidenfeld and Nicolson, London.
Exton-Smith, A. N. (1955) "Medical Problems of Old Age". John Wright and Sons, Bristol.

Cellular Specialization

17. The Neurone

J. D. W. TOMLINSON
Faculty of Medicine, Memorial University of Newfoundland, St. Johns, Newfoundland

I. Introduction

Nerve cells or neurones (the latter term being first employed by Waldeyer in 1891) are the basic anatomical and functional units of the nervous system. They are among the most highly specialized of cells and knowledge of their morphology and functioning has proved more difficult to determine than is the case for most other cell-types, although great advances have been made in recent times through the advent of the electron microscope and sophisticated physiological and biochemical techniques.

Nerve fibres were recognized by Leeuwenhoek at the end of the 17th Century, whilst what is now termed the nerve cell body or perikaryon was observed by Dutrochet in 1824 and described by Ehrenburg in 1833. However it took some ten years or so before the all-important fact, that the cell body and nerve fibre are continuous with each other, was established by the work of Remak, Schwann, Wagner and Kölliker; later Deiters, in 1865, was able to distinguish definitively between the two types of nerve cell processes now termed dendrites and axons. A more detailed early history of the neurone can be found in van der Loos (1967).

A. NEURONE THEORY

A long and often bitter controversy occurred over the question of whether the nerve cell bodies, together with their processes, formed distinct and separate entities, or whether they were arranged as a reticulum or network joined together by protoplasmic or fibrillary connections. There were distinguished protagonists on either side of the argument—Golgi and Nissl as "reticularists" and Cajal and Waldeyer for what became known as the "neurone theory". The latter concept has gradually prevailed, though definitive proof was obtained less than twenty years ago from evidence obtained by electron microscopy (see Palay, 1956). Thus each neurone is a separate, and indeed unique, cell, consisting of a cell body or perikaryon containing a nucleus and cytoplasm, together with their respective constituent organelles, and possessing one axon and at least one, and usually many more, dendritic processes (Fig. 1); such cells and processes are often in close contiguity, but never continuity, with other nerve cells, being separated from them by a surrounding plasma membrane approximately 7–8 nm thick (Peters *et al.*, 1970) which is somewhat thinner than that usually found in other cells.

B. CLASSIFICATION AND NUMBERS

Neurones probably exhibit a greater pleiomorphism than any other cell type, both in the morphology of their perikarya and in the number, arrangement and length of their processes. They have been classified (Shantha *et al.*, 1969) according to the number of processes (unipolar, bipolar and multipolar) length of axon (Golgi type I with axons up to a metre in length, Golgi type II with axons of under 1 mm), or according to their functional activities (sensory, motor, excitatory or inhibitory).

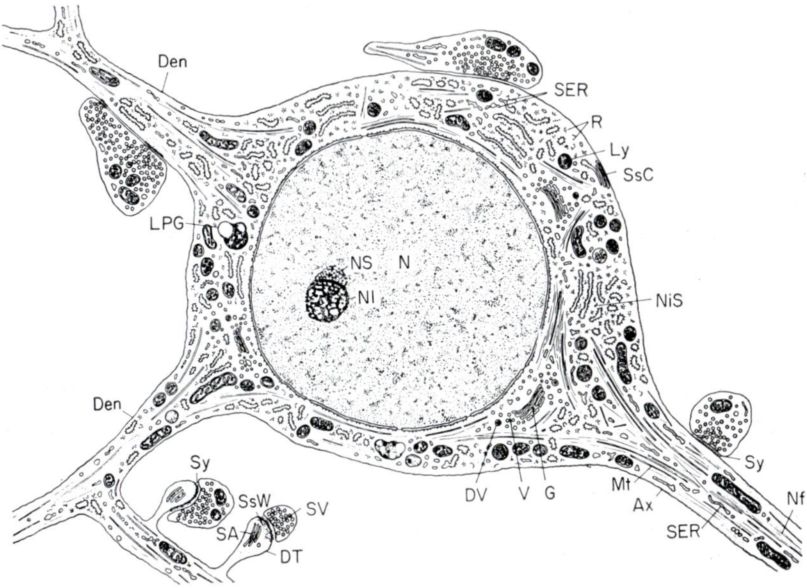

Fig. 1. Electronmicrograph drawing, showing principal features of a multipolar neurone Ax, axon; Den, dendrite; DT, dendritic thorn or spine; DV, dense-cored vesicles; G, Golgi apparatus; LPG, lipofuscin granule; Ly, lysosome; Mt, microtubules; N, nucleus; Nf, neurofilaments; Nl, nucleolus; NS, nucleolar satellite; NiS, Nissl substance; R, ribosome clusters; SA, spine apparatus; SER, smooth endoplasmic reticulum; SsC, Sub-vesicles. From Lentz, 1971. "Cell Fine Structure", W. B. Saunders Co., Philadelphia, London, Toronto.

The number of neurones possessed by a particular organism is extremely variable, and seemingly related to the range of behaviour of that organism. Some nematode worms have only one or two hundred nerve cells, the octopus brain some 200 million (though its "arms" contain almost twice as many as its brain), whilst the human brain has been estimated to contain some twenty thousand million neurones.

Exactly at what phylogenetic stage in the animal kingdom nerve cells first appear is controversial. It has been maintained, on histochemical and other grounds, that they exist in the sponges, whilst others doubt that such cells are in fact neurones.

Indeed, the criteria used to decide whether a particular cell is or is not a nerve cell are far from being absolute. Neurones should "look like nerve cells", display affinity for methylene blue and silver stains, and possess a process, the axon, which can influence other cells, not necessarily only in the immediate neighbourhood of the cell body, by the conduction of an impulse or signal along it. In electron micrographs the sites of such influence or control should display certain specialized characteristics such as vesicles and/or membrane modifications.

C. HISTOGENESIS

Some neurones, such as those contained in certain hypothalamic nuclei, display the phenomenon of neuro-secretion and it has been suggested that neurones may have first appeared as such secretory or regulatory cells, and only later evolved into rapid impulse conductors, usually exhibiting chemical transmission at axonal terminals (Horridge, 1968).

It is beyond the scope of this chapter to discuss the histogenesis of nerve cells in any detail, but some points may be mentioned.

After the formation of the vertebrate neural groove and tube and the establishment of ependymal and mantle layers, the classical description was that the primitive nerve cells or neuroblasts arose from rounded germinal cells in the ependymal (or luminal) layer, whilst the more columnar epithelial cells gave rise to spongioblasts or primitive glial cells. Doubt was early expressed as to whether this "dualist" theory was correct. In recent years, the employment of drugs such as colchicine (which blocks mitosis in metaphase), the technique of microspectrophotometry for DNA, and the use of tritiated thymidine combined with radioautography and electron microscopy have all produced evidence to suggest that it is not. The present concept (see Langman, 1968) is that in the early stages of development there exists only one type of cell, termed the neuroepithelial cell. These extend from the internal luminal surface of the neural tube, where they are "anchored" by terminal bars, through to the external membrane, i.e. they constitute both ependymal and mantle layers. These cells synthesize DNA in the outer mantle zone, and then contract down, round out and divide mitotically in the inner zone, following which the daughter cells elongate again, but still remain anchored at the luminal surface by terminal bars. Neuroblasts differentiate from the neuroepithelial cells, from which they are distinguished by their large round nuclei and pale cytoplasm. They have been shown to exhibit a so-called "transient dendrite" which temporarily links them to the internal layer, but this soon disappears. Once differentiated, they do not thereafter show evidence of DNA synthesis.

Two possible methods of their "release" have been postulated: either the anchoring terminal bars rupture, or division of the appropriate neuroepithelial cells occurs in a plane parallel to the luminal surface, so that one daughter neuroblast is "freed" from the ependymal zone. Mitotic orientation suggesting this latter method has been described, but it seems probable that the former may be the more usual method of neuroblast release.

II. Nucleus

A. NUCLEAR MEMBRANE

Each mature neurone (except some sympathetic ganglion cells) possess only one nucleus, which is usually centrally placed and surrounded by a nuclear envelope shown in electron micrographs to consist of two membranes, each 7 nm thick. The inner membrane is smooth whilst the outer exhibits a wavy appearance. The two appear to join at intervals to form "nuclear pores", and thus the apparent double membrane is probably a single one surrounding a "perinuclear cistern". This latter is said to communicate in places with the endoplasmic reticular cisternae, and thus the nuclear envelope has been regarded as a "specialization" of endoplasmic reticulum (see Peters *et al.*, 1970).

The structure of nuclear pores is probably similar in whatever cell they are found; they are circular in shape and when cut tangentially many display a central granule or "pore plug" 20 nm in diameter. This has been suggested to be RNA-containing material on the way from nucleus to cytoplasm. The pore boundary or annulus consists of two rings of eight peripheral electron dense granules 18 nm in diameter, one set on the nuclear and the other the cytoplasmic side of the pore. Roberts and Northcote (1970) have recently commented that "what the various granules represent, beyond the knowledge that they are ribonucleoprotein (Mitres, 1969) or how the pore operates in transport mechanisms remains obscure".

B. NUCLEOPLASM

The nucleoplasm (karyoplasm) is gelatinous in living nerve cells, and electron microscopy shows the contained chromatin mainly in the form of fibrillary strands about 20 nm in diameter, to be widely and irregularly dispersed throughout it though in the smaller neurones some irregular clumping occurs in various parts of the nucleus. In addition, granules varying in size from 20–35 nm are scattered in the matrix (Fig. 1).

C. NUCLEOLUS

Nerve cell nuclei usually contain at least one nucleolus, and occasionally two or more: up to ten have been seen in a single section through nuclei of rabbit retinal cells. In mammals they vary in size up to about 5 μm, and as a general rule the larger and more adult the neurone the larger the nucleolus it possesses. In man nucleoli of volumes up to 60 μm^3 have been described; in fish neurones they may be 100 times as large.

Electron microscopy shows the nucleolus to contain masses of closely packed filaments together with granules 15–20 nm in diameter, comprising the pars fibrosa and pars granulosa respectively (Fig. 2). In neurones, in contrast with certain other cells, these two are closely mingled with each other. Clear areas are present between the electron-dense elements which give many nucleoli a vacuolated appearance. Both filaments and granules contain ribonucleoprotein. The majority of the nuclear RNA is located within the nucleolus, where its concentration reaches 1–2%.

Though usually situated centrally within the nucleus the nucleolus often migrates to the nuclear membrane. Such migrations may be for the purpose of discharging synthesized material (possibly RNA) from the nucleolus into the perinuclear cytoplasm. In addition, or alternatively, such movement may serve the purpose of some reverse exchange across the membrane e.g. the nucleolus may receive ATP (Shantha *et al.*, 1969); Lindsay and Barr (1955), studying the behaviour of the nucleolus in neurones of the cat's hypoglossal nucleus during depletion and restoration of Nissl material following electrical stimulation of the hypoglossal nerve did not describe such nucleolar movement, but Rees (1971) demonstrated it in the rats hypoglossal nucleus following section of the nerve.

D. NUCLEOLAR SATELLITE

Barr and Bertram (1949) described, in female cat neurones, an intranuclear mass, closely associated with the nucleolus, which was absent in male cells. It was soon found to be identifiable in female cells from many species, including man. This body is the heterochromatic X chromosome and electron microscope studies show it to be composed of a mass of coiled fibrils, tightly wrapped round each other, closely attached to the surface of the nucleolus (Fig. 2).

E. ACCESSORY BODY OF CAJAL

In some, but by no means all, neurones stained with silver one or occasionally more spherical argyrophilic bodies can be seen within the nucleus. First described by Cajal, these measure 0·5–1·0 μm in diameter, and are situated in the nucleoplasm between the nucleolus and the nuclear membrane. It has been found in both male and female cells and tends to shrink and also to move further away from the nucleolus during chromatolysis. The former two properties serve to distinguish it from the nucleolar satellite, from which it also differs in possessing no detectable DNA. Indeed, in suitable preparations stained with protargol and then with methyl green the two structures have been demonstrated to co-exist in a single neuronal nucleus.

Haggar and Barr (1968) have demonstrated that accessory bodies

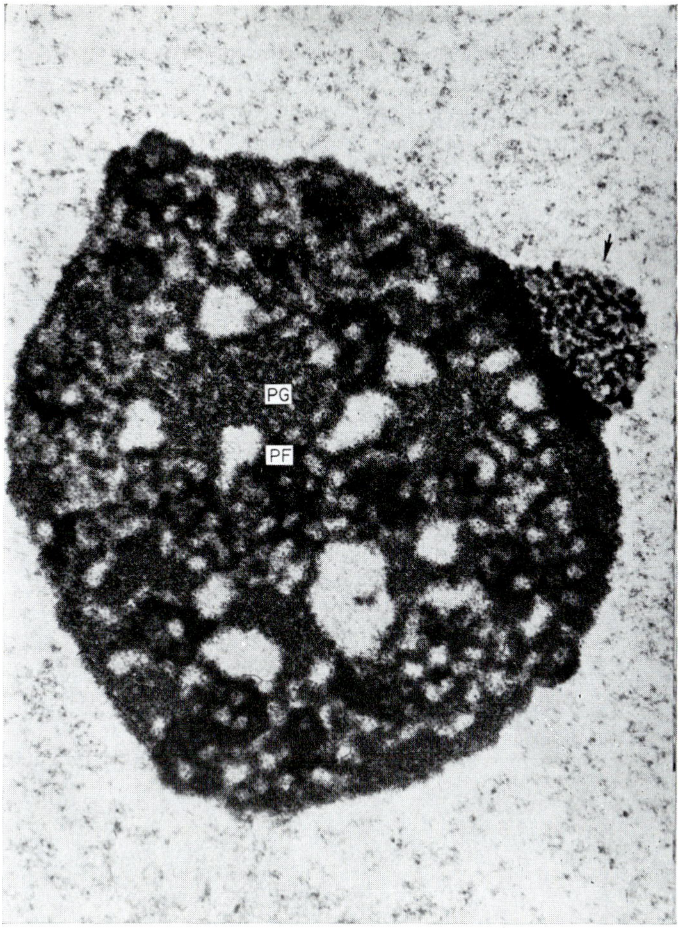

Fig. 2. Electronmicrograph of nucleolus in a female rat spinal neurone. PG, pars granulosa; PF, pars fibrosa. Note vacuolated areas within the nucleolus. The arrow indicates the nucleolar satellite. × 33 000. From Peters, Palay and Webster 1970.

possess proteins containing tyrosine and arginine and also lipoproteins. It may be derived from the nucleolus itself, but its function is not yet known.

F. NUCLEAR RODLETS

The description of rodlets within the nuclei of certain nerve cells in some species dates back to the end of the last century. They were termed "batonnets intranucléaires" by Cajal (1909–11). The problem of their nature, derivation and fine structure has been recently discussed by a number of authors. Siegesmund *et al.* (1964) were in no doubt that the bundles of closely packed fibrils of 5–7 nm described by them in electron microscope studies on rabbit, squirrel and monkey neurones following osmium fixation were the intranuclear rodlets seen in light microscopy by earlier workers (Fig. 3). It is however difficult to reconcile these rod-like structures with those reported by Colonnier (1965) in neurones of cat visual cortex fixed by glutaraldehyde prefusion. These appear not only as definite infoldings of nuclear membrane, complete with pores, but also to be quite often complex structures, single, double, bi- and tripartite, or even coiled, and often in close relationship with the nucleolus. The intranuclear rodlets of Masurovsky *et al.* (1969) consist of bundles of 5–7 nm fibrils associated with granulo-fibrillar bodies 0·4–0·8 μm in diameter.

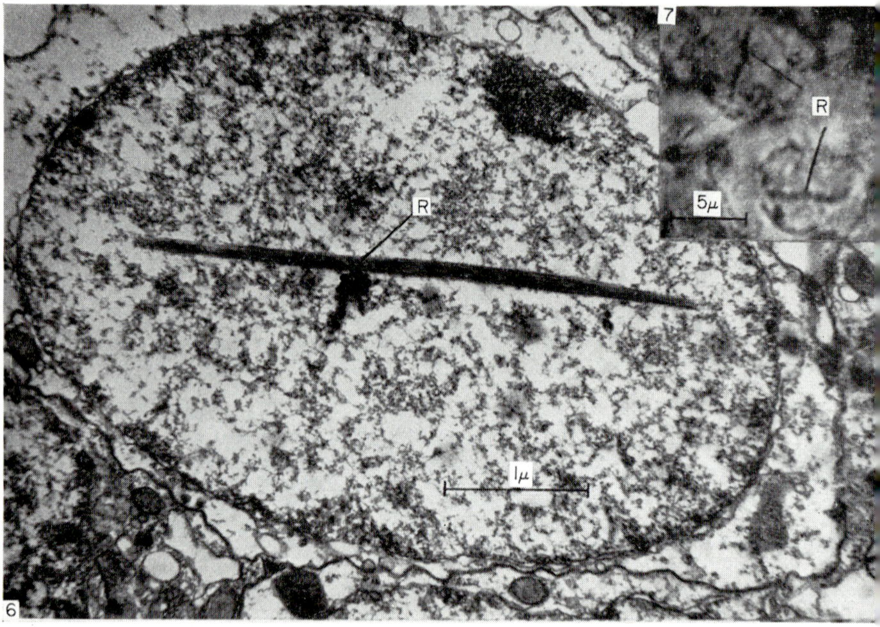

Fig. 3. Electronmicrograph of rabbit cerebellar granule cell showing intranuclear rodlet (R). ×25 000. From Siegesmund, Dutta and Fox (1964).

Until more is known about these structures their functional significance must remain unknown. Indeed whether the various nuclear inclusions described as "rodlets" by different authors are in fact the same structures is far from certain.

III. Cytoplasm

A. CELL MEMBRANE

The nerve cell, including its processes, is surrounded by a plasma membrane, approximately 8 nm in total thickness, which in high-power electron micrographs can be shown to be composed of three layers, two dark sandwiching a lighter third. At certain sites, notably at synapses, this membrane presents distinctive modifications (see Chapter 18). In light microscopy, "boutons terminaux" mark these sites (Fig. 4).

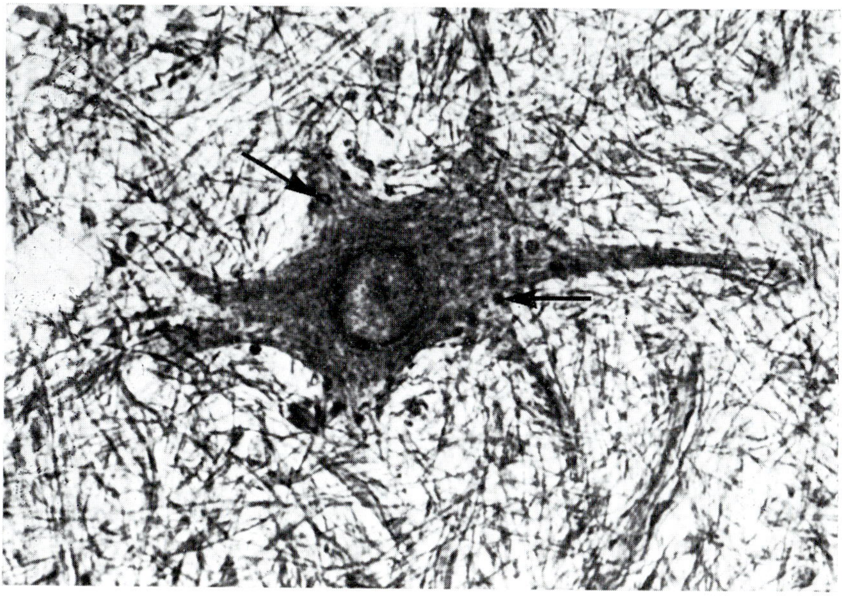

Fig. 4. Cat spinal motoneurone, silver impregnation, showing "boutons terminaux" (arrowed). ×480.

The cytoplasm is gelatinous in nature, and shows little tendency to flow out of the cell when the plasma membrane is ruptured. It contains numerous organelles few of which, however, are peculiar to neurones.

B. NISSL SUBSTANCE

Nissl bodies are one of the most striking features of nerve cells when stained with basic dyes such as methylene blue or cresyl violet, though different neurones exhibit marked variations in the size and distribution of these bodies. They are particularly prominent in large spinal motoneurones (Fig. 5) (cf. other large cells such as Purkinje and the giant cells of Dieters nucleus which have small Nissl bodies) whilst in most autonomic and small sensory neurones the Nissl substance is finely granular and evenly distributed throughout the cytoplasm. It should be emphasized that, although small neurones probably invariably have widely dispersed fine Nissl granules, large nerve cells by no means always possess large masses; indeed in the giant Mauthner neurone found in fishes it is hardly demonstrable, so widely is it scattered.

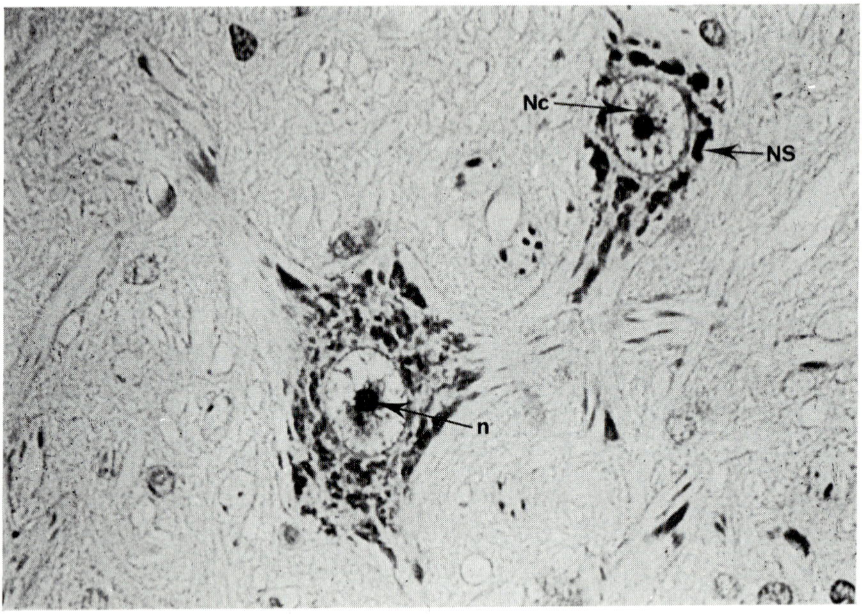

Fig. 5. Human spinal motoneurone, cresyl violet stain. NS, Nissl substance; n, nucleolus; Nc, nucleus. × 480.

For a number of years after their description in the last decade of the nineteenth century it was considered by numerous workers that Nissl granules were precipitative fixation artefacts. However in the early 1930's Benley and Gersh showed they could be demonstrated in freeze-dried neurones, and Beames and King that they could be displaced to the periphery of the cell by high-speed centrifugation. More recently Deitch and Murray (1956), by use of phase-contrast microscopy, demonstrated that they could be seen in living neurones. That Nissl substance is a distinct constituent organelle of living neuronal cytoplasm can hardly now be doubted. It was shown independently in 1940 by Caspersson (using ultra-violet absorption techniques), and by Brachet and Kunitz (studying ribonuclease reaction), that Nissl substance was largely composed of RNA, and the presence of the four ribonucleotides was demonstrated in rabbit motoneurones by Edstrom and Hydén in 1954.

Electron microscopy has revealed that the Nissl substance is composed of a collection of intercommunicating cisternae, arranged in layers, which form the rough or granular endoplasmic reticulum (Fig. 6). The external surface of the cisternal membranes have ribosomes lying in contact with them, though the majority of ribosomes are arranged as polysomes in rows, spirals, clusters or rosettes situated in the matrix between contiguous cisternae.

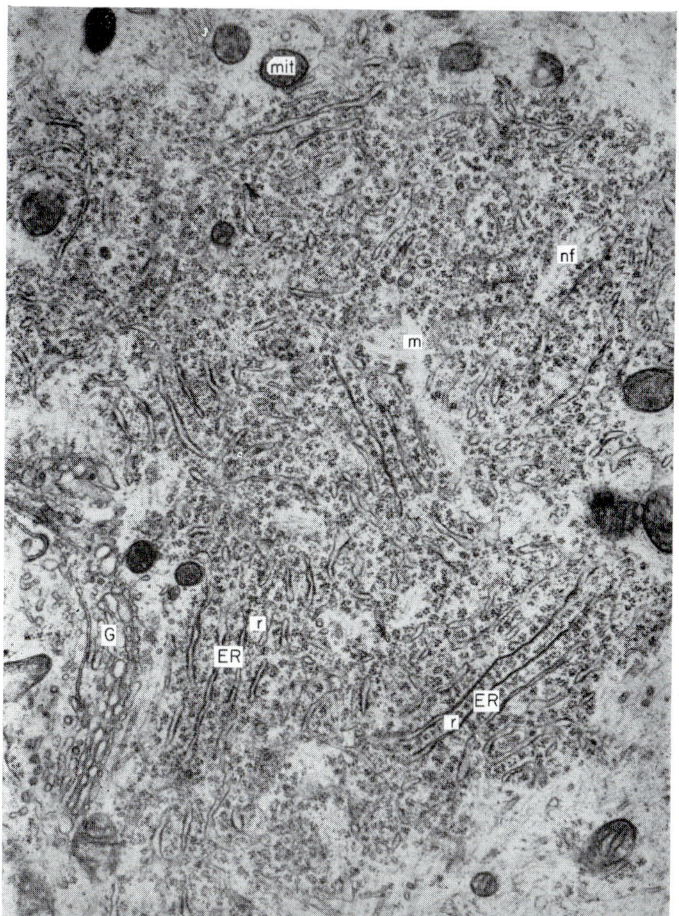

Fig. 6. Rat anterior horn cell, showing mitochondria (mit); microtubules (m); neurofilaments (nf); Golgi apparatus (G). ER is the granular endoplasmic reticulum of Nissl substance, between the cisternae of which can be seen collections of ribosomes (r). ×29 000. From Peters, Palay and Webster, 1970.

Continuity between this rough endoplasmic reticulum and the smooth or agranular type has often been demonstrated in nerve cells, and it has been emphasized that the Nissl bodies should be regarded not so much as discrete entities in neuronal cytoplasm, but rather as nodal points in the all-pervading endoplasmic reticulum of the cell where the principal protein synthesizing apparatus is concentrated (Peters *et al.*, 1970). These authors suggest that the Nissl ribosomes synthesize proteins which are used more for local utilization i.e. maintenance and renewal than for "exportation".

The arrangement pattern of the polyribosomes varies in different neurones. In the giant cells of Deiters nucleus they are less commonly attached to the endoplasmic reticular membranes than they are in either Purkinje cells or motoneurones and most of them, whether attached or free, consist of rows or clusters of only 4–6 ribosomes—many less than obtains in the polysomes of Purkinje cells. Since it appears that the size of polyribosomes is related to the type of protein molecule synthesized, such differences suggest that certain neurones may produce some proteins in some way distinctive to themselves (see Sotelo and Palay 1968).

Radioautography has enabled the protein metabolism of individual neurones to be examined, and shows that they have a very high rate of protein synthesis compared with other cells. This increases with the cells functional activity which results in increased nucleotide and protein metabolism. Protein synthesis occurs in the perikaryon, the nucleus itself, and basal dendrites, but not in the axon (or if so, in exceedingly small amounts, see p. 374).

In ganglion cells, for example, the highest concentration of newly synthesized protein (two minutes after injection of labelled precursors) is found in the Nissl substance in the perikaryon and in the nucleolus. "Sedentary" proteins are said to remain in the perikaryon for up to two weeks, whilst "migratory" proteins pass within 24 h from the cell body into the axon and dendrites. Part of the protein formed by the Nissl substance ribosomes passes to and accumulates in the Golgi apparatus, which may provide the protein content of primary lysosomes, whilst another fraction may play a part in the formation of neurofilaments and microtubules (see Droz, 1969). It has recently been shown that administration of ACTH increases protein synthesis of rat and mice spinal motoneurones by 25%.

It has been estimated that large nerve cells replace approximately 30% of their protein every 24 h. Most of this is concerned with maintenance and renewal of its protoplasmic constituents, not only in the perikaryon but also in the dendrites and axon.

C. GOLGI APPARATUS

The Golgi apparatus, commonly (though erroneously) stated to have been first described in Purkinje cells of the Barn Owl, appears in light-microscopic silver preparations (and, more recently, by the thiamine pyrophosphatase reaction) as a reticular complex of branching cords and droplets which, particularly in smaller neurones, closely surround the nucleus. Larger neurones tend to exhibit a rather more dispersed Golgi distribution.

The history of the Golgi apparatus has been extremely controversial. Its presence, early appreciated, in cells of both plant and animal kingdoms has led to many diverse views as to its nature. It has been recently pointed out that well over 2 000 publications are at present extant on the subject, and over a hundred different names have been attached to the complex, or to structures considered to be either the whole or part of it. The variations in its morphology both in the same cell type and also in different cells would indicate that it is a plastic structure whose appearance is indicative of the state of activity of the cell at any given moment. There would appear to be little evidence at present to indicate that the Golgi complex in nerve cells differs significantly either structurally or functionally from that in other cells.

Electron microscopy shows that the neuronal Golgi complex consists of closely packed large cisternae usually arranged in curved stacks some 5–7 layers thick (Fig. 1) (other cells may have 3–12 layers). This arrangement serves to distinguish the Golgi apparatus from the rest of the smooth endoplasmic reticulum. Cisternae of one level are seen freely to anastomose with one another but rarely with the layer above or below, though fenestrations of the walls of cisternae are said to be a prominent feature. Small round or ellipsoid vesicles are associated with the cisternae, the majority of these being smooth walled, though some are "coated"—the so-called alveolate vesicles are known to contain the elementary neurosecretory granules of cells that possess this faculty.

D. LYSOSOMES AND RELATED PARTICLES

Lysosomes are a prominent feature of the cytoplasm of neurones (Fig. 1). They do not appear to differ either structurally or cytochemically from those described in cells of other tissues. Most neuronal lysosomes are under 0·5 μm in diameter though some are as large as 2 μm. Their shape is most usually round or oval, but other forms are often encountered. In electron micrographs they are shown to be membrane bounded consisting of a finely granular dense matrix which may contain osmiophilic droplets, vesicles, membranous material or other inclusions.

The lysosomal system is discussed in Chapter 7. Its inter-relationship with the Golgi complex and other cytoplasmic inclusions, such as residual bodies, will not be discussed here but it can be stated that autophagic vacuoles (lysosomes involved in digestion of portions of the cells own cytoplasm or organelles such as mitochondria) are not conspicuous in normal nerve cells, but do increase in number if the cell is damaged, e.g. by axonal section or following a period of anoxia.

Why a long-lived cell such as the neurone possesses so many lysosomes (does this reflect the metabolic requirements of a cell with very extensive processes?) is far from being adequately explained at present. For detailed information on the Golgi complex in general, and neural lysosomes in particular, reviews by Novikoff (1967) Beames and Kessel (1968) and Koenig (1969) should be consulted.

E. LIPOFUSCIN GRANULES

Lipofuscin granules (Fig. 7) are usually regarded as "wear and tear" structures, as their numbers increase with the age of the neurone. Such granules are found throughout the cytoplasm of the cell body and in the basal dendrites but not in the axon. Their size varies from 1–3 μm, the larger tending to clump close together, and are often lobulated. In unstained preparations, viewed by light microscopy, they appear yellow or brown, due to contained lipofuscin pigment, and in electron micrographs as membrane-bounded dense structures, often containing vacuoles and small dense particles or even presenting a whorl-like laminated appearance. Different neurones in the same animal show variations in the number of lipfuscin granules; they are less common in cerebellar granule and Purkinje cells than in spinal motoneurones.

The most commonly held view is that the lipofuscin granules represent a coalescence of ageing autolysosomes where the contained pigment material is derived from the breakdown of unsaturated lipids: they have been shown to exhibit acid phosphatase activity which becomes peripherally situated as more lipofuscin pigment accumulates in the granules.

F. MELANIN PIGMENT

Neurones in certain nuclei of the brain, notably the substantia nigra and locus coeruleus, contain granules of this pigment in their cytoplasm (Fig 8). Cells from the area compacta of the former contain so much pigment that their internal morphology is largely obscured.

The site from which this pigment is produced is probably the Golgi apparatus, though the actual mechanism of its production is not known. In Parkinsonism there is an almost complete removal of melanin from nigral cells, and it is likely that the pigment is in some way involved in dopamine metabolism.

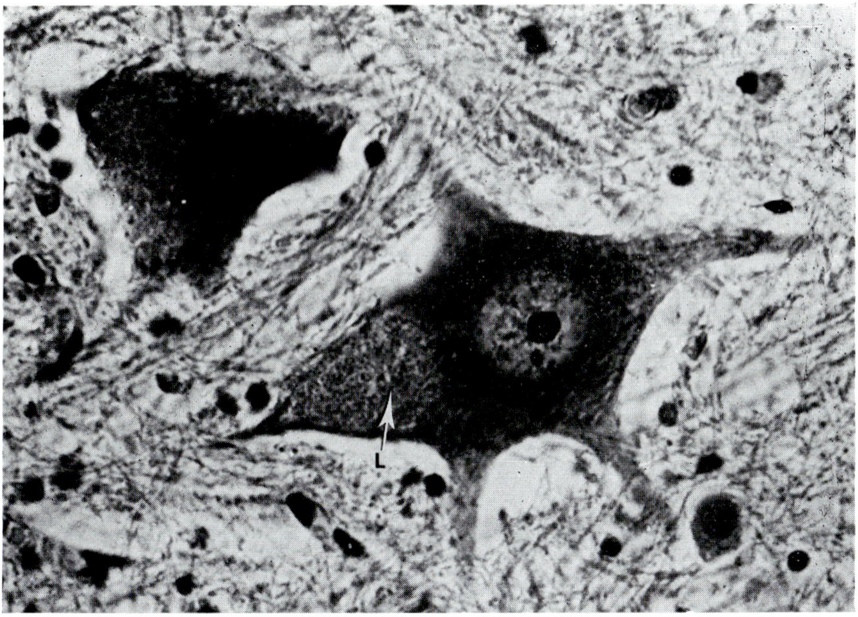

Fig. 7. Human spinal motoneurone, showing collection of lipofuscin granules (L). H & E stain, ×400.

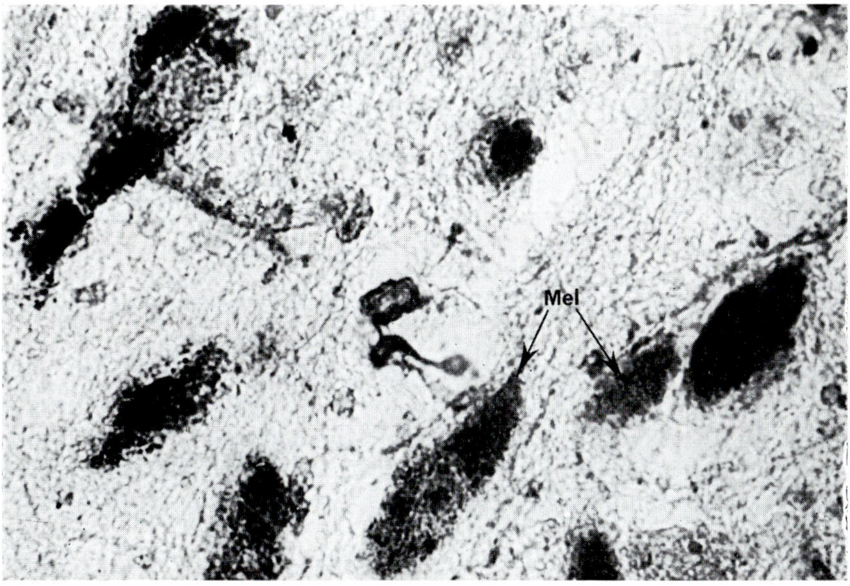

Fig. 8. Distribution of melanin pigment granules (Mel) in neurones of human substantia nigra. Holzer stain, ×480.

It may be stated that in albinos there is usually no reduction in the melanin contained in their nerve cells, so presumably the mechanism of neuronal melanin formation differs from that of melanocytes.

G. MITOCHONDRIA

The mitochondria in nerve cells do not appear to differ greatly from those in cells of other tissues (Fig. 6). They are most commonly approximately 1 μm long and 0·1 μm in diameter, but exhibit considerable variation in size, e.g. in the distal portions of dendrites they may reach 20 μm in length : branching is not uncommon.

Their morphology in a particular neurone is almost certainly dependent on the functional state of the cell. For instance, marked morphological lability has been demonstrated after barbiturate administration and tissue culture observations have shown that mitochondria are continuously changing their shape, size and location in the cell. They tend to surround, but rarely invade, the Nissl substance and, apart from their function as sites of oxidative phosphorylation, they have been shown, by leucine–^{3}H injections, to be sites of protein synthesis (Chapter 6). One of the characteristics of the mitochondria of nerve cells is said to be that they contain a lesser number of dense matrix granules (? hydroxyapatite) than those in, for example, muscle and liver cells, though the significance of this finding is uncertain. In addition, their cristae are commonly longitudinally rather than transversely orientated.

H. CENTROSOME

The presence of a centrosome in neurones such as spinal ganglion cells has been known for many years. The possession of this structure by adult nerve cells is surprising, though it must be said that it is more frequent and larger in young neurones than in older ones, and may represent a non-functional remnant which tends to degenerate with age.

As to the question whether mature nerve cells are able to divide the almost universal belief is that they cannot. If in certain circumstances they could do so it is difficult to conceive what would happen with regard to their dendrites and axonal processes, and the intimate connections of these with other nerve cells. Amitotic division, or appearances to suggest this, have been described in adult neurones from certain species of animals, by various authors, but whether this results in degeneration and death of the cells concerned has not been determined.

Until further evidence is available the general consensus that neurones are incapable of division once morphological and functional inter-relationships are established must prevail.

I. NEUROFIBRILS, NEUROFILAMENTS AND MICROTUBULES

For over a hundred years a meshwork of threadlike fibrils has been recognized in silver-stained preparations of nerve cells (Fig. 9). The presence of these fibrils has aroused considerable speculation as to whether they are fixation artefacts or, if an integral part of the living neurone, what functional role or roles they might serve.

Examination of neurones by different methods, such as *in vitro* phase-contrast microscopy of living cells, silver impregnation and and electron microscopy, have produced observations that are difficult to equate. Though fibrillary strands can be seen in some cells by the first technique, they appear to be somewhat thicker than the fibrils demonstrable in silver preparations. In addition, cells that do not exhibit a fibrillary stranding when examined *in vitro* have been found to manifest neurofibrils after subsequent fixation and silver staining, though the reverse seems not to obtain.

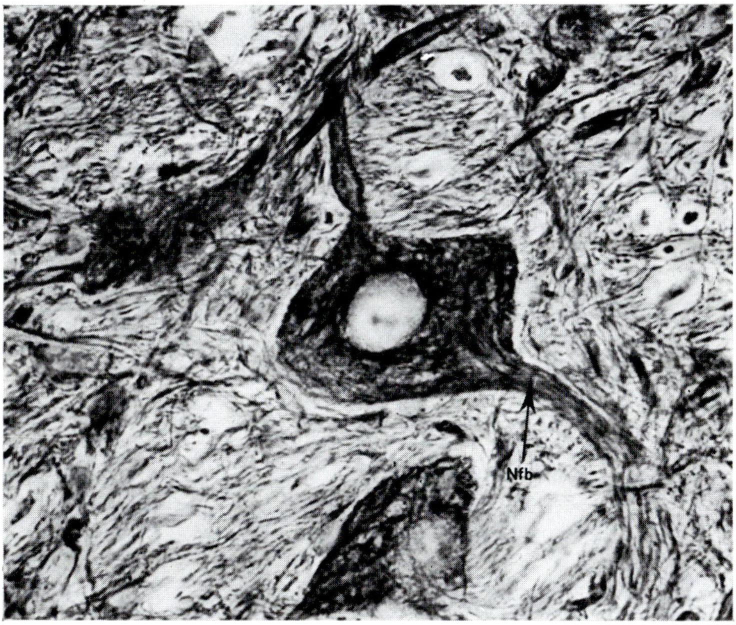

Fig. 9. Neurofibrils (Nfb) in human medullary neurone. Silver impregnation, × 480.

Electron microscope studies show that two distinct fibrillary entities exist in the perikaryal cytoplasm, between the more conspicuous elements such as Nissl substance, Golgi complex and mitochondria: they are also present in the axon and dendrites. These are the microtubules and neurofilaments, the former being circa 25 nm and the latter 10 nm in diameter (Fig. 10). Microtubules probably exist in all cells and have an electron dense wall 6–10 nm thick around a lighter centre, which sometimes has a dense dot within it. Their walls are composed of globular molecules arranged in up to 13 strands helically wound around the "core".

In high-power electron micrographs there is a suggestion that the neurofilaments also have a tubular structure, the walls of which appear to be made up of large numbers of spherical structures. These have

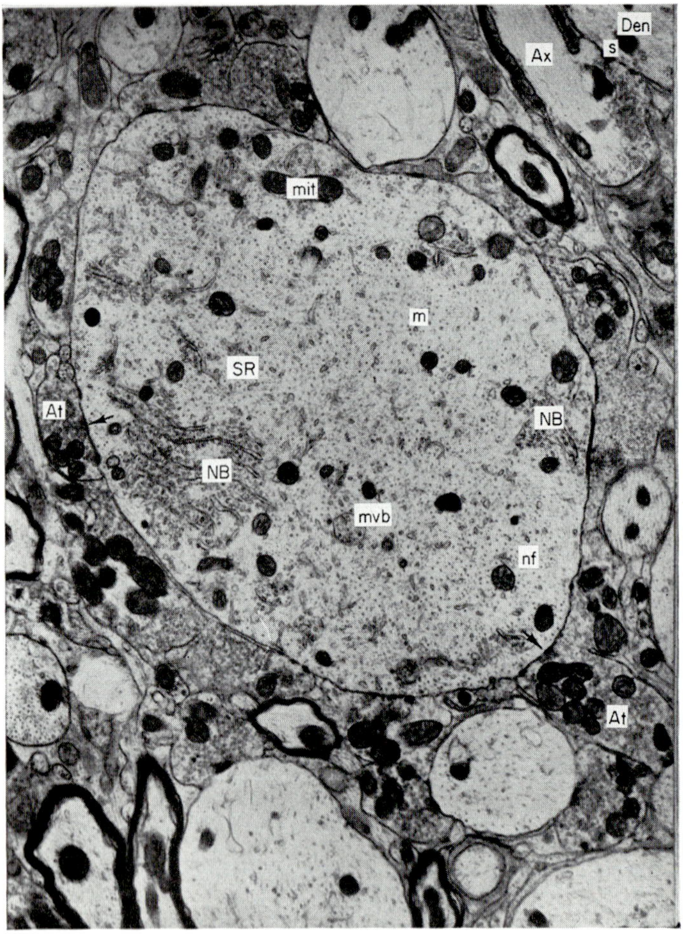

Fig. 10. Rat anterior horn cell dendrite, showing microtubules (m), neurofilaments (nf), Nissl body (NB), mitochondria (mit), smooth endoplasmic reticulum (SR), and multivesicular body (mvb). Note also the synapse (S) between an axon (Ax) and dendrite (Den). At lower right and mid left axon terminals (At) synapse on large dendrite (arrows). ×22000. From Peters, Palay and Webster, 1970.

been shown to be protein monomer subunits having a molecular weight of about 80 000 which bond closely together to form a cylinder with the globular subunits arranged helically around a central core of 3 nm diameter. The subunits are probably formed in association with the polyribosomes and subsequently bond together to form the definitive neurofilaments, which are especially prominent in axonal processes and also extend into the terminal boutons where they often present a coiled appearance.

It should be emphasized that microfilaments very similar to the neurofilaments of neurones have been demonstrated in many other types of cells.

It is obvious that these very small elements, microtubules and neurofilaments, could not individually be identical with the neurofibrils of light microscopy. They do, however, display a tendency to congregate in a roughly linear pattern, and it has been postulated that neurofibrils of light microscopy are aggregates of one or both of the E.M. elements. It has been stated that only neurofilaments are argyrophilic, but in some neurones it has been found that large apical dendrites may stain quite intensely with silver, although electron microscope studies show them to contain numerous microtubules but few neurofilaments.

IV Nerve Cell Processes

A. INTRODUCTION

Although as previously mentioned, Deiters was the first to distinguish between "protoplasmic processes" and the "cylinder axis" of nerve cells, the terms "dendrite" and "axon" were probably originally employed by His and Kölliker respectively at the end of the last century.

In general, distinction between the two types of processes can be made on a number of criteria, but absolute differentiation, in neurones from different species, when morphological, biochemical and physiological aspects are taken into consideration, is not always possible.

For example, if the cell body may be considered to be the "nodal" point of the neurone then dendrites are receiving processes which conduct impulses *to* the cell body, whilst axons conduct impulses *away* from it. The latter often possess a myelin sheath, the former never. Yet difficulties in this respect at once arise. In pseudo-unipolar neurones (such as those in spinal ganglia) the process from a peripheral sensory receptor is often considered a dendrite yet possesses a myelin sheath. As in such neurones a single process extends from the cell body and then divides into two (peripheral and central) at a T-junction, it has been proposed that in order to obviate complex morphological ambiguities an axon "may be defined as that single process of a neuron that is differentiated to conduct impulses (in a non-decremental manner) away from the dendritic or receptor portion of the neuron" (Peters, 1968). Although this definition to some extent begs the question as to what is "dendritic" it would seem to be in closer accord with known facts than to consider the myelinated peripheral "limb" of such processes as a dendrite. During development of chick embryo spinal ganglion neurones the peripheral process differs from the central in being initially larger in diameter and exhibiting markedly increased growth activity, but later development results in both limbs becoming of equal calibre. Section of the peripheral process results in chromatolytic changes in the cell body, that of the central process does not.

B. DENDRITES

In general morphological terms dendrites of adult neurones are usually multiple and fairly short (about 1 mm.) processes (Fig. 11) which diminish in diameter the further they extend from the cell body; they branch to form dendritic "trees" or patterns that are relatively constant (and often characteristic) for a particular type of nerve cell.

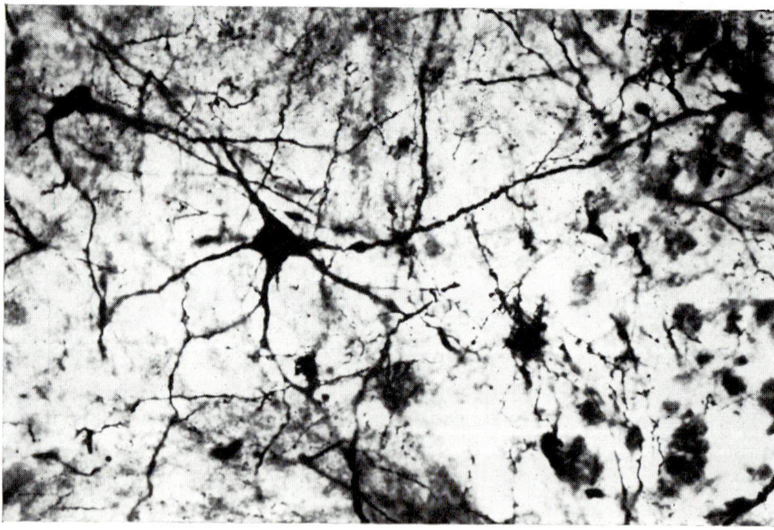

Fig. 11. Golgi-Cox preparation, cat spinal cord neurone. × 160.

Dendritic patterns are of bewildering variety and complexity. Invertebrate nerve cells manifest elaborate dendritic arrangements which differ from those exhibited by vertebrate neurones, (see Bullock and Horridge, 1965) but will not be considered here.

In vertebrate neurones dendritic patterns have been classified, on the basis of Golgi studies, by Ramón-Moliner (1968), into three broad categories:

A. Generalized (isodendritic)
B. Specialized (allodendritic)
C. Highly specialized (idiodendritic)

Isodendritic nerve cells manifest dendrites that tend to be straight and tapering, poorly equipped with spines and that have branches usually longer than the parent stem.

Idiodendritic neurones are those that present a very characteristic dendritic pattern, so that their provenance can be immediately determined from examination of a single specimen.

Allodendritic neurones are intermediate between the above two types, in that examination of several cells is required to identify their origin.

Examples of nerve cells with a dendritic morphology that fall into category A are present in the hypothalamus, subthalamus, nucleus solitarius and globus pallidus: into category B, lateral and paramedian reticular nuclei, striatal and cortical pyramidal cells, and into category C Purkinje cells of the cerebellum, mitral cells in olfactory bulb and retinal ganglionic cells.

Cytoplasm

In electronmicrographs the cytoplasmic appearance of dendritic processes (Fig. 10) differs according to their size and distance from the cell body. In general there is a proximo-distal gradient in the endoplasmic reticular structure. In the larger proximal dendrites the rough endoplasmic reticulum is very similar to that of the perikaryon, but in smaller dendrites the cisternae and free ribosomes become fewer, and the smallest dendrites possess hardly any of these constituents. The smooth endoplasmic reticulum is represented by long cisternae and vesicles but is not well developed. Microtubules are very prominent, and become longitudinally oriented parallel to each other in a regular pattern. Neurofilaments are present, but are far fewer in number relative to microtubules. Mitochondria are also usually oriented longitudinally, and more numerous in the smaller dendrites than in the larger.

Spines

It is onto dendrites that the majority of synapses occur, either directly or onto protuberances of the dendritic stems, the "spines" or "thorns", which commonly present a neck and terminal swelling or bulb (Figs. 1 and 12). Not all dendrites possess these, and proximal dendrites are usually devoid of them. It has been estimated that in some neurones they may account for nearly half the total surface area of the nerve cell, less its axon. They exhibit considerable differences in size and configuration—many are 1 μm or less in length, but may reach 5 μm. Those on the spiny branchlet dendrites of cerebellar Purkinje cells are commonly 1 μm × 0·7 μm and regularly arranged

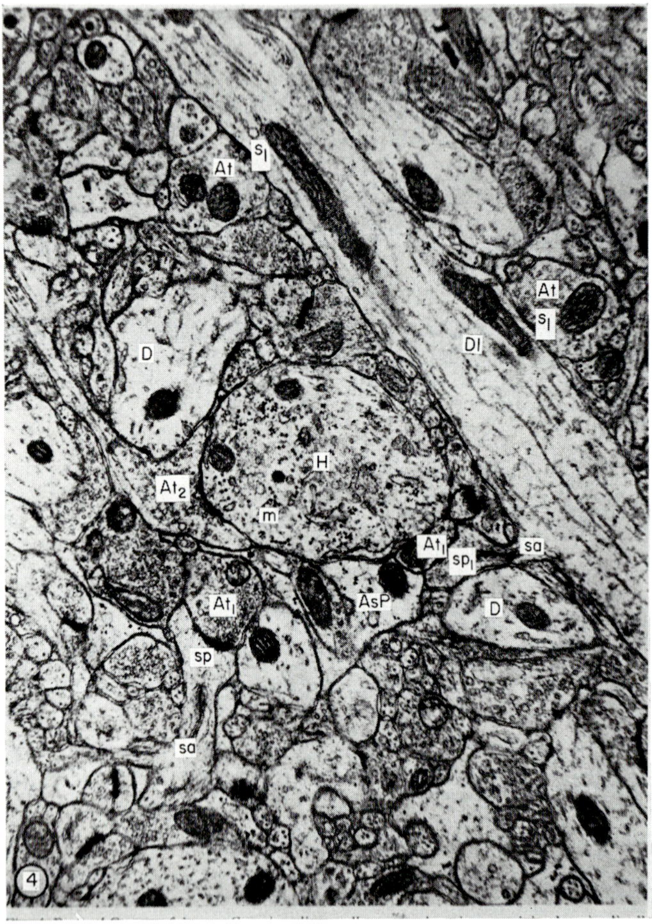

Fig. 12. Rat cortex. Longitudinally oriented dendrite (Dl) with axon terminals (At) synapsing upon it (S_1). Dendritic spines (SP and SP_1) containing a spine apparatus (sa) also have axon terminals (At_1) synapsing on to them. The axon hillock (H) presents microtubules (m) and ribosome clusters in its cytoplasm, and an axon terminal (At_2) synapsing on it. Two transversely sectioned dendrites (D) are also seen. ×28 000. From Nauta and Ebbeson, 1970.

(about 5 spines per 3 μm of dendrite): estimates of the total number of spines on the dendrites of these cells vary from 40 000–100 000 per cell.

Dendritic spines contain neither microtubules nor neurofilaments, but those of cerebral cortical neurones can often be shown to possess a laminated structure, consisting of a number of elongated cisternae with electron-dense layers sandwiched between them: some species contain more than one such "spine apparatus" (Fig. 1) the functional significance of which is unknown.

In certain parts of the nervous system (e.g. olfactory bulb, retina and thalamus) the existence of dendro-dendritic synapses has been described, one of the EM dendritic profiles being presynaptic and possessing synaptic vesicles. These latter usually occupy less than half the profile. Profiles containing such vesicles, which have been interpreted as dendritic, have never been traced as being continuous either with an axonic initial segment or an axon. A theory of the possible functioning of such synapses has recently been presented by Ralston (1971); his paper should be consulted for details.

C. AXONS

These are usually the thinnest, smoothest and longest process of a nerve cell and in general maintain a constant diameter throughout much of their extent, though occasional swellings can be observed. They vary in length from some 200 μm to over a metre. When collaterals arise from the main stem they typically do so at right angles, and are usually of comparable diameter to the parent axon.

The majority of axons arise directly from the cell body, though origin from the base of a dendrite is not uncommon as, for example, is sometimes the case in pyramidal cells of the cerebral cortex.

Axon Hillock

The origin of the axon of a multipolar neurone is from the *axon hillock*—an area of the perikaryon often distinguished in light microscopy by absence of basophilic Nissl substance. Electronmicrographs (Fig. 12) show that there is a reduction in rough endoplasmic reticulum in the cytoplasm of the hillock region though some free ribosomes are present. The most striking feature, however, is that the microtubules in this area channel towards the axon, become linearly oriented and cluster together in bundles.

Initial Segment

The hillock merges with what has become known as the "*initial segment*" or pre-axon. This may be distinguished from the main portion of the axon on a number of counts. It is unmyelinated, contains clusters of longitudinally arranged microtubules interspersed with similarly oriented neurofilaments and some clear vesicles and, in addition, possesses an accumulation of electron-dense material approximately 20 μm thick contiguous to, but not in actual contact with, the inner surface of the axolemma.

Although it is not universally accepted that axo-axonic synapses occur on the initial segment, workers who have described such synapses state that when they are present the electron-dense layer is absent at such sites, and electronmicrographs would seem to support this contention (see, for example, Fig. 52, Peters *et al.*, 1970). Normally, the electron-dense layer extends up to the commencement of the myelin sheath.

It should perhaps be mentioned that the initial segment of the axonal processes of sensory cells such as those in dorsal root ganglia exhibit neither the clustering of microtubules nor the electron-dense layer described above, and it has been suggested that such differences may be connected with the fact that they are not associated with the origin of propagated action potentials, though any such correlation between structure and function in the present state of knowledge remains purely speculative.

It is one of the unfortunate limitations of Golgi preparations that only the initial segment of an axon is usually demonstrable, as once the myelin sheath is present it prevents adequate impregnation of the axon contained within it.

Main Portion

The main portion of the axon, distal to its initial segment but proximal to its terminal portion, contains no ribosomal particles, but does exhibit profiles of smooth or agranular endoplasmic reticulum, represented by membrane-bounded vesicles, tubules and cisternae, which are interspersed amongst the neurofilaments and microtubules (Fig. 13). Communications between cisternae have been described, and it is possible that they constitute a continuous closed system throughout the extent of the axon.

In spite of the fact that ribosomes have not been demonstrated in the axon distal to the initial segment, it has been shown that a giant Mauthner axon in fishes contains, in total, three to four times as much RNA as the cell body (though the W/V concentration is only 1/40th). The base ratio of such RNA is similar to ribosomal RNA. Its presence in axons, not only in those of fishes but also numerous other species, has not yet been fully accounted for, e.g. it is not clear whether it is produced locally in the axon, imported from contiguous glial cells or transported down the axon from the soma, but its presence indicates that the mechanism for protein synthesis does exist in axons (see Lasek, 1970). Whether this is provided *in toto* by the mitochondria is a moot point not yet resolved.

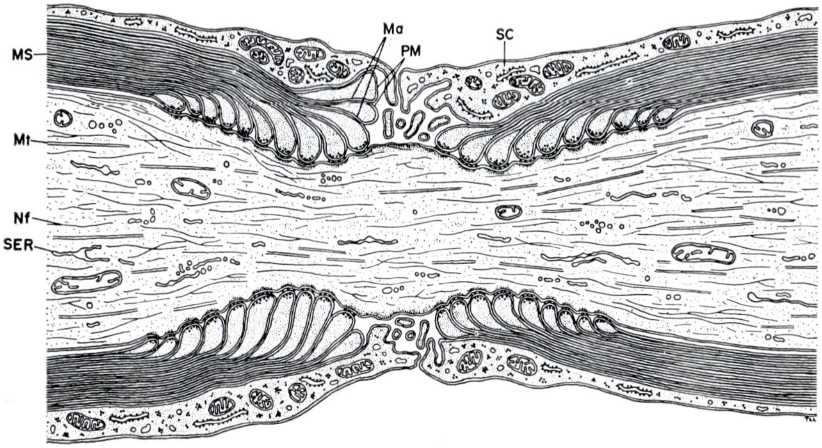

Fig. 13. Electronmicrograph drawing of node of Ranvier. MS, myelin sheath; Ma, mesaxon; Mt, microtubules; Nf, neurofilaments; PM, Plasma membranes; SC, Schwann cell cytoplasm; SER, smooth endoplasmic reticulum. From Lentz, 1971.

Axons contain numerous mitochondria, which tend to be long and thin with the long axis parallel to the axon. The cristae within axonal mitochondria are usually similarly oriented. Where synapses occur there is usually an increased number of mitochondria, and at such sites they are usually shorter than elsewhere in the axon, and closely associated with collections of vesicles. Most of these are about 30–50 μm in diameter though other sizes are not uncommon, e.g. 80 μm complex vesicles with clear centres and vesicles with dark centres. These latter are usually considered to be monoamine-containing synaptic vesicles. (See Chapter 18 for details.)

It is interesting that in axons of sympathetic neurones microtubules are always more numerous than neurofilaments, as is true also in growing axons. When the latter develop a myelin sheath neurofilaments become more numerous.

Lysosomes are not numerous in axons, though some are found throughout their length, and it is assumed that they are involved in normal turnovers of axon constituents. They become more numerous after axonal injury.

Nodes of Ranvier

Nodes of Ranvier occur at intervals along myelinated axons in both the central and peripheral nervous systems. The internodal intervals are greater in large diameter fibres than in smaller, and also increase as a particular axon grows in size. The nodes are sites between one myelin segment and the next, and it is here that myelinated fibre collaterals originate. The length of the nodes is of the order of 0·5–1 μm. At the nodes the surrounding myelin sheath thins out, and the lamellae that constitute it terminate in a regular order, the inner one first, the outer one last (see Chapter 19). As the lamellae terminate pockets of paranodal cytoplasm appear between contiguous lamellae. In longitudinal sections the appearance of "separateness" of these pockets is misleading; in fact the contained cytoplasm is a continuous spirally-wound core from the outside to the inside of the sheath (Fig. 13).

Near the node the plasma membrane of the inside of the sheath is more closely opposed to the axolemma than is the case in the internodal region. At peripheral nerve nodes the axon is covered by thin processes of the Schwann cell, but in the central nervous system the axolemma is bare, and close to its inner surface there is an accumulation of electron dense material some 20 μm thick similar to that present in the initial axonal segment. If a synapse occurs at such a node, the axon protrudes as a knob that contains synaptic vesicles: the dense material is then absent.

A number of authors have recently described a modification in the outer layer of the paranodal axolemma—some six to eight periodic thickenings each approximately 15 μm in length—the significance of which is obscure.

Axonal Flow

Much has been written on the concept that there is a constant flow of "something" from the nerve cell body along the axon since the first suggestion by Scott (1905) and its actual demonstration by Weiss nearly thirty years ago.

Two basic points have been emphasized by Weiss (1969). The first is that what is commonly referred to as "axonal flow" is a manifestation of the general neuroplasmic activity of the cell and, whilst conspicuous in the axon, a similar process is involved in the dendrites, though here the activity appears to be more of a seepage than a flow. The second is that "axonal flow" and "intra-axonal transport" are not synonymous terms and should be clearly differentiated. It is pointed out that in axonal flow one is dealing with movement of the axoplasm *itself*, whereas axonal transport is a much faster activity occurring *within* the axoplasm. The effector mechanism for the former must be present throughout the length of the axon (as has been shown by "damming up" by constriction at more than one site) and may be in the form of a microperistaltic wave.

The proximo-distal shift in this type of axonal flow is of the order of 1 mm. or so per day, but in recent years it has been shown that intra-axonal transport of, for example, protein occurs at a much faster rate than this. Recently Karlsson and Sjöstrand (1971), using ^{3}H leucine, have described four velocities of protein axonal transport, of the order of 2, 6–12, 40 and 150 mm per day. The two fastest they consider to be of light protein particles, possibly bound to synaptic vesicles and undergoing rapid turnover at the axon terminals, the 6–12 mm/day flow is associated with mitochondrial migration, and the slowest is concerned in the transport of soluble proteins.

In sympathetic adrenergic neurones noradrenaline is thought to be synthesized in the cell body and transported down the axon, and it has been shown, by means of the histochemical fluorescence method, that the terminals of such axons present intensely fluorescent varicosities, which are considered to be the site where the noradrenaline storage granules function. These are thought both to transport and synthesize noradrenaline and protect it from the action of monamine oxidase, and it has been estimated that the rate of proximo-distal transport of these granules is of the order of 3–10 mm per hour (Dahlström, 1969).

It seems possible that such fast intra-axonal transport is functionally related to the neuronal microtubules, and Schmitt (1969) has recently put forward a hypothetical sliding-vesicle mechanism to explain how this might occur. It is suggested that the microtubule subunits have binding sites for protein subunits on the outer surface of the vesicle or particle to be transported, and that there is in addition an ATPase binding site, which, when activated in some unknown manner, splits ATP and thus provided energy for transportation of the vesicle along the length of the microtubule.

Axons known to manifest fast transport mechanisms are rich in microtubules, and colchicine (which disrupts microtubule subunit structure) markedly inhibits transport of noradrenaline granules in adrenergic neurones, but does not appear to affect slow transport systems.

At present the fastest intra-axonal transport rate reported is that of neurosecretory granules in preoptic neurones of goldfish, which has been estimated at 2 mm per minute (Jasinsky *et al.*, 1966). For a recent review of axonal transport mechanisms and further references see Dahlström (1971), who makes the point that, at least in mammalian neurones, impulse conduction seems to have little or no influence on axonal transport rates. Although metabolic changes occur in axons when a nerve impulse passes along them, correlation of functional activity of this kind with possible changes in the structural constituents of axons remains a problem for the future.

Whilst there is so much still to be explained on the more easily demonstrable centrifugal axonal flow mechanisms, a possible flow in the opposite direction (i.e. towards the cell body) described by a number of authors must remain an even more obscure phenomenon.

V. Electrical Properties of the Nerve Cells

It is not possible here to do justice to the many important discoveries in this subject that have accrued since the early part of this century, more particularly in the last twenty years. Hodgkin (1964), Katz (1966) and Brazier (1969) should be consulted for further and more detailed information. For a review of the functioning of invertebrate nerve cells see Treherne and Moreton (1970).

A. Resting Potential

The nerve cell membrane, although postulated for many years, was only factually demonstrated following the advent of the electron microscope in the early 1950s. A potential difference—the trans-membrane or resting potential—exists across this membrane, whether it be that belonging to the nerve cell body, its dendrites or axon.

In squid giant axons, which have been extensively investigated with intracellular electrodes, this potential is of the order of −60 −70 mv; i.e. the inside is negative to the outside by this amount; in other neurones values of up to −90 mV have been demonstrated.

The concentration of potassium and sodium ions differs markedly on the two sides of the membrane—there is approximately 10 times as much sodium outside as there is inside and 30–40 times as much potassium inside as out.

It was the hypothesis of Bernstein (formulated in 1902 with regard to muscle but subsequently applied to nerve cells) that the potassium ion was the one responsible for the resting potential, only K^+ passing through the membrane. This produced an equilibrium system to which the theoretically derived Nernst equation could be applied. Thus

$$E = \frac{RT}{F} \log \frac{[K^+]\,o}{[K^+]\,i}$$

where E is the resting potential, R the gas constant, T the absolute temperature, F the Faraday and $[K^+]o$ and $[K^+]i$ the concentrations of K^+ outside and inside the membrane respectively.

Over the last thirty years modification of Bernstein's original hypothesis became necessary, as the cell membrane has been shown to be permeable to K^+, Na^+ and Cl^-, none of these being in equilibrium: there is a continual diffusion of Na^+ into and K^+ out of the soma and processes. Thus the resting potential is a *diffusion* potential, and its value may be given by the Goldman equation (assuming the electrical potential gradient through the membrane to be linear):—

$$E = \frac{RT}{F} \ln \frac{PK^+[K^+]o + PNa^+[Na^+]o + PCl^-[Cl^-]i}{PK^+[K^+]i + PNa^+[Na^+]i + PCl^-[Cl^-]o}$$

where P is the membrane permeability coefficient i.e. a measure of the ease with which the ions concerned pass through the membrane. This is not a constant as it may vary with changes in ionic concentrations. It will be appreciated that a change in the value of P for any of the ions will effect a change in the membrane potential.

As the resting nerve cell internal ionic composition is stable, yet continuous diffusion of Na^+ into and K^+ out of it through the membrane takes place, active "pumping" of these ions must occur to maintain such stability. This membrane pump requires an energy source which has been shown to be ATP, the necessary change ATP →ADP being through ATPase present in the membrane.

B. ACTION POTENTIAL

When an electric current is applied to the nerve cell membrane its flow decreases the resting potential—i.e. it "depolarizes" the membrane. If this reaches a certain critical level (see later Section "Exitation and Inhibition") the permeability to sodium ions (PNa^+) is suddenly increased, they pass into the cell and reverse the resting potential, so that the inside now becomes positive relative to the outside. This phenomenon is the action (or spike) potential, and Hodgkin and Huxley (1945) showed that in squid axons in which the resting potential was −45 mV the action spike reached +40mV, a change of 85 mV, i.e. the action potential is considerably larger than the resting potential. It was this fact, demonstrated by Hodgkin and Huxley in 1939, that caused the modification of Bernstein's theory and the development of the present hypothesis of nerve membrane potentials.

The inflow of sodium ions (the actual number is too small to alter the intracellular chemical content detectably) occurs during a period of approximately 0·2 msec, and is followed by an outflow of potassium ions which restores the resting membrane potential to its original value. The total duration of the spike potential is of the order of 1 msec. Following the action potential there is a recovery period during which the pump mechanism expels sodium. This is coupled with an inward pumping of potassium ions, which is largely dependent on the effectiveness of the sodium pump.

C. CONDUCTION AND REFRACTORY PERIODS

If an axon is repeatedly stimulated by two successive same-strength stimuli at shortening intervals then the amplitude of the second action potential will progressively decrease until, when the interval is brief enough, the second stimulus will no longer produce an action potential at all. The stimulus interval when the second potential produced is less than the first is within the relative refractory period and, when it is no longer produced, the second stimulus is said to be within the absolute refractory period: this is variable, but of the order of 1 msec.

The action potential that occurs at the point of stimulation travels down the axon as a propagated nerve impulse. This may be described as a patch of reversed polarity passing along the axon without a decrease in amplitude. In unmyelinated nerve fibres it is a continuously progressive event but in myelinated fibres, where the sheath acts as an insulator, the impulse "jumps" the internodal portion from one node of Ranvier to the next—"saltatory conduction"—and the conduction rate of the impulse in such a fibre is thus much increased.

The speed of conduction and the amplitude of a propagated impulse is independent of the strength of the stimulus—the all-or-nothing principle propounded by Adrian nearly 60 years ago.

The length of the axon involved in, or "occupied" by a single action potential (of approx. 1 msec duration) will depend on its speed of conduction—in axons conducting for instance at 10 msec, it will occupy approximately 1 cm of the axon at any given moment.

D. MEMBRANE CONDUCTANCE

Brief mention should be made of what has been termed the membrane conductance (g) of permeable ions. This has been studied by a technique known as the "voltage clamp". In this two silver electrodes are placed longitudinally, one within the axon, the other outside the membrane, and connected to a battery; this produces an artificial membrane potential. It has been found that when the clamping voltage is applied there is a flow of current into the axon, reaching a maximum after 0·5 msec. This declines until after 1·5 msec current flows out to the cell, to reach a value that remains steady for as long as the clamping voltage is maintained (curve *a*, Fig. 14). When the sodium in the extracellular fluid is replaced by impermeable choline, no current influx occurs, but the "outflowing" current reaches the same value as above (curve *b*).

The influx of current is due to Na^+ and the outflow to K^+, and by subtracting curve *b* from curve *a* a curve *c* for sodium current can be obtained. From such experiments it has been found that the conductance of sodium ions increases to a maximum at 0·5 msec and then returns to normal after approximately 3 msec; conductance of potassium is progressive and slower, reaching a maximum at 5 msec. The sodium and potassium conductances change with alterations in membrane potential, and have been shown to be functions of both time and membrane potential. Equations have been derived to describe such changes and the predicted behaviour of a nerve fibre has been found to be in close correlation to that actually obtaining. For details Hodgkin and Huxley (1952) should be consulted.

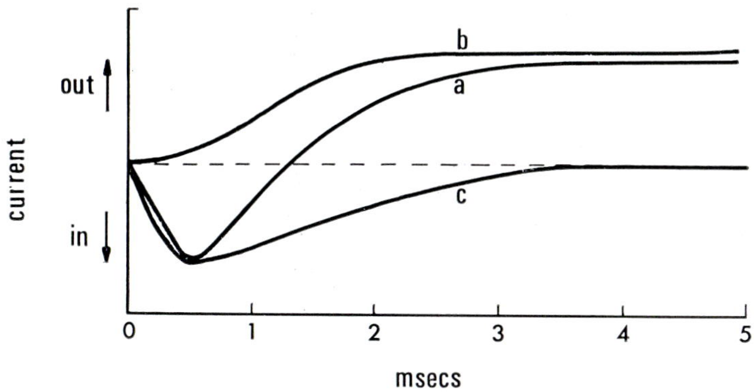

CURRENT CURVES IN VOLTAGE CLAMP EXPERIMENT

Fig. 14. Current curves following voltage clamp. For explanation, see text.

E. EXCITATION AND INHIBITION

In the physiological state the propagated spike potential is generated in the region of the axon hillock and initial axonal segment. The membrane of the nerve cell body and dendrites have a higher threshold than that of the above region; it requires 2–4 times the current to produce a spike potential on stimulation of the soma as it does to produce one from stimulation of the initial segment.

When a nerve impulse arrives at a synapse it may produce either an excitatory or inhibitory effect on the activity of the post-synaptic neurone. If excitatory, the impulse produces a localized non-propagated long-duration depolarization of the membrane—the excitatory post-synaptic potential (EPSP)—brought about by the release, from the pre-synaptic axon terminal, of a quantum of chemical transmitter such as acetylcholine, which effects a change in the permeability of the membrane. A single volley has been estimated to release 10^{-15} g of acetylcholine on a single cell.

Microelectrodes with the tip inside the cell are used to record EPSP: a single pre-synaptic stimulus will, after a short 0·5–1 msec latency, alter the membrane potential only slightly, say from −70 mV to −65 mV. This will slowly decay to the original level over a period of approximately 8–10 msec. If several stimuli are employed the EPSP's produced will, by electrotonic spread summate either spatially or temporarily. If this summation reaches a critical value—the threshold of the nerve cell—an action potential will be generated from the region of the axon hillock. The potential change necessary for this to occur is variable, but in spinal motoneurones is of the order of 10–25 mV.

Nerve cells, however, may be inhibited as well as excited, and there are two different ways in which this may occur, though the actual mechanism is similar in each case. Post-synaptic inhibition occurs by release of an inhibitory transmitter (such as ?GABA) at terminals on the soma or dendrites of the post-synaptic cell; pre-synaptic inhibition occurs when the inhibitory axon synapses on the terminal of an excitatory pre-synaptic axon. Microelectrode recordings show that, in inhibition, the membrane potential is increased i.e. a hyperpolarization of the membrane occurs. This change—the inhibitory post-synaptic potential (IPSP) is of the same order and has a similar time course to the EPSP. It is probably brought about by selectively changing the membrane permeability of Cl^-, allowing this ion to enter the cell, and making the inside more negative. In the case of pre-synaptic inhibition, the change in the membrane potential of excitatory terminals may be sufficient to prevent them from producing sufficient EPSP's to engender an action potential in the post-synaptic neurone.

The many thousand synapses that may be present on the soma and dendrites of a single neurone provide a state whereby the resultant activity of that cell, in producing or not producing an action potential and conducted impulse, is dependent on the interplay of inhibitory and excitatory impulses impinging on it. It is this which accounts for the plasticity so characteristic of the activity of nervous tissue.

A short account of methods available for studying how the nervous system may process information is given by Walter and Gardiner (1970).

Acknowledgements

I am grateful to Dr R. Sarnaker for the preparations from which Figures 4, 5, 7, 8, 9 and 11 were taken, and to the various authors and publishers named for permission to reproduce Figures 1–3, 6, 10, 12 and 13.

References

Barr, M. L. and Bertram, E. G. (1949). A morphological distinction between neurones of the male and female, and the behaviour of the nucleolar satellite during accelerated nucleoprotein synthesis. *Nature, Lond.* **163**, 676.

Bearns, H. W. and Kessel, R. G. (1968). The Golgi apparatus: structure and function. *Int. Rev. Cytol.* **23**, 209.

Brazier, M. A. B. (1969). Electrical activity of the nerve cell. Electrical activity of the nerve fibre. *In*: "The Structure and Function of Nervous Tissue", Vol. II. (Bourne, G. H., ed.). Academic Press, New York and London.

Bullock, T. H. and Horridge, G. A. (1965). "Structure and function in the nervous system of Invertebrates". Freeman, San Francisco.

Cajal, S. R. (1909–11). "Histologie du Système Nerveux de l'homme et des Vertébrés". Vols. I and II. Maloine, Paris.

Colonnier, M. (1965). On the nature of intranuclear rods. *J. Cell Biol.* **25**, Part 1, 646.

Dahlström, A. (1969). Synthesis, transport and life span of amine storage granules in sympathetic adrenergic neurons. *In*: "Cellular Dynamics of the Neuron". (Barondes, S. H., ed.). Academic Press, New York and London.

Dahlström, A. (1971). Axoplasmic transport (with particular respect to adrenergic neurons). *Phil. Trans. Roy. Soc. Lond. B*, **261**, 325.

Deitch, A. D. and Murray, M. R. (1956). The Nissl substance of living and fixed spinal ganglion cells. A phase contrast study. *J. biophys. biochem. Cytol.* **3**, 839.

Droz, B. (1969). Protein metabolism in nerve cells. *Int. Rev. Cytol.* **25**, 363.

Hodgkin, A. L. (1964). "The Conduction of the Nervous Impulse". Thomas, Springfield, and Liverpool Univ. Press.

Hodgkin, A. L. and Huxley, A. F. (1952). A quantitative description of membrane current and its application to conduction and excitation in nerve. *J. Physiol.* **117**, 500.

Horridge, G. A. (1968). The origins of the nervous system. *In*: "The Structure and Function of Nervous Tissue", Vol. I. (Bourne, G. H., ed.). Academic Press, New York and London.

Jasinsky, A., Gorbman, A. and Hara, T. J. (1966). Rate of movement and redistribution of stainable neurosecretory granules in hypothalamic neurons. *Science, N.Y.* **154**, 776.

Karlsson, J. O. and Sjöstrand, J. (1971). Axonal transport of proteins in the optic nerve and tract of the rabbit. *Acta. Neuropath. Berl.* **5**, Suppl. 5, 207.

Katz, B. (1966). "Nerve, Muscle and Synapse". McGraw-Hill, New York.

Koenig, H. (1969). Lysosomes in the nervous system. *In*: "Lysosomes in Biology and Pathology". (Dingle, J. T. and Fell, H. B., eds), pp. 110–162. North-Holland, Amsterdam and London.

Langman, J. (1968). Histogenesis of the central nervous system. *In*: "The Structure and Function of Nervous Tissue", Vol. I (Bourne, G. H., ed.). Academic Press, New York and London.

Lasek, R. J. (1970). Protein transport in neurons. *In*: "International Review of Neurobiology", **13**, 289. (Pfeiffer, C. C. and Smythies, J. R., eds). Academic Press, New York and London.

Lindsay, H. A. and Barr, M. L. (1955). Further observations on the behaviour of nuclear structures during depletion and restoration of Nissl material. *J. Anat. Lond.* **89**, 47.

Masurovsky, E. B., Benitez, H. H., Kim, S-U, and Murray, M. R. (1968).

Development and nature of intranuclear rodlets and associated bodies in chicken sympathetic neurons. *J. Cell. Biol.* **39**, Abstract 209, 86a.

Nayyar, R. P. and Barr, M. L. (1968). Histochemical studies on the accessory body of Cajal in neurones of the cat. *J. Comp. Neurol.* **132**, 125.

Novikoff, A. B. (1967). Lysosomes in nerve cells. *In*: "The Neuron", (Hydén, H., ed). Elsevier, Amsterdam.

Peters, A. (1968). The morphology of axons in the central nervous system. *In*: "The Structure and Function of Nervous Tissue", Vol. I. (Bourne, G. H., ed.) Academic Press, New York and London.

Peters, A., Palay, S. L. and Webster, H. de F. (1970). "The Fine Structure of the Nervous System". Harper and Row, New York, Evanston and London.

Ralston, H. J. (1971). Evidence for pre-synaptic dendrites and a proposal for their mechanism. *Nature, Lond.* **230**, 585.

Ramón-Moliner, E. (1968). The morphology of dendrites. *In*: "The Structure and Function of Nervous Tissue", Vol. I. (Bourne, G. H., ed.). Academic Press, New York and London.

Rees, E. L. (1971). Nucleolar displacement during chromatolysis. A quantitative study on the hypoglossal nucleus of the rat. *J. Anat. Lond.* **110**, 463.

Roberts, K. and Northcote, D. H. (1970). Structure of the nuclear pore in higher plants. *Nature, Lond.* **228**, 385.

Schmitt, F. O. (1969). Fibrous proteins and neuronal dynamics. *In*: "Cellular Dynamics of the Neuron". (Barondes, S. H., ed.). Academic Press, New York and London.

Scott, F. H. (1905). On the metabolism and action of nerve cells. *Brain*, **28**, 506.

Shantha, T. R., Manocha, S. L., Bourne, G. H. and Ariëns-Kappers, J. (1969). The morphology and cytology of neurons. *In*: "The Structure and Function of Nervous Tissue", Vol. II. (Bourne, G. H., ed.). Academic Press, New York and London.

Siegesmund, K. A., Dutta, C. R. and Fox, C. A. (1964). The ultrastructure of the intranuclear rodlet in certain nerve cells. *J. Anat. Lond.* **98**, 93.

Sotelo, C. and Palay, S. L. (1968). The fine structure of the lateral vestibular nucleus in the rat. *J. Cell Biol.* **36**, 151.

Treherne, J. E. and Moreton, R. B. (1970). The environment and function of invertebrate nerve cells. *Int. Rev. Cytol.* **28**, 45.

Van der Loos, H. (1967). The history of the neuron. *In*: "The Neuron". (Hydén, H., ed.). Elsevier, Amsterdam.

Walter, D. O. and Gardiner, M. F. (1970). Some guidelines from system science for studying neural information processing. *Int. Rev. Neurobiol.* **13**, 343.

Weiss, P. A. (1969). Neuronal dynamics and neuroplasmic ("axonal") flow. *In*: "Cellular Dynamics of the Neuron". (Barondes, S. H., ed.). Academic Press, New York and London.

18. The Synapse

E. G. GRAY

Department of Anatomy, University College London, England

I. Historical

The synapse, in current terminology, may be defined as a composite of the specialized structures found in a region where part of one neuron lies in close relationship with another and across which excitation or inhibition is transmitted. Synapses are also commonly recognized between receptor cells and neurons (also neuron-receptor feedback contacts) or between neurons and effector cells. Sherrington (1897) introduced the word synapse, derived from the Greek word "to clasp". He was much more concerned with the synapse as a functional region rather than morphological entity, just as nowadays physiologists explore "synapses" with micro-electrodes and pharmacologists with drugs. However, the electron microscopist now commonly uses the term synapse to denote a morphological configuration, often without any direct knowledge of its function. The morphological clues he uses are derived from certain synapses where structure and function can be analysed hand-in-hand and over the years his use of such clues has been seen to be amply justified.

In the first half of the 19th century neurohistologists were mainly concerned with recognizing the different parts of the neuron, the perikaryon, dendrites and axon (myelinated or unmyelinated) (see van der Loos, 1967) and by the turn of the century the various staining techniques, for example reduced-silver methods, methylene-blue-vital, Golgi and Nissl methods (nowadays still indispensable) were in use (see Nauta and Ebbesson, 1970) and the best oil-immersion objectives were available to the more fortunate investigators. Synaptic knobs or boutons could be demonstrated in various ways with the various techniques (Fig. 1) but their apparent absence in many regions of the central nervous system (see section II Pre-synaptic neurofilaments and microtubules) made interpretation difficult. Also, the reality and nature of the numerous small spines that could often be seen on dendrites in Golgi preparations (see section VII) were highly problematical.

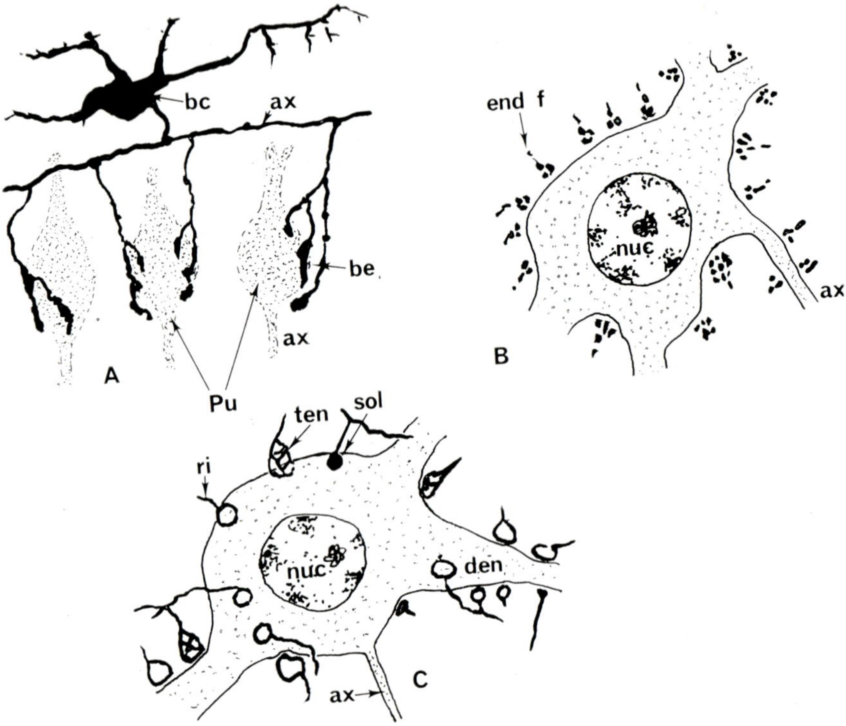

Fig. 1. A Golgi preparation—cerebellar cortex. The entire basket cell is impregnated: its axon branches and forms the basket endings at the bases of the Purkinje cells.
B The axonal endfeet on a spinal cord ventral horn cell are revealed by staining their mitochondrial content.
C Neurofibrillar boutons on a spinal cord ventral horn cell. The silver stain really only shows up the bundles of neurofilaments within their cytoplasm.

A fundamental problem that was much debated at the time was whether the nervous system was, like other tissues, composed of discrete cells (i.e. neurons), with axon and dendrites of finite length, or whether it was a complex net (i.e. reticulum) consisting of a web of processes all in cytoplasmic continuity. The principal protagonist of the neuron doctrine, which became generally accepted, was Ramón y Cajal: Gerlach and Golgi were amongst the reticularists. The controversy was not really settled until much later with the introduction of the electron microscope. As we shall see, neurons are quite independent units and there is no cytoplasmic continuity across the synapse (but clearly this could not have been decided with the limited resolution of the light microscope). (See Addendum a, p. 412.)

We owe much of our knowledge of the organization and connexions of the nervous system to Cajal (see for example his master works of 1911) the outstanding Spanish neurobiologist working in the latter part of last and the early part of this century. Curiously though, his circuitry of the CNS never allowed for inhibitory links. All his diagrams were based on excitatory pathways between the various neurons (see Eccles, 1969), although at about the same time Sherrington was developing his physiological concepts of excitation and inhibition in reflex arcs.

Optimal resolution had already been reached with the light microscope and although the elucidation of connexions (coupled with various experimental and histochemical techniques) in mammals and throughout the animal kingdom continued through the first half of the 20th century, because of the limited resolution, neurohistology, in a sense, stagnated. Physiological explorations, on the other hand, with the development and use of microelectrode techniques, forged ahead and on the pharmacological front strong evidence was produced to show that synaptic transmission in the vertebrate peripheral nervous system was chemical—in the autonomic nervous system acetylcholine or noradrenaline, and at the motor end-plate acetylcholine.

In the central nervous system, on the other hand, synaptic transmission was thought to be electrical (for various reasons, including the failure to demonstrate acetylcholine) rather than chemical. However, with the development of intracellular recording techniques (particularly by B. Fatt and B. Katz on muscle fibres) in the early fifties, Eccles and co-workers found good evidence that synaptic transmission in central neurons was chemical. Such evidence is often indirect however, for although the post-synaptic component (perikaryon or dendrite) can be easily impaled with an intra-cellular electrode tip, the presynaptic knob is usually too small for this. The squid giant fibre synapse, however, discovered by J. Z. Young in 1939, allows intracellular electrodes to be placed close to either side of the synaptic cleft, and it is from this synapse that much of the evidence for chemical transmission has been obtained. Since then electrophysiological and pharmacological evidence has strongly supported the view that as in the peripheral nervous system synaptic transmission at most synapses in the vertebrate (and as far as is known, invertebrate) central nervous system is chemical. Subsequently, however, a number of interesting examples of synapses with electrical transmission have been detected, mostly in the lower vertebrates and invertebrates.

Although the advanced electrophysiological and pharmacological techniques in present day use tell us much about the different sorts of excitatory and inhibitory mechanisms there is still much doubt about which transmitter is active at which synapse, although several different substances (see section II) are likely candidates.

The electron microscope has revolutionized the morphological approach to synaptic organization with the approaches mentioned above. In the first half of this century Ramón y Cajal made outstanding contributions to the morphology and connexions of the nervous system and it is a sad irony that the electron microscope (EM) was first used on biological material in 1934, the year Cajal died. It was not until the early fifties, however, that the EM and the required thin sectioning and fixation techniques were sufficiently developed to study synaptic fine structure. The surface membrane outlining the shape of the bouton could be clearly seen for the first time and the presence of a well defined synaptic cleft finally ruled out the possibility of cytoplasmic continuity across the synapse. Swarms of characteristic synaptic vesicles could be seen lying close to the presynaptic membrane, fulfilling physiological predictions that the transmitter substance would be found sequestered in small capsules of uniform size (see Katz, 1966). As we shall see, the fine structure of the synapse (central or peripheral) fitted well into the physiological criteria of a chemical synapse.

Then in the late fifties electrophysiology showed that after all there were a few electrical synapses in the central nervous system. These were found first in crustacea and then fishes and electron microscopy of the same synapses showed that, as expected, the synaptic membranes were apparently fused together. This correlation of structure with function added an exciting chapter but, as we shall see, (sections III and IV) at the same time revealed an intriguing anomaly.

At present we are concerned with the exciting task of differentiating excitatory from inhibitory synapses (section IV) and, with the improved perfusion fixation techniques now available (and experimental sectioning of fibre tracts), unravelling the connexions of nervous systems throughout the animal kingdom. EM autoradiography, histochemistry, fractionating techniques and molecular biology combine to make a multidisciplinary attack on all fronts. Indeed, there is so much information waiting to be won that one is often hard pressed to decide which problem to tackle first!

It is Editorial policy to restrict references in this book to a minimum. The reader should consult Gray and Guillery (1966), Gray (1969, 1970), Peters *et al.* (1970), Eccles (1964, 1969) and Katz (1966) for a more detailed review and references. (See Addendum b, p. 412.)

II. The Chemical Synapse

A. General Morphology

The chemical synapse is the best known type of two-component synapse and consists of a presynaptic process contacting a postsynaptic one. Motor end-plates can be included in this category and it is from these that much physiological evidence has been obtained. The vast majority of synapses have an onion-shaped, club or spheriodal presynaptic knob (Fig. 2) ranging from 0·5μm to several microns in diameter. The knob might arise as a terminal dilatation of a slender unmyelinated axon or might be one of a series of bead-like varicosities strung out along an unmyelinated axon. Alternatively, the knob may arise as a terminal expansion directly from a myelinated axon (Fig. 12). Certain presynaptic knobs have quite different shapes: the mossy fibre endings of the cerebellar cortex, for example, have an elongated irregular outline and may contact several dendrites of the granule cells. Large cup-shaped endings occur on the cell bodies of the avian ciliary ganglion and similar but more complex endings are found in the trapezoid nucleus—the chalices of Held.

Axonal presynaptic knobs can also occur at nodes of Ranvier in the CNS. Localized regions of perikarya or dendrites can very occasionally be observed with the morphological features of a presynaptic component. Indeed, as we shall see, almost any part of a neuron, with the exception of the initial segment, can show such morphology. The postsynaptic component can be either perikaryon (axo-somatic contact), dendrite (axo-dendritic on dendrite trunk or spine), initial segment (axo-IS) or another presynaptic knob (axo-axonal). The axonal region at the node of Ranvier is apparently never postsynaptic to a synaptic knob. The above remarks apply mostly to vertebrate synapses. In the invertebrates most neurons are unipolar and a complex of processes arises from the main trunk. These can often but by no means always be classified either as dendrite or axon. Synapses occur on these processes and seldom on the cell bodies. Detailed description of invertebrate synapses is beyond the scope of this article.

The synapse shown in Fig. 2 contains the features usually found at a synapse known to have chemical transmission. The surface membranes usually appear as a single dense line but at high magnifications and in suitable preparations the trilamellate (unit membrane) structure

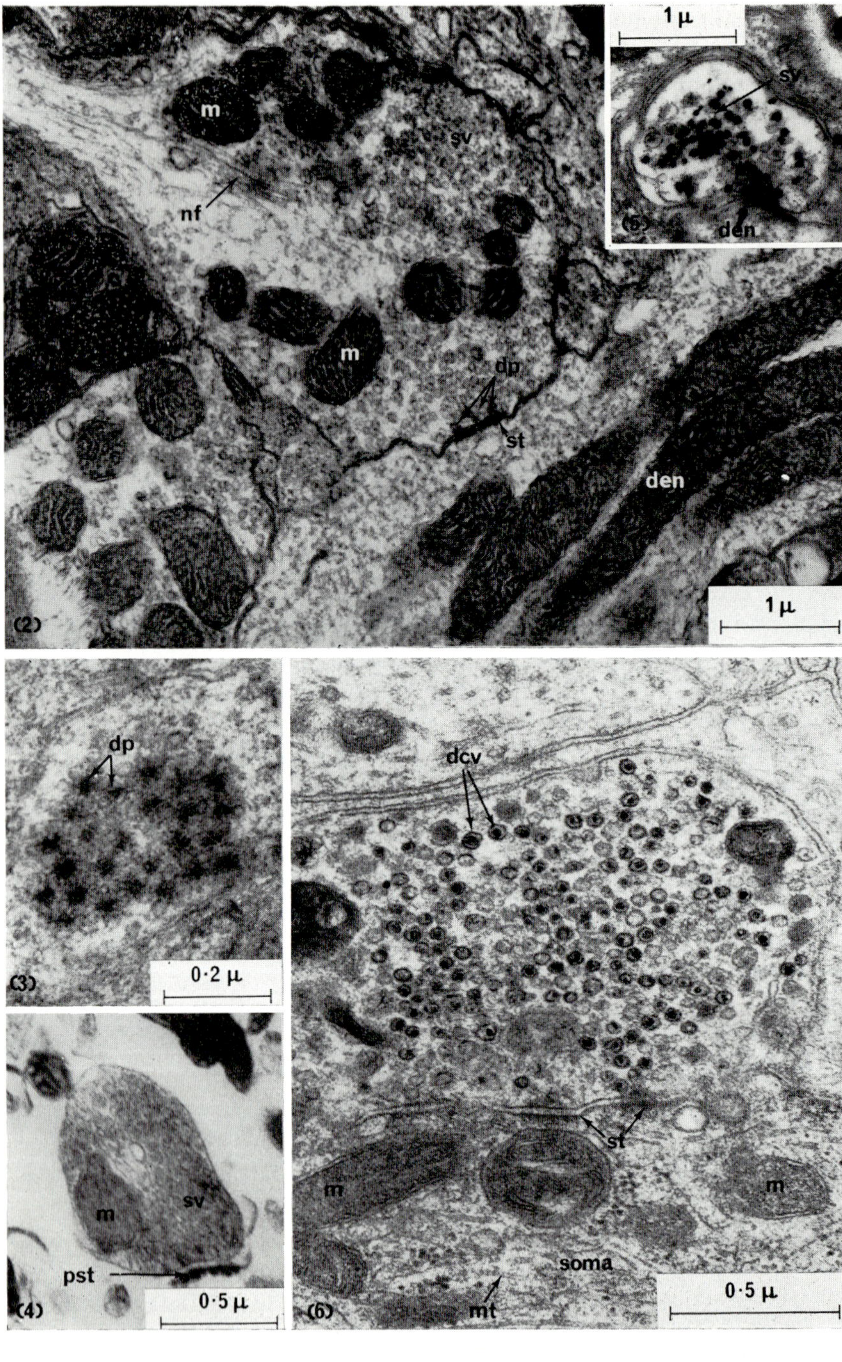
m
sv
nf
m
dp
st
den
1 μ
(2)
sv
den
(5)
1 μ
dp
0·2 μ
(3)
sv
m
pst
0·5 μ
(4)
dcv
st
m
m
soma
mt
0·5 μ
(6)

can be seen. Although these neuronal membranes have the special property of conducting action potentials, the electron microscope as yet has revealed no differences in these membranes from those of non-neural cells in general. The cytoplasm within the presynaptic bag contains numbers of synaptic vesicles of rather uniform size (40–50 nm in diameter). Their size varies in different types of synapse. They usually lie in aggregations close to the presynaptic membrane that lines one side of the synaptic cleft. They are thought to contain the transmitter substance, which is liberated into the synaptic cleft upon arrival of an impulse (see below), or, in smaller amounts, spontaneously.

One or more mitochondria (section II, Synaptic mitochondria) lie in the presynaptic cytoplasm usually further away from the synaptic cleft than the vesicles. These supply ATP for metabolic energy requirements: there is little evidence that they are specifically related to the transmitter mechanism. Other presynaptic organelles include the coated or alveolate vesicles (section II, Synaptic vesicles and transmitter mechanisms) and sometimes present are neurofilaments and/or microtubules (see II, Pre-synaptic neurofilaments).

The presynaptic membrane that lines the synaptic cleft may have patches or spots of dense material (protein?) adherent to its cytoplasmic surface and projecting into the cytoplasm (Fig. 2). These dense projections show especially clearly after phosphotungstic acid staining and, in vertebrates, when viewed *en face* they can be seen to be regularly spaced (~ 90 to 100 nm centre to centre) in a hexagonal pattern (Fig. 3). Their role is far from clear, but they may serve to guide the vesicles to specific points between them on the pre-synaptic membrane (see section II, Synaptic vesicles, for further discussion). The post-synaptic membrane may also be lined with a variable amount of dense material along its cytoplasmic surface—the post-synaptic thickening and vague strands of material bridge the synaptic cleft, probably constituting the bases for the adhesion that can be shown to exist between pre- and post-synaptic membranes.

The synaptic cleft is 20–30 nm across, and in this purely chemical synapse the membranes never lie closely apposed to each other in any

Fig. 2. An axo-dendritic synaptic contact in the brain stem (osmium fixed, PTA stained).
Fig. 3. The regularly arranged pre-synaptic dense projections seen face on (osmium fixed, PTA stained).
Fig. 4. A synaptosome (pinched-off nerve ending) ultracentrifuged from a sucrose homogenate of guinea pig cerebrum (osmium fixed, PTA stained).
Fig. 5. Cholinergic ending in the superior cervical ganglion. The synaptic vesicles have been impregnated by the osmium-zinc iodide method (lead stained). By kind permission of Dr. A. Matus.
Fig. 6. Adrenergic ending containing dense-cored synaptic vesicles in Auerbach's plexus (aldehyde-osmium fixed, lead stained). By kind permission of Dr. G. Gabella.

part of the cleft (as distinct from electrical synapses, section III). Diffusion of the transmitter across such a small space would only involve a matter of microseconds, and the synaptic delay measured at chemical synapses must represent the time taken for the mobilization at and release of the transmitter from the pre-synaptic membrane. When the transmitter arrives at the outside surface of the post-synaptic membrane it reacts with receptors and the resulting conformational changes momentarily open "gates" through which ions flow and result in depolarization or hyperpolarization in the post-synaptic cell. The nature of the receptors is only just beginning to be studied and they are too small to be visualized by present day electron microscopes.

B. SYNAPTIC VESICLES AND TRANSMITTER MECHANISMS

Like *Amphioxus*, if the synaptic vesicle hadn't been discovered it would have had to have been invented. In the early fifties Katz, del Castillo and Fatt showed that at the motor end-plate the transmitter acetylcholine is liberated in discrete or quantal amounts. When the nerve is not stimulated, only one or a few quanta are liberated and these are detected by measurements of miniature end-plate potentials. When an impulse comes along a large number of quanta are liberated. In the mid-fifties several independent groups described synaptic vesicles with rather uniform size (40–50 nm) in the pre-synaptic cytoplasm of the end-plate and it immediately became apparent that the vesicles might contain the transmitter, each vesicle being about the right size to contain one quantum of aceylcholine (about a thousand molecules). Since then (pre)synaptic vesicles have been described at most vertebrate and invertebrate peripheral and central synapses and in correlation there is also evidence for quantal release at peripheral and central synapses.

Since then an enormous amount of research has gone into attempts to show that the vesicles contain the transmitter (see Hubbard, 1970) and there is still much to be done. Attempts to isolate a pure fraction of synaptic vesicles for direct biochemical analysis were pioneered by Hebb and Whittaker. Later I joined forces with Whittaker and we studied various brain fractions and showed that a sub-fraction could be produced which was rich in pinched off nerve endings—the synaptosomes (Fig. 4) and which, on biochemical analysis, was shown to be rich in brain transmitters (acetylcholine, noradrenaline, glutamic acid, γ-amino butyric acid (GABA), dopamine.) More recently work has been aimed at extraction of the vesicles from the synaptosomes and preparing them as a pure sub-fraction. These can be shown to be rich in transmitters, and although it has been argued that the transmitter could be bound artifactually to the vesicle during extraction there is good evidence to indicate that this is not so. Nowadays, the isolation and study of synaptosomes is a much exploited technique in laboratories throughout the world.

Direct histochemical localization of the transmitter within the synaptic vesicle provides the most convincing evidence. Unfortunately, this is not feasible so far in cholinergic endings, because acetylcholine fails to be localized by the fixatives used at present. Akert's zinc iodide method shows up dense material inside vesicles of cholinergic endings (Fig. 5), but is also positive in vesicles of non-cholinergic endings. Our best direct evidence, so far, is from monoaminergic endings. Noradrenaline reacts to form a precipitate with osmium tetroxide fixative and many of the synaptic vesicles thus show dense cores (Fig. 6). Further, the dense cores in these vesicles can be made to disappear when the animal is treated with reserpine (which depletes noradrenaline) or the empty vesicles can be charged with dense cores when the endings are soaked *in vitro* in a medium rich in noradrenaline. Also, autoradiographic studies using tritiated noradrenaline show radioactivity related to the dense-cored synaptic vesicles. Vesicles in other monoaminergic endings where the transmitter is either dopamine or serotonin (5-hydroxytryptamine) also contain synaptic vesicles, some with granules, but less is known about them. (See Addendum c, p. 412.)

An additional advantage in the study of monoaminergic endings is that they can be studied with fluorescence microscopy, for adrenaline, after aldehyde gas treatment, forms a complex substance with a specific fluorescence colour. Constrictions in adrenergic nerves cause a proximal dilatation which fluoresces with u.v. light and can be seen with the electron microscope to contain accumulations of dense-cored vesicles. Many of these probably originate in the Golgi apparatus and move down the axon by axoplasmic flow. Whether cholinergic synaptic vesicles also originate in the Golgi apparatus is not known (see below).

Whereas the catecholamines, noradrenaline (autonomic nervous system, pineal and hypothalamus) and dopamine (caudate nucleus) and the monoamine, serotonin (5-HT-forebrain) can be related to endings with granulated synaptic vesicles, correlation with the other types of transmitter found in the CNS is less certain. These include acetylcholine (be it excitatory or inhibitory—autonomic nervous system, neuromuscular junctions and various regions in the CNS) and the amino-acids glutamic acid (probably excitatory-forebrain) and gamma amino-butyric acid (GABA-forebrain and cerebellum) and glycine in the spinal cord (these latter inhibitory). All these endings probably have the clear or non-granular type of synaptic vesicle. The shape of the vesicle (round or flat) related to excitatory or inhibitory mechanisms is discussed in section IV.

Electrical synapses contain characteristic pre-synaptic vesicles, yet at these synapses there is no evidence for a chemical mode of transmission (see section III). Thus, at present one cannot be dogmatic about synaptic vesicles always containing transmitters.

The mechanism of discharge of transmitter from the vesicle into the synaptic cleft (either spontaneously or as the result of a depolarization of the axon) still remains unknown. Calcium is essential for this mechanism. The effects of prolonged stimulation still remain controversial and it is not easy to demonstrate a depletion of synaptic vesicles. This would suggest that rather than the membrane of the synaptic vesicle opening into and fusing with the pre-synaptic surface membrane and releasing the transmitter into the cleft by a process of exocytosis (reverse pinocytosis), the wall of the synaptic vesicle may remain intact. Rather, a region of the vesicle touching the pre-synaptic membrane may undergo a transient reorganization of the lipo-protein structure in conjunction with the subjacent pre-synaptic membrane. This would result in a leak of the transmitter into the cleft (see Hubbard, 1970 for detailed discussion). (See Addendum d, p. 412.)

The mode of formation of synaptic vesicles still remains unsolved. In adrenergic endings many of the vesicles probably form at the Golgi apparatus and move down to the terminal by axoplasmic flow. The mode of formation of agranular vesicles in cholinergic and other non-adrenergic neurons is less well understood. Possibly some form at the Golgi apparatus, but there is evidence that they form locally at the synapse or end-plate by pinching off from the surface membrane lining the walls of the presynaptic bag (see Gray and Willis, 1970), when presumably they become charged with transmitter. The vesicle forms by an invagination and is surrounded by a spherical shell of pentagonal and hexagonal protein subunits (Fig. 17). These are of unit size and may therefore determine the quantal size of the synaptic vesicle. The vesicle plus shell is known as the complex, coated or alveolate vesicle. This moves into the pre-synaptic cytoplasm and the vesicle escapes from the shell but parts of the shell adhere to the surface of the synaptic vesicle. Those vesicles congregated against the pre-synaptic membrane can be seen lying around and touching the pre-synaptic dense projections (section II). It is possible in fact that the pre-synaptic dense projections are derived from pieces of the shells that have adhered to the synaptic vesicles. However, since the dense projections are regularly arranged on the pre-synaptic membrane (Fig. 3) they could be quite independent structures. (See Addendum e, p. 413.)

With the exception of cholinesterase, little is known about the enzymes that deactivate the various transmitters and their localization either intracellularly or in the synaptic cleft. Cholinesterase splits acetylcholine into choline and acetic acid, the former being retaken-up into the nerve ending. Its presence has been studied in most detail at motor end-plates, but its activity can also be localized with histochemical techniques at central synapses by both light and electron microscopy. The work of Shute and Lewis (Shute and Lewis, 1967; Lewis and Shute, 1967) is preeminent in this field. Although the enzyme can be shown to be localized to the synaptic cleft, techniques are not yet sufficiently refined to show the shape and location of its molecules. They probably lie in the outer surface of the post-synaptic membrane mixed with the receptor molecules (p. 459). Indeed some think that receptor and deactivator molecules may be the same or parts of the same molecule.

The presence of cholinesterase around central synapses usually but not invariably indicates that the transmitter is acetylcholine. The best evidence for this is the demonstration of the formative enzyme choline acetyltransferase in the presynaptic knob. This is a more difficult problem and is only just beginning to be attacked at the histochemical and fine structural level.

C. PRESYNAPTIC NEUROFILAMENTS AND MICROTUBULES (NEUROTUBULES)

In neurofibrillar silver-stained preparations of the CNS, the appearance of boutons as ring-shaped, tennis-racket or reticulated knobs (Fig. 1C and Fig. 8) puzzled the classical neurohistologists and equally bewildering was the appearance of such boutons on soma and dendrite in some places, for example, spinal cord, and not others. In particular, the failure to demonstrate boutons in the cerebral cortex (the axons just seemed to end blindly) proved exceedingly embarrassing to synaptologists concerned with elucidating cortical connexions. We (Boycott, Gray and Guillery, 1961) elucidated this problem with the electron microscope. Briefly, there are two main sorts of cytoplasmic fibrous proteins in neurons, the "solid" thread-like neurofilaments ~ 10 nm in diameter and the "hollow" microtubules (see Chapter 17) 20–25 nm in diameter. The neurofilaments are positive with the reduced silver stains and the linear deposits of silver grains along their length appear as dense strands, termed neurofibrillae, when viewed with the limited resolution of the light microscope. The microtubules, on the other hand, are negative with the silver methods. Neurofilaments and microtubules can both occur in various regions of the neuron. They are common in axons but do not usually enter the pre-synaptic cytoplasm (as in Fig. 2). In certain synapses the neurofilaments do enter the pre-synaptic bag, however, and when they do they form a ring-shaped bundle or a reticulated or compact mass (Fig. 7). These stain with silver and so

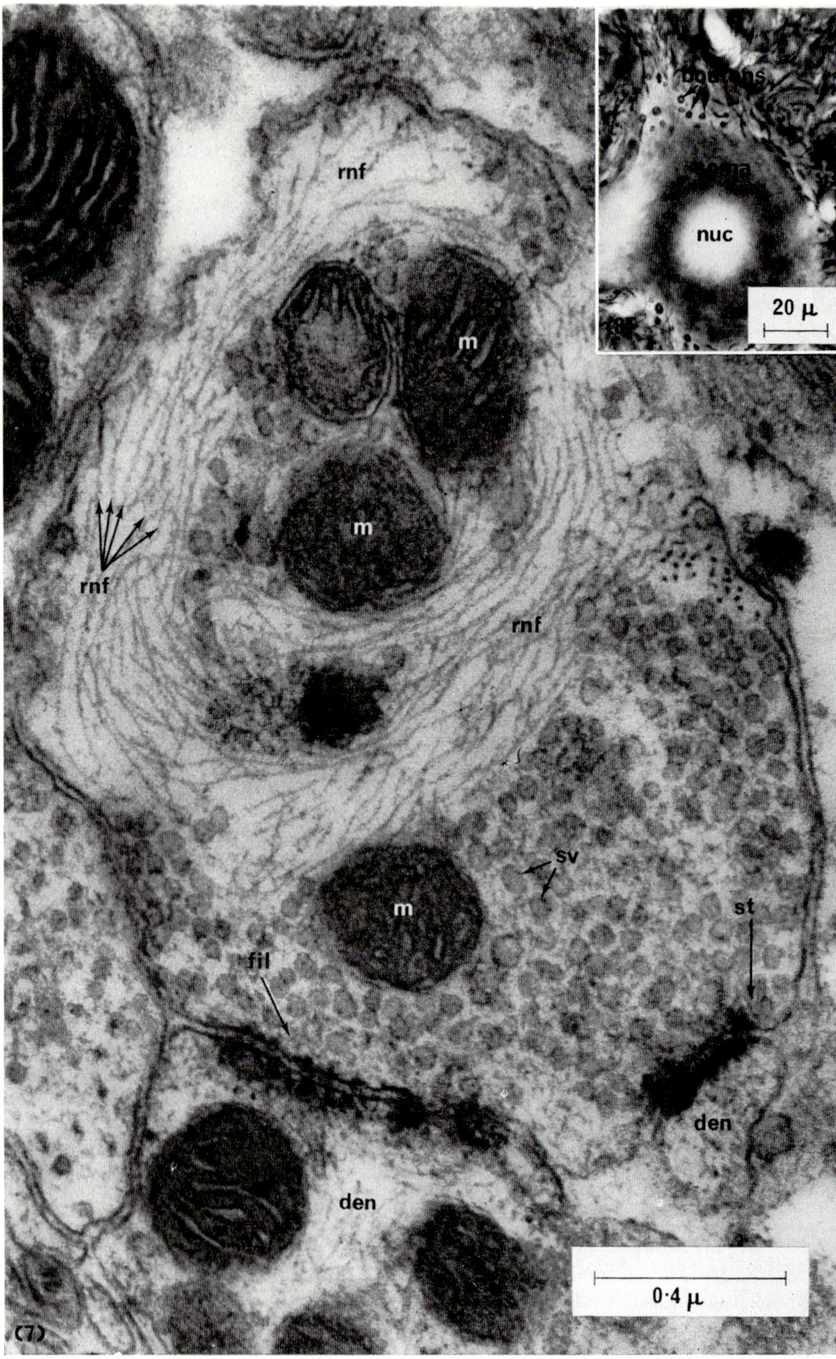

Fig. 7. Synaptic knob in the lateral geniculate nucleus containing a ring of neurofilaments (osmium fixed, PTA stained) (from Gray and Guillery, 1966).
Fig. 8. Ventral horn cell of the spinal cord of cat impregnated by the block Cajal silver method to show the boutons.

we see why some boutons appear as rings and others as reticulated or solid blobs (Figs. 1C and Fig. 8). Usually, however, the neurofilaments stop short at the junction between the pre-terminal axon and pre-synaptic bulb. In the cerebral cortex usually only the coarser axons contain neurofilaments whereas the finer ones contain only micro-tubules. No neurofilaments run into the pre-synaptic bags (e.g. Fig. 21) explaining why, in the cerebral cortex, no boutons can be demonstrated with the neurofibrillar silver methods, although with the EM we can see that the cortex is in fact crammed with conventional synapses (Fig. 21). (See Addendum f, p. 413.)

Why neurofilaments run only into certain endings is unclear at present—usually they run only into the larger ones. Also after axon section certain endings develop neurofilaments (see section VI). At present amino acid and structural analysis of isolated neurofilaments is in progress. Little is known about their function. They may play a part in neuron recognition and laying down of specific connections during ontogeny. It has been suggested that microtubules and also possibly neurofilaments may play a part in movement of synaptic vesicles and other particles down the axon by a rolling or sliding mechanism involving a surface active ATP on the tubule or neuro-filament.

D. SYNAPTIC MITOCHONDRIA

Most synaptic knobs contain one or more mitochondria (e.g. Figs. 2 and 7), showing the typical mitochondrial substructure. The early workers used mitochondrial stains to show up the end-feet on the ventral horn neurons of the spinal cord (Fig. 1B). The end-feet are particularly rich in mitochondria (the neurosomes of Held). Their main purpose is no doubt the production of ATP for the various metabolic activities in the ending, but they may also play other roles. Mito-chondria in the lateral geniculate nucleus have a pale or dark matrix depending on which type of ending they are in. In lizards the mito-chondrial content in synaptic knobs increases when the lizards are kept in the cold.

Mitochondria were once thought to play a direct part in production of the transmitter substance, but this has been discounted in more recent work. Some evidence indicates that ATP might be a transmitter at some terminals in Auerbach's plexus in which mitochondria would be the obvious source. Mitochondria may be involved, indirectly, how-ever. For example, they supply acetyl-CoA which with choline will from acetylcholine under the influence of choline acetyltransferase. Mitochondria also play a catabolic role, in, for example, monoamin-ergic endings, where the transmitter is either dopamine, noradrenaline or 5HT. Monoaminoxidase (MAO) of mitochondrial origin probably plays a part in the enzymatic destruction of the transmitter. Also, mitochondria play a part in the removal of the inhibitory transmitter GABA (see section II, Synaptic vesicles and transmitter mechanism) by means of their GABA-transaminase.

The postsynaptic component may or may not have mitochondria near the synaptic cleft. Most synapses in the cerebral cortex, corpus striatum and cerebellar cortex are found on dendritic spines rather than on the dendritic trunks. It is a curious fact that mitochondria are absent from these spines, the main organelle being the spine apparatus (see section VII).

E. OTHER PRESYNAPTIC ORGANELLES

Glycogen granules (Figs 13 and 14) may be present: also multivesicular bodies. Sometimes dense-cored vesicles 150 nm or more in diameter also occur. Whether these are lysosomes remains to be seen. Some receptor synapses have curious "ribbons" or spherical spiky bodies (see section V).

III. Electrical Synapses

Unlike the synapses with the purely chemical mode of transmission in the electrical synapse the synaptic cleft becomes reduced or obliterated either extensively or at localized patches (arrows, Figs. 11 and 12). When the trilamellate unit membrane structure can be seen (with the appropriate technique) the membranes come together at these points to give in section a three-lined appearance. This has become known as a tight junction (or closed contact, quintilamellate zone, nexus or in some instances zonula occludens) (Fig. 11, inset), but as we shall see, the more recent term "gap junction" is more appropriate. There is excellent correlation between their occurrence and the physiological detection of electrical transmission at the relevant synapses (absence of synaptic delay and refractoriness to pharmacological interference). Gap junctions also occur at the intercalated discs between cardiac muscle cells, between certain smooth muscles, between neuroglia (Fig. 14, ntj) and at certain epithelia, all of which can be shown to be electrically coupled.

Most synapses in the mammalian central nervous system are without gap junctions (the lateral vestibular nucleus and possibly the primate retina excepted) and taken in conjunction with the physiological evidence (see section I) it is probably true to state that almost all mammalian CNS synapses have a purely chemical mode of transmission. Electrical synapses are more common in the fish (Figs. 11 and 12, right) and frog CNS (brainstem and cord) and they also occur in

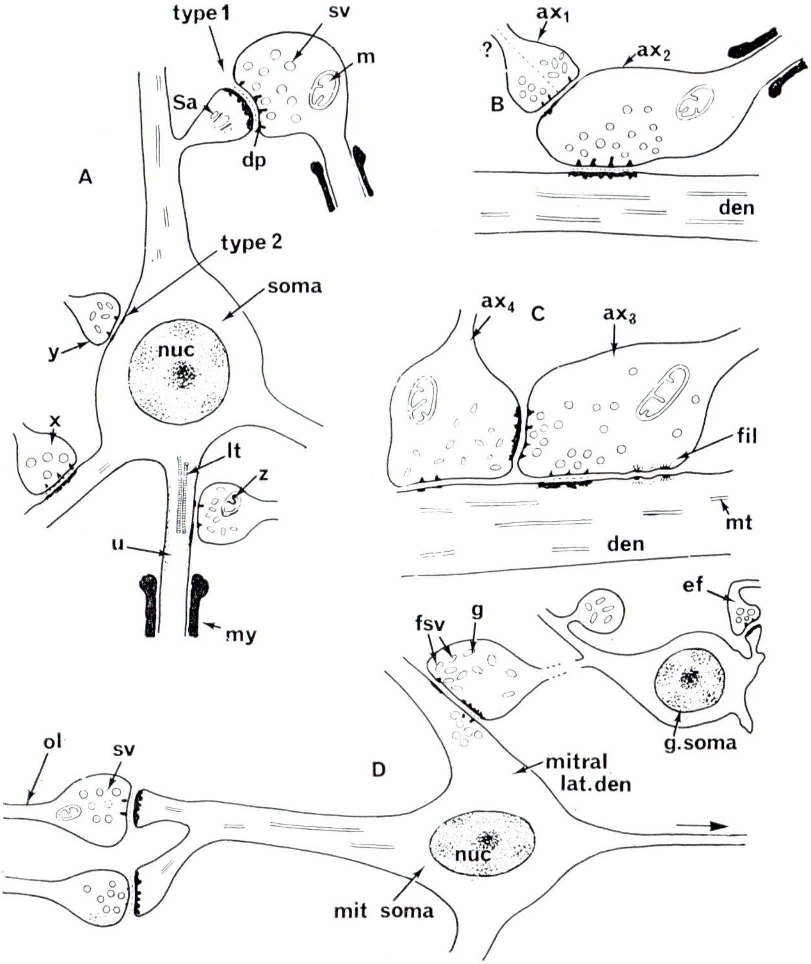

Fig. 9. Diagrams of various types of synapse.

various groups of invertebrates. The vast majority of synapses throughout the animal kingdom have no gap junctions though and so are presumably all chemical.

Most synapses with gap junctions have the characteristic presynaptic vesicles, yet physiological investigation of such synapses shows no evidence for chemical transmission (see section II). Part of the cleft near the gap junction may (Fig. 10, I) or may not (Fig. 10, H) show the characteristic asymmetrical thickenings against which the synaptic vesicles form aggregations. When it does, the synapse is called a "mixed"

have flat vesicles (see Colonnier, 1968); also those (z) lying on the axon initial segment (recognized by its undercoating of dense material, u and bundles of linked microtubules lt). Thus, it seems probable that chemical excitatory synapses are mostly on dendritic spines, whereas inhibitory synapses are located mostly on cell bodies and initial segments. Colonnier (1968) has shown that most of these cortical spine synapses have round vesicles (Fig. 16) whereas those on the soma have flat ones (Fig. 18). The stellate cells of the cerebral cortex tend to lack dendritic spines and have varicose dendrites. Both types of synapse are located on trunk and cell body. In the cerebral and cerebellar cortex the synapses are probably all of the simple two component (i.e. non-serial) category and this fits in with the electrophysiology. (See Addendum h, p. 413.)

Serial synapses are more complex arrangements and consist of three components—an axon knob presynaptic to a second axon knob, which in turn is presynaptic to another dendrite (Fig. 9, B) or soma (axo-axo-dendritic or axo-axo-somatic). Briefly, this arrangement is thought to be responsible for presynaptic inhibition in the spinal cord. Ax_2 (Fig. 14—compare Fig. 9, B) is chemically excitatory to the dendrite. Ax_1 is also excitatory and when it fires it lowers the resting potential in ax_2 and experiments have shown that this results in a dramatic decrease in output of the transmitter from ax_2 to dendrite (den). Thus inhibition is achieved presynaptically and no inhibitory transmitter is involved, only excitatory ones (see Eccles, 1969) One would therefore expect that after aldehyde initial fixation both knobs (ax_1 or ax_2) would have spherical synaptic vesicles (section IV). Observations on ax_1 are controversial, however. There is a parallel situation in crustacea and here ax_1 has small flat synaptic vesicles. This fits in though, for here there is an inhibitory transmitter (GABA) and this achieves depression of output of excitatory transmitter (glutamic acid) by increasing the Cl^- conductance. In the octopus (Fig. 10, E) msf endings contact amacrine (am) processes, which in turn contact spines (sp) of large neurons. Whether this constitutes a presynaptic inhibitory type of control remains to be seen. (See Addendum i, p. 414.)

A third variety of serial synapse occurs in the mammalian lateral geniculate nucleus (Fig. 9, C). The optic axon knob ax_3 is in all probability excitatory to a dendrite (and has spherical vesicles). It is also presynaptic to a second (interneuron) axon knob ax_4 which is in turn presynaptic to the dendrite. Ax_4 has flat vesicles, however, suggesting that it is chemically inhibtory. Thus, ax_3 could directly excite the dendrite, and also excite ax_4 and so reduce the output of inhibitory transmitter to the dendrite. This then would appear to be a pre-synaptic excitatory effect resulting from inhibition of release of an inhibitory transmitter. (See Addendum i, p. 414.)

A type of reciprocal synapse is illustrated in Fig. 9, D and Fig. 15 in work from Price and Powell (1970a, b and c). In the olfactory bulb, the endings (Type 1) of the olfactory nerve fibres (ol) are excitatory and have spherical vesicles. They contact the main dendrites of the mitral cells. Each mitral cell has lateral dendrites, which are contacted by grape-like gemmules (g) from granule cells (amacrine cells since they have no axons). These contacts are reciprocal synapses, dendrite-to-gemmule excitatory (Type 1 and spherical synaptic vesicles) and gemmule-to-dendrite inhibitory (Type 2 and flat vesicles) and probably constitute a negative feedback control. Efferent (centrifugal) fibres from the brain also end on the granule cells (ef). These probably excite the granule cells (Type 1 and spherical vesicles) which in turn inhibit the mitral cells through the gemmules. Thus there is both local and central inhibitory control.

Other categories include synapses at central nodes of Ranvier (either Type 1—round vesicle or Type 2—flat vesicle, so they can probably be either excitatory or inhibitory, Fig. 10, F) or somato-dendritic (Fig. 10, G). Indeed, if we interpret our morphology correctly, then it would seem that almost any part of the surface of a neuron can function as a pre-synaptic zone : i.e. modified to eject transmitter (an exception is the initial segment of a neuron); and also as a post-synaptic site; i.e. having receptors to receive and transduce transmitter (an apparent exception is the bare axon area of a node of Ranvier).

Electrical synapses (Fig. 10, H and I) have been described above (section III). These are always in the 2-component non-serial category.

Finally, there is a variety of synaptic structures found between receptor cells and the second order neurons, for example in the retina, cochlear and vestibular systems of the VIIIth nerve and taste buds. Usually pre-synaptic vesicles are present and the synaptic cleft is nowhere fused to form a "tight" or "gap" junction. Synaptic membrane thickenings are usually present. Thus, the morphology indicates a chemical mode of transmission, but studies on the transmitter involved at these synapses have scarcely begun. In the vertebrate rods and cones (Fig. 10, O) a curious synaptic ribbon (sr, actually a disc) occurs and numerous vesicles aggregate round its surface. In the electric receptor cells of certain fishes (Fig. 10, N) there is a presynaptic spherical body (b) with flanges that contact parallel rods (ro) on the presynaptic membrane. Thus tunnels are formed each just wide enough to accommodate a single file of synaptic vesicles. These ribbons and spherical bodies both possibly play a role in directing the vesicles towards the synaptic cleft, but experimental evidence for this is lacking. (See Addendum j, p. 414.)

VI. Experimental Degeneration and Pathology

When an axon is severed from its cell body the distal part usually undergoes Wallerian degeneration and there may be also changes in the proximal part or the cell body that will not be considered here. In the peripheral nervous system the axon may regrow to the end organ (sensory or motor) but in the central nervous system of mammals the axon does not usually regrow although it may do so in the CNS of lower vertebrates. Little is known about the electron microscopy of the re-establishment of synaptic connections after a lesion, although the initial degenerative processes are beginning to be elucidated.

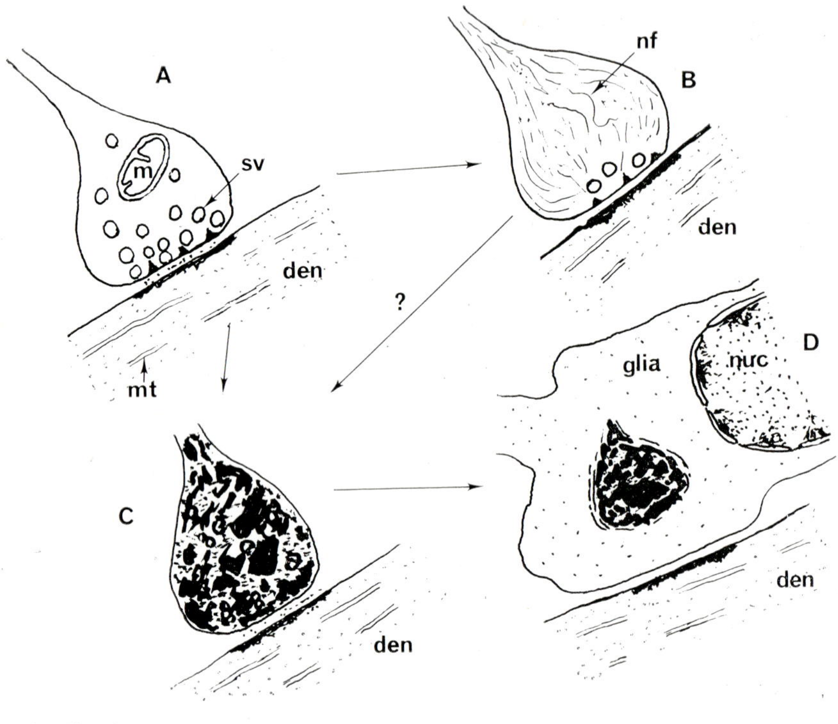

Fig. 19. Diagram to illustrate the two sorts of degeneration in a pre-synaptic knob after axon section.

These can take two main forms (see Guillery, 1970). When, for example, the optic axons are sectioned, the endings in the lateral geniculate nucleus lose their synaptic vesicles and the end-bulb (Fig. 19, A) becomes filled with neurofilaments (B) and, as would be expected, takes up silver stains (sections II, pre-synaptic neurofilaments and microtubules (neurotubules)) and can be seen with the light microscope. This process takes several days. In the cerebral cortex, thalamus, substantia gelatinosa and certain other regions there is a second, much faster type of degeneration. Within a day or two after axon section, the entire contents of the ending (C) become electron opaque due to the degradation of the lipo-proteins and other materials and the ending is phagocytosed by a neuroglial cell (D—usually an astrocyte). Whether the ending undergoing the first, neurofilamentous type of degeneration also eventually goes "dark" and is phagocytosed has not yet been investigated. Also, possibly as a third category, in the cerebral cortex for example, the degenerating ending will lose its content of synaptic vesicles and become clear rather than dense. (See Addendum k, p. 414.)

Such changes have not yet been firmly established in pathological conditions. A glance through a text-book of neuropathology shows that observations are mostly concerned with changes in the neuronal cell bodies (chromatolysis, shrinkage or sclerosis, vacuolation, appearance of inclusion granules, neurofibrillar changes, etc.) or in the axons (alterations or disappearance of axonal tracts, demyelination with sparing of axons, etc.), or in the neuroglia or the blood vascular system (including white cell infiltration from the blood). Such changes are caused by hereditory factors, protozoal, bacterial, viral or fungus invasion, metabolic disturbance, chemical poisons, autoimmune responses (demyelinating diseases) mechanical or radiation damage and often by factors unknown. Observations on changes of synapses are few, no doubt partly because of the low resolution of the light microscope and partly because of the difficulties of staining boutons in many parts of the CNS (section II, pre-synaptic neurofilaments etc.). The electron microscope is now opening up a new field in synaptic pathology, but one of the difficulties always to be borne in mind is what is a primary effect on the bouton or what is a secondary effect resulting from a remote lesion or other pathological change in the cell body, axon, surrounding neuroglia or indeed in any of the other structures of all the intimately related units that make up the PNS and CNS.

VII. Plasticity at Synapses and Learning Mechanisms

The modification of behaviour patterns in the light of past events is a ubiquitous and self-evident phenomenon yet tantalizing for we still know very little about the morphological and physiological alterations (the engram) that underlie the phenomenon of learning. Since we are still on insecure ground when it comes simply to interpreting the structures we see at the synapse with the electron microscope in terms of function, it is hazardous, to say the least, to read too much into any *changes* we can detect taking place at synapses.

A related problem is that when we visualize a synapse with the EM in an adult brain, it might appear to have all the features we associate with a mature synapse. Yet we have no way of telling whether that particular synapse is functional or latent—as if it were still waiting to be switched on, or whether it has ceased to function either temporarily or permanently. Thus, if we count the connections we see with EM in the cerebral cortex the result is not really meaningful till we have some way of deciding whether the synapses are active or not.

Most people would agree that learning leads to a modification in the firing propensity or pattern in synapses in the cerebral cortex. Few are prepared to speculate more precisely on the nature of these changes. Possibly the nucleic acid systems are changed so that the enzymes they programme are changed, and the change in enzymes brings about a change in the synaptic membranes during turnover. Possibly the membrane properties at the synaptic cleft change directly in the learning mechanism without the intervention of the protein synthesizing machinery. Possibly the mechanism for transmitter formation, storage (in the synaptic vesicles) and release is modifiable, or maybe it is the receptors on the postsynaptic membrane that may perhaps only come into being or undergo a change in efficiency in the process of learning.

Since learning is probably a special property of the cerebral cortex then shouldn't we expect to see something different with the EM when compared with, for example, synapses of the spinal cord? When I examined the mammalian cortex with the EM in 1959 I was struck by the fact that more than 90% of the synapses are formed on the tips of dendritic spines (Fig. 21) that can be revealed in Golgi preparations (Fig. 20) and many spines contained an organelle I termed the spine apparatus (section V, and Fig. 22). In the spinal cord and most other sub-cortical regions most synapses occur directly on dendrite or soma. The spine apparatus is unique to the mammalian forebrain. Since it is absent from the dendritic spines of the forebrains of the other vertebrate groups, however, the spine apparatus cannot be an essential part of the learning mechanism for members of these vertebrate groups are competent learners albeit with a more limited repertoire than mammals. Nevertheless, we need to know much more about the function of the dendritic spine and the spine apparatus it contains. Some hold that the spines are specific loci for different groups of incoming axons: i.e. spines play a part in the initial wiring of the circuitry during ontogeny. More recently, it has been suggested that the spines are silent in the sense that they might act as filters to exclude (perhaps the spine apparatus is involved in this) noise resulting from miniature synaptic potentials in the parent dendrite. Such noise might interfere with the synaptic activity at the tip of the spine (perhaps with

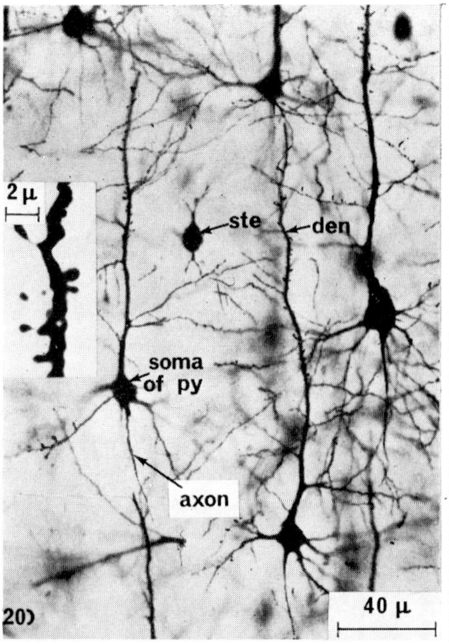

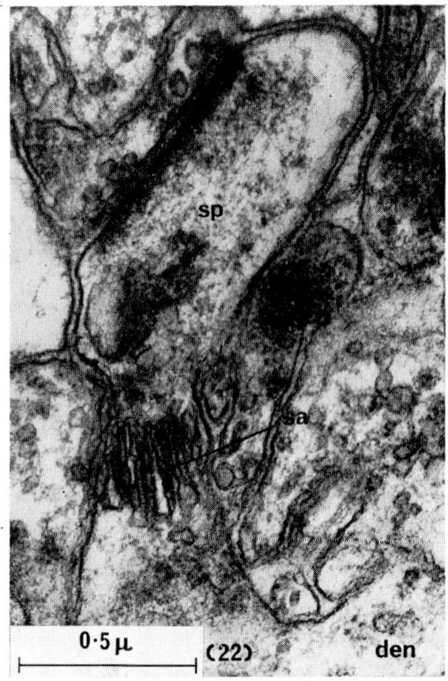

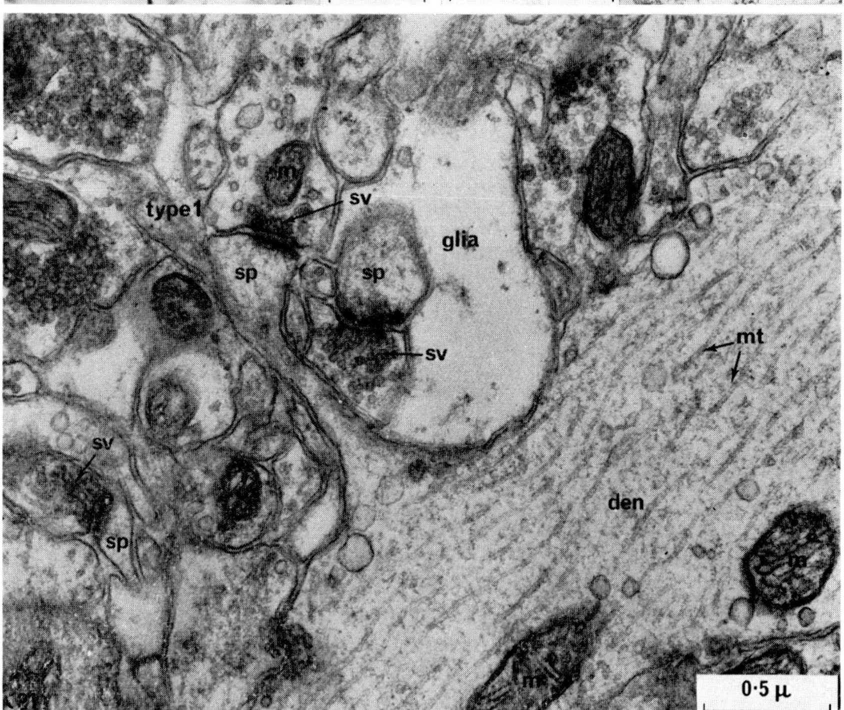

Fig. 20. Golgi preparation of the cerebral cortex of the rat (inset) high magnification (also light microscopy) to show the dendritic spines.

Fig. 21. A dendritic spine synapse in the cerebral cortex of the rat. No spine apparatus is visible in the spine (out of plane of section?). (Method as Fig. 12.)

Fig. 22. A dendritic spine synapse in the cerebral cortex of the rat. The spine contains a spine apparatus. (Method as Fig. 12.)

its timing mechanisms or subtle changes involved in laying down the engram).

At any event the dendritic spines are undergoing extensive study at the moment. They are known to be sites of synaptic contact (probably most are excitatory, section V) and since they can be easily seen with the light microscope they can be measured and counted. There is much evidence that spines can undergo "plastic" changes in that they come and go when their inputs are interfered with. This phenomenon has not yet been detected with the electron microscope and is at present under investigation.

ADDENDUM.

The text for the above chapter was written some time ago. The following remarks are made in an attempt to up-date the more obvious anachronisms. Reference to this addendum has been made in the appropriate places in the text.

(a) Since there is now evidence for ion-flow channels spanning the gap junctions of electrical synapses between certain neurons, this could perhaps be regarded as cytoplasmic continuity (see Addendum, g).

(b) See also Pappas, G. D. and Purpura, D. P. (1972). " Structure and Function of Synapses". Raven Press, New York and Pfenninger, K. H. (1973). Synaptic Morphology and Cytochemistry *Prog. Histochem. Cytochem.* 5, 1–86.

(c) Studies on aminergic synapses are progressing rapidly. For example the ability to take up and reutilise noradrenaline (involving a membrane pump) is well recognized. Also false transmitters such as 5- and 6- hydroxidopamine can be taken up into the granulated vesicle. The latter false transmitter results in peroxide formation which is poisonous and eventually destroys the nerve ending. This is being used as a powerful tool in experimental degeneration techniques. It is also now known that a formative enzyme (dopamine – β – hydroxylase) and a protein (Chromogranin) and ATP also occur within the vesicle and are liberated from it together with the transmitter during stimulation.

(d) The concept of the presynaptic grid (see Pfenninger, K., Akert, H., Moor, H. and Sandri, C. (1972) The fine structure of freeze-fractured presynaptic membranes. *J. Neurocytol* 1, 129–49), has been developed thanks to the relatively new EM technique of freeze-fracture-etching. Briefly, the dense projections are triagonally arranged and are surrounded by shells of hexagonally placed synaptic vesicles. "Synaptopores" – small pits in the presynaptic membrane—occur where the vesicles touch the membrane. These are thought to represent channels for flow of transmitter out from the vesicle into the synaptic cleft. They seem to be more frequent at stimulated synapses. Whether these pores are temporary or permanent and whether the membrane of the synaptic vesicle becomes incorporated or not in the presynaptic membrane are matters at present under intensive investigation: as is the related problem of membrane recycling and the postulated reformation of the synaptic vesicle by endocytosis. Stimulation has now been observed to result in depletion in the numbers of synaptic vesicles and horseradish peroxidase is being used as a marker to demonstrate endocytosis (see Addendum, e).

(e) Recent work has led to alternative theories (see Gray, E. G. (1972) Are the coats of coated vesicles artifacts? *J. Neurocytol.* 1, 363–82 and Gray, E. G. (1973). The Cytonet *etc. Brain Res.* 62, 329–335.) The dense projections may be the attachment points of a proteinaceous presynaptic cytonet, in the interstices of which lie the synaptic vesicles. Possibly the net contains actin-like protein and plays a role in propelling the synaptic vesicle into the grid on the presynaptic membrane. Myosin-like protein could be located in the wall of the synaptic vesicle. The coats of the coated vesicles may be condensations (artifactual?) of the cytonet.

(f) Colchicine is now used as a tool for the depolymerisation of microtubules to study their role in moving organelles along the axon (an aspect of axoplasmic flow). Growth cones at the tips of growing axons contain a marginal filamentous network which is thought to be "contractile" and responsible for the ruffling movements and pseudopod (microspike) forinations of the growth cone. Cytochalasin B inhibits these movements and is said to affect the microfilamentous network. Whether or not actomyosin-like proteins are involved has not yet been decided. Nor is it known whether the microfilamentous network of the growth cone is the same material as the cytonet of the adult synapse.

(g) Gap junctions are being studied intensively at present, especially with the EM freeze-fracture technique. Pegs spanning the gap are seen to be related to intramembranous particles. The pegs and particles are thought to form the walls of the ion-flow channels. Gap junctions have now been demonstrated at most electrical synapses and between muscle and other types of cells which show electrical coupling. Blood vessel endothelia and certain epithelia have true tight junctions between them, serving as extracellular seals.

(h) Serial synapses, probably involving presynaptic dendrites and also gap junctions have recently been described in the primate motor cortex. Also several studies aimed at elucidating the differences between pyramidal and stellate cells have been made. Large stellate cells have spines and like the pyramidal cells send their axons for long distances, probably to end in excitatory contacts. The small stellate cells have probably only local circuitry and make inhibitory contacts on pyramidal cell bodies and on other stellate cell bodies and dendritic varicosities. Incoming axons of extrinsic origin are probably all excitatory.

(i) GABA is now thought to be responsible for presynaptic inhibition in the vertebrate spinal cord and flat vesicles have been observed in the first component of a serial synapse. At present there is dispute as to the nature of the effect of GABA on the post-synaptic membrane.

(j) Gray, E. G. and Pease, H. (1971). *Brain Res.* 35, 1–15 have recently suggested a new interpretation concerning the functioning of the ribbon synapses.

(k) More recently other types of degeneration, including synaptic vesicle swelling, have been described. The origin and nature of the enzymic mechanisms that bring about the degenerative changes are far from understood at present. Lysosomes, as such, are not usually present, and diffuse acid phosphatase activity has been demonstrated in the presynaptic bag.

Acknowledgements

I am grateful to Professor J. Z. Young, F.R.S., for his continuing support, to Miss H. Pease for technical help and to Mr. R. Boddy for printing the photographs.

List of Abbreviations

am	amacrine cell
ax	axon or axon ending
b	dense body
bc	basket cell
be	basket cell endings
cv	complex vesicle
dcv	dense-cored vesicles
den	dendrite
dp	presynaptic dense projections
ef	efferent fibres
endf	synaptic endfoot
fil	filamentous contact (special type of synaptic contact)
fsv	flat synaptic vesicles
g	granule cell
glia	neuroglial cell

gln	glycogen in neuroglial cell
gly	glycogen granules
lt	linked tubules of initial segment
m	mitochondrion
mit lat den	mitral cell lateral dendrite
mit soma	mitral cell soma
msf	axon endings from median superior frontal lobe
mt	microtubule
my	myelin sheath
nf	neurofilaments
ntj	gap junction between neuroglial cell processes
nuc	nucleus
ol	olfactory axons
pst	postsynaptic thickening
Pu	Purkinje cell bodies
py	pyramidal cell
ri	ring-shaped bouton
rnf	ring of neurofilaments
ro	presynaptic rods
sa	spine apparatus
sh	shell of complex vesicle
sol	solid bouton
sp	dendritic spine
sr	synaptic ribbon
st	synaptic thickenings of synaptic cleft
ste	stellate cell
sv	synaptic vesicles
ten	tennis racket (reticulated) bouton
tj	gap junction
u	undercoating of axon initial segment
x, y, z	see text

References

Boycott, B. B., Gray, E. G. and Guillery, R. W. (1961). Synaptic structure and its alteration with environmental temperature. *Proc. Royal. Soc. Ser. B.* **154**, 151–172.

Cajal, S. R. (1911). "*Histologie du système nerveux de l'homme et des vertébrés*" *Vols I and II* Maloine, Paris.

Colonnier, M. (1968). Synaptic patterns on different cell types in the different laminae of the cat visual cortex. An electron microscope study. *Brain Res.* **9**, 268–287.

Eccles, J. C. (1964). "*The Physiology of Synapses*". Springer Verlag, Berlin.

Eccles, J. C. (1969). "*The Inhibitory Pathways of the Central Nervous System*". Thomas, Springfield, Illinois.

Gray, E. G. (1959). Axosomatic and axodendritic synapses of the cerebral cortex: an electron microscope study. *J. Anat.* **93**, 420–433.

Gray, E. G. (1969). Electron microscopy of excitatory and inhibitory synapses a: brief review. *Progr. Brain Res.* **31**, 141–155 Elsevier, Amsterdam.

Gray, E. G. (1970). The fine structural characterisation of different types of synapse. *Progr. Brain Res.* **34**, 149–160 Elsevier, Amsterdam.

Gray, E. G. and Guillery, R. W. (1966). Synaptic morphology in the normal and degenerating nervous system. *Int. Rev. Cytol.* **19**, 111–182.

Gray, E. G. and Willis, R. W. (1970). On synaptic vesicles, complex vesicles and dense projections *Brain Res.* **24**, 149–168.

Guillery, R. W. (1970). Light- and electron-microscopical studies of normal and degenerating axons, in "Contemporary Research Methods in Neuroanatomy" (W. J. H. Nauta and S. O. E. Ebbesson eds.) pp. 77–105, Springer Verlag, Berlin.

Hubbard, J. I. (1970). Mechanism of transmitter release. *Progr. Biophys. Mol. Biol.* **21**, pp. 35–124.

Katz, B. (1966). "Nerve, Muscle and Synapse". McGraw-Hill, New York.

Lewis, P. R. and Shute, C. C. D. (1967). The cholinergic limbic system; projections to hippocampal formation, medial cortex, nuclei of the ascending cholinergic reticular system, and the subfornical organ and supra-optic crest. *Brain*, **90**, 521–540.

Van Der Loos, H. (1967). The history of the neuron. In "The Neuron" (H. Hyden Ed.) pp. 2–47, Elsevier, Amsterdam.

Nauta, W. J. H. and Ebbesson, S. O. E. (eds.) (1970). "Contemporary Research Methods in Neuroanatomy". Springer Verlag, Berlin.

Peters, A., Palay, S. L. and Webster, H. de F. (1970). "The fine structure of the nervous system". Harper and Row, New York.

Price, J. L. and Powell, T. P. S. (1970a). The morphology of the granule cells of the olfactory bulb. *J. Cell Sci.* **7**, 91–123.

Price, J. L. and Powell, T. P. S. (1970b). The synaptology of the granule cells of the olfactory bulb. *J. Cell. Sci*, **7**, 125–155.

Price, J. L. and Powell, T. P. S. (1970c). An electron-microscopic study of the termination of the afferent fibres to the olfactory bulb from the cerebral hemisphere. *J. Cell Sci.* **7**, 157–187.

Sherrington, C. S. (1897). In "A Text Book of Physiology" (M. Foster ed.) 7th Ed. Vol. 3 Macmillan, London.

Shute, C. C. D. and Lewis, P. R. (1967). The ascending cholinergic reticular system: neocortical, olfactory and subcortical projections. *Brain* **90**, 497–520.

Uchizono, K. (1965). Characteristics of excitatory and inhibitory synapses in the central nervous system of the cat. *Nature, London* **207**, 642–43.

Young, J. Z. (1939). Fused neurons and synaptic contacts in the giant nerve fibres of cephalopods. *Phil. Trans. Royl. Soc. London Ser. B.* **229**, 465–503.

19. Satellite Cells of Neurons

H. J. GAMBLE

St. Thomas' Hospital Medical School, London, England

The major morphological characteristics of the neuron, the elongated cytoplasmic processes which extend from its nucleated part, or cell body, are related to the high development of its properties of irritability and conductivity. They allow a response to a suitable stimulus to be conducted over considerable distances which are, in some cases, to be measured in metres.

In these circumstances, where large parts of the cell are far from its trophic centre, it is not surprising that a system of satellite cells should be provided for the support of the neuron and be necessary to its proper functioning, if not to its continued existence. In the central nervous system the satellite cells (oligodendrocytes and astrocytes) provide physical support, as their name neuroglia (nerve glue) suggests, and there is evidence that they also play an important part in the nutrition of the neuron. In the peripheral nervous system, physical support of nerve processes is largely provided by collagenous connective tissues and the role of the satellites (Schwann cells) is here to "insulate" the nerve processes and to aid their nutrition. Special modifications of satellite cells of both central nervous system and peripheral nervous system provide myelin sheaths upon the larger axons, which, as well as insulating, greatly enhance the speed of conduction of the nerve impulse along the ensheathed axon. This role of ensheathing axons, by enormous and complex expansions of the plasma membrane, causes some neuronal satellites to become extremely specialized cells.

I. Development of Schwann Cells

Schwann cells, the satellites of the neuronal elements of the peripheral nervous system derive, in part, from elements of the basal lamina of the neural tube migrating peripherally in association with the development of ventral spinal nerve roots. Other Schwann cells, however, derive from the neural crest, migrating centrally in association with dorsal nerve roots and peripherally with nerve trunks.

It has been known since the later part of the nineteenth century that spinal nerve roots first appear in the embryo as two series of slender strands, linking the basal part of the neural tube to somites and the alar part of the neural tube to neural crest. The nature of these strands, which mimic ventral and dorsal roots respectively, was uncertain and they were called "protoplasmic". At later stages in development they were seen to contain nuclei and were described as becoming "fibrillated".

The classical tissue culture experiments of Harrison (1907) showed that embryonic nerve cells are capable of growing axonal processes of considerable length and led to the belief that the "protoplasmic" strands of earlier investigators consist, at first, of bundles of growing axons which later are invaded by migrating Schwann cells. Recent electron microscope studies of human fetal nerves suggest that Schwann cells may be present earlier than axons (Cravioto, 1965) perhaps as pioneers providing suitable surfaces over which axons may subsequently grow (Gamble, 1966). As will be seen below (p. 433) when the repair of injured peripheral nerve is considered, axonal regrowth and maturation occurs only after satellite (Schwann) cells have provided an ordered series of surfaces. It is possible that this aspect of nerve repair mimics the earlier processes of nerve growth in the embryo: that peripheral nerves and nerve roots at first consist of bundles of undifferentiated satellite cells, somewhat elongated axially, and invested communally by a basement membrane which separates the neurectodermal elements from surrounding mesodermal cells and fibrils.

Neuroblasts developing in the basal part of the neural tube extend their axonal processes into the ventral roots, infiltrating between adjacent satellite cells and, apparently, providing them with the stimulus to commence their characteristic differentiation. This is the formation of progressively more extensive and more complex cytoplasmic processes while establishing and then maintaining an intimate relationship with the invading axons (Fig. 1). Neuroblasts of the neural crest similarly extend axonal processes into dorsal roots and with similar effects upon Schwann cells. Axons at this stage of their development are of *c* 0·1—0·5 μm diameter.

Prior to the arrival of axons close to their surfaces, Schwann cells contain a centrally placed nucleus and organelles are rather evenly distributed through the cytoplasm. After axonal sprouts have arrived,

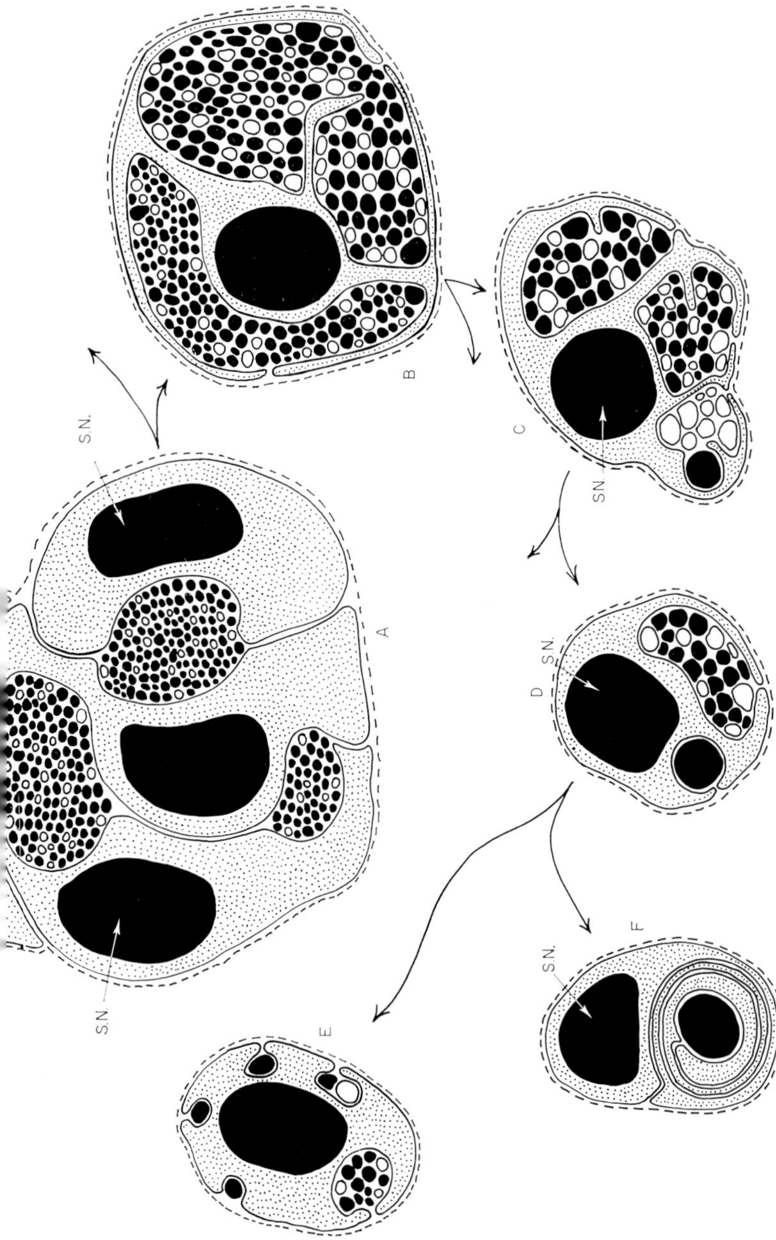

Fig. 1. Diagrammatic representation of transverse sections through fetal peripheral nerves at progressively more advanced stages of development leading to the formation of a small bundle of unmyelinated axons in E and to the formation of a myelinated axon in *F*. S.N.: Schwann cell nucleus. Dense profiles represent persisting axons. Empty profiles represent incipiently degenerate axons. - - - represent basement membrane.

mitochondria and ribosomes (both free and associated with endoplasmic reticulum) become polarized in the satellite cell close to the axons and the now eccentrically placed nucleus is covered only by a thin layer of cytoplasm on its other aspects. These changes must be assumed to be related to the more active role now undertaken by the Schwann cell, most particularly in relation to the production of a far more extensive plasma membrane.

As development proceeds more and more axons invade the satellite cell bundle and although many axons probably degenerate and disappear, many others remain and the satellites become deeply grooved to receive the (Fig. 1B).

As more grooves form and deepen, their margins come together to ensheath bundles of axons wholly. The meeting of the two margins forms a short "mesaxon", often marked by a desmosome-like thickening of the apposed plasma membranes (Gamble, 1966). The satellite cell now has the form of an elongated, deeply fluted column ensheathing as many as five hundred axons arranged in a few large bundles (Fig. 2). Concurrent with these events each bundle of satellite cells becomes broken up into a number of separate Schwann cell/axon complexes as basement membrane extends between adjacent cells (Figs. 1A and B), and collagen fibrils are laid down in the interstitial spaces. The basement membrane of the Schwann cell is of some importance in the repair processes which can occur in injured peripheral nerve (see p. 433).

As development proceeds, mitotic divisions occur in the satellite cells and these are so ordered that the ensheathing of the associated axons is maintained, with some of the axons ensheathed by each of the daughter cells. As a result the number of axons ensheathed by a satellite cell progressively declines. Degeneration and disappearance of axons probably accentuates the decline (Fig. 1 A–E). Among the surviving axons, some become much larger than their fellows and come to occupy separate grooves in the ensheathing satellite cells (Figs. 1 C and D). It is probable that these are the axons destined for further growth and subsequent myelination, but at their earliest appearance they differ only in size from axons which will remain small and unmyelinated.

At a slightly later stage of development (in human fetuses of *c* 8 cms C.R. length), as the result of further mitotic divisions, two populations of satellite cells are found. In the one kind, several axons are ensheathed, some singly, others in small bundles (Fig. 1 E). In the second kind a single axon only is ensheathed and frequently the mesaxon which marks the line of its invagination from the cell surface is elongated and shows a tendency to spiral (Fig. 1 F): the axon is usually of 0·5–1·0 μm diameter. No other features distinguish the two populations of Schwann cells. The invagination of bundles of small, unmyelinated axons persists in adult man only in the olfactory nerves. Elsewhere unmyelinated axons are invaginated separately although several may obtain their individual sheaths from one Schwann cell.

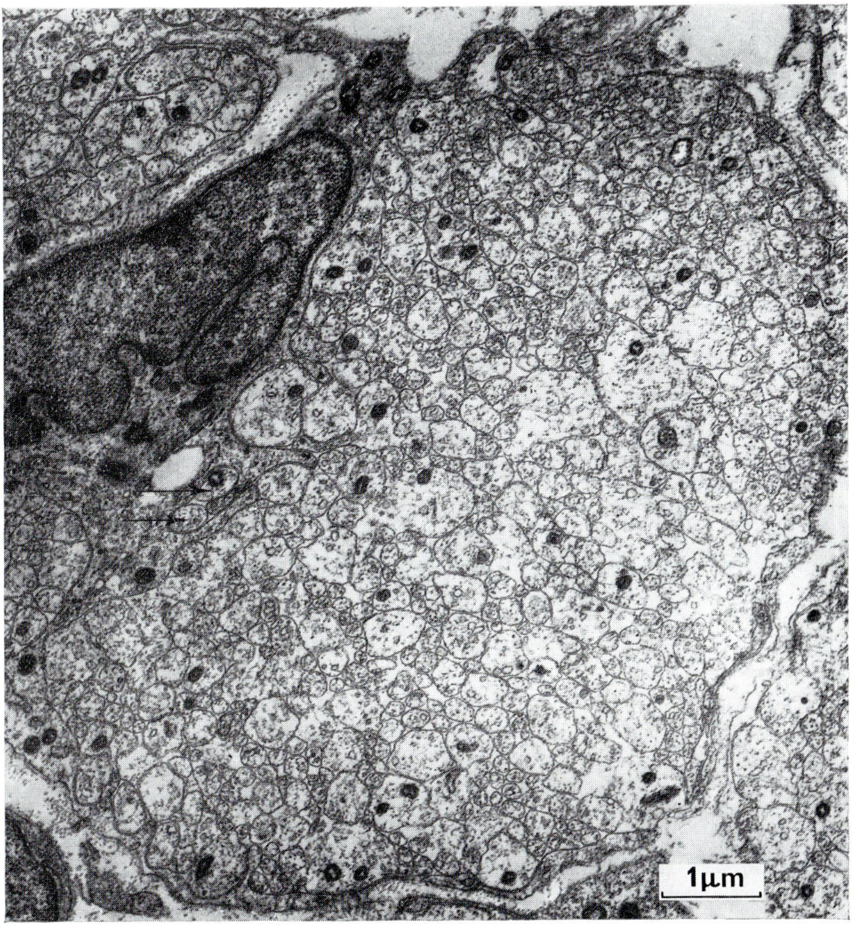

Fig. 2. Electron micrograph from transverse section of ulnar nerve of 5 cm. crown-rump length human fetus. A Schwann cell, with its nucleus toward the left, enwraps some 500 small axons. Most of these axons lie in one large bundle but two of them (arrowed) have been separated from their fellows by small overlapping processes of Schwann cell cytoplasm. From Gamble, H. J. (1966). *J. Anat.* **100**, 487–502, by kind permission of the publishers.

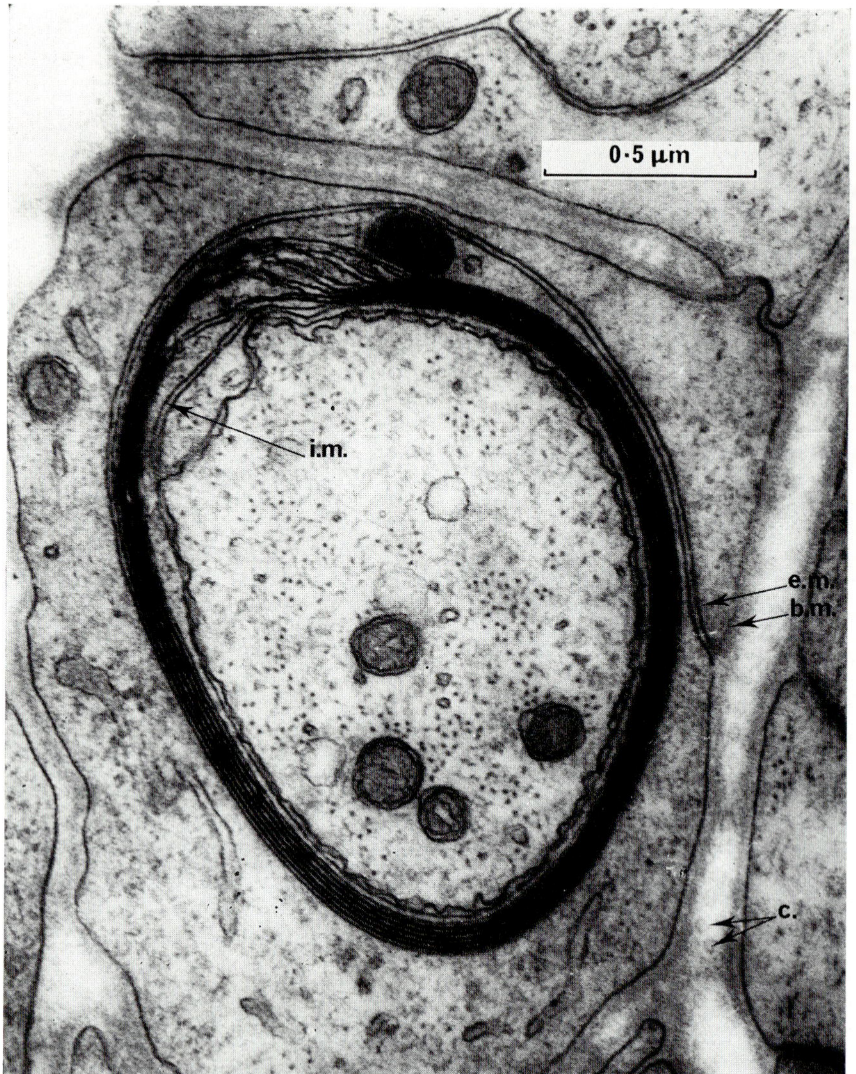

Fig. 3. Electron micrograph from transverse section of sural nerve from young human adult. A small axon is ensheathed by some six spiral lamellae of myelin and the "mesaxon" which forms the spirals is visible in its external and internal parts which extend from the cell surface to the spiral *e.m.* and from the spiral to the axon *i.m.* The basement membrane of the Schwann cell *b.m.* and endoneurial collagen fibrils *c.* may also be seen.
From Eames, Rosemary A. and Gamble, H. J. (1970). *J. Anat.* **106**, 417–435, by kind permission of the publishers.

Elongation and spiralling of a mesaxon marks the commencement of the process of Schwann cell differentiation which leads to the formation of the myelin sheath. This was first observed by Geren (1954) in the nerves of the chick embryo and subsequent studies leave little doubt that the same process occurs in all vertebrates. At first the spiralling involves a sheet of Schwann cell cytoplasm bounded as everywhere else by plasma membrane and separated, one layer from the next, by a narrow interval of extra-cellular space extending through the mesaxon from periaxonal space to the outer aspect of the Schwann cell (Fig. 1 F).

At this stage of development the myelin is loose but later, and by a means unknown, two changes occur together to produce compact myelin: the cytoplasm disappears from the flattened, spiralled sheet so that the inner aspects of the two bounding layers of plasma membrane meet and fuse; the extracellular space within the mesaxon (that is between the spiralled layers of Schwann cell plasma membrane) is also reduced and may disappear and the outer layers of two successive lamellae of plasma membrane also appear to meet and fuse. This apparent or real fusion is similar to an enormously extensive *zonula occludens* and the membrane so formed corresponds with the minor dense lamella of the adult myelin. The membrane fusion occurring as the result of the disappearance of cytoplasm leads to the formation of the major dense lamella of adult myelin. The spiralling is quite regular and the repeating period in the peripheral nervous system is *c* 17–18·5 nm (Figs 3 and 4).

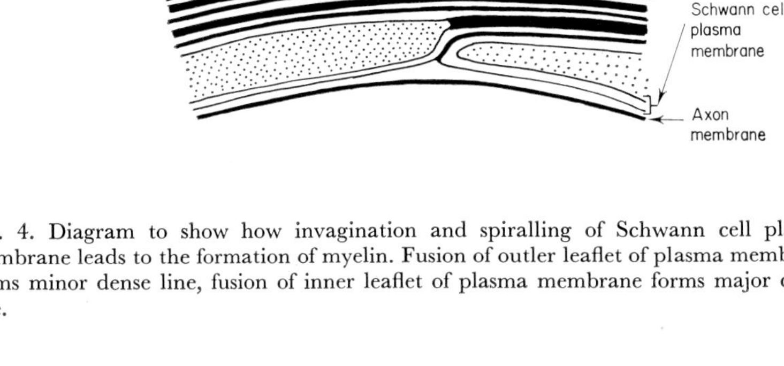

Fig. 4. Diagram to show how invagination and spiralling of Schwann cell plasma membrane leads to the formation of myelin. Fusion of outler leaflet of plasma membrane forms minor dense line, fusion of inner leaflet of plasma membrane forms major dense line.

Similar events occur in the Schwann cells proximal and distal to that described above so that the axon becomes invested along almost its whole course in the peripheral nervous system by segments of myelin (each the territory of a Schwann cell) separated each from the next by a myelin-free gap, the Node of Ranvier. At their terminal branching, the telodendria, many myelinated axons break up into unmyelinated branches, although even these maintain a Schwann cell investment.

Compaction of the myelin does not occur throughout the whole longitudinal extent of the Schwann cell. At each end of the myelin segment a small quantity of Schwann cell cytoplasm remains (Fig. 5 A and C) so arranged that each successive layer (with occasional exceptions) from within outward is slightly longer than that preceding, and by its bending inward comes to lie as close as 2·5–3·0 nm from the axon; the axon besides, may be marked by a series of transverse bands on its outer aspect (Peters *et al.*, 1970). The outermost turn of the Schwann cell alongside the node contains local concentrations of mitochondria and bears a number of micro-villus-like projections (called "finger-like" processes) which turn inward to make contact with the nodal axon membrane, interdigitating with similar processes from the next Schwann cell in the series, from the other side of the node (Fig. 5 A and B). It is important that although the Schwann cell ends at the node, its basement membrane extends across and is continuous with the basement membrane of the next Schwann cell.

By the methods of light microscopy, oblique clefts of Schmidt-Lantermann may sometimes be seen running through the myelin (on both sides of the axon), and apparently dividing it into minor segments where the bevelled end of one segment is received into the reciprocally bevelled end of the other. Electron microscopy has revealed that these apparent defects in the myelin are due to the presence of Schwann cell cytoplasm extending as a helix from the axonal to the outer surface of the sheath, each split in the major dense line (to enclose cytoplasm) being slightly offset from that next within or without in the helix (Fig. 5 A and E). With the classical myelin stains the cytoplasm was uncoloured and so was interpreted as a defect rather than as a change in the nature of the sheath.

It has been suggested that the clefts of Schmidt-Lantermann are not fixed and immutable but may represent shifting channels of cytoplasm (extending through the myelin from its outer to its inner aspects) allowing the nuclear region of the cell, with its Golgi apparatus closely associated, to influence more distant parts of the cell. If the clefts do indeed shift from time to time then clearly there would be opportunity for "maintenance" of the myelin sheath, by removal of components of it when it temporarily bounds Schwann cell cytoplasm. For further

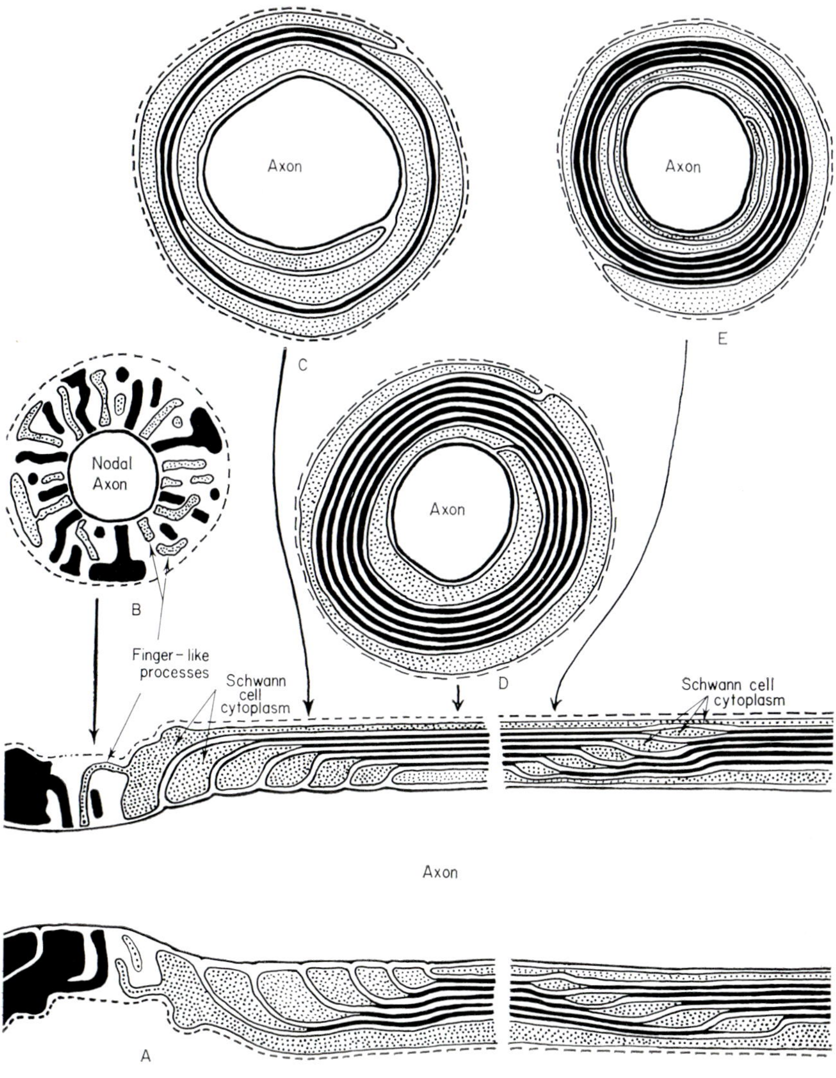

Fig. 5. Diagrammatic representation of myelinated nerve fibre of peripheral nervous system. A: in longitudinal section. B: in transverse section through node of Ranvier. C: in transverse section immediately adjacent to node of Ranvier. D: in transverse section through compact myelin. E: in transverse section through cleft of Schmidt-Lantermann. Note that basement membrane, represented by dotted lines, continues across node of Ranvier.

discussion of this concept, and of the nature and biochemistry of the membranes constituting the myelin sheath, reference may be made to Elfvin (1968).

Some consideration of the dimensions of Schwann cells of myelinated nerve fibres may be of interest. When myelin formation has advanced sufficiently to be recognizable by light microscopy, nodes of Ranvier are some 200–250 μm apart, and this dimension may be represented as the length of the Schwann cell. The fibre shown in Fig. 3 has six lamellae of myelin and their mean diameter is of the order of 1·2 μm. If it were possible to unroll the myelin sheath and so spread the cell out flat, it may be calculated that its breadth would approximate to 30 μm. Cytoplasm would be present along both sides, the outer side containing the nucleus, and the inner which had lain along side the axon: a narrow strand would traverse each end, having lain paranodally, i.e. alongside the node at either end.

After some growth and maturation of the myelin sheath, the internodal length might have increased to 1 000 μm with 60 lamellae of myelin (the repeating period of myelin has been estimated at 11·5–13·0 nm in fixed tissues and at 17·0–18·5 nm in fresh tissue) forming a sheath the mean diameter of which approximates 10 μm. If, again, it were possible to unroll the myelin sheath, the breadth of the Schwann cell may be calculated to approximate to 2 000 μm. Cytoplasm would again be limited to the margins of the cell, together with narrow strips traversing its breadth if clefts of Schmidt-Lantermann had been present. Thus although the increase in cell volume might be relatively small, the increase in surface area would have been of the order of 330 times.

The length of the Schwann cells of unmyelinated nerve fibres is not known. The ends of the cells cannot be seen by light microscopy and their length is sufficient to make serial sectioning for electron microscopy difficult.

In fetal nerves, as has been seen, some hundreds of unmyelinated axons may be ensheathed by one Schwann cell. Beyond the limits of that cell the further fate of those axons is unknown; they may pass into the territory of one or of several other Schwann cells.

In nerves of the human child, and in those of many laboratory animals, a Schwann cell may ensheath as many as twelve or fifteen axons, providing a separate groove for each. Nearer their terminations, e.g. in the skin or associated with blood vessel walls, Schwann cells are usually associated with rather fewer axons so that some dispersal must have occurred, with the axonal complement of one Schwann cell being passed on to two or three others.

Particularly in cutaneous nerves of the adult human Schwann cells of unmyelinated nerve fibres are of a far more complex shape, and this paradoxically, allows easier interpretation of the junctions between successive cells. In such nerve fibres cytoplasmic processes branch laterally from the nucleated part of the Schwann cell, and then bifurcate to form a parallel array of processes elongated in the axis of the nerve (Fig. 6). Many of the processes are grooved to provide a sheath for an axon but others appear to be merely supernumerary while still others may be grooved to provide sheaths for bundles of collagen fibrils (Fig. 6). Tracing a small process of Schwann cell cytoplasm with its ensheathed axon through serial sections by electron microscopy has revealed something of the structure of the junctional regions between successive Schwann cells.

There are no true nodes of Ranvier on unmyelinated nerve fibres. The axon is nowhere uncovered and often one ensheathing Schwann cell process is itself ensheathed for a short distance by the overlapping process from the next Schwann cell. The inner of the two processes may exhibit an increase in its electron density but no other specialization has been regularly observed in this region. At these junctions the basement membrane of the one Schwann cell becomes continuous with that of the succeeding Schwann cell (Fig. 7). It is likely in nerve trunks, and certain nearer their terminations, that the complement of axons ensheathed by one Schwann cell is passed on to two or more others and that the Schwann cells together form a complexly branching network (Eames and Gamble, 1970).

The terminal and pre-terminal parts of nerve fibres of the peripheral nervous system generally maintain the separation of the neural element from contact with collagen-containing tissues. They achieve this by the presence of a sheath of Schwann cell cytoplasm (perhaps incomplete) or by investment by basement membrane, or by the presence of both. Where these become defective, their place is usually taken by epithelial cells of sensory end organs or by contact with muscle or gland cells.

Close to their endings, unmyelinated axons occupy only shallow grooves upon the surfaces of Schwann cell processes but are covered by the investing basement membrane. Such endings are sometimes called "naked" and occur close to the epidermis, in the cornea, in some glands and among smooth and cardiac muscle fibres. Sometimes the axonal element contains synaptic vesicles, when it is assumed to be an efferent ending. In the absence of such vesicles its functional standing is obscure. The terminals of many myelinated nerve fibres are of essentially similar structure, the myelin sheath having given way to a simpler Schwann cell investment of the final branches. Thus, the terminal branches of the myelinated afferent fibres of muscle spindles may, within the spindle capsule, be invested by little more than basement membrane, while in Golgi tendon organs the basement

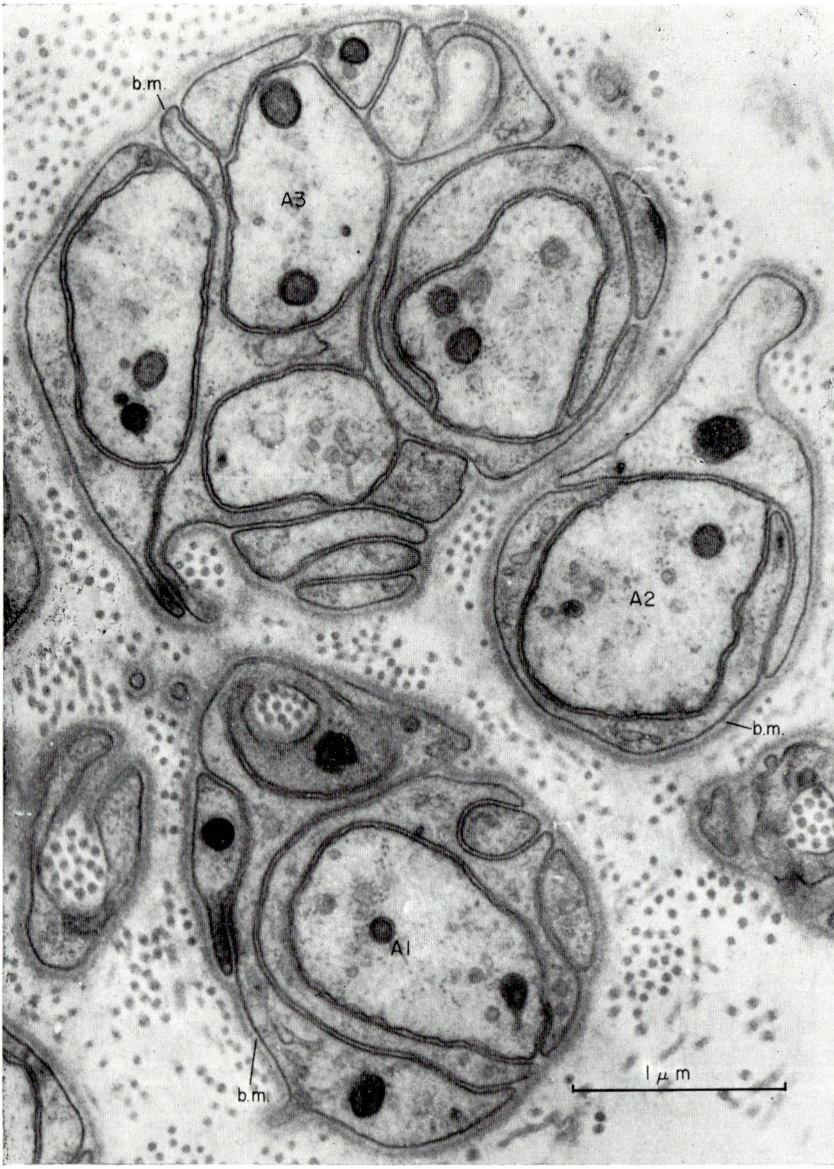

Fig. 6. Electron micrograph from transverse section of sural nerve from young human adult. Unmyelinated axons are variously ensheathed: A1 by a single Schwann cell process. A2 by two Schwann cell processes and A3 by five Schwann cell processes. Note that some processes of Schwann cell cytoplasm ensheath little bundles of collagen fibrils, and all Schwann cells are themselves invested by basement membrane *b.m.* From Eames, Rosemary A. and Gamble, H. J. (1970). *J. Anat.* **106**, 417–435, by kind permission of the publishers.

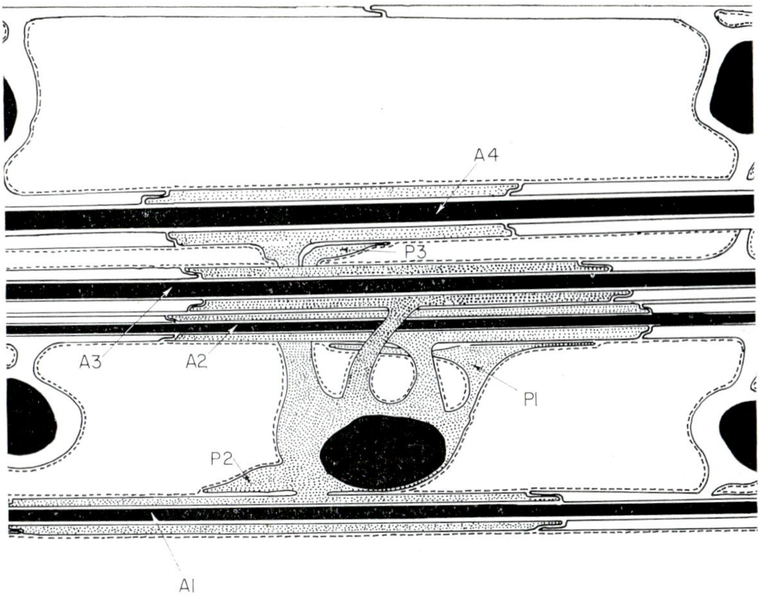

Fig. 7. Diagrammatic representation of Schwann cells of unmyelinated nerve fibre bundle. The cytoplasm of one Schwann cell is shaded and it may be seen to form 7 processes; of these, 3 (P1–P3) are supernumerary but each of the others bifurcates and provides a sheath for a segment of each of four axons (A1-A4). At the limits of the shaded Schwann cell the axons are passed on to ensheathing processes of more than one Schwann cell.

membrane itself may be defective in parts so that the axon terminal makes direct contact with collagen fibrils. Their relationship however, is quite exceptional (Merrilees, 1962).

In other highly organized sensory endings of the peripheral nervous system various types of investment of axon terminals are found. In the sensory endings in the inner ear, axons are ensheathed by supporting elements of the epithelium except where they contact its sensory elements. Rather similar arrangements are found in taste buds and in the carotid body. In Meissner's corpuscles, Schwann cells everywhere invest the fine axonal terminals and so separate them from the lamellae of cells and collagen that make up the bulk of the organ. The inner core of the Pacinian corpuscle provides a complexly arranged immediate environment for the axon terminal but it is uncertain whether the core cells are of Schwann cell or of perineurial origin.

At the endings of somatic efferent fibres (i.e. those supplying skeletal muscle) the myelin sheath ends at a hemi-node, i.e. where the proximal Schwann cell forms compact myelin and bears the usual paranodal apparatus of finger-like proceses etc. *vide supra*, but its successors form no myelin and provide a simple investment of the terminal branches, excepting only at their distal faces which are apposed to the muscle fibre. Here the basement membrane investment remains intact, being fused with that of the muscle fibre (Chapter 20). Visceral efferent fibres to smooth and cardiac muscle may be similarly separated from the muscle cells they innervate, but sometimes they make direct contact without intervention of basement membrane. Visceral efferent nerve fibres to glands sometimes make direct contact with the plasma membrane of the gland cell although surprisingly few have been shown to do so by electron microscopy. In the medulla of the suprarenal gland a relatively small number of very complexly branched Schwann cells provide cytoplasmic ensheathing processes upon widely divergent axons, except where these terminate upon the glandular cells. In exocrine glands, where a system of ducts imposes some sort of order upon the secreting elements, axons ensheathed by Schwann cells pursue more ordered courses through the gland but they are not often seen forming synaptic contacts with its cells.

The cell bodies of neurons which lie in the peripheral nervous system, i.e. in the ganglia of dorsal nerve roots and in the larger ganglia of the autonomic nervous system, are invested by a thin capsule of satellite cells, more or less complete, and by an investing layer of basement membranes. Except where the nucleus is present, these satellite cells are extremely thin and curved to follow the outline of the neuron; almost certainly they are Schwann cells of rather simple form. The satellite cells invest a dorsal root ganglion cell almost completely, leaving only one small opening through which its single axonal process may emerge. Perforations occur in the capsules of the multipolar neurons of autonomic ganglia to allow entry of axonal processes to synapse upon the neuron as well as to allow the emergence of its own several processes. The presence of axonal terminals upon the neuronal surface loosens the satellite cell investment which, besides, may be incomplete allowing contact between adjacent neurons. There may also be areas where only basement membrane intervenes between neuronal surface and endoneuronial connective tissues.

In the smaller ganglia of the autonomic nervous system and in its peripheral plexuses, e.g. Auerbach's myenteric plexus, Schwann cells (as satellites) make only sporadic "contact" with the neuronal cell bodies which, for the greater part, are covered by synapsing terminals of other neurons, or by basement membrane. At the other extreme, the bipolar neurons of the spiral ganglion of the auditory nerve are provided with a multilayered capsule of satellite cells so thin and closely packed as to resemble a loose myelin sheath, an arrangement which, presumably, promotes the rapid conduction of the nerve impulse from the dendritic pole across the cell body to the axonal pole.

During the passage of a nerve impulse along an unmyelinated nerve fibre active depolarization of the axonal membrane leads to a temporary increase in the concentration of potassium ions in the narrow extracellular space between the axon and the investing Schwann cell. So far as is known no active depolarization occurs in the Schwann cell membrane, but changes in the concentration of potassium or other ions in the interstitial fluid might passively lead to variation in the potential difference across the membrane, and so to variation in the concentration of ions in the intracellular fluid of the Schwann cell. The Schwann cell might thus play a supporting role in the propagation of the nerve impulse, perhaps by taking up some excess potassium ions and so helping to maintain a stable concentration in the extracellular fluid surrounding the axon.

In myelinated peripheral nerve fibres active depolarization is apparently limited to the nodal axonal membrane. Here the axon is set in a relatively extensive extracellular space and surrounded by specialized parts of the Schwann cells, namely the finger-like processes of cytoplasm which project from the mitochondrion-rich paranodal cytoplasm. The exact nature of Schwann cell involvement in saltatory conduction is obscure but the structures observed suggest the presence of oxidative metabolism and perhaps the active transport of solutes between the extracellular fluid and the intracellular fluid of the Schwann cell. The function of such activity might again be the maintenance of a suitable ionic composition in the extracellular fluid of the node, i.e. the region where this is most likely to be disturbed by repetitive activity in the axon.

It is a very persuasive argument in favour of the concept of axonal dependence upon its Schwann cells that an axon can function several metres from its parent cell body, that cell body's own dimensions being measurable in tens of microns only. There is indirect evidence that Schwann cells are involved in the nutrition of the axon from certain pathological conditions where a primary disease of Schwann cells leads to local demyelination with consequent disorder of transmission of nerve impulses and, sometimes, the destruction of the axon. More direct evidence is lacking at present, however. There is no direct evidence bearing upon the functions of the satellite (Schwann) cells of ganglion cells of the peripheral nervous system. As elsewhere, however, their situation suggests functions concerned with the nutrition and insulation of the neuronal cell body.

It will have been seen that neurons of the peripheral nervous system and axons of central origin which enter into the peripheral nervous system are so invested by Schwann cells and their basement membranes, or by special ensheathing cells in certain epithelia that, almost without exception, they make "contact" only with each other, with effector cells or with specialized sensory cells. It seems that they can grow and function properly only in the special environment thus provided, shielded from the fluids of connective tissue spaces and, presumably, nourished by their satellites. Besides these functions their satellites, when spun into myelin sheaths, promote their ability to conduct the nerve impulse rapidly, without unwanted spread to neighbouring nerve fibres.

II. Injury and Repair in Peripheral Nervous System

When a myelinated axon of the peripheral nervous system is interrupted, as by cutting or severely crushing it, that part lying distal to the injury undergoes degenerative changes which lead to changes in the associated Schwann cells. These changes in the axon and associated myelin sheath are known as Wallerian degeneration and include the following features. That part of the Schwann cell which is the myelin sheath is sloughed from the nucleated part and breaks up into successively smaller segments, each a closed cylinder containing degenerating axonal material. The debris of the degenerated myelin and axon is later and rather slowly removed by the action of macrophages which either invade the nerve through the walls of its blood vessels or are themselves transformed Schwann cells.

The nucleated part of the Schwann cell, having sloughed the associated myelin, rounds up and begins to undergo mitotic division. Its basement membrane remains as an intact tube, so that the products of degeneration and the now more numerous Schwann cells are closely confined. As the myelin/axon debris is removed so do the Schwann cells come to form an elongated band of cells (of von Büngner) extending from the site of injury to the terminations of its arborization, and confined within the branching tube of basement membrane which leads to these terminations. Macrophages are occasionally observed traversing the basement membrane tube but whether to enter or to leave it is not easily determined.

Proximal to the injury, Schwann cells are little affected although changes similar to those occurring distally will extend as far as the node of Ranvier immediately proximal to the injury.

If continuity of the proximal and distal parts of the nerve fibre is preserved (as basement membrane and connective tissues may be after a crush injury) or is reformed (as after transection and re-suturing) by proliferated Schwann cells migrating to bridge a gap, then sprouting from the proximal stump of the axon may lead on to regeneration and repair. In this process the Schwann cells of the band of von Büngner are known to provide surfaces over which axonal buds will grow (as in the development of fetal nerves) and by which they will later become ensheathed. Probably they provide metabolic support as well. The maintenance of the Schwann cells in ordered rows leading to suitable terminations is determined by their containment in the original tube of basement membrane, which despite its ill-defined nature, is clearly of fundamental importance in the promotion of nerve regeneration. Given that an axon enters a suitable band of von Büngner, e.g. that a motor fibre enters a tube leading to a motor ending then good recovery of function is likely, with a new myelin sheath forming from the now more numerous Schwann cells. The internodal length of the new myelin is likely to approximate to that of fetal nerve but its diameter will be little less than that originally present and the nerve impulse will travel at very nearly its original speed. Obviously the chance of an axon entering the right tube is reduced where a nerve is cut and then sutured together. Enough of the thousands of axons involved often do bridge the gap into tubes sufficiently suitable that, with some re-education, some useful restoration of function may be expected.

Regeneration of nerve fibres and restoration of a high order of function often follow the degenerative changes consequent upon the interruption of unmyelinated fibres of the peripheral nervous system. The histological changes accompanying the processes of degeneration and regeneration have been less extensively studied and documented in these than in myelinated fibres due to the difficulty of resolving such fine axons and such slender Schwann cell processes by light microscopy. More recently, electron micrographs have proved difficult to interpret once the normal relationship of axons and Schwann cells is disturbed. It is known, however, that very little, if any, Schwann cell proliferation occurs and it is obvious that in the absence of myelin there is less debris to be removed by the activities of macrophages.

Apart from the more dramatic forms of nerve injury occasioned by gun-shot wound or fracture of an adjacent bone, minor injuries are frequent especially in the more superficial nerves and their cutaneous branches. Minor cuts and prolonged pressure are often sufficient injury to set in train the processes of degeneration and, given reasonably good fortune, regeneration. There are many neurological disease processes where Schwann cells degenerate, or at least fail to maintain their segments of the myelin sheath, without an accompanying degeneration of the ensheathed axon. In such circumstances nerve conduction may persist in some fibres but at a reduced velocity and the information received into the central nervous system may be badly misinterpreted.

Interpretation of sensory input does not depend merely upon the types of sensory nerve ending being stimulated but also upon the impulses being carried centrally from the endings at conduction velocities appropriate to each. Afferents from some sensory endings normally conduct more rapidly than those from others. If the conduction velocities of some elements become abnormally slowed, the interpretation of the whole sensory input may be misinterpreted unless and until re-education occurs in the central nervous system. Such conditions belong to the realm of pathology and will not be further pursued here.

III. Development of the Neuroglia

At their first appearance the neurectodermal cells of the neural plate are of columnar form. The rapid division of cells during the thickening and grooving of the plate which lead to the formation of the neural tube is accompanied by a cyclical change of position in the nuclei and by changes (as yet not fully determined) in the shape of the cells as a whole. Nuclei in premitotic configuration are found only superficially, migrating towards the luminal surface to undergo mitosis after which some of the daughter cells migrate toward the periphery. Some cells, known as spongioblasts, maintain their radial

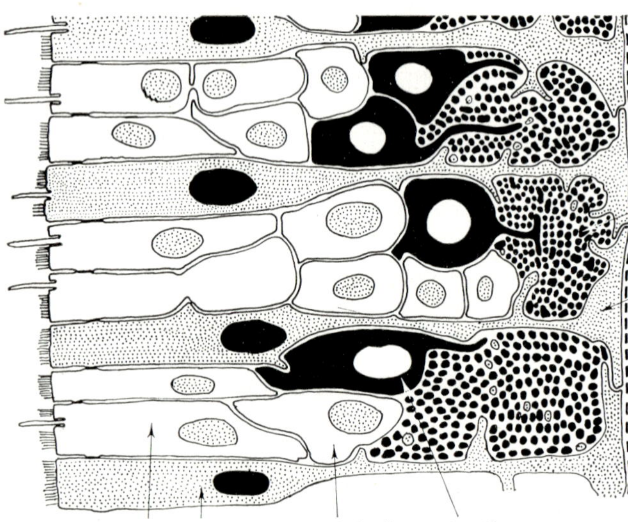

Fig. 8(A). Diagrammatic representation of transverse section through wall of neural tube of foetus to show spongioblasts extending from luminal to outer surface while trapping cells and processes in mantle and marginal layers respectively.

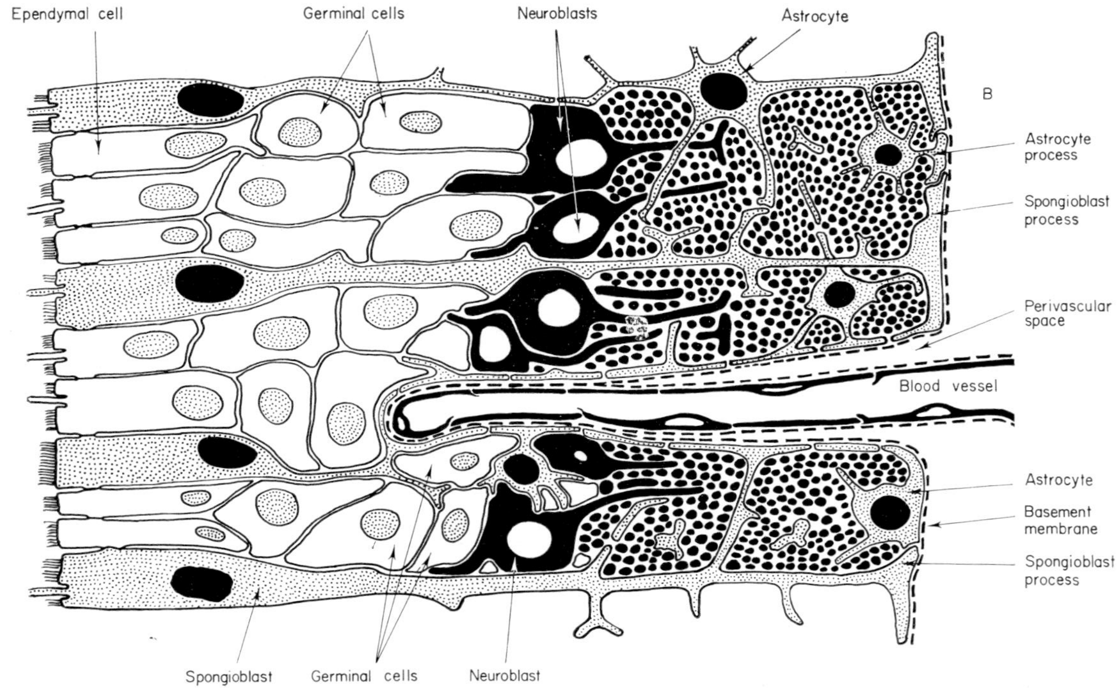

Fig. 8(B). Diagrammatic representation of transverse sections through wall of neural tube of fetus to show later stage in development where astrocytic processes have taken over some part in the formation of the external limiting membrane and also form the outer wall of a space surrounding an invading blood vessel. No cells identifiable as oligodentrocytes are shown at this stage.

extent from the luminal to the outer surface of the neural tube where their expanded extremities meet edge to edge to form an external limiting membrane (Fig. 8). The migrated daughter cells of mitotic divisions which originally took place close to the lumen occupy the spaces between the peripheral processes of the spongioblasts and are known as germinal (also neuroepithelial or matrix) cells. They later undergo differentiation to become neurons or neuroglial cells. Another population of cells remains at the luminal surface maintaining an ependymal lining to the tube.

The ependyma which lines the neural tube and which persists as the lining of the central canal of the spinal cord and of the ventricular cavities of the brain is, for the most part, a simple columnar epithelium bordering fluid-filled spaces. These cells are not in any meaningful sense the satellites of neurons. In the early fetal human spinal cord however, some ependymal cells temporarily ensheath unmyelinated axons and to this extent truly are satellite cells (Gamble, 1969). In some other situations ependymal cells may have rather specialized endocrine functions, but these lie outside the scope of this chapter (Knowles and Anand Kumar, 1969).

From an early stage in development the external limiting membrane of the neural tube is covered by a basement membrane. This, like the basement membrane surrounding Schwann cells, intervenes between cells of neurectodermal origin (i.e. the flattened extremities of spongioblasts) and the cells and fibrils of mainly mesodermal origin from which pia and arachnoid mater will later develop.

The accommodation of successive migrations of germinal cells into the outer layer of the neural tube, and their growth and differentiation there is permitted by changes in the spongioblast network, so that while an ependymal layer is maintained at the inner face of the tube, astrocytes come to provide the cytoplasmic processes of the external limiting membrane. The precise mechanism of this change is obscure, and conflicting views are discussed by Langman (1968).

Possibly the original spongioblasts divide *in situ*, one daughter cell remaining ependymal in situation, the other migrating outward and becoming an astrocyte. It is possible, alternatively, that the outer end of the spongioblast degenerates and is simply replaced by an astrocyte differentiated from an adjacent germinal cell. By whatever means, however, astrocytes take over the formation and maintenance of the external limiting membrane and by forming the interface between neurectoderm and mesoderm, mimic one of the characteristics of the Schwann cell of the peripheral nervous system.

In its earlier stages of development the central nervous system is wholly avascular, being dependent for its nutrition upon diffusion from plexuses of capillary blood vessels set in the surrounding mesoderm. Little information is available concerning the early vascularization of the brain but the human spinal cord is first invaded by blood vessels at the 7–8 mm C.R. length stage when capillaries penetrate from the ventral aspect. Almost immediately thereafter there is a second invasion from plexuses situated dorso-laterally and the invading vessels come to supply, respectively, those regions in which ventral and dorsal horns of grey matter will differentiate.

The invasion is accomplished by active sprouting from leptomeningeal vessels and mitotic figures are conspicuous in the endothelial cells. Clearly the entry of vessels must disrupt the external limiting membrane but contact of endothelial cells (and the basement membrane that covers their outer aspect) with cells and processes of the neural tube is limited by the provision of a sheath of cytoplasmic processes derived from spongioblasts and astrocytes. As at the outer surface of the cord, this is covered by a basement membrane (Fig. 8B). Around the smaller vessels a single basement membrane may be common to endothelial cells and the neuroglial processes that confront them. Around larger blood vessels the basement membrane is usually in two layers, one proper to the endothelial cell (and sometimes split to enclose a pericyte), the other proper to the astrocyte process, with collagen fibrils present in the perivascular space.

Essentially the same relationship between endothelial cells and astrocytes persists throughout life although it is possible that oligodendrocytes come to play some part in forming the neurectodermal outer wall of the perivascular sheath while in the walls of the larger vessels of the adult C.N.S., smooth muscle cells are lodged in the basement membrane proper to the endothelial cells (Jones, 1970).

Early astrocytes, and their parent spongioblasts, are characterized by a pale and watery cytoplasm apparently lacking organelles except in the immediate vicinity of the nucleus which perhaps indicates no more than inadequate fixation. Their nuclei show some clumping of chromatin but the glycogen granules and the bundles of filaments which characterize the cytoplasm of both protoplasmic and fibrous astrocytes are not present in the immature cells which are identifiable as much by their situation as by their fine structure.

The germinal cells which largely fill the interstices of the framework provided by ependymal cells, spongioblasts and maturing astrocytes, are the stem cells of two major categories, those cells destined to become neurons and those destined to become neuroglial cells. There is evidence that neuroblasts possess a conspicuous nucleolus throughout their successive divisions and subsequent development into mature neurons, while the nucleus as a whole displays little clumping of the chromatin.

The precursors of the glial cells, by contrast, show clumping of the chromatin through all their stages of multiplication and differentiation and do so whether these lead to the formation of additional astrocytes or to the formation of oligodendrocytes. The changes which occur in the further development of glial elements of the neural tube are related to and dependent upon concurrent changes in the neuroblast/neuron population. It is convenient to discuss these latter developments first.

A cytoplasmic, axonal process grows out from the cell body of a neuroblast to mark the beginning of its transformation into a neuron. The axonal sprout traverses the web of glial processes in the outer part of the neural tube and after further elongation either perforates the external limiting membrane to enter the primordum of a ventral nerve root (or comparable cranial nerve) or bends to run longitudinally in the outer part of the tube. Here, with many similarly orientated axons and a web of glial processes, it contributes to the formation of the marginal zone which is the precursor of the white matter of the spinal cord and of parts of the brain stem. Deep to the marginal layer dividing germinal cells and differentiating neuroblasts and glioblasts, together with their shorter cytoplasmic processes, constitute the mantle layer.

In the early marginal layer a relatively small number of glial processes traverse the bundles of increasingly numerous axons, so that axon bundles are large and axonal contact with glial processes is only occasional and even then for only a short distance: none of the axons receives individual ensheathing and all are of small diameter (*ca.* 0·1–0·5 μm). The neuronal cell bodies in the mantle layer make extensive contacts with glioblasts and some contacts with astrocytic processes but are not, apparently, in any sense encapsulated by them. Additional cytoplasmic outgrowths from neuronal cell bodies branch and form the dendrites but little information is available about their relationships with glial processes at this stage of development.

More or less concurrently with neuroblast differentiation, glioblasts of the mantle layer begin to differentiate into oligodendrocytes. Their nuclei resemble those of astrocytes but their cytoplasm is usually considerably more electron dense (due mainly to the presence of ribosomes) and their cytoplasmic processes finer and more freely branching.

Extension of the cytoplasmic processes into the marginal layer and their branching there break up the axonal bundles. Some axons remain with apparently casual, intermittent contact with astrocytic processes. Others, either singly or in small groups, become encircled by oligodendrocyte processes and the encircling of single axons is undoubtedly the prelude to the formation of a myelin sheath in many, perhaps all, instances. A slightly more mature oligodendrocyte may, by its several cytoplasmic processes, provide communal sheaths for several small bundles of axons while simultaneously providing a separate sheath for another single axon. Since these conditions have, as yet, been observed only in transverse sections of axon bundles which reveal width and breadth of the encircling oligodendrocyte process, the length of the process is not known. Other processes of oligodendrocyte cytoplasm extend through the neighbouring parts of the mantle layer but their relationships here with neurons, neuroblasts, etc. has not yet been elucidated (Gamble, 1969).

In the adult central nervous system glial cells greatly outnumber neurons although their density varies from one region to another and even between one area of cerebral cortex and another. In general they are small cells, at least so far as the perinuclear part of the cell is concerned, but their cytoplasmic processes may extend widely into adjacent regions. The interrelationships between glial cells and neurons has been studied extensively and good accord obtained between electron microscope and light microscopic appearances.

Adult astrocytes are known to light microscopy in two forms; fibrous astrocytes occurring mainly in the white matter and protoplasmic astrocytes occurring in the grey matter. By electron microscopy, both forms have been shown to contain glycogen particles and microfibrils composed of an alpha keratin and of 8–9 nm diameter. These fibrils are more numerous in the fibrous astrocytes where they often occur in dense-packed bundles; in these cells too the nucleus is frequently deeply indented while in the protoplasmic astrocyte it seldom is.

Cytoplasmic processes of the protoplasmic astrocytes often occur as extremely thin sheets and may be multilaminar but both types of cells provide segments of the external walls of perivascular spaces and contribute to the formation of the marginal glia. Protoplasmic astrocytes additionally may lie alongside the cell body of a neuron and, in part, encapsulate it; or situated more distantly, may spread processes over the neuronal surfaces effectively reducing the area of neuronal surface available for synaptic contacts.

Apart from an insulating function which these close relationships with neuronal surfaces seem to imply, there can be little doubt that astrocytes provide structural support for the central nervous system and there is evidence of adhesion between adjacent astrocytic processes (Peters *et al.*, 1970).

The consistently close relationship of astrocytic processes with blood vessels and with the basement membrane associated with both blood vessels and the external surface of the central nervous system suggests functions associated with fluid transport at these surfaces. A possible role in the maintenance of the blood-brain barrier (an incompletely understood mechanism by means of which large molecules are prevented from passing from blood into brain) has sometimes been attributed to the astrocytes although recent failures to demonstrate fenestrae in the endothelial cells of blood vessels of the central nervous system may suggest that the "barrier" may reside in these cells.

It is also true that large molecule tracers, which may pass readily between the endothelial cells of capillaries elsewhere fail to do so in the capillaries of all but a few regions of the central nervous system. Micropinocytic vesicles are extremely rare in brain capillary cells and while large molecule tracers in the blood may be taken up in small quantities there is no evidence of their later release upon the abluminal side (Reese and Karnovsky, 1967). The blood-brain barrier may thus reside in the endothelial cells of the brain capillaries. To exclude the astrocytes from active participation in the blood-brain barrier is not necessarily to exclude them from consideration in relation to other aspects of transport of metabolites between blood vessels and neurons. It is known that certain enzymes (e.g. nucleoside phosphatase and ATPase) occur at the perivascular end feet of astrocytes (perhaps in the covering basement membrane) and that astrocytes are involved in the maintenance of a constant flow of glycogen toward the neurons. It is also known that astrocytes respond rapidly to changes in the osmolarity of the blood circulating through brain capillaries so that they may act as buffers shielding neurons from these same changes which might otherwise have ill effects upon the maintenance of a nicely adjusted ionic balance at the neuronal plasma membrane.

It is worthy of note that astrocytic cytoplasm is frequently of "watery" or "empty" appearance in tissue prepared for electron microscopy and that intercellular spaces between the various elements of the central nervous system are extremely small, although to some extent variable according to the procedures of fixation and preparation which may be employed. The extensive literature on this subject has been reviewed recently by Davson (1967) in relation to the rather larger extent of the extracellular space which physiological experimentation suggested must be present in the nervous tissue. The greatest disparity seems to involve the "sodium space", calculated to comprise *c* 35% of the whole brain volume (and thought to coincide with the extracellular space, in other tissues) and the sum of the apparent extracellular spaces which can be measured between the plasma membranes made visible by electron microscopy: these have been estimated at *c* 10% of the total volume, being less than 10% where cells and processes are large and as much as 30% where processes are especially (and unusually) small. This great disparity has sometimes been explained in terms of the astrocytes functioning as an "intracellular extracellular space" and being themselves a part of the "sodium space". Some recent evidence points to astrocytes being rich in sodium (Siegesmund, 1969) although it is as yet uncertain whether the sodium occurs in a bound or free form within the cells.

Like the astrocytes, oligodendrocytes occur in both white and grey matter of the central nervous system. In the white matter, they frequently occur in rows, interfascicular with respect to bundles of axons. In the grey matter they may occur as the close satellites of neuronal cell bodies, but many are disposed, apparently at random, in the neuropil.

As their name suggests, oligodendrocytes bear fewer cytoplasmic processes than astrocytes and these processes are commonly slender strands where they branch from the perinuclear cytoplasm. Distally the processes from satellite oligodendrocytes often appear to embrace axons by expansions and spiralling, form myelin sheath segments. Similar processes from satellite oligodendrocytes often appear to embrace the neuron beside which they lie.

Apart from the number and disposition of their cytoplasmic processes oligodendrocytes differ from astrocytes in several ways. They are generally rather smaller with an eccentrically placed nucleus containing dense, clumped chromatin. Cytoplasm is scanty except at one pole of the nucleus where organelles are present including granular endoplasmic reticulum. The cytoplasm generally is dense due to the presence of many fine particles while free ribosomes are numerous. Microtubules of *c* 25 nm diameter are often conspicuous but microfibrils and glycogen granules occur only very rarely.

Satellite oligodendrocytes have been found most often associated with large neurons, such as those of Deiter's nucleus, and the larger pyramidal cells of the cerebral cortex. There is evidence that prolonged activity in neurons (as in the anterior horn cells during swimming) may lead to a temporary increase in the number of oligodendrocytes immediately adjacent to the active neurons and that this is accomplished by a temporary migration rather than by local proliferation (Kulenkampff, 1951). This migration has been interpreted as a means by which more numerous oligodendrocytes might give metabolic aid to the neuron and some evidence of this (or at least, of a metabolic interdependence) comes from the work of Hydén and Pigon (1960) who showed that increased cytochrome oxidase and succinoxidase in neurons is accompanied by a fall in the activity of the same enzymes in the satellite oligodendrocytes. Similar reciprocal changes in neuronal/glial metabolism of RNA and protein synthesis have been shown by Ezyházi and Hydén (1961); but direct evidence of direct neuronal dependence upon oligodendrocytes is at present lacking. Furthermore, recent electron microscopic studies have shown that slender sheets of astrocytic cytoplasm often intervene between neuron and oligodendrocyte and the nature of the triple interrelationship is obscure.

In white matter the role of the oligodendrocyte is better understood. Although evidence is to some extent circumstantial there can be no doubt that oligodendrocytes form myelin sheaths upon axons of the central nervous system and in a manner essentially similar to that by which Schwann cells form myelin in the peripheral nervous system.

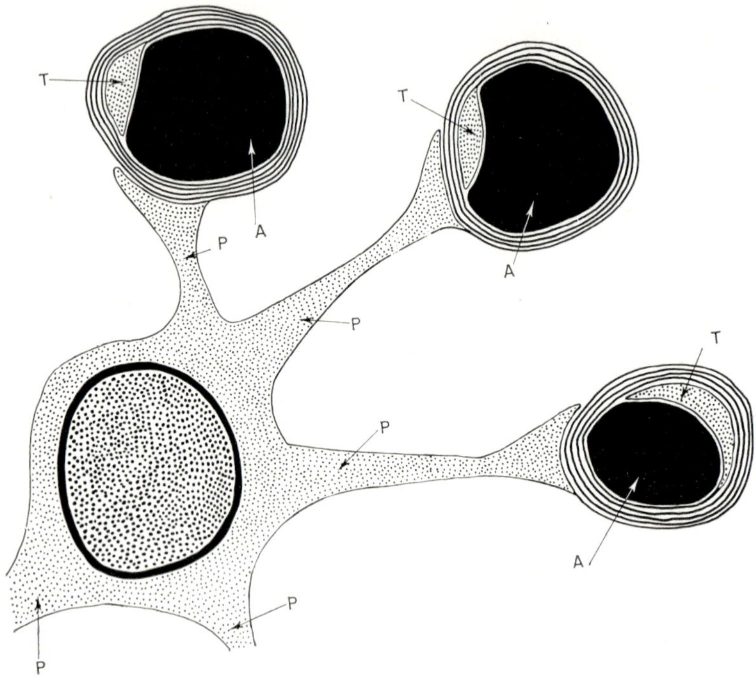

Fig. 9. Diagrammatic representation of oligodendrocyte from which five cytoplasmic processes *p* radiate. Three of these each form a short segment of myelin upon each of three axons *a*. A small quantity of oligodendrocyte cytoplasm lies as a "tongue" *t* alongside the axon.

There is a major point of difference, however: a single oligodendrocyte forms segments of myelin upon more than one axon (Fig. 9) while a Schwann cell never forms myelin upon more than one. The upper limit to the number of myelin segments formed by a single oligodendrocyte is not known.

In other respects structural differences between central and peripheral myelin appear to be minor. Thus oligodendrocyte cytoplasm is very scanty at both inner and outer aspects of the myelin sheath (except where a cytoplasmic strand extends to the perinuclear region of the cell) so that both inner and outer "mesaxons" are much reduced (see Fig. 9).

Perhaps more importantly, the repeating period of central myelin has been shown to be some 10% less than that of peripheral myelin, the difference being due to the width of the "minor dense line". Clefts of Schmidt–Lantermann have not been seen in central myelin and although the "major dense line" splits paranodally to enclose a helix of cytoplasm (just as do Schwann cells paranodally) neither finger-like processes nor local concentrations of mitochondria are present at central nodes of Ranvier. The nodal axon, consequently, is uncovered even by basement membrane and may make contacts with adjacent cell processes whether glial or neuronal; at some nodes, indeed, the axon may bulge to form the presynaptic element of a synapse, but in other respects is not related to conspicuously modified parts of its satellite cells.

Finally it is to be noted that myelin segments of the central nervous system are much shorter than those of peripheral nerve fibres.

Oligodendrocyte processes are often found in neuropil and among bundles of unmyelinated nerve fibres. Their disposition appears often to be quite random and at present their function is obscure.

With clear evidence that intrafasicular oligodendrocytes may form myelin sheath segments upon adjacent axons it is natural to suppose that the "insulation" of these axons is their primary function. The very close apposition of the successive lamellae of myelin must render difficult any diffusion from the general extracellular space to the periaxonal space (although there is some evidence that it may occur, Hirano and Dembitzer, 1969), while the absence of clefts of Schmidt–Lantermann similarly must reduce the possibility of transport through the medium of the oligodendrocyte itself, especially since the perinuclear part of the cell is joined only by slender cytoplasmic strands to its myelin-forming parts.

Given that central myelin appears so well adapted to an insulating role, it is surprising to find the nodal axon without any specialized covering. It appears to be in free communication with extracellular fluids and randomly related to neighbouring glial and neuronal processes. However, the occurrence of synapses at some nodes suggests that the relationships at other nodes may, with further study, prove to be less random than is now apparent.

From what has been written thus far it is obvious that where axonal processes enter or leave the central nervous system, as nerve rootlets, there will be a region where investment by glial cell changes to investment by Schwann cell. This region, the Obersteiner-Redlich zone, is marked by abrupt change upon each individual axon but the frontier as a whole is ragged with greater and lesser salients of glia-invested axons projecting among the Schwann cell-invested axons of the rootlet. As might be expected, the entering or leaving axons must perforate the marginal glia but the basement membrane that covers this is not itself perforated: it extends peripherally without interruption as the basement membranes of each of the axon/Schwann cell complexes that make up the nerve rootlet and so far as it is known invests these to their terminations.

It would thus seem probable that in the central nervous system astrocytes and oligodendrocytes together perform all of the functions which Schwann cells perform in the peripheral nervous system, namely that oligodendrocytes are concerned with insulation of myelinated axons while astrocytes are involved in their nutrition. Astrocytes also provide the physical support and strength which the fibroblasts and endoneural collagen provide, in greater measure, in peripheral nerves. Besides these features one may note the basement membrane which everywhere, and indeed as one continuous sheet, intervenes between cells of neurectodermal origin and the cells, fibrils and fluids derived from mesoderm. The survival of this membrane during Wallerian degeneration and its role in maintaining the orientation of Schwann cells and axon during regeneration imply greater durability than its appearance might suggest. It may share in the provision of physical support for the nervous tissues as well as involving itself in the nutritional activities of the astrocytes and Schwann cells to which it appears to belong.

A seemingly unorthodox view of the roles to be attributed to oligodendrocytes and astrocytes has been expressed by Blunt *et al.* (1972), and others. They believe, for example, that in the developing optic nerve of the kitten immature astrocytes perform a "myelin sheath supporting role" and they refer to an "association between an astrocyte precursor, the early astrocytoblast, and the deposition of myelin sheaths". The terminology is obscure and difficult to interpret since the "support" and the "association" are not defined; and even in the summary, where it is claimed that "results support the hypothesis that the myelin sheath depositing cell is an immature astrocyte (early astrocytoblast)" it is difficult to equate a mere "deposition" with the spiral wrapping of plasma membrane (whether of astrocyte or oligodendrocyte) which is the myelin sheath.

In addition to the astrocytes and oligodendrocytes which together constitute the macroglia, a much smaller and less numerous glial element (perhaps 5–10% of all glial cells) occurs as the microglial cell. Aparently because it is recognizable only after the invasion of blood vessels into the developing central nervous system, the microglial cell is often said to be of mesodermal origin.

By light microscopy the nucleus of the microglial cell is seen to be round or triangular and set in a scanty cytoplasm from which short, wavy and spined processes are given off. These cells become much more numerous in inflammatory or destructive lesions of the central nervous system when they act as macrophages and are called Gitter cells.

Electron microscopic investigations have only recently revealed cells which might correspond in size and form with the microglial cells of light microscopy. Again the nucleus is small and rounded or triangular, set in a scanty cytoplasm which may show many Golgi profiles, but little endoplasmic reticulum, few tubules and no fibrils. They are found most commonly in association with blood vessels and close to neuronal cell bodies. They have also been shown to be numerous in the subependymal plate of the lateral ventricle.

In some degeneration experiments upon the optic nerve microglial cells were found to be more numerous, to surround degenerating myelin and to contain numerous lipid inclusions, i.e. they had taken on the characters of small macrophages. Recent studies however have done little to elucidate the origin of the microglial cell : a mesodermal origin is little more than probable than an ectodermal.

In an earlier section dealing with Schwann cells, brief mention was made of their role in the processes of degeneration and regeneration which may follow upon the interruption of an axon (as by cutting or crushing) in the peripheral nervous system. Emphasis was laid upon the role of the Schwann cell basement membrane in imposing order upon the debris resulting from the degeneration and, later, upon the proliferated Schwann cells : it ensures correct alignment of these cells and so upon the regenerated axon that traverses them, from the lesion to a suitable termination.

In the central nervous system there is no system of basement membranes surrounding oligodendrocyte/axon complexes. As in the peripheral nervous system, axonal and myelin degeneration follow axonal section, and glial cells proliferate. A new axonal sprout may appear and attempt to grow but is at once faced by disordered debris and a "scar" of proliferated glial cells, mainly astrocytes. There is a barrier to new growth as well as a lack of preserved route. Regeneration is possible in theory but in practice is of no value, except in some of the lower vertebrates.

Much more might be written of glial cell involvement in injury to and disease of the central nervous system but such considerations belong to pathology and will not be treated here.

IV. Summary

It has been shown that two types of cell, Schwann cells and oligodendrocytes, of a common origin in neural ectoderm, may be so modified in the course of their development as to form relatively enormous expanses of plasma membrane. By the tight spiralling of this about an axon, or axons, and the virtual elimination of both extracellular space and cytoplasm, a myelin sheath (or sheaths) is formed and is maintained normally throughout life by a cell not conspicuously well equipped with organelles such as granular endoplasmic reticulum, Golgi apparatus etc.

There seems to be no doubt that the myelin sheath is essential to the saltatory progression of the nerve impulse and so responsible for its rapid transmission in myelinated nerve fibres.

Schwann cells also play a part in insulating peripheral axons and peripheral ganglion cells from the fluids of the endoneurial tissue spaces. They are also probably involved in the metabolic support of the cells and processes whose satellites they are.

Astrocytes of the central nervous system provide the physical support and resistance to stresses which collagen fibrils provide in the peripheral nervous system. They probably provide metabolic support also to the nervous elements as well as providing insulation from the tissue fluids surrounding the blood vessels of the central nervous system. The question of their involvement in the "blood-brain" barrier is undecided.

References

Blunt, M. J., Paisley, P. B. and Wendell-Smith, C. P. (1972). Oxidative enzyme histochemistry of immature neuroglia during myelination. *J. Anat.* **110**, 421–433.

Bunge, R. P. (1968). Glial cells and the central myelin sheath. *Physiol. Rev.* **48**, 197–251.

Cravioto, H. (1965). The role of Schwann cells in the development of human peripheral nerves. *J. Ultrastruct. Res.* **12**, 634–651.

Davson, H. (1967). "The Physiology of Cerebro-Spinal Fluid". J. and A. Churchill, London.

Eames, Rosemary A. and Gamble, H. J. (1970). Schwann cell relationships in normal human cutaneous nerves. *J. Anat.* **106**, 417–435.

Elfvin, L. G. (1968). The structure and composition of motor, sensory and autonomic nerves and fibres. In "The Structure and Function of Nervous Tissue", (G. H. Bourne, ed.), Vol. 1. Academic Press, New York and London.

Ezyházi, E. and Hydén, H. (1961). Experimentally induced changes in the base composition of the ribonuclear acids of isolated nerve cells and their oligodendroglial cells. *J. biophys. biochem. Cytol.* **10**, 403–410.

Gamble, H. J. (1966). Further electron microscope studies of human foetal peripheral nerves. *J. Anat.* **100,** 487–502.

Gamble, H. J. (1969). Electron microscope observations on the human foetal and embryonic spinal cord. *J. Anat.* **104**, 435–453.

Geren, B. B. (1954). The formation from the Schwann cell surface of myelin in peripheral nerves of chick embryos. *Exper. Cell Res.* **7**, 558–562.

Harrison, R. G. (1907). Observations of the living developing nerve fibre. *Anat. Rec.* **1**, 116–118.

Hirano, A. and Dembitzer, H. M. (1970). The transverse bands as a means of access to the periaxonal space of the central myelinated nerve fiber. *J. Ultrastruct. Res.* **28**, 141–149.

Hydén, H. and Pigon, A. (1960). A cytophysiological study of the functional relationship between oligodendroglial cells and the nerve cells of Deiters' nucleus. *J. Neurochem.* **6**, 57–72.

Jones, E. G. (1970). On the mode of entry of blood vessels into the cerebral cortex. *J. Anat.* **106**, 507–520.

Knowles, F. and Anand Kumar, T. C. (1969). Structural changes, related to reproduction, in the hypothalamus and in the pars tuberalis of the Rhesus monkey. *Phil. Trans. Roy. Soc. London Ser B* **256**, 357–375.

Kulenkampff, H. (1951). Das Verhalten der Vorderwurzelzellen der weissen Maus unter der Reig physiologischer Tätigkeit. *Z. Anat. Entwgesch.* **116**, 143–156.

Langman, J. (1968). Histogenesis of the central nervous system. In "The Structure and Function of Nervous Tissue". (G. H. Bourne, ed.) Vol 1. Academic Press, New York and London.

Merrilees, N. C. R. (1962). Some observations on the fine structure of a Golgi tendon organ of a rat. In "Symposium on Muscle Receptors" (D. Barker ed). Hong Kong University Press.

Peters, A., Palay, S. L. and Webster, H. de F. (1970). "The Fine Structure of the Nervous System. The Cells and their Processes." Hoeber Medical Division. Harper and Row, New York.

Reese, T. S. and Karnovsky, M. J. (1967). Fine structural localization of a blood-brain barrier to exogenous peroxidase. *J. Cell Biol.* **34**, 207–217.

Siegesmund, K. A. (1969). Sodium localization in the cerebellum. *J. Anat.* **105**, 403–413.

20. The Skeletal Muscle Cell

A. R. MUIR
Royal (Dick) School of Veterinary Studies,
Edinburgh, Scotland

It would appear simple to define a muscle cell as one which by contraction produces animal movement. In more general terms it might be defined as a cell that can convert chemical energy, in the form adenosine triphosphate (ATP), into the mechanical energy of contraction. However, other types of cell can convert chemical into mechanical energy, some carry cilia and flagellae which beat, others throw out and contract pseudopodia for phagocytosis and others progress by following projections of their cytoplasm. In all dividing cells chromosomes separate at anaphase and evidence is accumulating that all such movements are generated by conformational changes in proteins similar to the actin and myosin that have been so extensively studied in muscle. Myosin, very similar to that of muscle, can be extracted from the lowly slime mould. Further within a single cell the cytoplasm streams, the nucleus gently rotates and ions are transported across barriers; these activities could also be accomplished by similar conformational changes, but their elucidation may not be possible with present techniques.

So the conversion of chemical into mechanical energy is a universal feature of cells and a muscle cell is merely one that has developed the property of contractility. Muscle cells which produce the rapid contractions, so characteristic of vertebrate movement, have their contractile protein arranged to produce cross-striations, which can be seen in the living cell by direct microscopic examinations. Smooth muscle cells do not have these striations although they contain the same proteins (Chapter 22).

One form of striated muscle cell is not under voluntary control and this forms cardiac muscle. Most of the other muscle cells are under voluntary control, and so can be called voluntary striated muscle. These cells, usually attached to the skeleton, are also called skeletal muscle. Some are not attached to the skeleton, e.g. cutaneous muscles and sphincters, and others are not voluntarily controlled, e.g. oesophageal wall, so neither name is entirely satisfactory.

Skeletal muscle cells make up almost 50% of an adult man's weight and they contain over 80% of his body water. The skeletal muscle cells of domestic mammals and birds provide, as meat, much of our dietary protein.

As described below, skeletal muscle cells even in one species show considerable variation in length, diameter, arrangement of contractile proteins and enzymic properties. But it is profitable to consider first the structure of a single skeletal muscle cell such as would be found in one of the larger limb muscles of man.

I. Morphology of an Extrafusal Skeletal Muscle Cell

The word cell is used here to mean a large multinucleated syncytium forming an elongated, roughly cylindrical, contractile mass of cytoplasm which is surrounded by a single plasma membrane. It is usual to refer to this investing membrane as the sarcolemma and the contained cytoplasm as sarcoplasm. The notable feature of the sarcoplasm is the mass of tightly packed protein filaments, myofilaments; there are thick and thin myofilaments which overlap in a complex, but highly ordered, fashion to produce the characteristic cross-striations of the whole cell. Groups of myofilaments, in parallel, form structures large enough to be recognized by light microscopy; these are the myofibrils (Fig. 1). Current theories of contraction are based on the

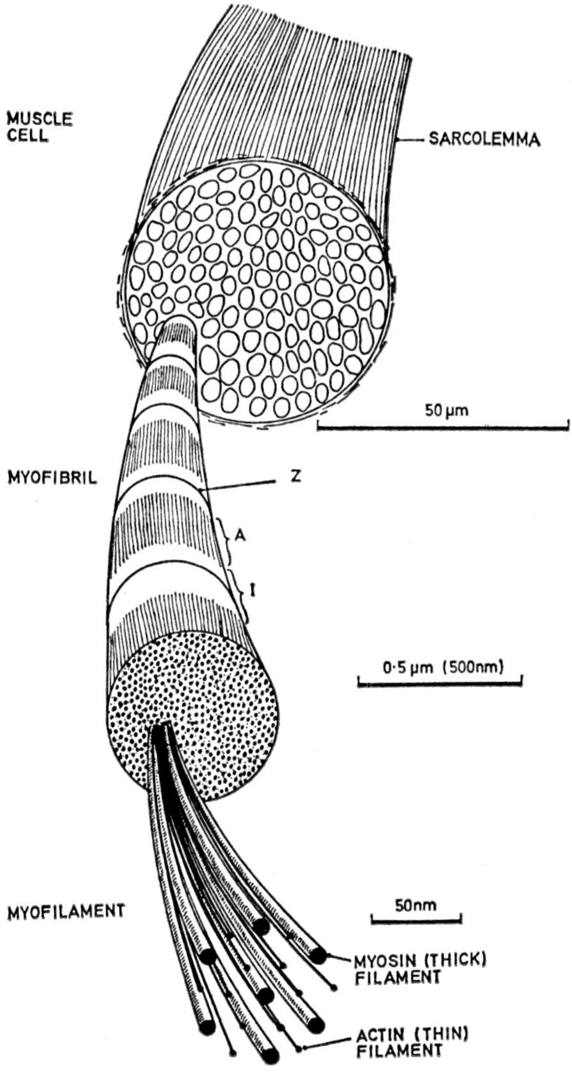

Fig. 1. Diagrammatic representation of the three dimensional structure of a muscle cell, showing how the myofilaments form myofibrils which, in turn, are contained within the cell. (Reprinted from Companion to Medical Studies.)

structural overlap between the thick and thin myofilaments, and change in length is associated with these filaments sliding past each other (Wilkie 1968). Other organelles are packed in the sarcoplasm between the myofibrils and these and many general features of skeletal muscle structure are summarized in Fig. 2.

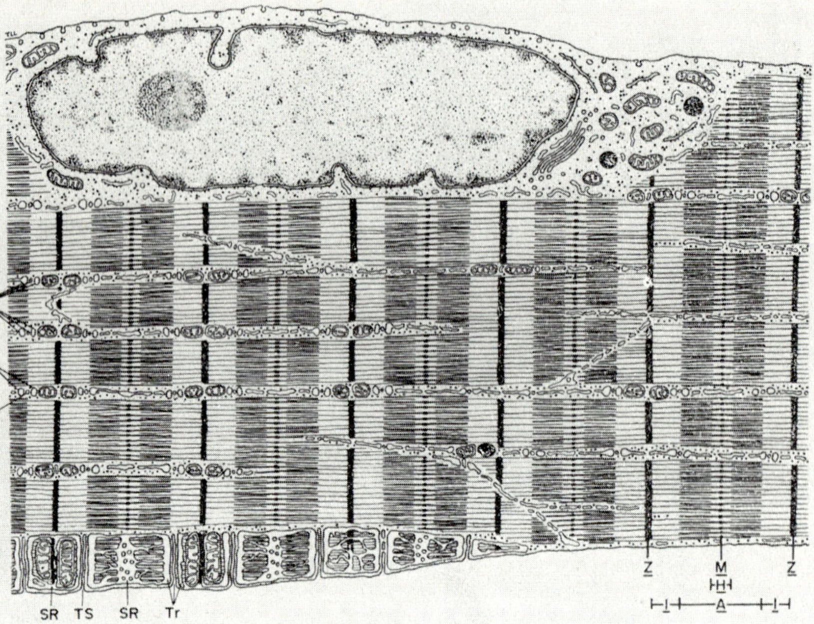

Fig. 2. Line drawing constructed from electron micrographs of longitudinal sections of skeletal muscle cells. A subsarcolemmal nucleus and associated organelles lies across the top of the picture. Most of the rest of the cell is filled by branching myofibrils; these are surrounded by sarcoplasmic reticulum SR. From the sarcolemma at the bottom, T-tubules TS penetrate the cell to come into contact with terminal cisternae Tr of the sarcoplasmic reticulum. (With permission from Lentz, T. L. (1971) Cell Fine Structure, Saunders, Philadelphia.)

A. LENGTH

The muscle cell is often called a fibre because of its great length but confusion is avoided if the term fibre is restricted to the contents of a fibrous connective tissue sheath which surrounds either a single cell or a longitudinal series of cells. This sheath, which runs the whole length of the fibre, is called endomysium. Groups of endomysial tubes are bound together in parallel by perimysium; the fascicles thus formed can be seen by the naked eye on the surface of a muscle. Epimysium is the connective tissue around the whole muscle. If the muscle cell runs from origin to insertion the endomysium blends with these collagenous attachments, but when more than one cell makes up a fibre, their endomysial sheaths are continuous with each other. This feature

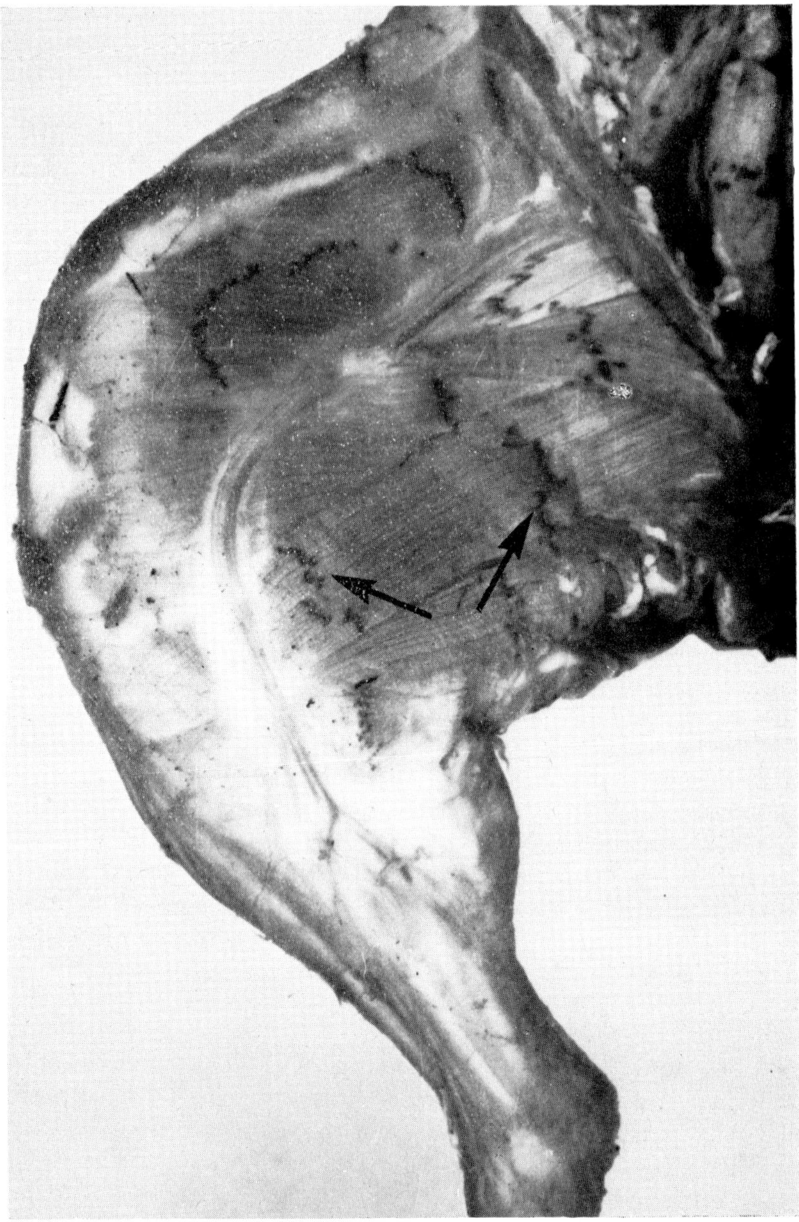

Fig. 3. The myoneural junctions of the muscles of the medial side of the rat's thigh are shown by staining for acetylcholinesterase. Most muscles have one band of myoneural junctions across the middle of their bellies, but anterior gracilis (arrows) has two bands (preparation by Dr. B. Mackay).

has made it difficult to determine the length of a single muscle cell either by dissection or serial sectioning. Single fibres which ran the whole length of a human sartorius muscle have been dissected but this does not mean that such a fibre contains only one cell.

As most mammalian skeletal muscle cells have only one nerve terminal, the demonstration of cholinesterase activity at these terminals on whole mounts of skeletal muscle can show the arrangement of component cells. Most of the muscles shown in Fig. 3 have one band of end plates traversing the middle of their bellies, so the likeliest interpretation is a single array of cells in the middle of the muscle. Other possible explanations of this observation involve overlapping of cells and end plates not near the mid-point of each cell's length. The anterior gracilis has two bands suggesting a tandem array; an interpretation which is supported by fibre counts taken on serial sections through its mid-point. As soon as cholinesterase can be demonstrated in the developing mammal, that is just before birth in small rodents, the pattern of distribution is the same as in the adult. Therefore the analyses of human neonatal material made by Schwarzacher (1959) are likely to apply in the adult.

After using this technique to determine the cellular arrangements in human muscle, it would appear that all short muscles, e.g. intercostal, extraocular and those of the hand and foot, and most large limb muscles, e.g. adductor magnus, are composed of cells which run from origin to insertion. Some muscles, e.g. sartorius, have their cells arranged in an irregular series as their end plates are scattered throughout their substance. Composite muscles, such as the rectus abdominis, have one band of end plates between each pair of fibrous intersections.

Obviously the length of a fibre is markedly affected by the state of contraction, when at rest the longest muscle cells probably run for about 10 cm before ending at an attachment or abutting against the end of another skeletal muscle cell.

Factors that control length

Muscle fibres by shortening must produce a complete movement at the joint over which they act. The necessary shortening can be assessed, on a cadaver, by placing the joint at the two extremes of its movement and then measuring the fully extended and fully contracted length of a fibre. Such studies show that a muscle fibre, or cell if one makes up the whole length, can shorten by 57% of its fully extended length.

The above observations can be expressed in another way; namely, a muscle cell grows until it reaches a length which will allow it to produce the full movement at the single joint where it acts. Such an expression suggests that if the range of movement is increased, the cells and fibres should grow in length. Crawford (1954) designed such an experiment, using the tibialis anterior of the rabbit, and by releasing the tendon from the extensor retinaculum, he produced a greater excursion for the point of insertion; the belly of this muscle grew in length. Conversely arthrodesis results in a shortening of the muscle bellies acting over the affected joint.

Similar, but more complex, considerations apply to muscles that act over more than one joint. Sometimes the fibres in such a muscle are not long enough even when fully extended to allow a full movement at each of the two joints over which they act; this is called passive insufficiency and is easily demonstrated by attempting to straighten one's knee when the hip is flexed. This is due to the hamstring muscles being unable to lengthen sufficiently. The connective tissue within these muscles is preventing their overextension, so passive insufficiency is also known as the ligamentous action of a muscle.

Active insufficiency indicates that the fibres are so short that their complete shortening can only be effective, at any one time, at one of the two joints it crosses. For example, gastrocnemius and plantaris, can forcibly plantarflex the ankle joint only when the knee is extended; with a flexed knee, the Achilles tendon is lax and a rather painful plantar-flexion is produced by the soleus and other deep muscles of the calf. Training, especially during childhood, can reduce the insufficiencies presumably by increasing the length of the contractile fibres; the high-kicking dancer having lengthened her hamstrings and the cossack dancer having increased the range of contraction of his gastrocnemii.

Especially during growth the length of a muscle fibre is variable and, if demanded, its cells will grow in length relatively more. Such growth occurs at the expense of the associated tendon and the final length of the fibre reflects the range of movement.

Effective length

Few muscle fibres run parallel to the direction of the movement of their tendon. Instead they are inserted at angles in a variety of uni-, bi-, multi- and circumpennate arrangements (Fig. 4). These patterns increase the number of fibres that can be packed into a given volume, and by increasing the sum of the cross-sectional area of the contained fibres, make the contraction of the muscle stronger (p. 484). However, the force is reduced by the cosine of the angle which the fibre makes with its tendon and there is a corresponding increase in the movement of the tendon which this angulated fibre produces (Fig. 5). So when considering the function of a muscle, the effective length of its fibres should be considered after determining its internal architecture.

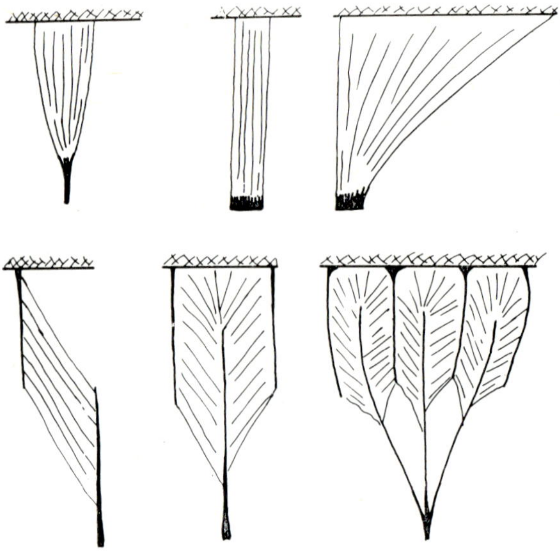

Fig. 4. The internal architecture of whole muscles is shown. In the upper row, fusiform, strap and fan types of muscle confer the maximum range of movement. In the lower row, unipennate, bipennate and multipennate forms increase the strength of contraction at the expense of the range of movement.

Consider two muscles in the human calf, plantaris and gastrocnemius, which have similar attachments and both act at the knee and ankle joints. One has a tiny short belly and the longest tendon in the body, while the other has two long fleshy bellies which produce the familiar contours of the calf. Yet, when dissected, it can be seen that the plantaris fibres, running longitudinally, have the same effective length as the relatively short oblique fibres of the gastrocnemius.

B. GIRTH

If a fibre is teased out from a muscle, it appears cylindrical, but in fresh frozen transverse sections, and after fixation, the profiles of fibres are irregular polygons; irregular, because of the close packing of fibres of different sizes. If the reticular connective tissue of the endomysium is impregnated with silver, a similar irregular pattern is obtained, although the muscle cells themselves have shrunk away from their sheaths during embedding. Capillaries, which run mainly along the long axes of the muscle fibres, are usually found at the corners of these interlocking polygons. Although it is difficult to be certain of an *in vivo* shape of a muscle cell, it is usual to consider them as cylinders and refer to their diameter; this can vary from 10 μm to over 120 μm in adult mammals.

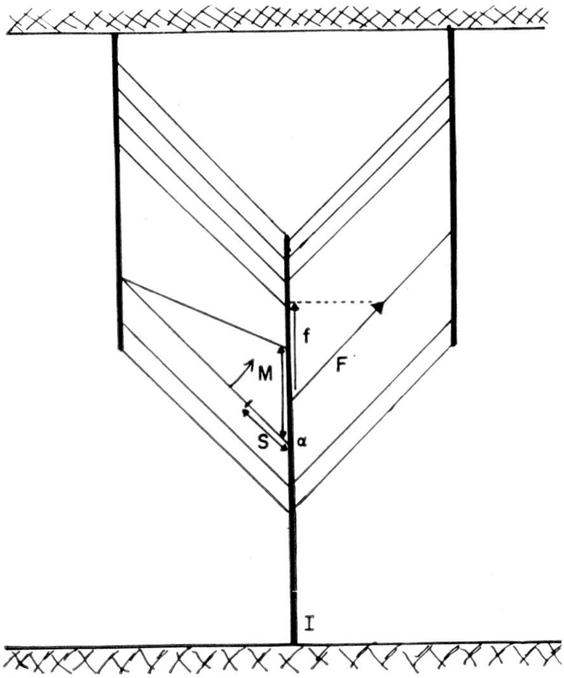

Fig. 5. The effects of a bipennate arrangement on force and range of movement are shown. To the right of the tendon of insertion I it is seen that the actual force f exerted by a single fibre F is F cos α, a diminution which is overcome by the large number of fibres present. On the left it is shown that an actual shortening S will produce a larger movement M.

The most effective method of determining these "diameters" is to measure the mean area of the fibres in fresh frozen truly transverse sections and convert this area into that of a circular profile before calculating the diameter. Measurements of diameter made on the shrunken fibres after wax embedding can only be used for comparisons on the same preparations.

Factors affecting girth

(a) Growth causes a steady increase after the fibre number has become constant.

(b) In the adult, or during growth, increase in certain types of exercise or paralysis of neighbouring muscles will cause hypertrophy (p. 502).

(c) Atrophy due to denervation, disuse or disease, is shown by a decrease in diameter.

(d) Type of fibre has an effect in that those fibres that are rich in myoglobin and mitochondrial enzymes, i.e. red fibres, tend to be more slender than the pale fibres which do not require such a rapid O_2 interchange with the related capillaries (p. 494).

(e) The size of the animal exerts little effect on the diameters of its muscle fibres. The higher metabolic rate in smaller animals demands a larger proportion of red fibres and so, for the reason given above, such animals may have smaller fibres.

As fibre number is determined before birth, much of the selection in animal husbandry has been applied to the rates of post-natal growth of muscle, and hence its fibres; this may explain the large diameter of pig muscle fibres. Perhaps future improvements in selection will be obtained by choosing species and breeds that lay down large numbers of fibres during embryonic development.

C. EXTERNAL SURFACE

Sarcolemma

This was first demonstrated by W. Bowman who caused a muscle to contract so vigorously that the myofibrils ruptured and left a clear delicate membranous tube containing two myofibrillar clots. Such a membrane appears structureless by light microscopy, but some reticular fibres adhere to its outer surface, and there is evidence of the attachment of the Z-discs on its inner aspect. It is the latter attachments which give the outer surface of a contracted fibre a serrated or scalloped appearance; each serration being opposite the sarcomere between a pair of Z-discs.

In section, the sarcolemma is seen to be composed of a plasma membrane less than 10 nm thick, which has the usual structural and functional properties of an excitable membrane (Chapters 17 and 22), and an outer basal lamina. The latter consists of an amorphous layer, about 30 nm thick, which lies apparently free of the plasma membrane since there is an intervening clear gap of 10–20 nm; the edges of the basal lamina are not sharply defined, and reticulin fibres blend with its outer aspect. Like the basement membranes of epithelia, the basal lamina consists of mucopolysaccharides and fine fibrils which are probably protein in nature and may be continuous with the microfibrils of the associated reticulin fibres. Certainly, the basal lamina is a developed structure and not merely a precipitation of connective tissue ground substance, since the lamina will persist in a plicated form after the contained fibre has atrophied or degenerated.

The plasma membrane shows the caveoli, which are usually thought to be involved in micropinocytosis, but there seems to be no experimental demonstration of such a function in skeletal muscle cells. Freeze-etched preparations have confirmed the presence of these caveoli. Modifications of the sarcolemma at the attachment of a tendon and the formation of T-tubules are better described in relation to force transmission and excitation-contraction coupling respectively (pp. 465 and 488).

Myoneural junction

A slight elevation of the muscle cell surface indicates the site of a myoneural junction. Practically all mammalian extrafusal fibres have only one such junction situated near the mid-point of the cell. The muscle cytoplasm in and near the elevation is packed with muscle mitochondria and contains some vesicles, but the main modifications affect the surface membranes. The sarcolemma is indented to receive the terminal ramifications of one of the branches of the axon supplying this motor unit (Fig. 6). The sides and bases of these indentations are thrown into deep subneural clefts; the surface pattern of these clefts resembles a finger-print. The motor nerve endings are separated from the muscle plasma membrane by the basal lamina and are thus described as epilemmal: in contrast, afferent endings on muscle cells in muscle spindles (Chapter 21) are insinuated in a hypolemmal position between the basal lamina and the plasma membrane.

The bulbous nerve endings are sheathed right up to their termination by Schwann cells, and each ending is packed with mitochondria and synaptic vesicles. Most of the description of a synapse in Chapter 18 applies to this specialized form of intercellular connection.

True acetylcholinesterase can be located by light and electron histochemistry in the subneural clefts, where, of course, it can hydrolyse the transmitted acetylcholine. The acetylcholinesterase seems to be an integral part of the muscle membrane since it persists for many months after the degeneration of the nerve terminals. Also, changes in the sensitivity of the sarcolemma to acetylcholine, applied in minute amounts, are one of the early signs of developing innervation, originally the whole surface is sensitive but the middle region becomes more sensitive about the time when the developing neurites reach the muscle (Diamond and Miledi, 1962). When the motor nerve is crushed the regenerating neurites follow the endoneurial tubes to the original end plates, but if a cut nerve is implanted into a denervated muscle, it can form new motor-end plates and so the neurites need not search out the original site of innervation (Gwyn and Aitken, 1966).

The single motor-end plate at the middle of a muscle cell is called an *en plaque* ending to distinguish it from a simpler form of *en grappe* ending, where the nerve meanders over the surface and makes numerous punctate contacts with the muscle. *En grappe* endings do not produce propagated impulses and the whole cell contracts only when many of the numerous contacts are activated. There is no subneural

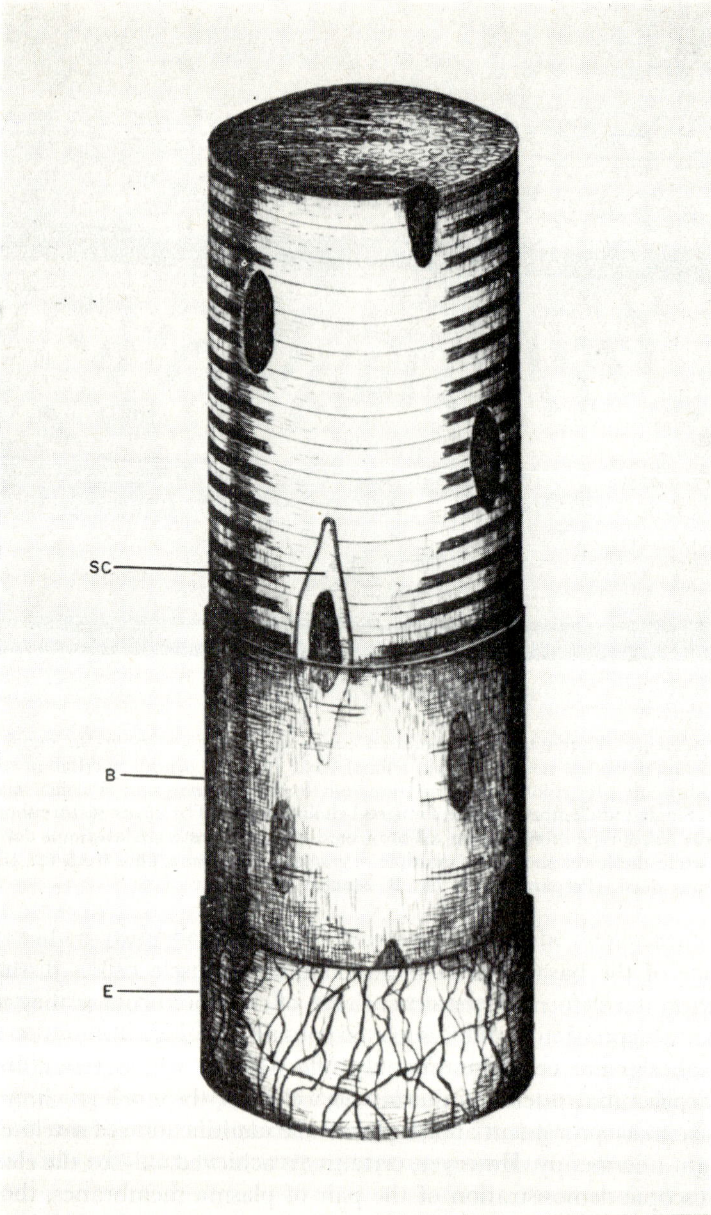

Fig. 8. A segment of a skeletal muscle fibre showing the endomysium E and basal lamina B. One satellite cell SC nestles in a sarcolemmal depression beneath the basal lamina. This length, 125 μm, would be expected to contain nine myonuclei and one satellite cell. (Reprinted from Muir (1970) in "Regeneration of striated muscle and myogenesis" (A. Mauro, S. A. Shafiq and A. T. Milhorat eds.) Excerpta Medica, Amsterdam.)

mitochondria are smaller and have fewer cristae than in the neighbouring muscle cell; free ribosomes are present, there are only a few cisternae of rough-surfaced endoplasmic reticulum, and the surface plasma membrane carries more pinocytic caveoli than the related muscle cell.

Number

As satellite cells can only be recognized with assurance in sections for the electron microscope, it is hard to assess the frequency with which they occur. Estimates vary from 5–20% of the nuclei within the basal lamina being assigned to satellite cells, with some suggestion that they are more numerous in growing muscle. A proportion of 10% means that there are about 400 satellite cells related to a muscle cell which is 5 cm long.

Distribution

Since their original demonstration by Mauro (1961), satellite cells have been reported in many mammalian, avian, reptilian and amphibian skeletal muscles. They do not occur in cardiac muscle of vertebrates, but a crab, which has a form of multinucleated skeletal muscle in its heart wall, has a complement of satellite cells. They are distributed along the whole length of the fibre, and do not appear to be more concentrated near either the motor end plate or the myotendinous junction.

Function

Most workers now consider that the satellite cells represent committed myoblasts in a resting, or reserve, state ready to multiply and fuse with the muscle cell in growth or regeneration (p. 501). However, it has been suggested that a satellite cell is merely an intermittent phase in the life cycle of a muscle nucleus; with such a nucleus creating surrounding cell membranes and producing a "satellite cell" in response to injury, or even the demand for nuclear division. These matters are discussed further on p. 505).

E. CONTENTS

Nuclei

Red muscles (p. 494) may have some nuclei in the centre of the fibre, but most mature extrafusal fibres have their nuclei immediately beneath the sarcolemma. The nuclei are ovoid, with their long axes along the fibre; as these axes measure 12–16 μm, and as there are between 50–100 nuclei per millimetre length of the fibre, a single transverse section will usually contain one or two sectioned nuclei. The syncytium of a muscle cell, 5 cm long, will contain some 4 000 nuclei. The nuclei have a dark chromatin which is uniformly distributed in the nucleoplasm, and a pair of nucleoli are usually evident.

Cytoplasm

(a) Sarcoplasm is the name given to the cytoplasmic matrix of the muscle cell. It contains the usual small molecular components of all cells and the special ones concerned with contraction, such as creatine phosphate which serves as an energy store (p. 492). Glycogen as single β particles provides an important store of carbohydrates. The protein myoglobin, which can store oxygen, is present in all muscles, but its concentration varies markedly between different muscles (p. 494).

(b) Centrioles have not been seen in a skeletal muscle cell.

(c) Golgi apparatus is represented by a few flattened cisternae of smooth endoplasmic reticulum close to a subsarcolemmal nucleus.

(d) Rough endoplasmic reticular cavities can be seen, but they are small and infrequent. Ribosomes are present.

(e) Lipofuscin accumulates at the poles of the nuclei during the liftime of the cell, and they can be a prominent feature in the muscle cells of old animals (Chapter 16).

(f) Lipid droplets.

(g) Mitochondria.

The fundamental features of these organelles have been discussed in Chapter 6 and their shape and distribution in skeletal muscle cells are affected by the metabolic pattern of each cell. So these aspects will be discussed below in relation to the definition of fibre types (p. 494).

(h) Microtubules, which cannot be spindle fibres because of the absence of centrioles, make their straight courses through the cytoplasm without any connections with surrounding organelles.

(i) Myofibrils.

(j) Sarcoplasmic reticulum, which is a modified form of smooth endoplasmic reticulum.

These last two contents are unique to and characteristic of muscle and so deserve separate discussion under the headings of Contraction and Excitation-Contraction Coupling respectively.

II. Structure of Myofibrils and Contraction

A. ORGANIZATION IN THE CELL

The cross-striations of muscle are due to the longitudinally oriented contractile myofibrils being arranged in register through the whole thickness of a cell. As each myofibril is 0·5–1·0 μm in diameter, this means that some 2 500 myofibrils are held in constant relation to each other throughout the contractile cycle in a 50 μm fibre. The repeating period along a single myofibril is the functional unit of contraction, the sarcomere, and it is limited by two Z-discs. At rest, the sarcomere measures 2·4 μm; hence, in a 5 cm long cell, there are over 20 000 sarcomeres along each myofibril.

There is no obvious reason why myofibrils should be kept in register, and they readily lose this orderly arrangement during certain preparative procedures. But the clearly marked anisotropic bands stretching across a whole fibre, under polarized light, testify to the efficiency of the *in vivo* retaining mechanisms. Fine fibrils, 7–8 nm in diameter (i.e. intermediate between the thick and thin myofilaments), course through the sarcoplasm; they are not associated with myofilaments and are especially prominent between the Z discs. Thus, these fibrils could hold the Z-discs in register, and we have already mentioned the attachment of the Z-discs of peripheral myofibrils to the inside of the sarcolemma. The combination of the Z-discs and these fibrils does therefore produce a perforated diaphragm which crosses the whole fibre and, through the plasma membrane, gains attachment to the endomysium and the other surrounding connective tissue. It is these surface attachments which produce the scalloped surface seen on contracted cells.

In many muscle fibres the Z-discs of adjacent myofibrils are actually continuous with each other, so forming more or less complete diaphragms across the fibre. Because this results in large fields of myofilaments rather than discrete myofibrils, this organization was called "felderstructur" by Krüger in distinction to "fibrillenstructur" (Fig. 9). These two types of structure may have functional implications (p. 495), but the distinction between them is not absolute, because of the filamentous attachments between adjacent myofibrils in "fibrillenstructur" and because widely separated fusions of Z-discs along the lengths of such a fibre are overlooked in electron micrographs.

Myofibrils occupy almost all the interior of a muscle cell, the narrow clefts between them containing mitochondria, lipid droplets, T-tubules and sarciplasmic reticulum (p. 488).

B. MYOTENDINOUS JUNCTION

The cell may end by tapering into a tendinous attachment, or by abutting against a sheet of collagenous tissue such as an intermuscular septum or periosteum, or it may join the end of another cell within a composite muscle fibre. In each case the sarcolemma at the end of the fibre is profoundly modified (Fig. 10). Deep tubular invaginations of the plasma membrane, up to 20 μm long, penetrate the cell between the myofibrils. Into the mouths of these invaginations run collagenous

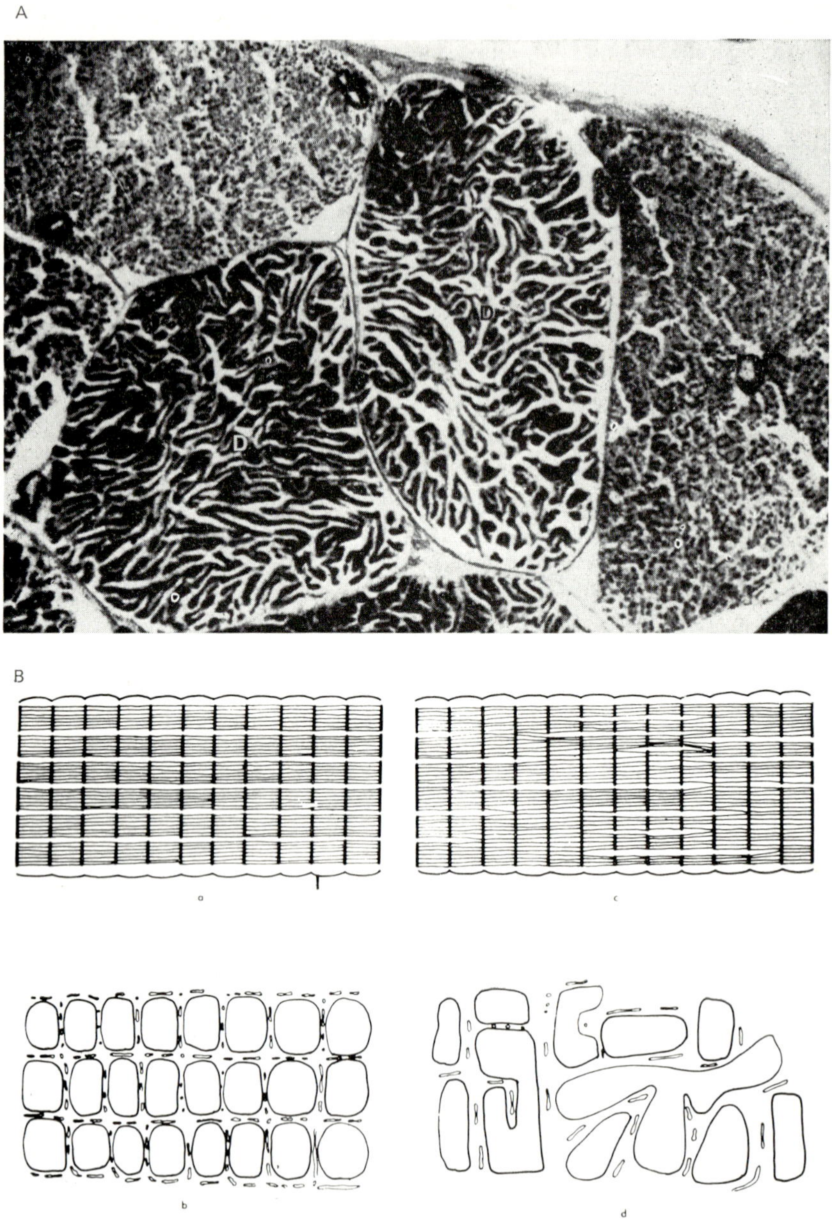

Fig. 9. (A) Felderstructur and fibrillenstructur shown in transverse sections of frog muscle. In felderstructur fibres D, the myofilaments clump together in irregular areas, whereas in the surrounding fibrillenstructur fibres the myofilaments form apparently discrete myofibrils, ×600. (With permission from Krüger (1952), Akad-Verlags., Leipzig.) (B) Diagrammatic representations of fibrillenstructur a, b and felderstructur c, d in longitudinal and transverse sections. (Designed by Miss A. MacNaughtan.)

fibres from the tendon before they blend with amorphous connective tissue in the lumen of the invagination. By digesting off the connective tissue with collagenase, and then examining the end of the muscle cell with the scanning electron microscope, a striking picture of these invaginations is obtained (Fig. 11).

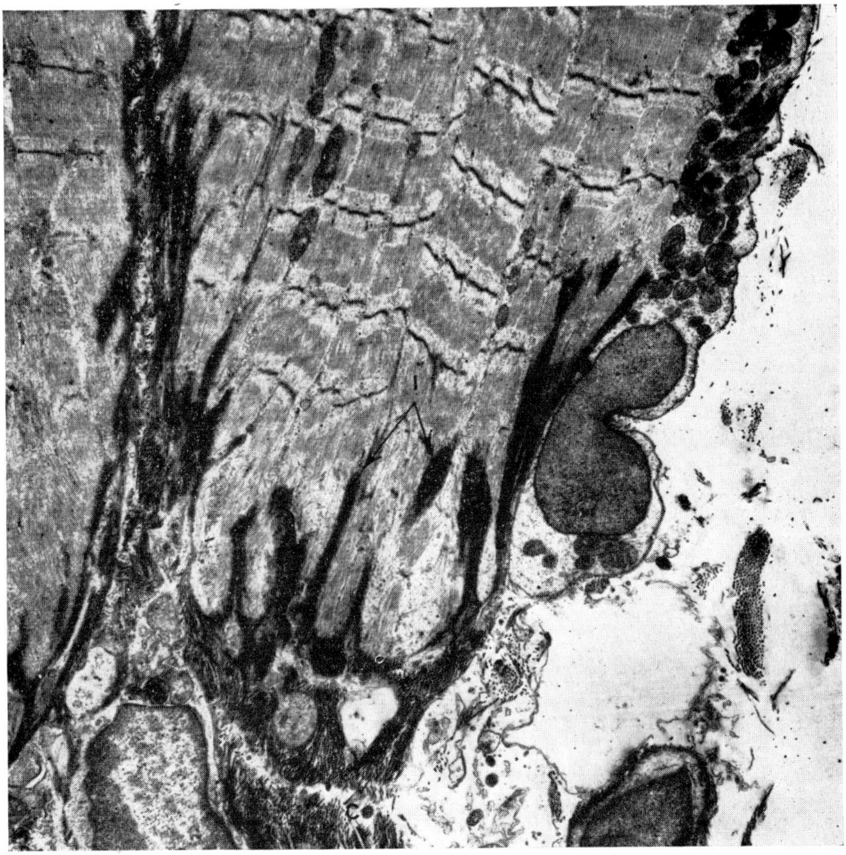

Fig. 10. The tapering end of a rat muscle fibre is shown in this electron micrograph. The collagen C of the tendon emerges from terminal invaginations, I ×4 700.

On the inside of the sarcolemma, the myofibrils usually approach the membranes of the tapering cell and its invaginations at oblique angles. The sarcomeres continue in regular series to within less than 5 μm from the end of the myofibril and then the myofilaments become less regularly arranged and are embedded in electron-dense material which is applied to the inside of the plasma membrane.

Acetylcholinesterase can be located at the myotendinous junction in lower concentration than at the motor end-plate. Its function is obscure, but it could represent a phylogenetic remnant of the enzymes associated with the innervation of lower vertebrate muscles, some of which have their myoneural junctions at the ends of the cells.

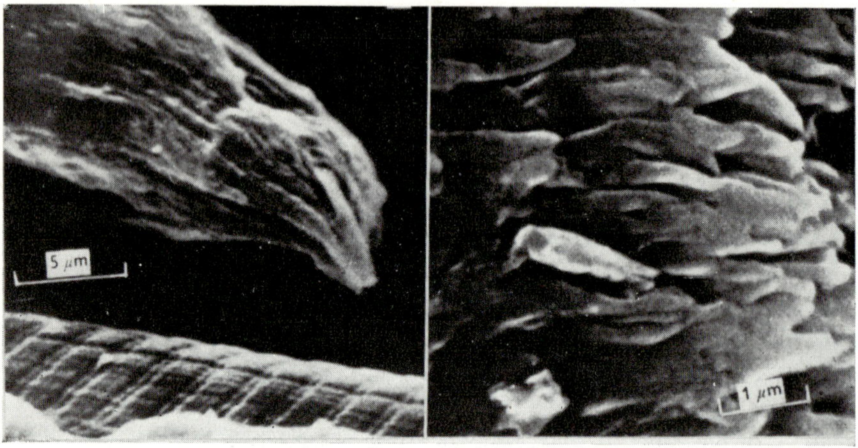

Fig. 11. Scanning electron micrographs of the ends of frog muscle fibres after collagenase removal of the surrounding connective tissue. On the left the tapering end of one fibre is seen above a surface view of an adjacent one. On the right, the complex invaginations at the end of a fibre are seen at higher power. (With permission, from Williams, J. C. P. and Boyde, A. (1968), *J. Anat* **103**, 597.)

Before describing the structure of the contractile unit of striated muscle or sarcomere (p.474), it is convenient to discuss the features of the known constituent proteins.

C. MUSCLE PROTEINS

If fresh skeletal muscle is minced and extracted with KCl solutions of high ionic strength, its proteins are extracted and, as these have different solubilities at different ionic strengths, various fractions can be precipitated. Great care and ingenuity is needed to obtain pure samples, and much confusion has arisen due to impurities leading investigators to believe that they had isolated a new important protein. However, modern methods produce three clearly different proteins, actin, myosin and tropomyosin, and two more which are also almost certainly involved in the process of contraction, troponin and α-actinin.

Actin

This protein has a molecular weight of 45 000 and with a high proline content, 5 out of each 100 residues, its polypeptide chain readily coils to form a globular molecule, 5·5 mm in diameter. This globular actin (G-actin) will combine with ATP and other G-actin molecules to produce long chains of globular molecules firmly bound together by ADP after the ejection of a phosphate radicle. These chains are known as fibrous or F-actin and, as they can be made *in vitro* and yet also form the main part of the thin filaments in the whole cell, they are now described in detail. Each F-actin consists of two strings of G-actin beads, wound around each other like a twisted necklace. The pitch of this twist is affected by the method of preparation, but is presumably determined by the site of adhesion between adjacent dimers of G-actin. One full turn of the superhelix need not be completed in a whole number of actin molecules, for example when we are dealing with 13 pairs, each measuring 5·5 nm along the axis, then a length of 70 nm is involved, and half of this, 35 nm, should equal the distance between cross-over points; F-actin made *in vitro* and examined on E.M. grids by negative staining (Fig. 12) gives a similar repeating period of between 30 and 40 nm.

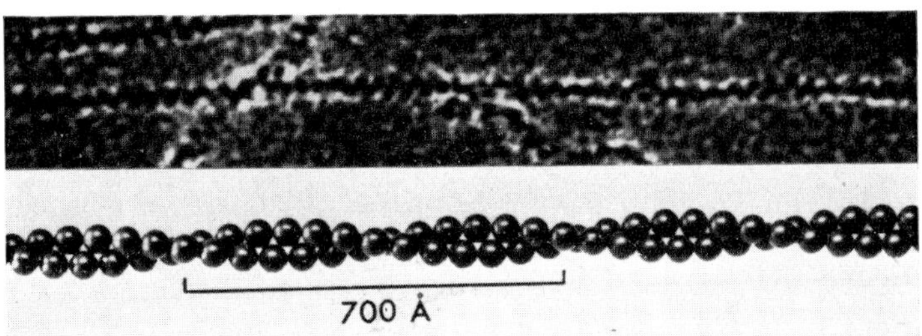

Fig. 12. A model of fibrous actin is shown below with 13 dimers per complete twist, which would measure about 70 nm in length. Above, a negatively stained filament of *F*-actin is shown in an electron micrograph, × 445 000. (With permission, from Hanson, J. and Lowy, J. (1965).)

The three dimensional molecular structure of G-actin is not yet known, but when it is resolved it can be expected to show three zones which will combine through ADP with other G-actins, the eccentricity of these zones determining the structure of F-actin. There should also be a site which can be recognized as responsible for combining with myosin, this site must also be eccentric because in an F-actin strand all these sites point to the same end of the twisted chain. This can be readily demonstrated by treating F-actin with the heavy meromyosin parts of myosin molecules (Fig. 13).

Myosin

These molecules are much bigger than actin as their molecular weight is about 500 000 and they are much more complex. It is extracted at lower ionic strength and, when purified and examined directly in the E.M., its molecules slightly resemble spermatozoa (Fig. 14). The "head" and a short length of "tail" can be separated from the remaining part of the "tail" by trypsin digestion and this produces heavy meromyosin (HMM) and light meromyosin (LMM) respectively.

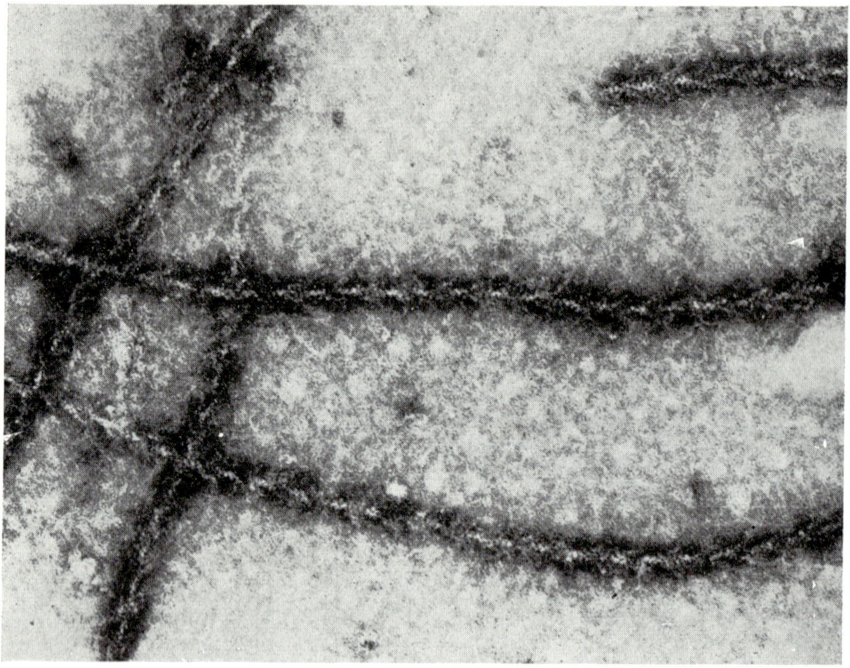

Fig. 13. Negatively stained filaments of F-actin treated with a solution of heavy meromyosin. On each filament, the heavy meromyosin molecules are attached obliquely, so giving the appearance of arrowheads which all point in the same direction. × 132 000. (With permission from Huxley, H. E. (1964), *Proc. Royal Soc. Ser. B.* **160**, 442.)

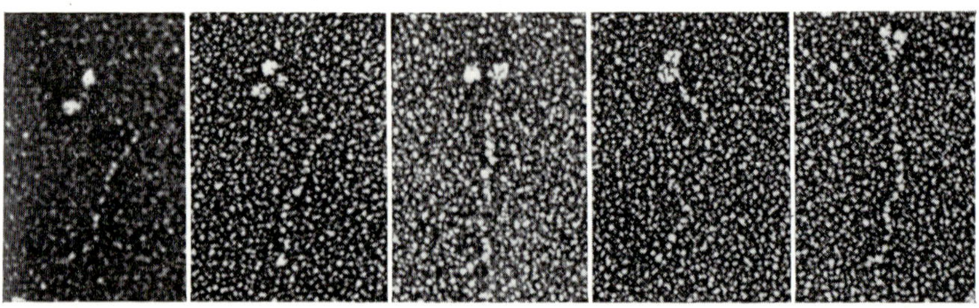

Fig. 14. Single molecules of myosin which were rotated while they were being shadowed for electron microscopy. × 153 000. (Reprinted, with permission, from Slayter, H. S. and Lowey, J. (1967), *Proc. Nat. Acad. Sci. USA* **58**, 1611.)

Light meromyosin contains no proline, and its properties indicate that it consists of two α-helically coiled polypeptide chains twisted about each other. Heavy meromyosin can be further digested into two fractions, the S_2 fraction (the short length of "tail") having much the same structure as LMM, while the S_1 fraction ("head") contains almost as many proline residues as actin, 3 per 100. It is the S_1 fraction that forms the globular heads, which are capable of binding so strongly with actin, and which on their own have slight ATPase activity; an activity that is greatly enhanced after their combination with actin.

The structure of the whole myosin molecule as revealed by a careful shadowing method for the electron microscope, which involves rotating the specimen while it is being shadowed, and incorporating the information which is derived from biochemical methods, is shown in Fig. 15.

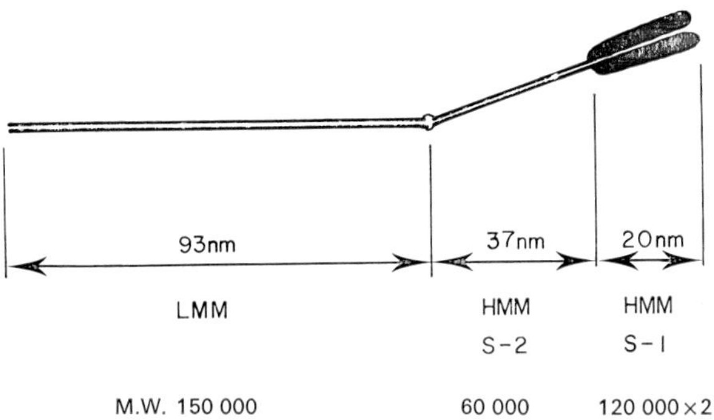

Fig. 15. Diagrammatic representation of a myosin molecule from data provided by Lowey *et al.*, (1969).

It is evident that each molecule has two heads, S_1 segments, joined to a straight neck, some 37 nm long, which then joins a 93 nm length of a similar double α-helix; the joint being a flexible one.

If the ionic strength of a myosin solution is lowered to 0·6 KCl, ordered molecular precipitates are formed (Fig. 16). These sequences can be interpreted as showing an affinity for the ends of the LMM portions, which at the outset joins two myosin molecules so that their HMM portions point away from each other. As more precipitation occurs, the fibrous myosin grows equally from this centre, where the polarity changes, by the affinity of the LMM portions to each other, now oriented in the same direction. In this way a strand is produced which possesses in the middle a region, nearly 200 nm long, which has no projecting heads and then regions, of varying lengths, with HMM projections. In electron micrographs these projections have so far been shown as blobs on the side of the filament; it is likely that

Tropomyosin

This protein, which has no ATPase activity, occurs in two slightly different forms, A and B. Tropomyosin A precipitates to form long fibrous strands of variable length and width; each molecule being double-coiled α-helix with little or no disturbance due to proline. In every way this protein resembles the LMM and the HMM-S_2 fractions of myosin. Tropomyosin B precipitates in the form of a quadrangular mesh with a side length of 20 nm (Fig. 17); from the corners of this mesh strands of precipitated tropomyosin are sometimes seen.

Troponin

This is a globular molecule, molecular weight about 80 000, which has been difficult to separate and purify. Troponin has the power to bind Ca^{2+}, and in its absence the combination of actin and myosin loses its characteristic sensitivity to low levels of Ca^{2+} concentration.

α-Actinin

This is also a globular protein which seems to be associated with tropomyosin B and is thought to be found in the Z-disc region.

It would be foolish to assume that the proteins described form a complete inventory of the special proteins that are synthesized by muscle cells and play an essential part in the development of a contractile force. One more is known to be associated with the M-line, which has cholinesterase activity, but it has not been isolated or characterized. There may be many more such proteins. However, the known ones are sufficient to explain much of the architecture of the contractile unit in striated muscle, the sarcomere.

D. THE SARCOMERE

The elucidation of the structure of the sarcomere has proceeded at an erratic pace since the early light microscopists described the cross-striation and polarized light properties of myofibrils in muscle. Early electron microscope work confirmed and amplified the descriptions of the banding and showed that each myofibril contained filaments oriented along its long axis. Thick filaments, 10–12 nm in diameter, restricted to the A-band, explain the anisotropy in polarized light. The isotropic (I) bands contain thin filaments, 5 nm in diameter, which have almost no effect on polarized light. The isolation of myofibrils and the device of preserving them by impregnating the cell with glycerol and so removing the other components, opened the way to more detailed electron microscope sectioning studies of the myofibrils (Huxley, H. E., 1957). Interference microscopy showed the constant length of the A-band during contraction (Huxley, A. F., 1957). Extraction of particular muscle proteins followed by phase contrast examination of myofibrils demonstrated that the thick filaments contained the myosin and the thin filaments, the actin. This interpretation was confirmed later by fluorescent antibody studies (Pepe, 1967). The structure of sarcomere components in relation to known protein structure also showed their distribution in the myofibril (Huxley, 1963). X-ray diffraction studies on whole muscle fibres indicated the paired filamentous structure and showed the separation between the filaments and their axial periodicity (Huxley, 1963). More recently these have been applied to contracting muscle, and these studies show that the axial periodicity of the filament remain constant during contraction; perhaps the strongest evidence in favour of the sliding filament theory (Elliot *et al.*, 1967). In parallel with these structural studies, independent experimental work has shown a close relationship correlation between the tension that a muscle can develop and the expected performance based on the structure of the sarcomere (Gordon *et al.*, 1966).

The following description of a sarcomere must for brevity be an amalgam of all these lines of investigation. Its documentation can be found in the articles quoted above, and this account merely shows the way that these special muscle proteins are organized to produce a functional unit that is amply large enough to be recognized by light microscopy in a living cell. In this account any measurements, made on fixed material, have been adjusted to the figures believed to obtain in life.

Z-Discs

These provide the attachments at the end of the sarcomeres for the thin filaments. Detailed studies of the structure of these discs (Knappeis and Carlsen, 1962) give slightly different interpretations which may depend on the amount of stretch at the time of fixation. All agree that the thin filaments are linked with their neighbours in the next sarcomere through a quadrangular lattice. Landon (1970) shows, with glutaradehyde fixation, that two such lattices, with spacings of 22 nm, overlap each other 50% out of register, thus giving the appearance of an 11 nm side length; the thin filaments of adjacent sarcomeres attach to these lattices. The apparent similarity between the pattern of the osmium-fixed Z-disc filaments and that of isolated tropomyosin B crystals has been taken to indicate that the skeleton of the disc is composed of this protein. Other elements must be present, α-actinin has been mentioned, and the permeating electron density of these discs requires a chemical explanation.

Thin filaments

These contain actin and, if isolated and examined by negative staining, they closely resemble F-actin threads. They differ however in three important ways; the thin filaments extend for 1·1 μm on either side of the two Z-discs, while F-actin threads are of indefinite length; if treated with HMM, the polarity of the thin filaments is shown to change at the Z-disc so that, when the myosin head has attached to the contained actin, it points away from the Z-discs, while on synthetic F-actin they are oriented in only one direction however long the filament has grown; and lastly, the thin filament contains other proteins,

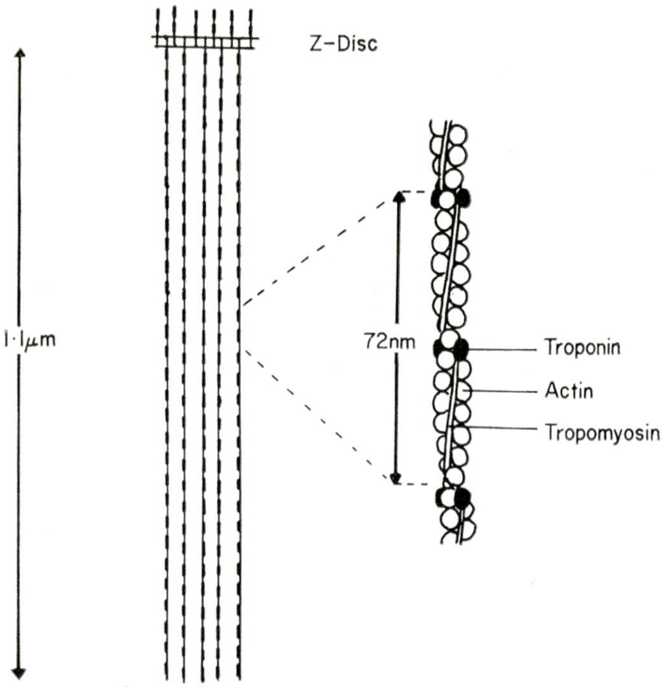

Fig. 18. Models for the fine structure of the thin filaments. From the Z-disc the thin filaments extend for about 15 complete twists of their F-actin component. One complete twist, on the right, shows a possible arrangement of component molecules as suggested by Ebashi *et al.*, (1969).

tropomyosin, the long strands of which may lie in the spiral grooves formed by the G-actin molecules, and troponin, which can be shown by immunologically labelled ferritin markers to be attached at 40 nm intervals along their length. These features of a thin filament are summarized in Fig. 18, where it is seen that each filament will contain about 15 complete twists of F-actin, each twist measuring 71 nm in length and having about 13 dimers of G-actin. So the whole filament has some 390 G-actin molecules, perhaps two 1·1 μm long tropomyosin double helices running in the two spiral grooves and globular troponin attached to produce 25 cross striations. The diameter of the thin filaments in fixed sectioned muscle is about 5 nm, a figure that is considerably less than the values measured on negatively stained fresh filaments or computed from a knowledge of the molecular structure of these filaments.

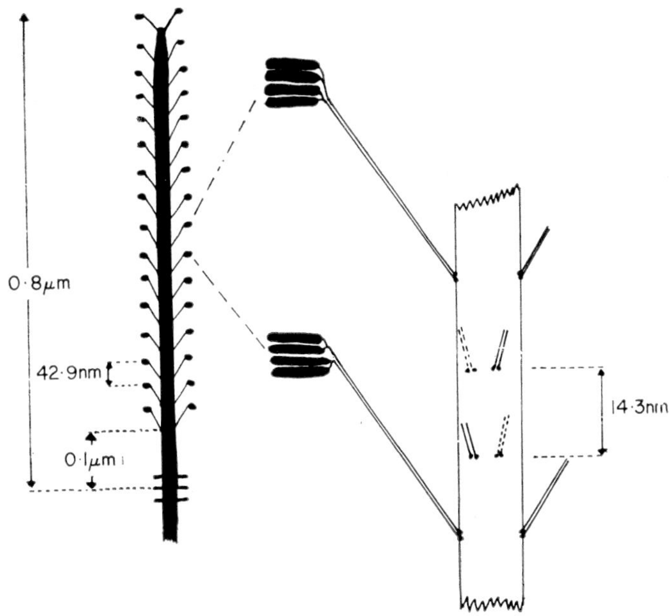

Fig. 20. Scale diagram of half of a thick filament constructed from data summarized by Bendall (1969). On the right the six pairs of projections, arranged in three planes at 120°, are shown.

Thick filaments

Containing myosin, these filaments resemble fibrous precipitated myosin *in vitro* in having radially projecting heads, a zone 200 nm wide free of heads in the centre, and a change of polarity on either side of the middle of this zone. As with the thin filaments, the natural article shows some differences from its manufactured copy. The thick filaments as separated from fresh muscle, by gentle homogenization in the presence of a calcium-chelating agent such as EDTA, are constantly 1·6 μm long, and they taper towards their ends (Fig. 19). The smooth central zone has three sets of projecting arms forming the M-bands (p. 480) and it also has an ovoid swelling, which contains a cholines-

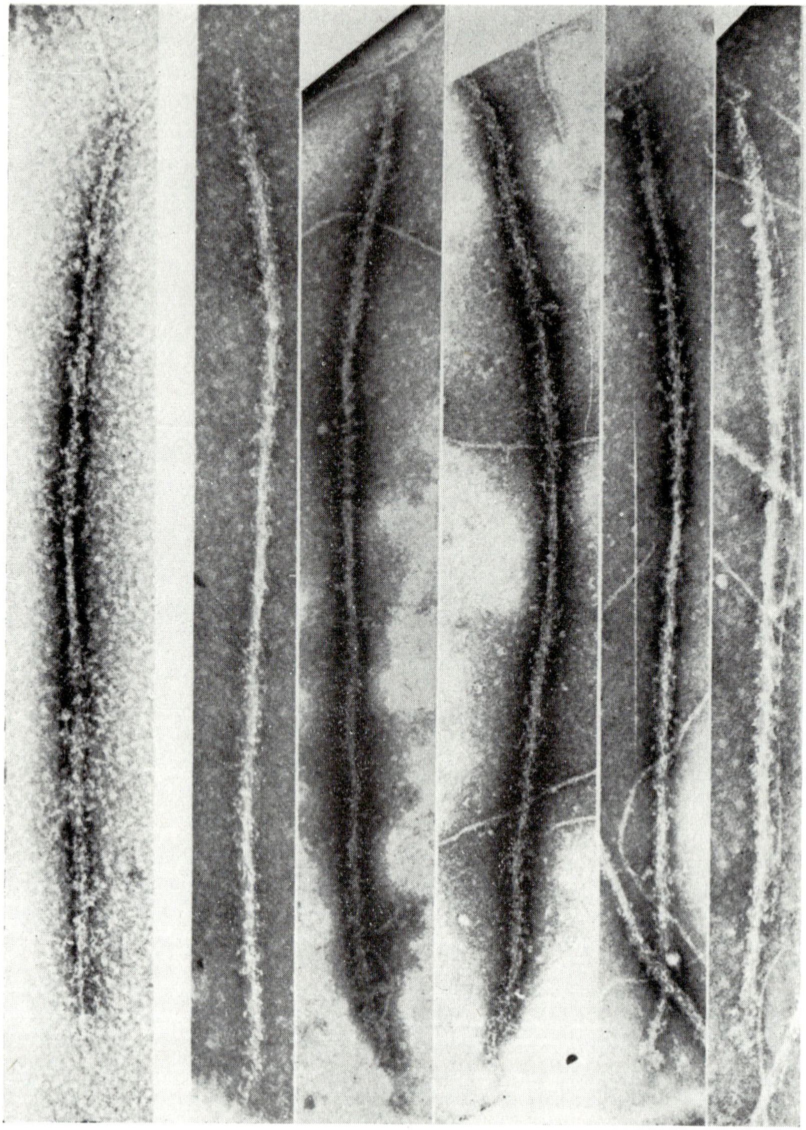

Fig. 19. The five thick filaments on the right were isolated from intact muscle, and the one on the left was formed by precipitating myosin. Note the central projection-free zones and the tapering ends, ×75 000. (With permission from Huxley (1963).

terase enzyme and may have other components. A scale diagram showing one half of a thick filament is shown in Fig. 20; 0·8 μm long, it has a central region 0·1 μm long which is free of projections and which houses the 93 nm length of the LMM portions of the first myosin molecules. Strands of tropomyosin may run the whole length of the thick filament, in association with the parallel arrangement of the LMM portions of all the myosin molecules. These portions of myosin aggregate so that pairs of HMM portions project in three planes at 120° intervals, that is in the direction of the neighbouring thin filaments (Fig. 21). The axial distance between successive pairs of projections is just over 14 nm so that the projections, which are in the same plane as the first pair, are 42·9 nm further along the axis. Sixteen such sets of six projections gives a thick filament, half of whose length is 100 + (15 × 42·9) = 743 nm, i.e. nearly 0·8 μm long. Thus each half of a thick filament has about 16 × 6 = 96 projections. We have

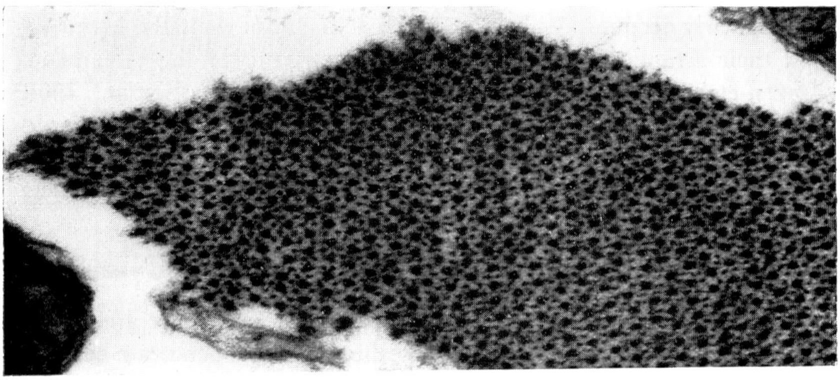

Fig. 21. Above, an electron micrograph of a transverse section of an A-band stained with phosphotungstic acid, ×44 100. The diagrams below indicate the changes in interfilament spacing which are associated with sarcomere lengths of 3.6 and 2.0 μm (see Elliott *et al.*, 1967).

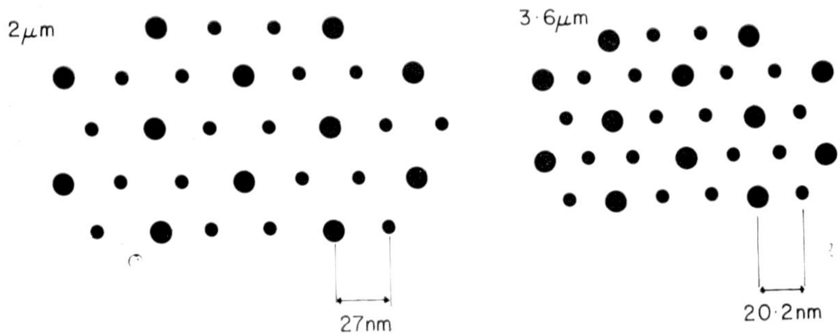

X-ray studies during contraction show that the interfilament distance varies with the degree of shortening. The distance from the centre of the thick to the centre of the thin filament being 20·2 nm at a sarcomere length of 3·6 μm and increasing to 27 nm when the sarcomere has shortened to 2·0 μm (Fig. 21). There is a similar increase (19·7–30·1 nm) in interfilament spacing for the thin filaments near the Z-disc; it is not known whether this is associated with an increase in the radius of the Z-discs, or whether there is a bending of the thin filaments at this point during contraction.

E. FUNCTIONAL CONSIDERATIONS

Method of Contraction

The detailed study of the architecture of a sarcomere, which has been outlined above, permits speculation on the changes which occur during contraction. The combination of actin with the S_1 segment of heavy meromyosin is profoundly affected by the presence of troponin, Ca^{2+} ions and ATP. These effects can be shown *in vitro* on isolated actin and myosin, and also on glycerinated myofibrils. The reactions of really pure actin are not affected by Ca^{2+}, because troponin confers calcium sensitivity on the filament. Most *in vitro* experiments and all work with natural filaments have been performed in the presence of troponin and the following results obtained. In the presence of Ca^{2+} at concentrations of $>10^{-4}$ M, ATP causes combination of actin and myosin and intense ATPase activity at the site of union with the liberation of energy in the form of contraction and heat. If Ca^{2+} is present and ATP absent, the two proteins combine firmly with each other; this is the state of rigor. In the absence of Ca^{2+}, ATP causes a repulsion between the proteins and this is the state of a relaxed muscle.

Consider a sarcomere at a length of 2·0 μm with its thin filaments being separated from each of the surrounding three thick filaments by 27 nm. From each thick filament 16 projections, each containing two double headed HMM segments, incline towards the thin filament. This means that the 0·7 μm overlapped length of thin filament, containing some 250 G-actin molecules, is within range of 96 heavy meromyosin molecules or 192 S_1-HMM segments. A flush of Ca^{2+} ions (p. 490) will cause many of these to fuse and react with the actin. A movement of the thin filament of 5·5 nm would bring the next G-actin molecule into a corresponding position, and such a movement could be produced either by a shortening or bending of the 20 nm long S_1 segment. So many interacting points are possible, that it is not even necessary for this small amount of movement to be produced by a single S_1 segment, as a vernier effect can be envisaged with the long series of interacting points between the two sets of filaments.

There is some indication that the movement is generated by bending rather than a molecular shortening. Insect muscle, in rigor, shows the cross-bridges between filaments bent away from the Z-discs, and we have seen that the HMM segments react with the thin filaments to produce arrowheads, which also point away from the Z-disc. In relaxed muscle the HMM segments probably lie along the sides of the thick filament, as a repeating period is obtained by x-ray diffraction, which disappears at the onset of contraction when, presumably, the segments all become active.

During an isotonic contraction, successive G-actin molecules will be moved by an HMM segment, with an ATP molecule being split to ADP at each section. If the cell is held so that an isometric contraction is produced, the same or closely related molecules will interact; ATP will be split at each flush of Ca^{2+}, but more of its energy is dissipated as heat. If the time interval between stimuli is short enough to prevent the removal of Ca^{2+}, a tetanic contraction will be produced.

Force of contraction

The contractile force generated by a single filament will be affected by temperature, hydrogen ion concentration, availability of Ca^{2+} and ATP; but with all these factors constant, it will be influenced mainly by the extent of overlap of the filaments. Figure 23 shows how increasing overlap should result in an increase in developed tension bringing more projections into play until the ends of filaments begin to interfere with each other, and then a further reduction to zero when the ends of thick filaments are crumpled and compressed by the Z-discs. Note that when the thin filaments project into the opposite half of the sarcomere they cannot react because of their reversed polarity.

Experimental proof that the force of contraction is in fact related to sarcomere length in the fashion expected was difficult to obtain. In intact muscles the connective tissue framework exerts a passive tension at longer sarcomere lengths. So the estimations had to be performed on single fibres and, because the terminal sarcomeres stretch less than those in the middle of the fibre, an elaborate apparatus had to be constructed to hold the middle of the muscle cell in a truly isometric condition. With these precautions, a length-tension curve was obtained which closely fits the expected values (Fig. 23). This correlation provides one of the strongest proofs of the validity of the sliding filament mechanism.

The force developed by a whole muscle does not follow that of a single fibre, as, with increasing sarcomere length, the connective tissue framework is stretched and it exerts a passive tension. The role of connective tissue varies between muscles, being especially important in

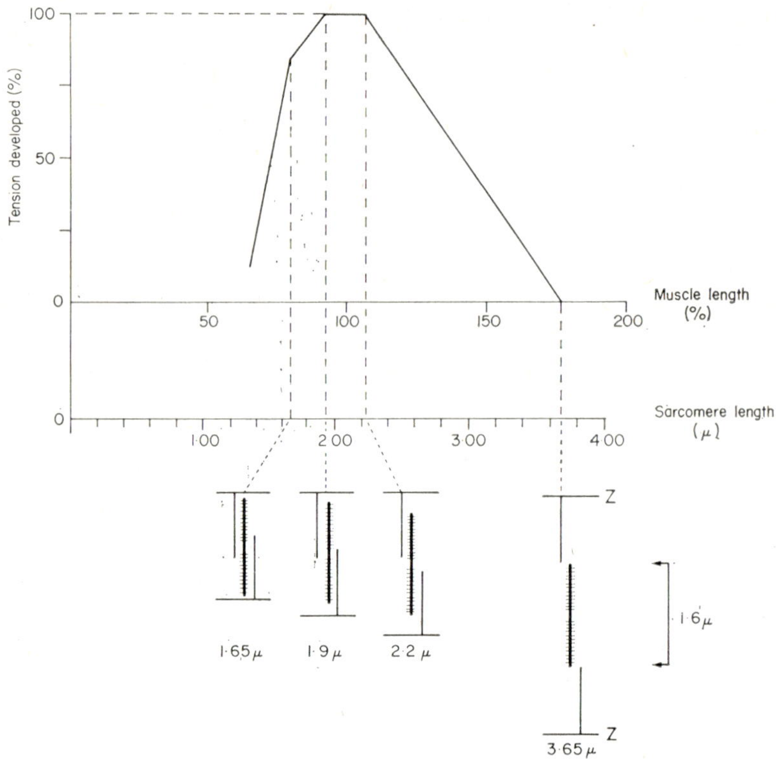

Fig. 23. The upper figure shows the tension produced by a single fibre plotted against the percentage of muscle length and against sarcomere length as observed in an interference microscope; this data was provided by Gordon Huxley and Julian (1963). The figure shows the lengths and relative positions of the 1·6 μm long thick filaments and the 1·0 μm long thin filaments at four selected sarcomere lengths; the lengths of the filaments were given by Page and Huxley (1963). (Reproduced, with permission, from Hanson and Lowy (1965).)

those muscles which act through passive insufficiency (p. 455). Most human muscles are not passively insufficient, so their greatest force is exerted when they are slightly less than half their maximum length. Consequently, the posture we assume for any strong effort involves placing all the important joints midway between full extension and flexion. Think of the pose of a sprinter on the blocks.

The force of contraction of a cell will depend mainly on the number of interacting filaments, and as these virtually fill the cells, this number is reflected by their cross-sectional area. Similarly, the cross-sectional area of all the fibres indicate the maximum force a whole muscle can produce; this has been measured and figures of between 2 and 4 kg/cm² of cross-section muscle obtained. Since there is no increase in the number of cells, it follows that the hypertrophy of an athlete's muscles, particularly those such as weight-lifters and gymnasts, consists of an increase in cell, and so fibre, diameter. In extreme hypertrophy, such as that induced experimentally by severing adjacent muscles, some fibre splitting has been reported, but normally the extra myofilaments can be accommodated merely by enlarging the existing cells. The design of whole muscle into pennate forms can be explained by referring to the need to produce the maximum number of interacting filaments.

The most effective means of measuring the contractile force which a muscle can produce is to create a so-called "active state". This is achieved by rapid direct stimulation of an isolated muscle, which thus generates a tetanic isometric contraction; as soon as this is generated a stop is released to allow a small shortening against a considerable load. In this way the force of the fully activated muscle can be measured as any elastic properties of the muscle have been fully stretched before the measurement is made.

Range of contraction

Readers may have noticed that the range of shortening, which the sarcomere can produce, 58% from fully extended, is very close to the range necessary to effect a full movement at the relevant joints, 57% as measured on cadavers. This is a remarkable correlation between observations made many years ago with a scalpel and ruler and recent ones which required x-ray diffraction and electron microscopy. It means that the number of sarcomeres a muscle cell acquires in series along its myofibrils is the distance its attachment must move in millimetres multiplied by 500; since each sarcomere can contribute about 2 μm of movement. If cells are in series along a fibre in a long flat, or fan shaped, muscle such as sartorius or pectoralis major, they are each contributing their quota of contraction to the extensive movement demanded of such muscles.

It should be noted that the arrangement of sarcomeres in series implies that increasing length does not affect force; it is not a tug-of-war as there is no rope, only one sarcomere pulling on the next which must contract with equal strength or it will itself be stretched. Indeed at the end of 20 000, in a long cell, the force exerted on the collagen through the myotendinous junction cannot be any greater than that which operates at each Z-disc. Also, remember that athletic training does not noticeably change the length of a muscle belly although it increases the girth.

Elasticity

The connective tissues of muscle contain a few elastic fibres and these are more numerous in postural muscles. This elastin and the elasticity due to the arrangement of collagenous tubes around the fibres are referred to as the "parallel elastic component". The "series elastic component" consists of the elasticity due to the tendinous attachments at the ends of the fibres and, probably, some intracellular elasticity. During contraction, the bent or shortened HMM-S_1 segments, suspended from their thick filaments by the inclined HMM-S_2 segments could confer elastic properties on the transient connections between the two sets of filaments.

A single twitch expends much of its energy in stretching the series elastic component, so the time course is prolonged and the maximum tension developed is only a fraction of that which the same fibre, or muscle, produces during a tetanic contraction. The force produced by the cell itself during a single twitch is the same as in each of a short series of tetanic contractions.

Speed of shortening

The applied load will determine the speed with which each succeeding actin molecule will come in range of a myosin. The fastest movement is obtained with no load, while one of about 4 kg/cm^2 prevents shortening and produces an isometric contraction. Between these two extremes there is a characteristic relationship between load P and speed of shortening v, which was defined by Hill (1938) as:

$$(P + a)\ v = b\ (P_o - P).$$

The terms a and b are constants and P_0 is the maximum isometric force that the muscle can produce.

The constants, a and b, vary widely as different muscles in different species have their own inherent speeds of shortening. It has been shown that the ATPase activity of extracted myosin is directly related to the speed of contraction. Many muscles contain cells that differ in their myofibrillar ATPase activity and these cells are scattered throughout the muscle belly. So a histochemical demonstration of such activity gives a mosaic pattern on a transverse, fresh-frozen section (Fig. 24). It is seen that the activity, at the unphysiological pH of 9·5, is graded into high and low; these would indicate fast and slow inherent speeds of contraction.

Apart from the inherent myofibrillar speed of shortening, there are variations due to the pattern of innervation and the deployment of internal membranes. Those cells with a slow response, which can produce a graded contraction, are innervated by *en grappe* endings and their myofibrils do not have M-bands.

Various measurements can be made to indicate the speed of contraction. Contraction time is the interval from the onset of tension to peak isometric tension, while half relaxation time is that elapsing from time of peak tension until half the maximum isometric tension remains. The ratio between the tension developed by a single isometric twitch and that of a short burst of tetanic activity also indicates speed because the slower fibres, which are less affected by the series elastic component attain a higher twitch: tetanus ratio.

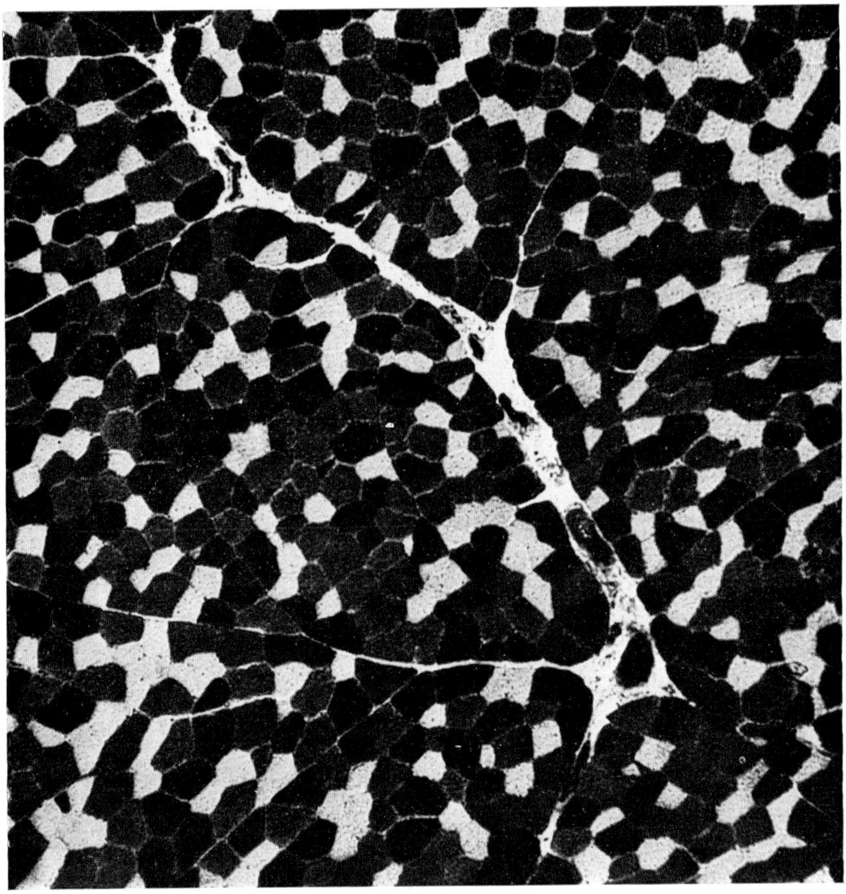

Fig. 24. A transverse frozen section of the pectoral muscle of a horse stained to show myofibrillar ATPase activity, $\times 80$. (Preparation by H. M. Gunn.)

Power developed

The product of the distance moved and the applied load indicates the power developed. The greatest power is developed when the speed of shortening is about one-third of the maximum speed. The characteristic equation above shows that at this speed of shortening the load must also be about one-third of the maximum isometric force. The design of bicycle gears, racing rowing boats, hand-pumps etc. endeavours to ensure that our muscle cells can work at this most effective load.

With no load or a load large enough to prevent shortening, no mechanical work is performed; notice that the word, work, has another colloquial meaning since a sustained but unsuccessful attempt to lift a large weight would be counted as work even though, in the physical sense, none is performed. After such an effort energy is liberated as heat and during the attempt some shortening and heat liberation occurs as the muscle cells stretch their elastic attachments; a truly isometric contraction can only be developed in a physiological laboratory.

III. Excitation and Excitation-Contraction Coupling

A. T-TUBULES

If the quanta of acetylcholine, which produce the excitatory post-synaptic potentials, are sufficient to depolarize the muscle plasma membrane at the motor end plate, a propagated impulse begins to spread towards the two ends of the cell. The speed of this propagation is affected by the diameter of the cell and the electrical capacitance of its membrane; for a cell with a 50 μm diameter it will be about 5 m/sec. Consequently the whole surface of a 5 cm long cell is depolarized in less than 5 milliseconds.

The ionic changes taking place at depolarization are the same as at any other excitable membrane (Chapters 17 and 22). As long as 1948 Hill predicted that a process not a substance must be concerned with coupling the surface membrane changes to the initiation of contraction. He calculated that the time needed for ionic concentrations, such as those obtaining at the surface, to penetrate by diffusion to the middle of the cell was too long to explain the onset of contraction in the central myofibrils. Consequently the rediscovery of a system of transversely oriented tubules, derived from the plasma membrane of the sarcolemma, which penetrate the whole cell between its myofilaments, has been taken to provide the predicted link between the surface and the internal myofibrils. These transverse or T-tubules were first seen after impregnation by gold salts by E. Verrati in 1902, but were ignored until they were shown by electron microscopy. In mammalian skeletal muscle, the T-tubules enter the fibre just on the A-band side of the plane of the A-I junctions of the sarcomeres, and they insinuate themselves between the myofibrils in this plane. Thus there are two T-tubules in relation to each sarcomere (Fig. 26).

It has proved difficult to demonstrate the actual continuity between T-tubule and surface membrane in mammalian skeletal muscle, as the fine T-tubules, 20 nm in diameter, readily break up into a line of spherical or sausage-shaped vesicles. However, their continuity in fish muscle, the ability of ferritin to penetrate the T-tubules of amphibian muscle, and the demonstration of an internally propagating stimulus to contract the half sarcomeres at that level after weak stimulation of the surface at the A-I junction, all imply a similar continuity. The latter evidence is particularly convincing because in other vertebrates, e.g. amphibia, the T-tubules are at the level of the Z-discs (Fig. 25) and correspondingly the site of maximum sensitivity moves to this level. In mammalian cardiac muscle the T-tubules are much larger, 0.1 μm in diameter, and their openings can be readily demonstrated in sections or on freeze-etched preparations of the surface.

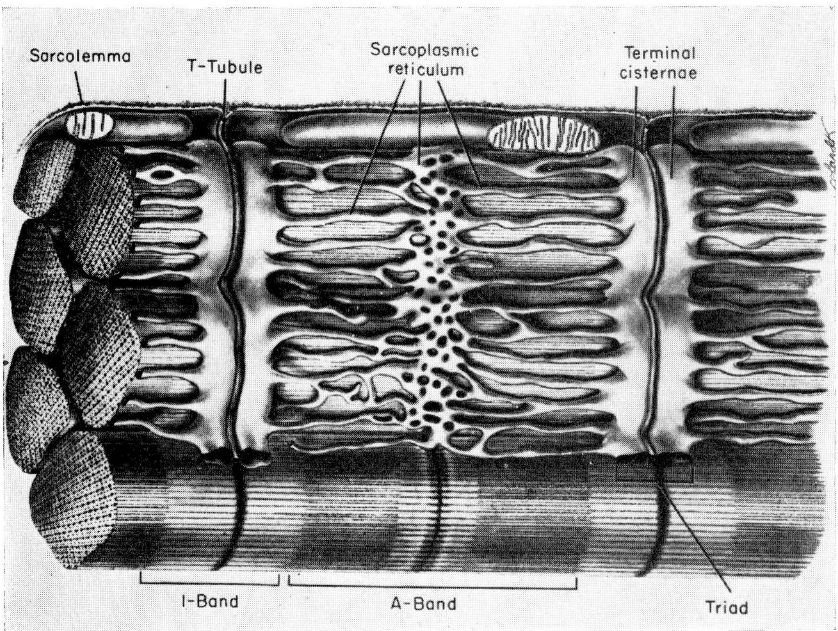

Fig. 25. Diagram showing the arrangement of T-tubules and sarcoplasmic reticulum in frog muscle. (With permission, from Fawcett, D. W. and McNutt, N. S. (1969), *J. Cell. Biol.* **42**, 1.)

B. SARCOPLASMIC RETICULUM

The smooth endoplasmic reticulum in skeletal muscle is arranged in a pattern quite as remarkable as that of the T-tubules. Called sarcoplasmic reticulum (SR), pairs of transversely disposed networks embrace separately the A and I-bands of the myofibrils. These networks are separated by the T-tubules at the A-I junction, and at these contacts the SR is dilated to form large terminal cisternae applied to the T-tubules. The combination of two terminal cisternae and an interposed T-tubule is sometimes referred to as a triad. There is no good evidence of continuity between the extracellular space in the T-tubule and the lumina of the SR, but there are signs that the intervening cytoplasm is modified because it shows a periodic electron density. A similar beaded appearance is observed if a terminal cisterna makes a contact with the surface membrane. The terminal cisternae may contain dense granular material and after oxalate precipitation there is some indication of a Ca^{2+} content. Otherwise there is little sign of structural specialization in different parts of the SR.

In those muscles which have T-tubules opposite the Z-discs, there is only one SR network applied to each sarcomere, but this forms a much finer mesh opposite the M-line (Fig. 25).

C. FUNCTION

The present theory is that the impulse at the surface causes electrically transmitted changes in the T-tubules, which then activate the SR through the connections with the terminal cisternae. The activated SR then releases Ca^{2+} into the sarcoplasm around the myofibrils. The Ca^{2+} diffuses into the myofibril and, because troponin is present, causes the HMM myosin and actin to combine, split ATP and produce a contraction. Relaxation is achieved by the active pumping of Ca^{2+} by the membranes of the SR, so that this cation is sequestered in the lumina of the SR and the concentration in the sarcoplasm is reduced to less than 10^{-6}M.

This theory stands mainly on the following evidence:

(a) The direct demonstration of non-propagating impulses, which will spread into the cell at the level of the T-tubule opening, whether this is at the A-I junction or at Z-disc level.

(b) The dependence on Ca^{2+} concentration for myosin-actin interaction can be shown *in vitro*. No other ion exerts the same effect at the low concentration of 10^{-4}M.

(c) In large crab muscle, the use of micropipettes to injections can show that only Ca^{2+} in low concentration will produce a transient contraction.

(d) Centrifuged homogenates of muscle produce a fraction which will cause the relaxation of isolated myofibrils or *in vitro* actomyosin combinations. This relaxing factor can be shown to be capable of removing Ca^{2+} from the medium if it is supplied with energy in the form of ATP.

(e) Sections of the centrifuge pellet of relaxing factor shows that it has been formed from smooth cytomembranes.

(f) The pumping back of Ca^{2+} to concentrations of 10^5 times that obtaining in the surrounding sarcoplasm requires work and so it may be significant that the SR membranes can be shown to possess ATPase activity.

(g) "Fast" muscle, that is those with short excitation-contraction delays have more profuse and more regularly arranged T-tubules and SR than "slow" muscles. Some fish use a high frequency sound to

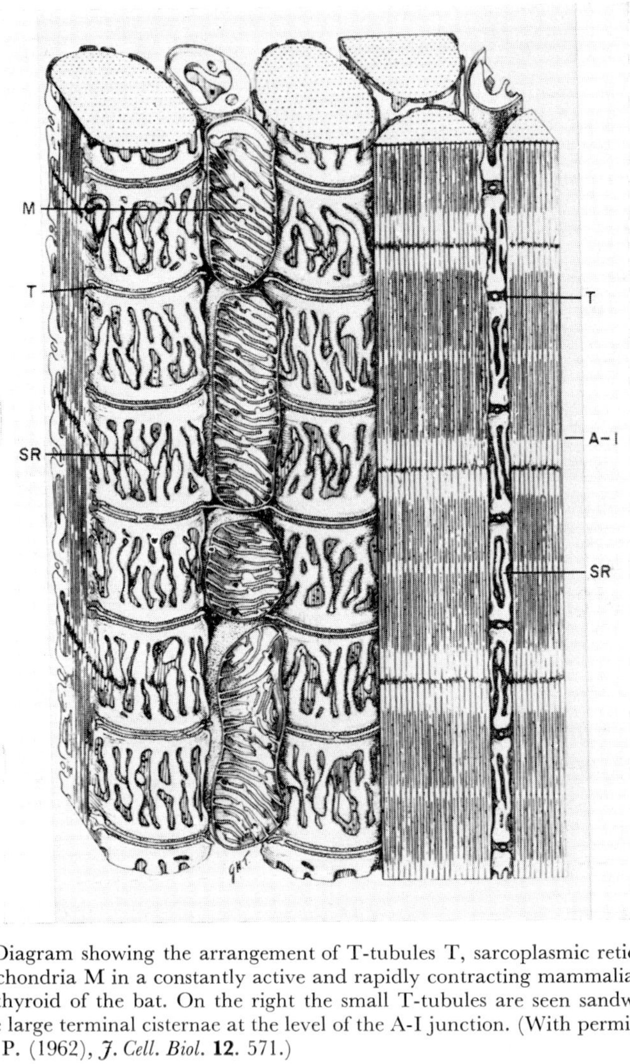

Fig. 26. Diagram showing the arrangement of T-tubules T, sarcoplasmic reticulum SR and mitochondria M in a constantly active and rapidly contracting mammalian muscle, the cricothyroid of the bat. On the right the small T-tubules are seen sandwiched between the large terminal cisternae at the level of the A-I junction. (With permission from Revel, J. P. (1962), *J. Cell. Biol.* **12**. 571.)

echo-locate their depth in muddy waters, and most bats use the same method to fly in the dark; the muscles involved, swim-bladder and larynx have a rich complement of T-tubule and SR (Fig. 26). A rapidly contracting insect muscle even has its SR penetrating between the myofilaments at the M-line level.

IV. Metabolism

The ADP produced at each contraction is rapidly restored to ATP by receiving a high energy phosphate moiety from creatine phosphate, which thus acts as the immediate energy store. Creatine phosphate levels are restored, during relaxation, by the same ATP-ADP transformation which operates in the myofilaments. So the basic source of energy is the phosphorylation of ADP, which can be achieved either aerobically by the citric acid cycle or anaerobically by the degradation of glycogen or glucose to lactate (Fig. 27).

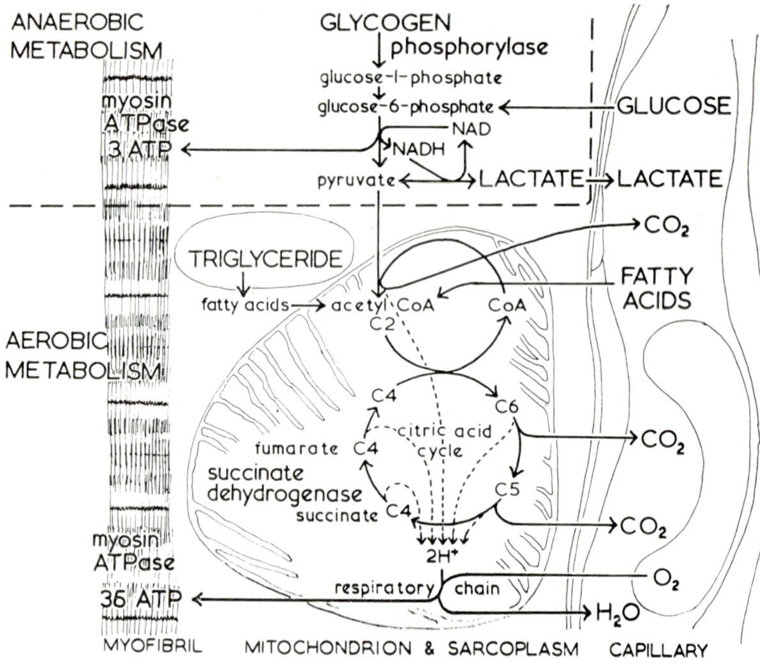

Fig. 27. An outline of aerobic and anaerobic metabolism in skeletal muscle. The activities of the three enzymes mentioned, myosin ATPase, succinate dehydrogenase and phosphorylase, can be used to characterize a fibre by histochemical methods. (Designed by A. S. Davies.)

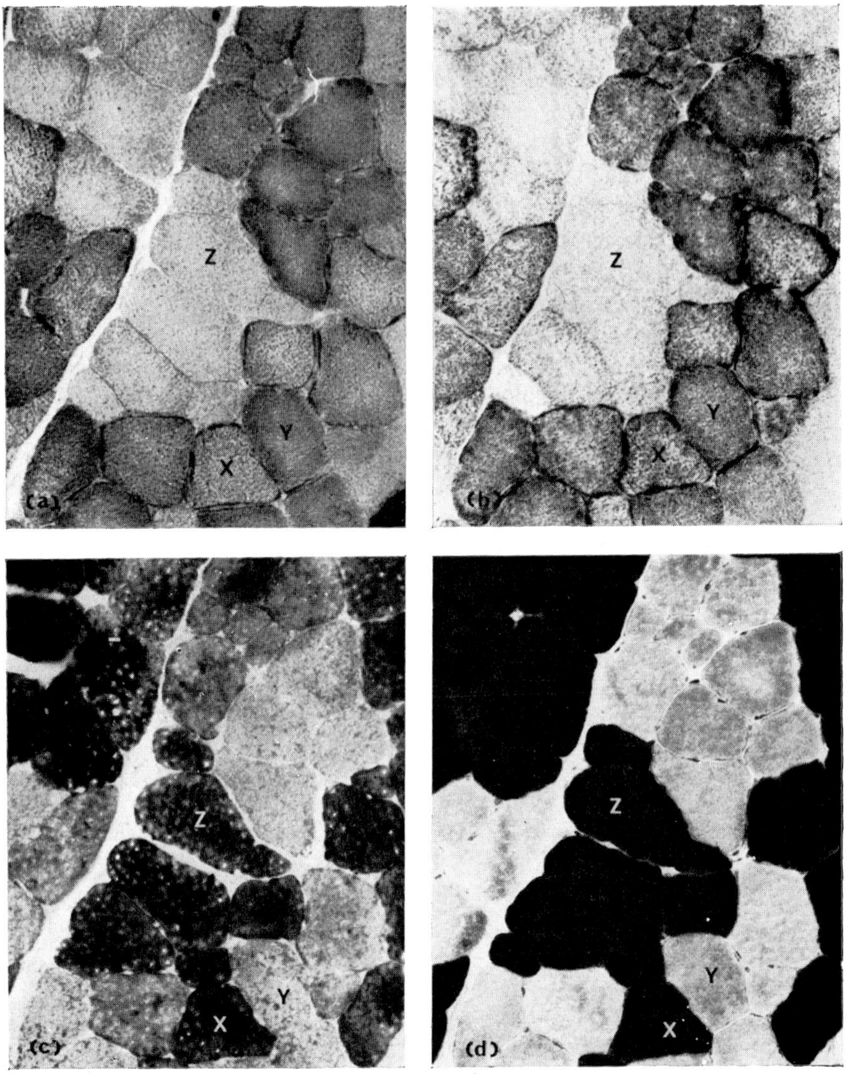

Fig. 28. Serial frozen transverse sections of pig diaphragm stained for (a) phospholipids by Sudan Black, (b) succinate dehydrogenase, (c) glycogen phosphorylase, and (d) myosin ATPase, $\times$ 152. Note that the cell X has a strong reaction in all sections, but those like Y which react strongly for phospholipids and succinate dehydrogenase usually are low in phosphorylase and ATPase: conversely cells like Z are high in phosphorylase and myofibrillar ATPase but low in phospholipids and succinate dehydrogenase. (Preparation by A. S. Davies.)

Like most other tissues, skeletal muscles use both aerobic and anaerobic pathways, but its cells are usually adapted to use predominantly one of these methods. In most muscles, cells using mainly aerobic and those using anaerobic methods are randomly mixed, so homogenization and enzyme analytical methods can only reveal an overall content. Structural and histochemical methods can however distinguish the metabolic characteristic of single cells and give, in transverse section, the mosaic pattern shown in Fig. 28. The features listed in Table 1 can all be assessed on serial fresh-frozen sections of the same cell and, in combination, they indicate whether the metabolism is dominated by the aerobic or anaerobic pathways. Many cells, in a given specimen, fall into these two categories but cells using both methods are present. In electron micrographs, the mitochondrial content and the elaboration of their cristae are the factors that show the metabolic pattern. If the mitochondria are sparse, they are ovoid, $1{\cdot}0 \times 0{\cdot}5$ μm, and present only near the end plate and scattered randomly amongst the myofibrils; when present in larger numbers, subsarcolemmal accumulations appear; and when numerous, they form pairs of bracelets surrounding the myofibrils at either side of the level of the Z-disc.

TABLE 1. Metabolic properties

	Aerobic	Combined	Anaerobic
1. Mitochondria	numerous	numerous	sparse
Phospholipids Mitochondrial enzymes e.g. succinate dehydrogenase cytochrome oxidase	+	+	−
2. Myoglobin	+	+	−
3. Colour	"red"	"red"	"pale"
4. Capillary blood supply	rich	rich	poor
5. Fibre diameter	lesser		greater
6. Intracellular lipids	more		less
7. Glycogen stores	usually less		usually more
Phosphorylase activity	−	+	+
UDPG—Glycogen synthetase activity	−	+	+

All the features in Table 1 can be functionally related to each other. Thus the colour classification into "red" and "pale" (Ranvier, 1873) is due to the content of iron-containing enzymes, such as mitochondrial cytochromes, and myoglobin. Such "red" cells receive a richer capillary blood supply which augments the same colour in a whole muscle containing many such cells. The tendency towards smaller diameter in "red" cells favours interchange with the capillaries. Lipid content varies with species and nutritional state. A phospholipid stain, such as Sudan Black on frozen aldehyde-fixed muscle sections, for light microscopy or an EM assessment will usually serve to show mitochondrial content, and thus indicate the oxiditative capacity of any particular muscle cell.

V. Classification of Muscle Cells

The foregoing descriptions have often included references to features which vary between cells, such as diameter, innervation method, profusion of T-tubules and SR, felderstructur and fibrillenstructur, absence of M-band, thickness of Z-disc, myofibrillar ATPase activity, redness and paleness and the important time course of contraction.

Confusion has arisen because of premature attempts to link all these factors in a functionally relevant fashion; in particular, the common association of "redness" and slowness, although true for postural muscles, which are constantly active and need not produce rapid movements, is demonstrably false in other situations. Indeed the very fast laryngeal muscles are often "red".

It would appear that metabolic pattern and contractile speed are independently variable, but that most of the factors can be shown to be associated with one or other of these two variables. Table 1 has shown the dependent linkages related to metabolism, and a similar correlation is attempted in Table 2 for contractile properties. Again,

Table 2. Contractile properties

	Slow	Twitch	
		Slow-twitch	Fast-twitch
1. Contraction time	300 ms	80 ms	20 ms
Twitch-tetanus ratio	0·6	0·25	0·15
2. Innervation	en grappe	en plaque	en plaque
3. T-tubule and SR	sparse	more profuse	very profuse
4. Myofibrils	felderstructur	felderstructur	fibrillenstructur
5. M-bands	absent	present	present
6. Myofibrillar ATPase activity	low	low	high

most of the correlations have functional implications; for instance a rich SR, embracing myofibrils, requires a fibrillenstructur organization. The relations between ATPase activity and speed of contraction does not appear to have been confirmed on single cells, but fast muscles contain a large proportion of high ATPase cells and the myosin extracted from such muscles has a higher activity than that from slow muscles.

Muscle cells also exhibit varying thickness of Z-discs, with thick discs being associated with small mitochondria-rich muscle fibres by Schiaffino *et al.*, 1970. However, the evidence for such an association does not seem to be strong and this feature could be related to contractile properties.

Even between species, let alone families, orders and classes of vertebrates, there are great variations in the contractile and metabolic properties of muscle cells. Quantitative histochemical methods are not available, and so it is impossible to compare accurately the properties of muscle cells in different species. There are, however, obvious qualitative differences, for example the smallest, reddest muscle cell in the cow is larger and contains fewer mitochondria per unit of volume than the largest, whitest cell in a mouse. These distinctions are probably due to the higher metabolic rate in the smaller warm blooded mammal.

Great interest has been aroused by the demonstration that the structure and function of muscle can be affected by its innervation pattern (Buller and Lewis, 1965). It appears certain that a slow-twitch fibre reinnervated by a nerve, which had been supplying a faster muscle, will develop faster properties. Long periods of cross-innervation are necessary, and often the nerves are transposed before growth is completed. Close (1969), for instance, transfers the nerves between the slow soleus and faster extensor digitorum longus muscles in three-week-old rats and shows that their contraction time has changed after 480 days. In such experiments, it is reasonable to accept the evidence showing that myofibrillar ATPase has been modified by the "trophic" influence of the nerve. In contrast, suggestions that a similar influence affects the aerobic-anaerobic metabolic pattern should be viewed with care. Such enzymes are affected by exercise and denervation, they cannot be quantitatively assessed between different cells and there is no independent physiological evidence, such as is obtained for speed of contraction. The possibility that the observed effects are merely an enzymic adaptation to a different functional state rather than the direct influence of the nerve, should not be overlooked. The structural ATPase of the myofibrils, on the other hand, is much less labile, and here after a long interval, the processes of turnover and growth could replace the myosin with a neurally induced variant.

This conclusion leads to the expectation that the cells of a particular motor unit will all have the same level of ATPase activity, and the mosaic pattern seen in transverse sections can be attributed to the fact that the muscle fibres of a motor unit are scattered throughout a muscle. Since the proportion of myosin ATPase low fibres increases in post-natal growth (Davies, 1972), particular motor neurones may be changing their trophic influence or a labile terminal innervation pattern must be postulated. Henneman and Olsen (1965) have shown that motor nerve fibres supplying the cat soleus and gastrocnemius innervate units with differing contraction times and that the slower conducting, smaller diameter nerve fibres supply slower twitch units than the larger fibres.

VI. Development, Growth and Regeneration

A. EMBRYONIC DIFFERENTIATION

Mammalian skeletal muscle cells arise from mesoderm, usually in the myotomes of the somites or in the branchial arches. At the outset mononucleated spindle-shaped cells, held close together by zonulae adhaerentes, multiply by mitosis; these myoblasts soon become oriented in parallel array to form recognizable muscle primordia. Into the limbs and other sites, such as the tongue, they may migrate as single unrecognizable cells to proliferate *in situ* and so form masses of myoblasts. Only the later demonstration of their nerve supply and phylogenetic analogy reveals the possible pathways taken by these early (highly speculative) migrations. Many consider that myoblasts are induced in the mesoderm of the limbs and tongue, and that the courses of the nerves are merely phylogenetic remnants.

Suddenly many of the myoblast aggregations become multinucleated myotubes and, for the first time, myofibrils are detected under the sarcolemma. As soon as myofibrils appear they contain the same size of sarcomere as in the adult. The surrounding sarcoplasm contains randomly oriented filaments, but there is evidence that the true myofilaments are assembled in the actual sarcomere (Larson *et al.*, 1969). The earlier stages of differentiation can be envisaged even though, as yet, they cannot be demonstrated. Imagine the following steps; activation of the DNA operons related to myofibrillar proteins, production of messenger RNA, development of ribosomes and polyribosomes, synthesis of the two polypeptide chains of myosin and those of actin, troponin, etc., coiling and assembly to form myosin and G-actin, coalescence to form filaments but with lengths precisely determined by tropomyosin. The sarcomere is the only complex structure, visible by light microscopy, whose molecular pattern of development can be so envisaged on the basis of *in vivo* and *in vitro* observations (p. 499).

The elongating myotubes acquire more nuclei; although division of myotube nuclei is never observed, and they all contain the 2n quantity of DNA. Their peripheral cross-striations become more evident as more myofibrils, in register, are added. As each developing myofibril reaches its adult thickness, new myofibrils appear, probably by the splitting of existing ones. The T-tubule and SR systems are relatively late in developing; in the cat, they appear just after birth.

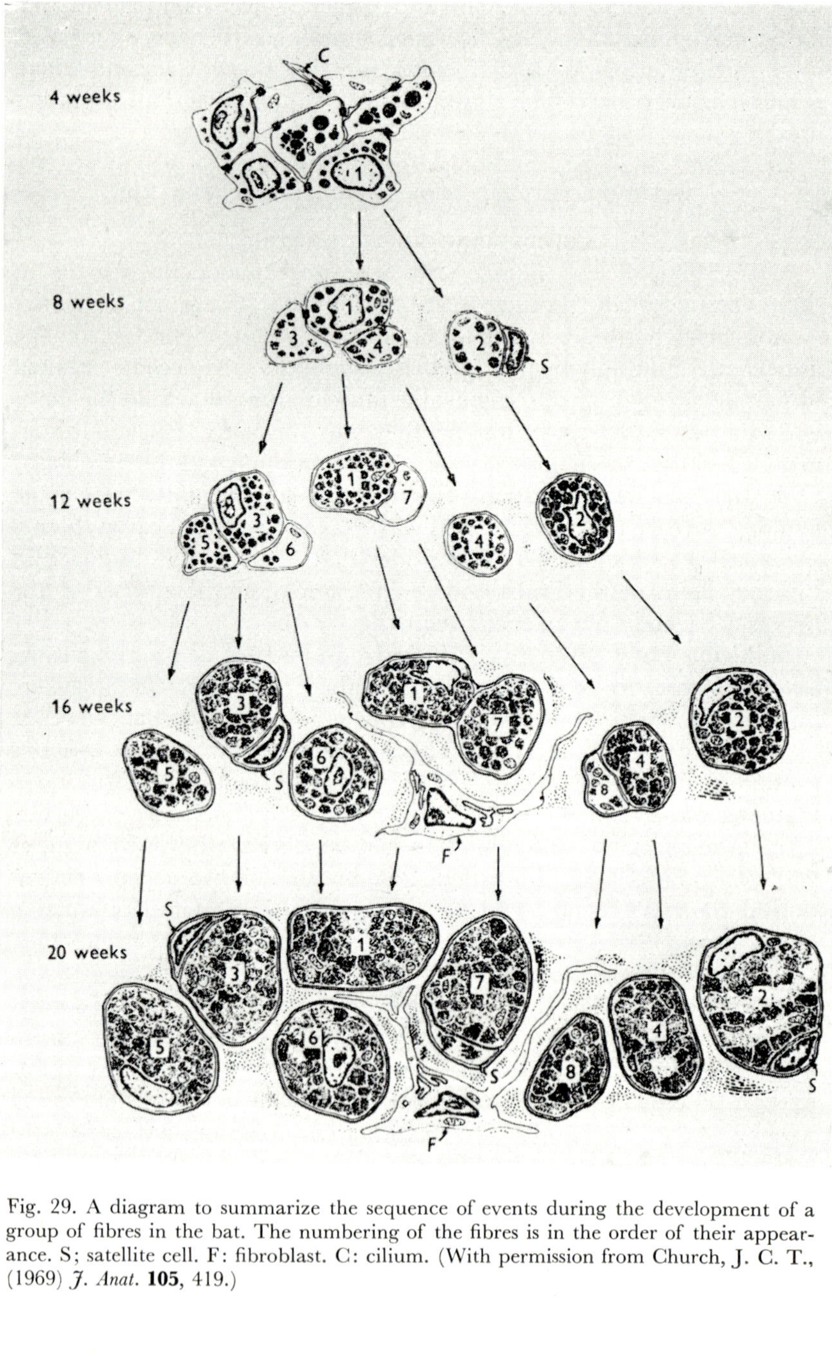

Fig. 29. A diagram to summarize the sequence of events during the development of a group of fibres in the bat. The numbering of the fibres is in the order of their appearance. S; satellite cell. F: fibroblast. C: cilium. (With permission from Church, J. C. T., (1969) *J. Anat.* **105**, 419.)

Separation of myotubes with the formation of endomysial and other tubes begins before the development of new myotubes ends. Once a basal lamina is defined, satellite cells can be recognized as mononuclear myoblasts, not containing myofilaments, within these sheaths. By division and fusion with myotubes, these cells contribute new nuclei to the developing muscle cells. Their division and subsequent fusion in this situation can also produce new myotubes which separate and form their own investing sheath of basal lamina (Fig. 29). Later development compounds proliferation of myofibrils and migration of nuclei to subsarcolemmal positions.

The number of myotubes, or endomysial sheaths which an adult muscle contains, is established in larger mammals before birth. Subsequent growth in length and diameter is adequate to explain the increase in muscle mass, although very few workers have had the courage to confirm MacCallum's (1898) counts of embryonic and adult human muscles. There is no evidence that muscle cells, originally in single array, slide past each other to produce series of cells in the adult; on the contrary, the pattern of motor-end-plate distribution, revealed as soon as cholinesterase becomes demonstrable, stays virtually the same throughout growth. Chapter 15 gives further details concerning formation of muscle in the embryo.

B. DIFFERENTIATION IN CULTURE

Myoblasts isolated from myotomes by mincing and trypsin digestion will divide rapidly in flat culture dishes. For some unexplained but interesting reason, the culture plate must be primed by coating with reconstituted collagen or by allowing fibroblasts to inhabit the plate before they are replaced by myoblasts. If satisfactory conditions are attained, the proliferating myoblasts will form myotubes and as they appear within hours, and as mitoses are not seen in myotube nuclei, it is evident that the myotubes have arisen by fusion of mononuclear myoblasts. As in the embryo, myofilaments appear only in the peripheral cytoplasm of myotubes.

The recognition of DNA synthesis, by brief pulse labelling of the culture with ^{3}H thymidine, and of myosin by fluorescent labelling, allows a detailed analysis of the life cycle of these myoblasts (Bischoff and Holtzer, 1969). As shown in Fig. 30, after the mitosis the cell may enter another division cycle or it may fuse to form a myotube or join an existing myotube. Because neither labelled nuclei, less than 8 hours after labelling, nor 4n DNA nuclei are found in myotubes, it can be concluded that myoblasts can only fuse in the G_1 stage of their cycle. Further, because all myoblasts in the G_1 stage do not fuse, it must be assumed that only a proportion develop a "fusion factor" at the time

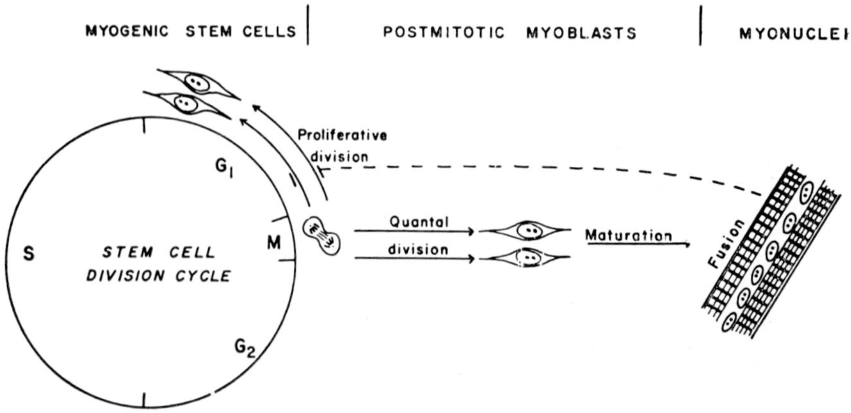

Fig. 30. A diagram to show the myogenic cell cycle which may continue through proliferative divisions, but at some stage a "quantal" division occurs and so the daughter cells then fuse to form a myotube. (Reprinted with permission from Bischoff, R. (1970). In "Regeneration of Striated Muscles," Excerpta Medica.)

of, or soon after, that division; otherwise all the stem cells would be lost.

It has been stated that cells that can synthesize DNA cannot synthesize myosin and *vice versa*, but the widespread distribution of myosin and the demonstration of mitoses in cardiac muscle cells reveal these as overstatements. Nevertheless, myonuclei in the syncytium never show mitosis or autoradiographic evidence of DNA synthesis; in this they conform with other syncytial nuclei in the placenta and osteoclasts, which also arise by fusion. If myofibrillar synthesis begins before fusion, and if the coalescence of two multinucleated myotubes can take place, many of the difficulties of interpretation are avoided. Perhaps the synthesis of "fusion factor" is coupled to that of the special muscle proteins so that a skeletal muscle cell is always ready to receive any neighbour who is engaged in the same business. These intriguing matters are being thoroughly studied at present (Holtzer, 1970).

C. REGENERATION

After muscle cell damage, the myofibrillar material coagulates into irreversible contractures. This response is elicited by ischaemic, physical, toxic and thermal agents, and subsequent events are not affected by the nature of the causative agent. Some cells are left in endomysial tubes after the contained muscle cell has withdrawn; these cells were shown in the original account of regeneration by W. Waldeyer in 1865 and they were seen during the direct observation of the process by

Speidel (1938). It is likely that these residual cells represent the unaffected satellite cells; although it has been suggested that satellite cells are not normally present and that they represent a response to injury. This suggestion would demand that an ordinary myonucleus should wrap itself up in a new plasma membrane at the time of injury.

Fig. 31. Two regenerating fibres are shown in this section of adult pig muscle, the one on the left is in the myotube stage, whereas the other is more mature with subsarcolemmal nuclei, ×500.

Before regeneration begins, the debris of the original cell is removed by macrophages, which enter from the circulating blood. This process takes 3–4 days, and then gradually the tube becomes filled with dividing and growing mononuclear basophilic myoblasts. These could also come from the circulation, or they could be derived from the transitory "satellite" phase of the life of a myonucleus, but it is more likely that they are the progeny of the pre-existing satellite cells. Evidence tending to exclude the possibility of a circulating origin includes the observation that whole body irradiation, with a protected damaged region, has no effect on repair while irradiation of the site of damage impairs recovery. It is difficult to exclude the possibility that syncytial myonuclei can reconstitute themselves into mononuclear dividing cells; but the discovery of satellite cells in control material, the absence of any evidence that syncytial cells divide, and the observation that satellite cells are present on the repaired fibre at the site of the original lesion, all indicate that satellite cells are the source of the myoblasts.

The subsequent stages of fusion, appearance of peripheral myofilaments and central nuclei, growth in diameter, migration of nuclei to the subsarcolemmal region, all mimic those steps in embryos and in culture (Fig. 31). The whole process is completed, under ideal conditions, in less than three weeks. The normal incidence of regenerating cells requires further investigation, because although they are found in about 5% of human post-mortem samples, such samples cannot indicate the rate in a healthy population. In other mammals, muscle damage seems more common, for example unaccustomed exercise in a horse will produce severe myoglobinuria due to massive destruction of muscle cells.

After injury, the effectiveness of muscle regeneration depends on local conditions such as blood and nerve supply, and the distance separating the damaged portions of the muscle. Scar formation obstructs regenerating cells, but does not prevent them attempting to bridge the deficiency.

Remarkably extensive regeneration has been achieved by Studitsky (1963), who excised the whole of the triceps surae in rats, minced the muscles and implanted the resulting soup in the original calf. Initially, short segments of endomysial tubes were randomly oriented until they became full of myoblasts, then they rotated into the lines of tension and, by passing through the myotube and muscle fibre stage, produced a new muscle. This had only one belly attached to the Achilles tendon, and although it received a small nerve supply, it did not function and did not reach more than a quarter of the original weight.

D. GROWTH

The relatively slow-growing human body is an exception, but other mammals show remarkable growth during the first few weeks of life. Much of this growth is skeletal muscle as this tissue forms a greater proportion of the body than any other tissue. The instantaneous daily growth rate of single muscle fibres during this early phase of growth is about 10%. This means that the actual amount of accretion is an even greater percentage, since the newly formed muscle will itself grow. All this growth is due to increase in size of cell, since the cellular number is established before birth (p. 499). As the cells grow with the growth of the whole body, they increase with their cross-sectional area increasing by the two-thirds power of the weight of the muscle. Gradually growth in length ceases, and any subsequent increase in muscle mass is directly reflected in growth in diameter. Such growth in girth is indistinguishable from hypertrophy. When a muscle atrophies because of splinter or denervation, it loses contractile material and the fibre diameter diminishes.

Increase in girth

The local factors which control increase in girth are work and load; such exercise-induced growth affects all constituents. Myofibrils increase in number so they continue to fill the larger cross-sectional area without themselves increasing significantly in size; consequently the embracing SR and associated T-tubules must increase. Recent experiments have shown that mitochondria increase in size and number, and this improvement in oxidative capacity probably matches the improved intrinsic circulation and general cardio-respiratory effects of athletic training.

At the molecular level, increasing girth does not pose any particular problems, since the process of myofibrillar production in the embryo and foetus involves the addition of new myofilaments in lateral alignment with existing ones.

Other general factors affect growth in diameter; breed is important, and in humans the mesomorphic types are largely determined by the extent of muscle development. Type of nutrition has its effects, either by increasing intracellular and interstitial lipid, or in protein deprivation by impairing cellular growth. Not all muscles are similarly affected in starvation, and the reasons for sparing some, such as the diaphragm, are probably functional. Vitamin E is necessary for muscle growth and maintenance as deficiency causes necrosis with subsequent attempts at regeneration. Androgenic hormones cause the spurt in muscle growth at puberty, and there are some muscles that are peculiarly responsive to testosterone. A detailed study of hypertrophy in such a sexually dependent muscle, the levator ani of the mouse, is given by Venable (1966).

Increase in length

It has been established above (p. 454) that length is determined by range of movement, but it is difficult to explain the actual mode of increase. As sarcomere length is constant, any elongation involves an increase in sarcomere numbers. Speidel (1938) counted the addition of cross-striations during the growth of myotomes in tadpole tails. He thought that the sarcomeres were added at the myotendinous junction, because he failed to see growth elsewhere along the fibre. It is difficult to envisage a new sarcomere being inserted in the middle of a functional myofibril. Further, the intermyofibrillar attachments at the Z-discs, particularly in the "felderstructur" fibres, suggest that such an insertion would affect all the myofibrils at that level; this should make it observable in a living fibre. At the ends of the cell the oblique attachment of the myofibrils to the invaginations of plasma membrane provide a possible mechanism for the assembly of a new sarcomere before the original myofibrillar attachment is loosened; alternatively, new myoblasts containing organized myofilaments could fuse with the end of the cell.

Attempts to label growth with wire markers and by measuring the separation of *en grappe* end plates indicate that the connective tissue and sarcolemma of a fibre grow interstitially. Because growth in length and girth are always associated and because no specific radioactive amino acid for myofibrillar proteins is available, it has proved difficult to obtain direct evidence on the site of myofibrillar growth. Recently Griffen and his colleagues have used radioactive ADP which is incorporated as a structural part of the thin filament (p. 469); subsequent autoradiographs of glycerol-extracted fibres clearly shows terminal growth (Fig. 32). This subject requires further investigation of its precise mechanism and controlling factors.

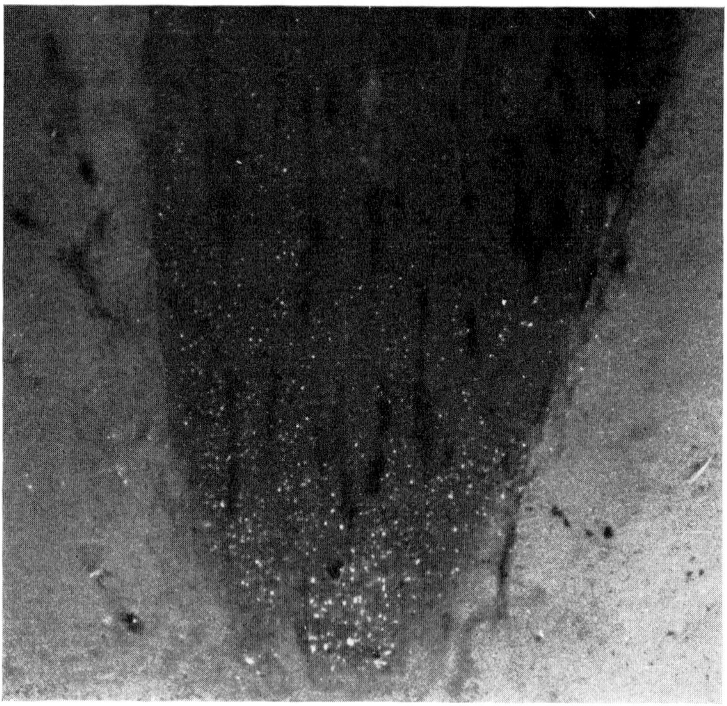

Fig. 32. In this autoradiograph of a longitudinal section of the termination of muscle fibres, in the mouse biceps brachii, the silver grains are glowing white. As radioactive adenosine had been given, the concentration of silver grains at the ends of the fibres indicates the sites of thin filament formation. (Photograph supplied by Dr. G. Goldspink; reprinted with permission from Nature.)

Increase in nuclei

The tibialis anterior muscle of a 30 g rat contains eight million muscle nuclei, but 70 days later when the animal weighs 86 g there are 12 million more such nuclei (Enesco and Puddy, 1964). Similarly, Moss (1968) shows that the pectoral muscle of New Hampshire cockerels, from 0–226 days after hatching, accumulates over 75×10^9 nuclei and that the rate of increase is directly proportional to the increase in cross-sectional area of the fibres. This means that the nuclei are related to a progressively larger volume of sarcoplasm because the cross-sectional area is only increasing in relation to the two-thirds power of the increase in mass. It follows that at the peak rate of growth of over 10% per day, the nuclei contained within the fibre must increase by about 5% during the same period.

Synthesis of DNA and mitoses have been detected in satellite cells by electron microscopy (Moss and Leblond, 1971). It is likely that all mitoses within endomysial tubes are occurring in mononucleated cells, whether they have been temporarily formed from myonuclei for this purpose, or whether they represent a continuing constituent of the fibre. As there are between 5 and 20 satellite cells for every 100 fibre nuclei, between 100% and 25% of these cells must divide daily to produce the cells which will fuse with the syncytium during the rapid phase of growth.

If we assume that satellite cells are a permanent feature of the fibre, and that they represent committed cells, which by division will produce myoblasts destined to fuse, we still have intriguing questions to answer. How does the growing muscle cell inform its attendent satellite cells that new nuclei are required? When a satellite cell divides, does it produce two myoblasts which fuse while other satellite cells are engaged in restoring the satellite population. Or does a satellite cell division produce two different cells, a non-fusing satellite cell and a fusing myoblast? These are general problems which apply equally to differentiating erythroid, epithelial and spermatogenic cells. The availability of tissue culture experiments and the techniques for identifying the contractile proteins, make muscle a convenient model for the investigation of these fundamental aspects of cell biology.

VII. Conclusion

The main aim of this brief review has been to show how, in this tissue, molecular structure, functional properties, histological features and topographical anatomy are interrelated. Similar relationships must apply in all other tissues, but at present striated muscle is unusual in that there is sufficient quantitative biophysical data to permit an attempt at such correlations. Another purpose has been to show the variety of cells which are grouped under this title, they are built of the same materials but adapted for their role in particular muscles of particular animals. Space has prevented reference to further adaptation of striated muscle to form vertebrate myocardium and Purkinje and nodal cells in mammals and birds. The most bizarre adaptation of skeletal muscle, the electric organs of fishes, also had to be omitted.

Finally, consideration should be given to outstanding problems in this intensely investigated field. The crucial question of how a muscle contracts remains unanswered, as there is little knowledge of the actual changes occurring in HMM-S_1 segments during the splitting of ATP. If these proteins could be crystallized, their three-dimensional structure could be elucidated by x-ray diffraction, and this alone might suggest a likely functional mechanism for either shortening or bending. Otherwise direct electron optical methods on contracted and relaxed filaments, using improved preparative techniques, will have to be employed; present methods of preparation do not permit such studies on small single protein molecules. The sarcomere, as currently conceived, is an unstable structure: in series along a myofibril any slight weakness in the sarcomere would be exploited by its stronger neighbours, and as the developed tension falls when length increases, a vicious circle would be produced. Although the attachments between Z-discs and their attachment to the sarcolemma confer some stability, it is surprising that no internal structure appears to limit extension. Is there a fibrous protein "myofibrillin" which prevents the disaster of stretching a sarcomere beyond 3·8 μm?

At the cellular level, the development of the synthesis of the characteristic muscle protein should continue to reveal features of differentiation. As the process can be studied in embryos, culture, growth, regeneration and hypertrophy, evidence may be obtained which has escaped detection in other more mobile and less co-operative cells. In particular, the stage when a precursor cell becomes a myoblast and when it loses the potential to perform any other function should be determined. The most intriguing problems of growth relate to its control and, for the growth of meat, these problems are not only academic. What stimulus controls the production of new myonuclei? Can muscle cells be induced to grow without a proportionate increase in myonuclei? Is the volume of muscle sarcoplasm supported by a single myonucleus similar in all animals at all stages of growth? What limits the number of myotubes to be formed? What controls their growth in diameter? How is growth in length controlled at the cellular level?

Similar problems arise for the control of the metabolic machinery of muscle cells. They vary, as described, between species, between muscles, and within single muscles. How much of the phenotype is determined by the genotype, and how much by local environmental factors such as vascularity, tensile forces and patterns of utilization? What is the range of effects which can be induced by the nerve supply? When muscle cells lie in series, do successive ones have the same contractile and metabolic structure?

Much of the interest remains in attempting to relate macromolecular structure to the function of a whole animal. Do all muscles utilize the whole range of contraction of a sarcomere? One-third of the weight of a quadruped passes through one forelimb, and this is borne by the serratus ventralis muscle. At rest, is this muscle stretched to its weakest point of active contraction? It is known that cardiac muscle sarcomeres do not normally stretch beyond 2·2 μm, so they operate only on the ascending limb of the length-tension curve; perhaps some skeletal muscles are similarly restricted. Little attempt has yet been made to synthesize the knowledge of muscle cell structure, the arrangement of cells in a muscle, and the performance of that muscle in a particular muscle action. Such a synthesis will place the knowledge of ergonomics on an established scientific base and may lead to improvement in athletic performance and the design of manually operated machines.

References

Adams, R. D., Denny-Brown, D. and Pearson, C. M. (1962). In "Diseases of Muscle", 2nd Ed. Kimpton, London.

Bendall, J. R. (1969). "Muscles, Molecules and Movements". Heinemann Educational Books, London.

Bertaud, W. S., Rayns, D. G. and Simpson, F. O. (1968). *Nature, London* **200**, 381.

Bischoff, R. and Holtzer, H. (1969). *J. Cell Biol.* **41**, 188–200.

Bourne, G. H. (1960). "Structure and Function of Muscle", Vols. I, II and III. Academic Press, New York and London.

Boyd, J. D. (1960). In "Structure and Function of Muscle" (G. H. Bourne, ed), Vol. I, pp. 63–85. Academic Press, New York and London.

Buller, A. J. and Lewis, D. M. (1965). *J. Physiol.* **178**, 343–358.

Caspar, D., Cohen, C., and Longley, W. (1969). *J. Mol.Biol.* **41**, 87–108.

Close, R. (1969). *J. Physiol.* **204**, 331–346.

Couteaux, R. (1960). In "Structure and Function of Muscle" (G. H. Bourne, ed.), Vol. I. Academic Press, New York and London.

Crawford, G. N. C. (1954). *J. Bone Joint Surg.* **36B**, 2: 294–303.

Davies, A. S. (1972). *J. Anat.* **113**, 213–240.

Diamond, J. and Miledi, R. (1962). *J. Physiol.*, **162**, 393–408.

Ebashi, S. and Endo, M. (1968). *Progr. Biophys. Mol. Biol.* **18**, 125–183.

Ebashi, S., Endo, M. and Ohtsuki, I. (1969). *Quart. Rev. Biophys.* **2**, 351–384.

Elliot, G. F., Lowy, J. and Millmann, B. M. (1967). *J. Mol. Biol.* **25**, 31–45.

Enesco, M. and Puddy, D. (1964). *Amer. J. Anat.* **114**, 235–244.

Gordon, A. M., Huxley, A. F. and Julian, F. J. (1966). *J. Physiol.* **184**, 170–192.

Griffin, G. E., Williams, P. E. and Goldspink, G. (1971). *Nature New Biol.* **232**, **27**, 28–29.
Gwyn, D. G. and Aitken, J. T. (1966). *J. Anat.* **100**, 111–126.
Hanson, J. (1968). *Quart. Rev. Biophys.* **1**, 177–216.
Hanson, J. and Lowy, J. (1965). *Brit. Med. Bull.* **21**, 264–271.
Henneman, E. and Olson, C. B. (1965). *J. Neurophysiol.* **28**, 581–598.
Hill, A. V. (1938). *Proc. Royal Soc. Ser. B.* **126**, 136–195.
Hill, A. V. (1965). "Trails and Trials in Physiology". Arnold, London.
Holtzer, H. (1970). In "Symposia of the International Society for Cell Biology" (H. Padykula, ed.), Vol. IX, pp. 69–88. Academic Press, New York and London.
Huxley, A. F. (1957). *Progr. Biophys.* **7**, 225–318.
Huxley, H. E. (1957). *J. Biophys. Biochem. Cytol.* **3**, 631–647.
Huxley, H. E. (1960). In "The Cell" (Brachet and Mirsky, eds.), Vol. IV, pp. 365–481. Academic Press, New York and London.
Huxley, H. E. (1963). *J. Mol. Biol.* **7**, 281–308.
Huxley, H. E. (1968). *J. Mol. Biol.* **37**, 507–520.
Katz, B. (1966). "Nerve, Muscle and Synapse". New York, McGraw Hill.
Knappeis, G. G. and Carlsen, F. (1962). *J. Cell Biol.* **13**, 323–335.
Landon, D. N. (1970). *J. Cell Sci.* **6**, 257–276.
Larson, P. F., Hudgson, P. and Walton, J. N. (1969). *Nature London* **222/5199**, 1168–1169.
Lowey, S., Slayter, H. S., Weeds, A. G. and Baker, H. (1969). *J. Mol. Biol.* **42**, 1–29.
MacCallum, J. G. (1898). *Bull. Johns Hopkins Hosp.* **9**, 208–215.
Mauro, A. (1961). *J. Biophys. Biochem. Cytol.* **9**, 493–495.
Mauro, A., Shafiq, S. A. and Milhorat, A. T. (1970). "Regeneration of Striated Muscle and Myogenesis". Excerpta Medica, Amsterdam.
Meigs, E. B. (1932). In "Special Cytology" (E. V. Cowdry, ed.), Vol. II, pp. 1089–1125. Hoeber, New York.
Moss, F. P. (1968). *Amer. J. Anat.* **122**, 555–563.
Moss, F. P. and Leblond, C. P. (1971). *Anat. Rec.* **170**, 421–436.
Needham, D. M. (1971). "Machina Carnis". University Press, Cambridge.
Page, S. and Hunley, H. E. (1963). *J. Cell. Biol.* **19**, 369–390.
Pepe, F. (1967). *J. Mol. Biol.* **27**, 203–225.
Perry, S. V. (1967). *Progr. Biophys. Mol. Biol.* **17**, 325–381.
Ranvier, L. (1873). *C. R. Acad. Sci.* **77**, 1030–1034.
Schiaffino, S., Hanzlikova, V. and Pierobon, S. (1970). *J. Cell. Biol.* **47**, 107–119.
Schwarzacher, H. F. (1959). *Acta Anat.* **37**, 217–231.
Speidel, C. C. (1938). *Amer. J. Anat.* **62**, 179–235.
Studitsky, A. N. (1963). In "Cinemicrography in Cell Biology" (G. G. Rose, ed.), pp. 171–200. Academic Press, New York and London.
Tiegs, O. W. (1953). *Physiol. Rev.* **33**, 90–144.
Venable, J. H. (1966). *Amer. J. Anat.* **119**, 271–302.
Wilkie, D. R. (1968). "Muscle". Arnold, London.

Further Reading

Apart from the references included in the text and those pertaining to the illustrations, the following should be consulted for additional information. They provide an adequate entrance to the relevant literature.

General accounts
Short: Wilkie (1968).
Medium: Bendall (1969).
Long: Bourne (1960).

Cytology
Meigs (1932).

Innervation
Tiegs (1953), Couteaux (1960), Katz (1966).

Molecular structure
Myosin: Lowey *et al.* (1969).
Actin: Hanson (1968).
Tropomyosin: Caspar *et al.* (1969).
Troponin: Ebashi and Endo (1968).

Contraction
Huxley (1960, 1968), Perry (1967).

Metabolism
Needham (1971), Schiaffino *et al.* (1970).

Physiology
Hill (1965).

Development and regeneration
Mauro *et al.* (1970), Boyd (1960).

Pathology
Adams, Denny Brown and Pearson (1962).

21 The Muscle Spindle

D. N. LANDON

Institute of Neurology, Queen Square, London, England

I. Introduction

The muscle spindle has been an object of great interest to both morphologists and physiologists since its identification in 1843 by Miescher in the abdominal muscles of the house mouse. These structures were subsequently observed and described by many histologists among whom was Kuhne, who coined the term "muscle spindle". By the 1880s it was generally agreed that amid the mass of normal striated muscle fibres, in many muscles from many species, it was possible to detect small bundles of diminutive fibres containing numerous nuclei in their central regions. The nuclear zone of each bundle was enveloped by a multilamellar sheath, very similar to the perineurial sheath surrounding peripheral nerve fibres, and this was invariably pierced by a small nerve fascicle containing at least one large myelinated nerve fibre which ended as a series of non-myelinated branches among and around the nuclear regions of the small muscle fibres.

Despite agreement on the morphology of muscle spindles, two schools of thought developed concerning their functional significance. One considered that these structures represented pathological regressive changes in groups of previously normal muscle fibres, the other that the small size of the muscle fibres in the spindle and their massed central nuclei were evidence that they were embryonic rests of immature myotubes, perhaps capable of further development if given an appropriate stimulus.

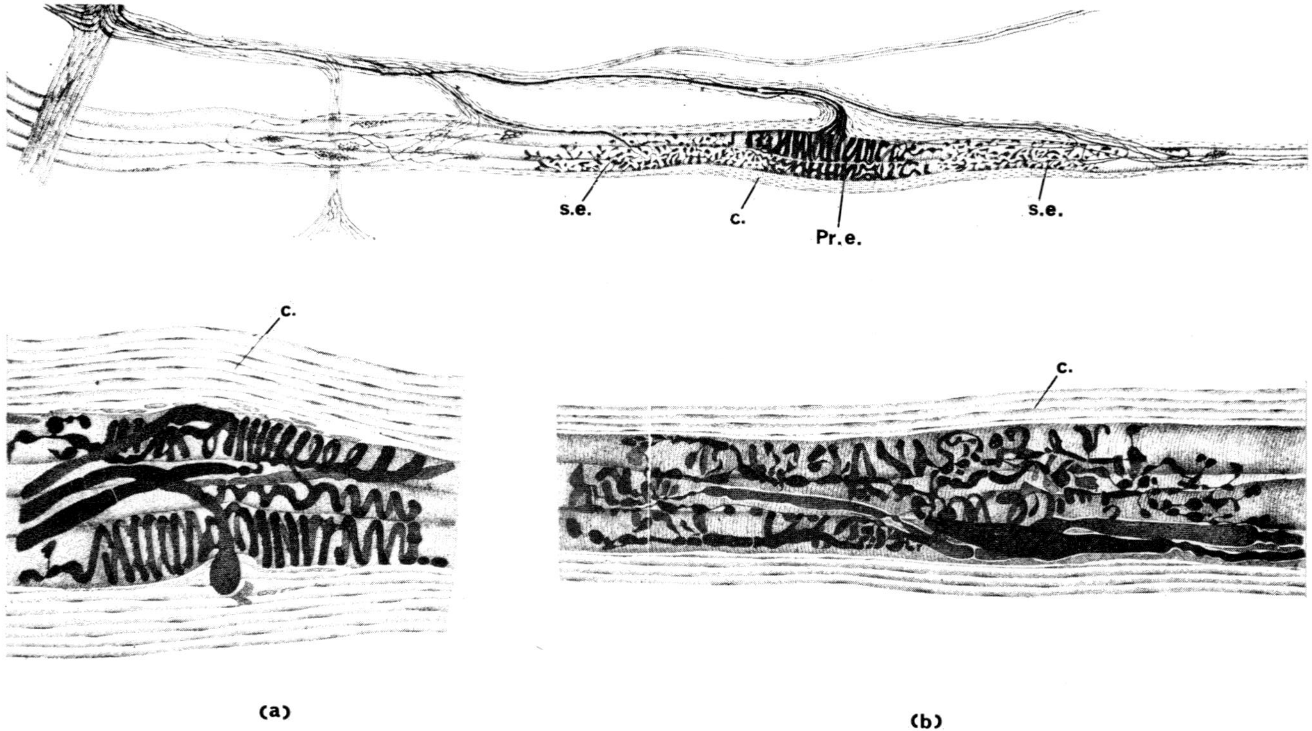

Fig. 1. A gold chloride impregnated preparation of a muscle spindle from a cat (Ruffini, 1898). A bundle of intrafusal fibres are shown enclosed by their mutilamellar capsule *c*. The annulo-spiral primary sensory nerve terminal surrounds the equator of the intrafusal fibres *pr. e* (and in (a) and (b)) and on either side, in a more polar position, are found the

In 1888 Kerschner proposed that the muscle spindle might be a complicated end-organ serving muscular sense. This view was supported by Ruffini, who between 1892 and 1898, using his gold chloride impregnation technique, provided elegant histological evidence of the extensive and complex innervation of the central regions of the spindle (Fig. 1). Experimental evidence for the sensory nature of this equatorial innervation was provided by Sherrington's demonstration that in the degeneration of muscle tissue that followed denervation by ventral root section only the muscle fibres of the muscle spindles and their equatorial innervation were spared, and Batten (1897), in a review of his own and earlier observations, was able to conclude that "the muscle spindle is the organ from which muscular afferent impulses are derived".

These deductions have been confirmed and extended by the observations of numerous later workers during the present century. They have shown that early in embryonic development, and sychronous with the entry of the first nerve fibres into developing skeletal muscles, small groups of myotubes are sequestered within extensions of the perineurial sheath covering branches of the muscle nerve, and that the subsequent differentiation and final morphology of these myotubes is greatly modified by the trophic influences of their extensive sensory innervation. It has also been demonstrated beyond any reasonable doubt that the muscle spindles formed in this way act as length-tension receptors in parallel with the normal "extrafusal" muscle fibres comprising the bulk of the muscle, and that the properties of the equatorial sensory region of the spindle can be modified by contractile activity of the small "intrafusal" muscle fibres, enabling the organ to play an important role in the servo-control mechanisms which govern integrated muscular activity.

The functional characteristics of mammalian spindles will be described below and this description is preceded by a brief account of their morphology, a necessary preliminary to any discussion of the significance of data derived from electro-physiological experiments.

II. Distribution

Muscle spindles are found throughout the skeletal musculature of terrestrial and amphibious vertebrates, but their numbers vary in different muscle groups, as does their spatial distribution within individual muscles. The ratio of muscle spindles to extrafusal muscle fibres is greater in the distal limb muscles and the axial muscles of the trunk than in the limb girdle musculature; the small muscles of the hand and the short suboccipital muscles being particularly rich in spindles, from which it has been inferred the the spindle density in a given muscle is an index of the degree of refinement of movement control of which that muscle is capable. The rotator muscles of the limbs, for example, have rather higher spindle densities than might otherwise be expected from their position along the proximo-distal axis of the limb. Cooper (1966) has shown that the number of muscle spindles in a muscle is probably related to the number of its motor units, and that for most muscles for which counts of spindle numbers are available the total spindle population is approximately 10% of the total number of myelinated nerve fibres entering that muscle.

Within an individual muscle the spindles are seldom uniformly distributed throughout its mass, but segregated to some extent within particular zones or fascicles, for instance to the muscle axis, in (say) the tibialis anterior muscle of the rat; to one fascicle, e.g. the deep slip of the medial head of the rat gastrocnemius; or to one end of the muscle, e.g. the proximal end in human extrinsic ocular muscles. Such instances exemplify a general rule that the majority of spindles in a muscle are located in that portion containing the greatest numbers of small, mitochondria rich, slow-twitch extrafusal muscle fibres (Chapter 20), and this association may well have both functional and ontological significance.

III. Structure

The muscle spindle consists of three distinct sets of structural elements—the capsule, the intrafusal muscle fibres, and the sensory and motor nerve terminals; there are significant differences in the form and arrangement of these components in the spindles of amphibians, reptiles and mammals, and the description which follows refers to the arrangement most commonly seen in mammals.

A. The Capsule

The capsule, the appearance of which has given the muscle spindle its name, is a fusiform many-layered sac, some 200 μm in diameter at its widest point, covering the central 3–4 mm of the intrafusal muscle fibres. The majority of muscle spindles have a single capsule, but tandem spindles can occur in which two or more capsules share some common intrafusal fibres, each capsule being associated with a zone of sensory innervation.

The capsule is divisible into two components, an outer which in all respects resembles the perineurial sheath of the peripheral nerve fascicle with which it is continuous, and an inner "axial sheath of Sherrington" which closely invests the intrafusal fibres in the equatorial regions of the spindle. The outer capsule consists of up to nine or ten layers of flattened cells with fine collagen fibrils between each layer, and is thickest at the equator of the spindle, the numbers of layers of cells progressively diminishing towards the poles until reduced to a single layer, from the end of which some of the intrafusal fibres pass out into the general endomysial tissue space.

In the equatorial region the space enclosed by the outer capsule is considerably larger than the contained intrafusal fibres and their associated nerve endings. This space appears to be fluid-filled and, since Sherrington's demonstration of its continuity with the lymphatic system by retrograde injection of indian ink into the regional lymphatics of the hind limb of a dog, it has been known as the "lymph space". The only confirmation of the presence of such a connection has been the observation by Jahn of lymphocytes within the capsular space of living frog spindles. Brezezinski (1962) has demonstrated histologically that the lymph space contains a metachomatically staining material which he has identified as hyaluronic acid, and ultrastructurally it commonly contains dispersed flocculent matter not seen in the endomysium. Several writers have ascribed a cushioning function to the capsule and its enclosed fluid contents, suggesting that it insulates the mechanically sensitive central region of the intrafusal fibres from stresses other than those delivered by tension along their axes, and the presence of an osmotically active material of high molecular weight within the capsular space may be an important factor in maintaining its normal distension.

The axial sheath of Sherrington is most apparent in the equatorial region and consists of a series of extremely flattened cells, having tight junctional contacts at their contiguous edges, forming a series of individual compartments for the intrafusal muscle fibres and their associated sensory and motor nerve terminals. The cytoplasm of these cells shows evidence of protein synthetic activity and they may be responsible for the production of the metachromatic material of the lymph space. Away from the equator of the spindle the compartments of the axial sheath progressively coalesce and the cells are reduced in number, few persisting beyond the end of the lymph space. Capillaries are seldom seen within the spindle capsule and those which do occur are in general in the equatorial region and may be the terminal branches of the vasa nervorum; elsewhere the capillaries run outside or between the outer layers of the capsule, and such a vascular arrangement may explain the selective resistance of spindle structures to haematogenously delivered local anaesthetics and toxins.

B. THE INTRAFUSAL FIBRES

The intrafusal muscle fibres are 10–30 μm in diameter and vary in number from three to fourteen or more, the spindles in small muscles, e.g. the lumbrical muscles, in general having relatively few intrafusal fibres. In most mammals, the rabbit being an exception, these muscle fibres are clearly distinguishable histologically on grounds of diameter, length and equatorial morphology into two classes, "nuclear bag fibres" and "nuclear chain fibres" (Boyd, 1962). The ratio of the numbers of "bag" fibres to "chain" fibres within an individual spindle varies between species and between different muscles, but a preponderance of "nuclear chain" fibres has been claimed to be associated with fine motor control (Granit, 1970).

The "nuclear bag" intrafusal fibres are usually twice the diameter of the "nuclear chain" fibres, and considerably longer, extending for several millimetres beyond the ends of the capsule into the endomysial tissue space. At the equator the myofilament content of the more polar regions is largely replaced by an aggregation of centrally placed myonuclei, two or more nuclei thick and up to 100 μm or more in length; the "nuclear bag". This zone is flanked by a myotube region of varying extent containing a single row of central nuclei, while further from the equator the myonuclei occupy their normal subsarcolemmal position. Not all the myofilaments are lost in the nuclear bag region, longitudinal continuity between the two contractile poles of the fibre being maintained by a few slender myofibrils which span the "bag".

The "nuclear chain" fibres are of smaller diameter, 10–15 μm, are usually coextensive with the capsule, and have a central nuclear aggregation in the form of a myotube. This consists of a central core of elongated nuclei surrounded by a coherent shell of myofilaments; the myofilament component occupying a much greater proportion of the cross sectional area of the intrafusal fibre than do the few persisting myofibrils which traverse the "bag" region of the nuclear bag fibre (Fig. 2).

Specialized contact areas have been observed between contiguous nuclear chain fibres in the cat (Corvaja *et al.*, 1967) and these may have provided the basis for previous histological reports of the branching or splitting of intrafusal fibres. However, at their ends within the polar extremities of the spindle capsule the nuclear chain fibres often divide into two or more short "tails" through which the fibre is inserted into a local aggregation of connective tissue in close relationship to the surface of an adjacent nuclear bag fibre or the capsule wall.

The two types of intrafusal fibre possesses further ultrastructural differences, the nuclear bag fibres having many of the characteristic features of slow amphibian muscle—few triads, sparse sarcoplasmic reticulum and absent M lines; whereas the nuclear chain fibres have prominent M lines and pseudo-H zones, greater numbers of larger mitochondria and frequent triads, and in most respects closely resemble

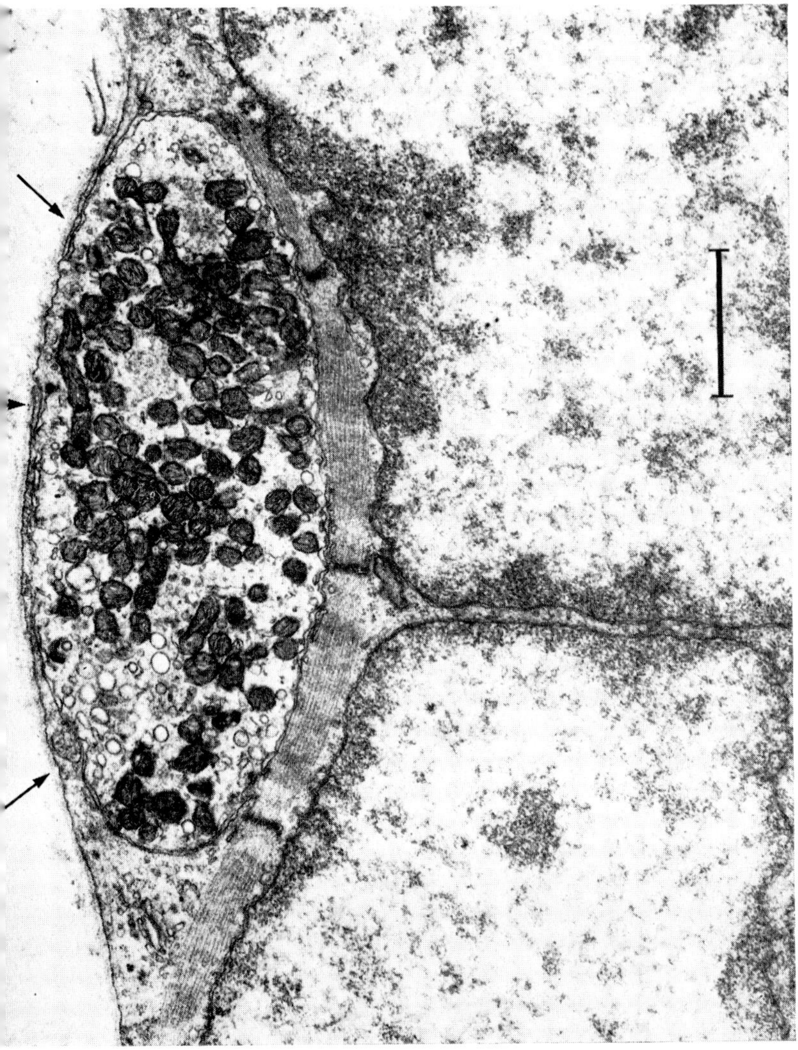

Fig. 2. A longitudinal section through the "bag" of a nuclear bag intrafusal fibre in a rat spindle. The primary sensory ending lies in a trough in the surface of the intrafusal fibre beneath its basal lamina, and is largely covered over on its external surface by thin "flaps" of sarcolemma (arrows) (scale = 1 μm).

mammalian, slow-twitch, extrafusal muscle fibres (Landon, 1966). Such histochemical evidence as is at present available appears to confirm the similarity of nuclear chain fibres to slow-twitch extrafusal fibres, i.e. those with a high content of oxidative enzymes (see Chapter 20), while the nuclear bag fibres with low activities of all enzymes possess the histochemical characteristics of immature extrafusal fibres in neonatal animals. Althought the muscle spindles of the rabbit have been noted to provide an exception to the "bag"—"chain" division seen in other mammals, in that they possess only "nuclear bag" intrafusal fibres, these fibres are in fact distinguishable fine structurally and histochemically into at least two types comparable to the "chains" and "bags" of other species.

A striking feature of the connective tissue surrounding the intrafusal muscle fibres is the presence of large elastic fibres which form an interconnecting network around the intrafusal fibres throughout the capsule and accompany the polar regions of the nuclear bag fibres on their extracapsular course to their insertion. It has been demonstrated in man and in the rat that the elastin is concentrated at the poles of the capsule and in the mid-equatorial region, with rather less in the intervening zones; and in the cat it has been shown that whereas the elastin in the polar regions of the capsule is predominantly associated with the "nuclear bag" intrafusal fibres, at the equator there is a 2 to 3 fold increase in elastic fibres around the "nuclear chain" fibres with a comparable fall in the numbers surrounding the "nuclear bag" fibres (M. H. Gladden, personal communication).

C. SENSORY NERVE TERMINALS

The sensory nerve terminals present in the equatorial region of the muscle spindle are divisible into two classes on the basis of the diameter of their parent axon and their form and distribution. The largest afferent fibres end by division at the last node of Ranvier into a number of non-myelinated branches each of which forms an "annulo-spiral" terminal that enwraps the equatorial region of an intrafusal fibre. These terminals are known as "primary endings" and are situated over the bag and adjacent myotube regions of the nuclear bag fibres, and the central zone of the myotube region of the nuclear chain fibres. A second group of sensory terminals are derived from one or more afferent nerve fibres of smaller diameter. Apart from a few rings and short spirals these terminals have a diffusely spreading arborescent morphology—the "flower sprays" of Ruffini; the majority of these "secondary endings" are located on the nuclear chain fibres polar to the central primary ending, only occasional branches pass to the para-equatorial regions of the nuclear bag fibres (Fig. 1).

The nerve terminals have no Schwann cell covering and closely enwrap the intrafusal muscle fibre, lying beneath its basal lamina in a trough or gutter in its surface, and on the nuclear bag fibres the edges of this trough are frequently prolonged into sarcoplasmic lips partially enclosing the external aspect of the nerve terminal (Fig. 2). The fine structure of the primary terminal differs from that of its parent axon in a manner similar to that seen in other mechanoreceptors, the cytoplasm consisting of fluffy amorphous electron dense material with a few centrally placed clusters of neurotubules and neurofilaments, and large numbers of small mitochondria and small vesicles of from 20–100 nm diameter. No basal lamina intervenes between the adjacent nerve and muscle cell membranes, unlike the arrangement at the motor end plate, and the only observed structural connections between the two membranes have been occasional zonulae adhaerentes.

These morphological features suggest that the sensory terminal is a distinct entity with biophysical properties that differ from those of its parent axon, and further that it is securely tethered to the surface of the intrafusal muscle fibre and must therefore comply with changes in that fibre's longitudinal dimensions (Fig. 3).

The secondary sensory terminals have a similar close relationship to the surface of the "nuclear chain" intrafusal fibres, lying beneath their basal laminae but without any obvious trough for their reception. However, unlike the primary terminals, they are divisible into two distinct components: the first consists of relatively large receptor terminal bulbs and spirals with similar ultrastructural characteristics to the coil of the primary terminal, the second of their connecting elements the "stalks" of the "flower spray", which in their internal structure resemble small non-myelinated nerve fibres (Figs. 4 and 5).

D. MOTOR INNERVATION

The details of the arrangement of the motor innervation to the intrafusal fibres of muscle spindles have been the subject of some controversy (Boyd, 1962; Barker, 1966). While it is agreed that in mammals the intrafusal fibres are, with few exceptions, supplied by the smallest myelinated nerve fibres in the muscle nerve, ranging from 2 to 8 μm in diameter and known as the γ efferents, it has not been possible on morphological grounds alone to determine whether one particular fusimotor nerve fibre innervates one class of intrafusal fibre exclusively, or whether "bag" and "chain" fibres commonly share terminals from a single axon. However, there are now both electrophysiological and functional grounds for preferring the first arrangement and these will be given below.

In addition the observations of Barker *et al.* (1970) have clearly

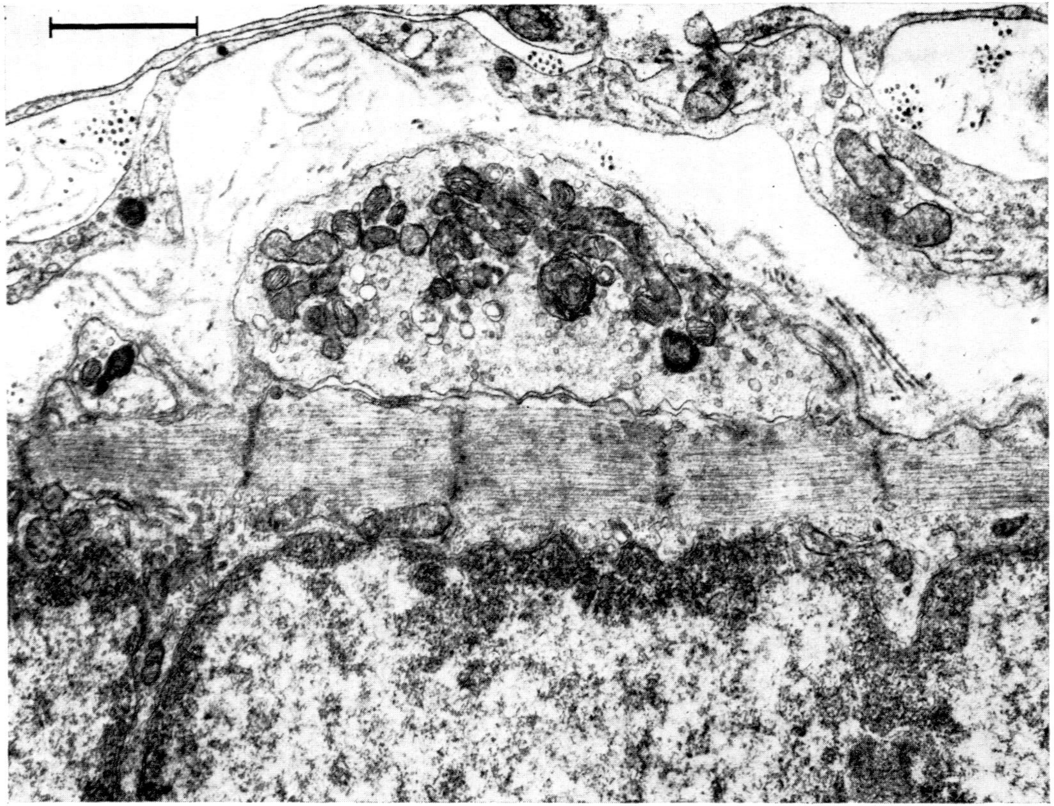

Fig. 3. A longitudinal section through the nuclear region of a contracted nuclear chain intrafusal fibre in a rat spindle. The primary sensory terminal lies in close contact with the muscle fibre surface beneath its basal lamina, located only by small folds of sarcolemma. The appearance of the cytoplasm is typical of such a receptor, containing "fluffy" amorphous material with many small mitochondria and numerous vesicles of varying

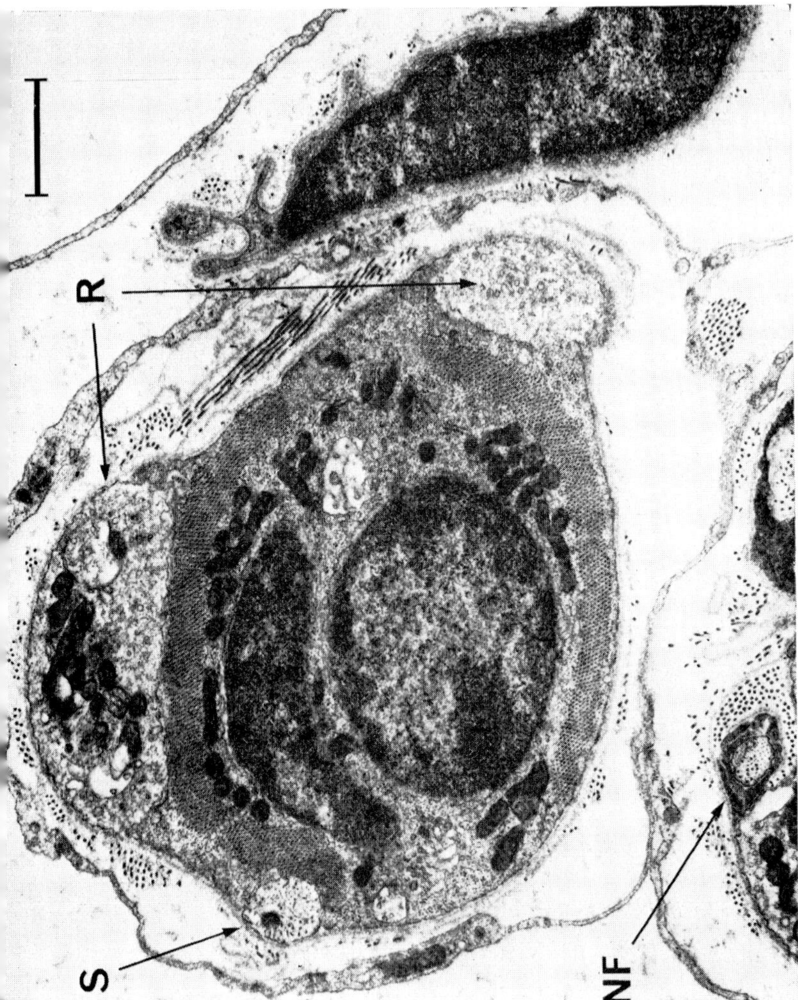

Fig. 4. A transverse section through a nuclear chain intrafusal fibre of the rat, in the region of the secondary sensory ending. The receptor terminals *R* have similar cytoplasmic features to those of the primary ending, but the "connecting stalks" of the "flower spray" *S* resemble adjacent non-myelinated nerve fibres *NF*. They lack a Schwann cell covering and run along the surface of the intrafusal fibre beneath its basal lamina. (scale = 1 μm).

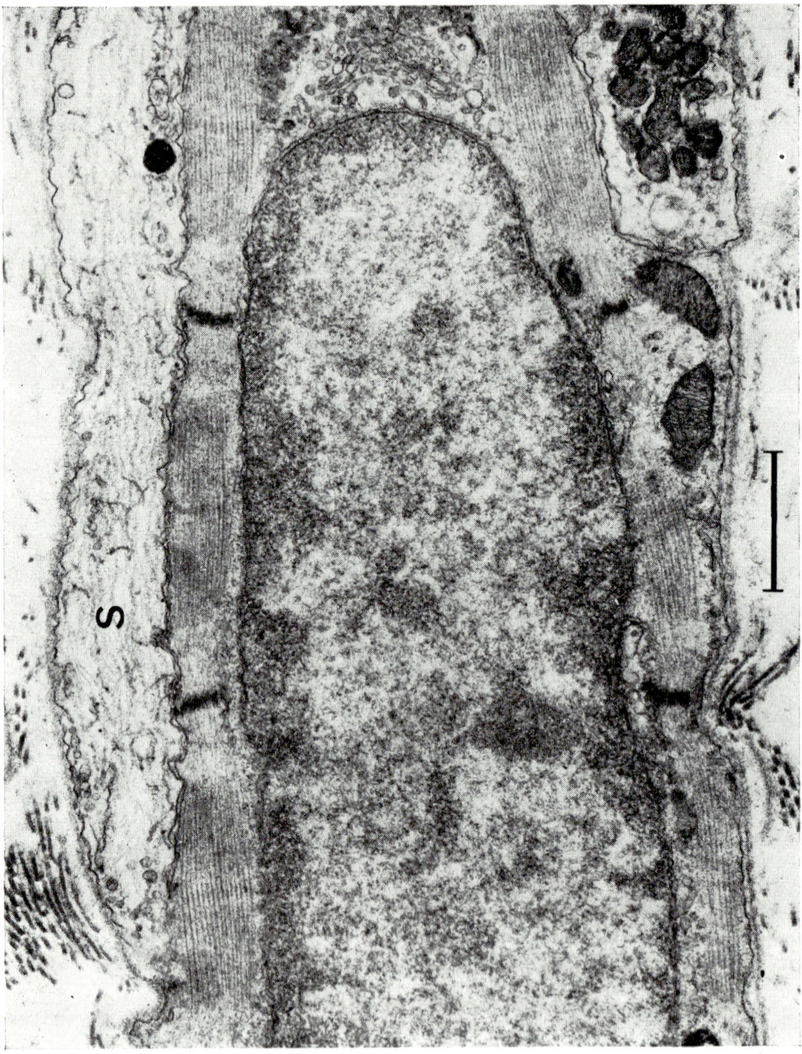

Fig. 5. A longitudinal section through the secondary sensory region of a nuclear chain fibre in the rat. At the top of the figure one of the connecting stalks *S* of the secondary terminal is seen running along the surface of the intrafusal fibre, its internal structure resembling that of a non-myelinated nerve fibre (scale = 1 μm).

shown that, in the cat at least, the motor terminals themselves are divisible into three classes on morphological grounds, and that the form of a particular terminal is to some extent related to the type of intrafusal fibre on which it lies and its position along the axis of that intrafusal fibre relative to the equatorial sensory innervation.

The motor terminals found closest to the equator of the spindles occur predominantly on the nuclear chain fibres and have been termed "trail" endings. They are derived from the smallest diameter γ efferent nerve fibres and consist of fine axon terminals without obvious end bulbs having extensive longitudinal ramifications in the juxta-equatorial regions of the nuclear chain intrafusal fibres. Despite having cholinesterase activity the sarcoplasm and sarcolemma underlying the terminal show no evidence of specialized sole plate formation, the only fine structural features that distinguish such endings from branches of the secondary sensory terminals, with which they overlap, are the presence of a partial Schwann cell sheath on their external aspect, and of the muscle fibre basal lamina between the adjacent nerve and muscle cell membranes.

Towards the poles of the spindle two kinds of "plate" terminals occur and both of these are found more commonly on nuclear bag than nuclear chain fibres. One type, the P.1 plate, may be found anywhere between the polar end of the capsule and the extreme extracapsular poles of the nuclear bag intrafusal fibres, and closely resembles a small extrafusal motor end plate. It occupies a restricted area of sarcolemma and has an obvious sole plate with synaptic gutters and deep branching subsynaptic clefts. These P.1 plate endings are supplied by relatively large diameter axons of which some, and possibly all, are branches of members of the β group of efferent nerve fibres whose principal distribution is to the slow-twitch, extrafusal muscle fibres.

The second class of plate ending is the P.2 plate (Barker *et al.*, 1970). This is also found most commonly on nuclear bag fibres, either within or beyond the pole of the capsule, but where both classes of plate ending are present on the same intrafusal fibre, the P.2 plate occupies the more equatorial position. It consists of an elongated area of knob-like terminals overlying a thin sole plate having shallow unbranched subsynaptic clefts, and is in many respects very similar to the *en grappe* terminals on the extrafusal muscle fibres of lower vertebrates. The nerve fibres from which these P.2 plates arise come from the larger diameter members of the γ efferent group, and in general each fibre supplies a single ending.

The innervation of a typical mammalian muscle spindle may be summarized therefore as consisting of two morphologically distinct types of sensory receptor, the "primary" and "secondary" endings, derived from large diameter group Ia and smaller group II afferent nerve fibres; and possibly three types of motor terminal, P.1 and P.2 "plate" and "trail" endings, supplied by the β, and large and small γ efferent nerve fibres respectively (Fig. 6).

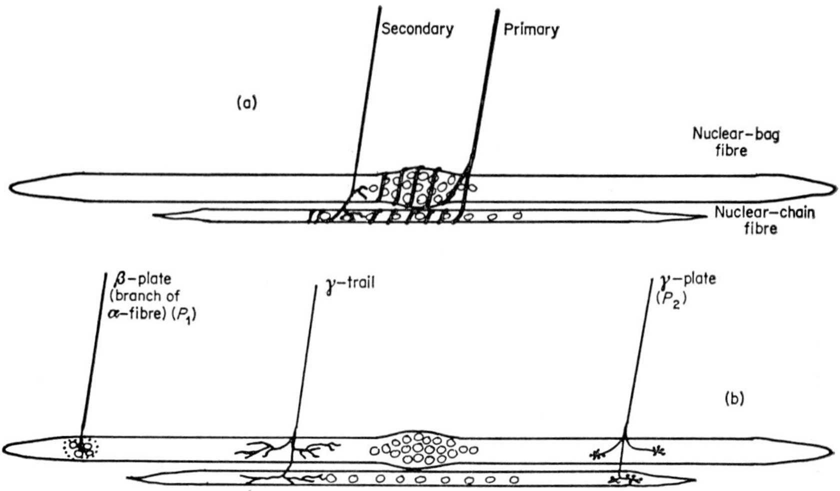

Fig. 6. (a) Simplified diagram of a muscle spindle, containing one nuclear bag fibre and one nuclear chain fibre, to show the pattern of sensory innervation.
(b) The same spindle showing the motor innervation according to Barker. The motor nerve fibres do not necessarily contact the intrafusal fibres on the sides shown in the diagram, which is intended to illustrate the distribution of the three types of motor terminal along the length of the intrafusal fibres (from Crowe, 1968).

V. Functional Properties

The first steps in the elucidation of the functional activity of the muscle spindle were provided in 1933 by B. H. C. Matthews, who demonstrated a number of its basic electrophysiological properties. He showed that the spindle is a muscle stretch receptor of great sensitivity adapting only slowly to maintained stretch, and that it was capable of responding both to change of length and rate of change of length, the velocity sensitivity of the nerve ending being signalled by an increasing rate of afferent discharge with increasing velocity of stretching. Matthews also confirmed previous predictions that the afferent discharge from the spindle would be silenced by contraction of the muscle within which it lay due to its inparallel relationship to the extrafusal muscle fibres.

These valuable early results of the application of electrophysiological methods to the study of muscle spindles, in particular the technique of recording from single nerve fibres, were subsequently greatly extended by other investigators. Cooper (1961), relying upon the differing conduction velocities of the large afferent fibre from the primary annulo-spiral terminal and the smaller diameter Group II afferents from the secondary sensory terminals, was able to show that the two classes of terminal differed in their responses to applied stretch. She found that although their sensitivities to maintained stretch were very similar the primary ending was responsible for the major part of the velocity sensitivity of the spindle which had been demonstrated by Matthews (Fig. 7). A diagrammatic summary of the responses of the

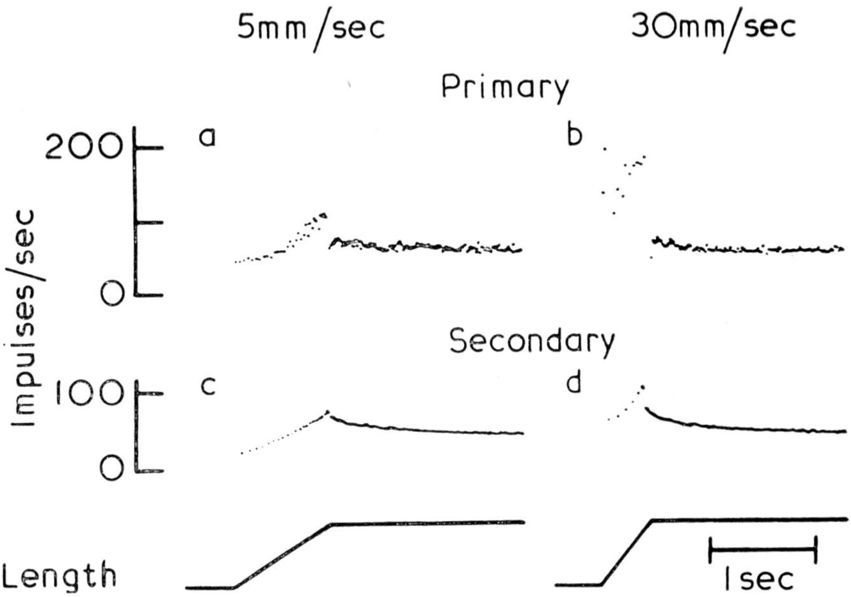

Fig. 7. A comparison of the response of typical primary and secondary endings to constant velocity stretches. The records show the instantaneous frequency of firing of the endings determined for successive action potentials, each of which contributes one spot to the display (from Brown *et al.*, 1965).

two kinds of afferent ending to a number of different stimuli has been provided by Matthews (1964) (Fig. 8). "The primary ending appears to respond to the combined stimuli of instantaneous values of the length of the muscle and the velocity at which that length is changing, while the secondary ending is, to a first approximation, insensitive to velocity and responds only to the length of the muscle. Thus linear stretch combines length and velocity stimuli, the brief tap is a pulse with a large velocity component but a small displacement component, a sinusoidal stretch has an appreciable velocity but a small length component and a release combines a negative value of velocity with a positive but declining value of length" (Matthews, 1971).

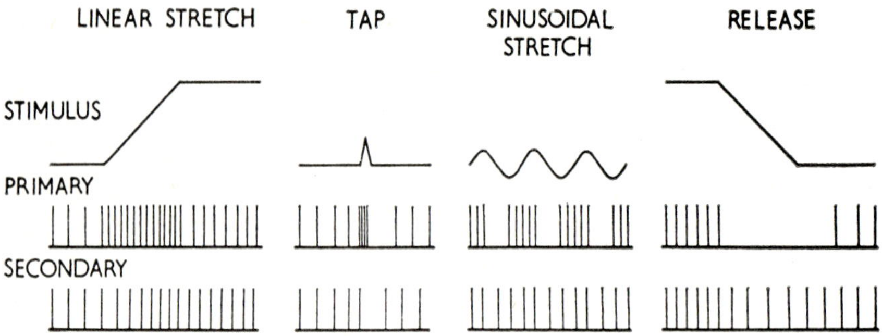

Figure 8. Diagrammatic comparison of the responses of "typical" primary and secondary endings to various stimuli, according to Matthews (1964). The responses are depicted as if the muscle were under moderate initial stretch, and as if there were no fusimotor activity.

The differing behaviour of the two types of afferent ending has been ascribed by Matthews to the fact that the terminals lie on structurally different regions of the intrafusal fibres which he suggests may have significantly different visco-elastic properties (Fig. 9). The sensitivity of the primary ending to velocity of stretching may be due in part to the underlying presence of the nuclear region of the intrafusal fibre which, particularly in the nuclear bag fibre, has fewer myofibrils than the remainder of the fibre and therefore possibly a lower viscosity and proportionately greater elasticity when stretched. On the other hand the secondary sensory endings lying on the contractile paraequatorial regions of the nuclear chain fibres, may be expected to have the same mechanical properties as the remainder of the intrafusal fibre, and this in turn may account for their higher threshold to stretching. Other factors that may affect the sensitivity of the sensory terminal to stretch in the intact spindle will be discussed below.

Subsequent studies by P. B. C. Matthews and his collaborators have demonstrated further points of difference between the behaviour of the primary and secondary sensory endings in the decerebrate cat. Whereas the rate of firing of the secondary ending is directly proportional to the extent of the applied stretch for stretches of both large and small amplitude, the response of the primary ending is non-linear for large stretches but linear for stretches of less than 100 μm amplitude. Within the latter range the terminal shows great sensitivity to small length changes but, since the response is linear, a reduced velocity sensitivity when compared to the behaviour of the same ending subjected to a large stretch. A further interesting, and probably advantageous property of this highly sensitive component of the response is that it cannot be abolished by a large initial maintained stretch, the primary ending resetting itself at the new muscle length to continue to give a large response to further small amplitude stretches. As Matthews has pointed out this high sensitivity of the primary ending to small stimuli at differing initial muscle lengths makes it a valuable detector of small length-tension changes, with the effect of rendering the stretch reflex particularly good at resisting small mechanical disturbances.

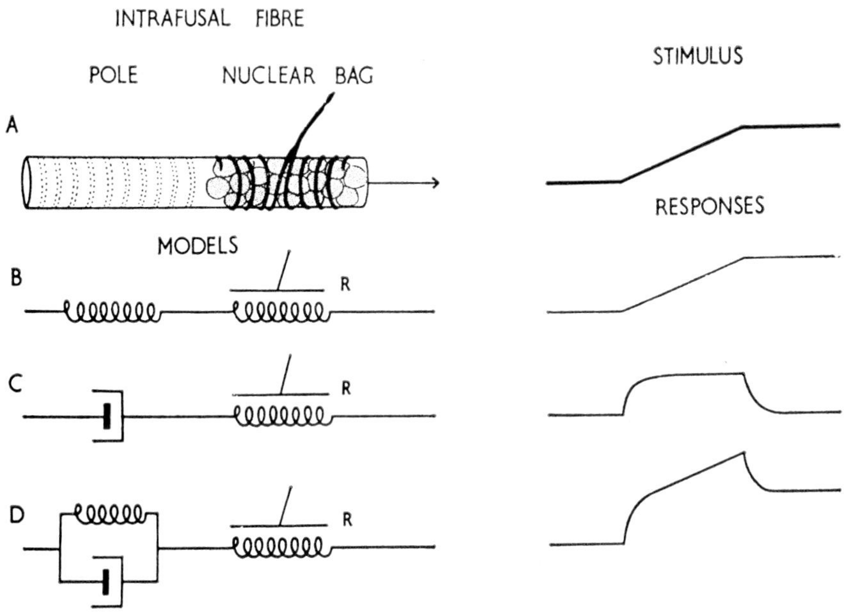

Fig. 9. A series of simple mechanical models illustrating how the insertion of a viscous component (shown as a dash pot) into the intrafusal fibre, in series with an elastic equatorial region, can introduce velocity sensitivity into the afferent response from the receptor *R* (from Matthews, 1964). The response of model *D* is closest to that of the primary sensory ending.

The intrafusal muscle fibres receive a motor as well as a sensory innervation and, as described earlier, this is derived from the smallest group of muscle efferent fibres, the γ efferents, with at least occasional contributions from collaterals of β fibres innervating extrafusal muscle fibres. This specific intrafusal innervation has been deduced from both anatomical studies (v.s.), and electrophysiological experiments involving single motor fibre stimulation with simultaneous recording from single afferents. Jansen and Matthews (1962) proposed that the motor innervation of the spindle might possess the further refinement of independent control of the two distinct anatomical classes of intrafusal muscle fibre demonstrated by Boyd (1962), and suggested that such an arrangement would permit the separation of the two functional components of the primary afferent response, i.e. its response to change in length and to velocity of change in length. Matthews (1962) subsequently demonstrated that individual γ efferent fibres were indeed distinguishable into two classes on the basis of the effect of their stimulation on the response of the primary afferent terminal to simple "ramp" stretches (Fig. 10). Both kinds of γ efferent excite the primary ending under isometric conditions, but whereas the "dynamic" efferent fibre increases the velocity response at the beginning and end of the applied stretch, the "static" efferent fibre reduces the velocity response but greatly enhances the primary discharge at the new sustained length. In addition it has been shown that if one fusimotor fibre supplies several spindles,

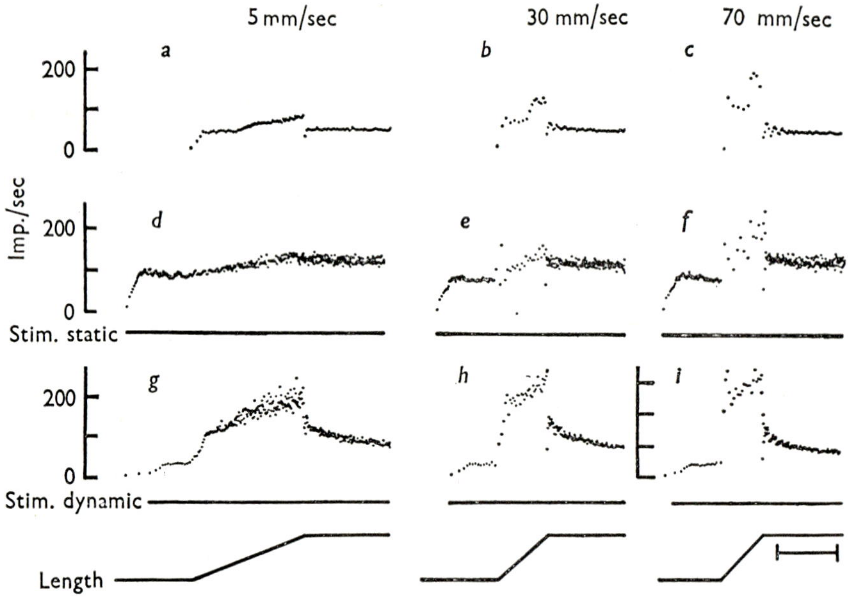

Fig. 10. Records showing the differing responses of a single primary ending to stimulation of the two kinds of fusimotor fibre at differing velocities of stretch. As in Fig. 7 each action potential is represented by a spot, the vertical displacement of which is proportional to the reciprocal of the time elapsed since that which preceded it. The top row of records shows the response of the ending to a 6 mm stretch, at the velocities indicated, in the absence of fusimotor stimulation. The lower two rows of records show the response of the same ending subjected to identical stretches during static and dynamic fusimotor stimulation at 70/sec. (From Crowe and Matthews, 1964.)

as may occur, it induces the same "dynamic" or "static" effect on the primary efferent discharge of each, indicating that this effect is either a specific property of that nerve fibre or that each efferent exclusively innervates a functionally distinct class of intrafusal fibre. Since the "dynamic" and "static" γ efferents are indistinguishable with regard to their conduction velocities, and there is no reason to suppose that they operate through the release of different transmitter agents at their terminals, the second alternative appears the more likely. Matthews (1962) made the further suggestion that if the two morphological varieties of intrafusal fibre do indeed receive a separate and distinct innervation, the ending on the nuclear bag fibres would be responsible for the "dynamic" component of the primary afferent response and those on the nuclear chain fibre for the "static" component.

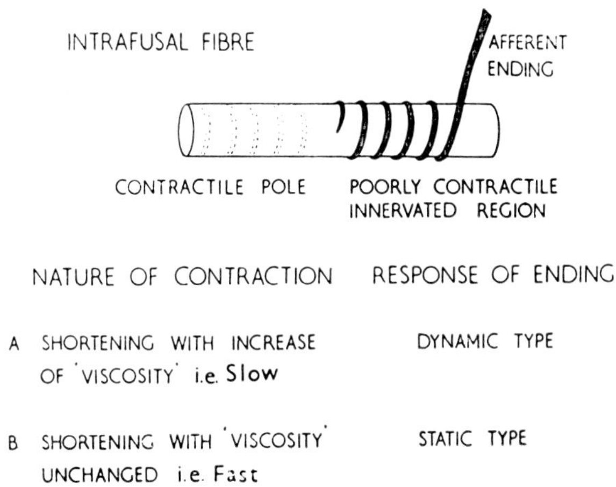

Fig. 11. A diagrammatic illustration of the possible origin of static and dynamic fusimotor actions mediated by different intrafusal fibres. (From Brown and Matthews, 1966).

Matthews (1964) later proposed that a possible explanation for the mechanism by which the "static" and "dynamic" efferents produced their respective effects upon the primary afferent discharge might lie in a difference in the balance of viscoelastic properties within the contractile regions of the two kinds of intrafusal fibre. By analogy with observations made on extrafusal fibres of a slow muscle, soleus (Crowe and Matthews, unpublished data), which show a much larger viscous contribution to the tension developed during stretch than do the fibres of faster muscles, he suggested that the "dynamic" component of the afferent discharge would be derived from that portion of the primary annulo-spiral ending associated with the physiologically "slower", and therefore more viscous, class of intrafusal fibre; the expected properties of such a fibre being a high tension under increasing stretch with adaptation to a maintained increase in length (Fig. 11). An alternative view, that the "dynamic" and "static" fusimotor fibres may exert their specific effects through a gradient of fast/slow and corresponding viscous/elastic properties within each individual intrafusal fibre, has been argued by Crowe (1968) and the reader is referred to this article for details.

Although much has been made of the possible functional specificity of the range of different motor terminals demonstrated by Barker, it is likely that attempts to equate specific modes of motor activity with morphologically distinct varieties of motor terminals may well prove misleading, if it is the properties of the endings *per se* which are considered to determine the time-course and mechanical characteristics of the intrafusal fibre contraction. A possibly more fruitful explanation for the observed structural differences may be found in the differing morphogenetic influences to which the terminals and intrafusal fibres are subject during their ontogeny. Zelená (1957) has shown that in the embryonic development of the muscle spindle the formation of an extensive annulo-spiral sensory terminal precedes the first effective contact of motor nerves with the intrafusal muscle fibres by an appreciable interval, and she has demonstrated that this initial, purely sensory, innervation has a decisive effect upon the structural organisation of the equatorial regions of the intrafusal fibres during their development, and thereafter is capable of maintaining the structural integrity of the central regions of the intrafusal fibres by some unknown trophic mechanism for long periods after de-efferentation has produced total atrophy of the extrafusal muscle fibres (Zelená, 1962). As the precise form adopted by individual motor nerve terminals is likely to be strongly influenced by the structural and functional characteristics of the intrafusal muscle fibre at their point of contact, it is probable that the observed differences between "plate" and "trail" motor endings are primarily a reflection of the declining influence of the sensory terminal upon the structure and properties of the intrafusal fibre, and its capacity to respond to motor innervation by the development of discreet motor end-plates of the extrafusal type, with increasing distance from the spindle equator.

Matthews' two propositions, that the intrafusal fibres receive independent motor innervation and that the nuclear bag fibres possess slower and therefore more viscous contractile properties than the nuclear chain fibres, and are thus responsible for the "dynamic" component of the afferent discharge, have received substantial confirmation and support from subsequent experimental studies. It has been shown that in rat lumbrical spindles the chain and bag fibres receive an entirely separate innervation, while direct observation of muscle spindles subjected to motor stimuli has shown that the nuclear bag fibres do indeed contract more slowly than do the nuclear chain fibres and that in the cat they are usually independently innervated. Electrophysiological evidence has been obtained that the dynamic effect upon the afferent discharge is mediated through a slower intrafusal contractile mechanism with a lower tetanic fusion frequency than is the "static" effect. It has also been shown that while depolarizing agents such as succinylcholine applied directly to spindles can cause a prolonged contracture of nuclear bag fibres, leaving nuclear chain fibres apparently unaffected, the same agent delivered systematically to the intact animal produces a powerful dynamic effect upon the primary afferent discharge. Lastly the inability of the "dynamic" efferents to cause a discharge of the secondary sensory afferents, although these terminals are strongly excited by the "static" efferents, can be conveniently correlated with the more or less exclusive localization of the secondary sensory terminals to the nuclear chain fibres. These, and other points, are treated at greater length by Granit (1970) and Matthews (1972).

At the present time it appears that Matthews' suggestions that each intrafusal fibre has only one mode of contraction and that they are independently innervated ("dynamic" efferents supplying nuclear bag fibres and "static" efferents nuclear chain fibres) provide the surest foundation upon which to construct hypotheses of spindle function. This position is unlikely to be significantly affected by either a small degree of cross-innervation or the presence of a gradient of contractile properties within one intrafusal fibre, of the kind suggested to occur in the nuclear bag fibres of dog spindles by Banker and Girvin (1971), since twitch response originating in the poles of such an intrafusal fibre can only be transmitted to the sensory terminal through the medium of the morphologically "slow" para-equatorial regions.

Three other aspects of the anatomy of muscle spindles which may relate to their functional activity deserve further mention. The first is the presence of the elastic fibres which form a conspicuous component of the intracapsular structure of the spindle and extend out into the endomysial spaces in parallel with the poles of the nuclear bag intrafusal fibres. They have received relatively little attention from those concerned with analyses of the mechanical properties of spindles but, as Granit (1970) suggests, they may play a significant role both in assisting the elastic recoil of the spindle on release from tension, and by the addition of a parallel elastic component to the intrafusal fibres during stretch.

The second point concerns the apparent insertion of the ends of the nuclear chain fibres into the sides of the adjacent nuclear bag fibres at the extremities of the capsule. Boyd has reported such an arrangement in the cat and it is a common feature in rat spindles; indeed an attachment of this kind, either to the nuclear bag fibres or to the capsule itself must of necessity be present if stretch of the muscle as a whole is to be transmitted to the nuclear chain fibres within the spindle. Cooper (1961) proposed that under these circumstances contractile activity of the nuclear chain fibres would result in unloading of the central regions of the nuclear bag fibres with which they lie in parallel. This phenomenon was observed by I. A. Boyd in a cat spindle and suggests that the reduction in the dynamic component of the primary afferent response to stretch which occurs during "static" fusimotor stimulation (Fig. 10) may be due to similar "unloading" of the velocity sensitive equatorial region of the nuclear bag fibre. It must be said, however, that Crowe and Matthews (1964) consider such an effect to be unlikely.

Lastly the mechanisms by which changes in length and tension of the intrafusal muscle fibres are converted into trains of propagated action potentials in the afferent nerve fibre deserve some comment. It is generally assumed that mechanical deformation of the polarized axolemma of the sensory terminal results in the production of local "ion sinks" and that these give rise to locally propagated receptor potentials. The convergence and summation of a sufficient number of such potentials arriving at the terminal heminode of the myelinated afferent fibre along its unmyelinated terminal branches causing the generation of a spike discharge which will thereafter be conducted proximally in a saltatory fashion. Any morphological features of the sensory terminal structure which will tend to increase the magnitude or extent of the mechanical deformation of its cell membrane with stretch will obviously increase the efficiency of the system, and therefore its sensitivity as a transducer of mechanical stimuli.

The close anatomical relationship of the primary sensory terminal to the intrafusal fibres has been described and this must be presumed to result in the precise following by the annulo-spiral sensory coil of every small dimensional change of the intrafusal fibres. It is possible that the effects of such changes are enhanced by the presence of the "lips" of sarcoplasm that partially enclose the superficial surface of the terminal in the equatorial regions of the intrafusal fibres, these "lips" concentrating the strains in the intrafusal fibre induced by stretch upon the small remaining proportion of the terminal surface unrelated to the

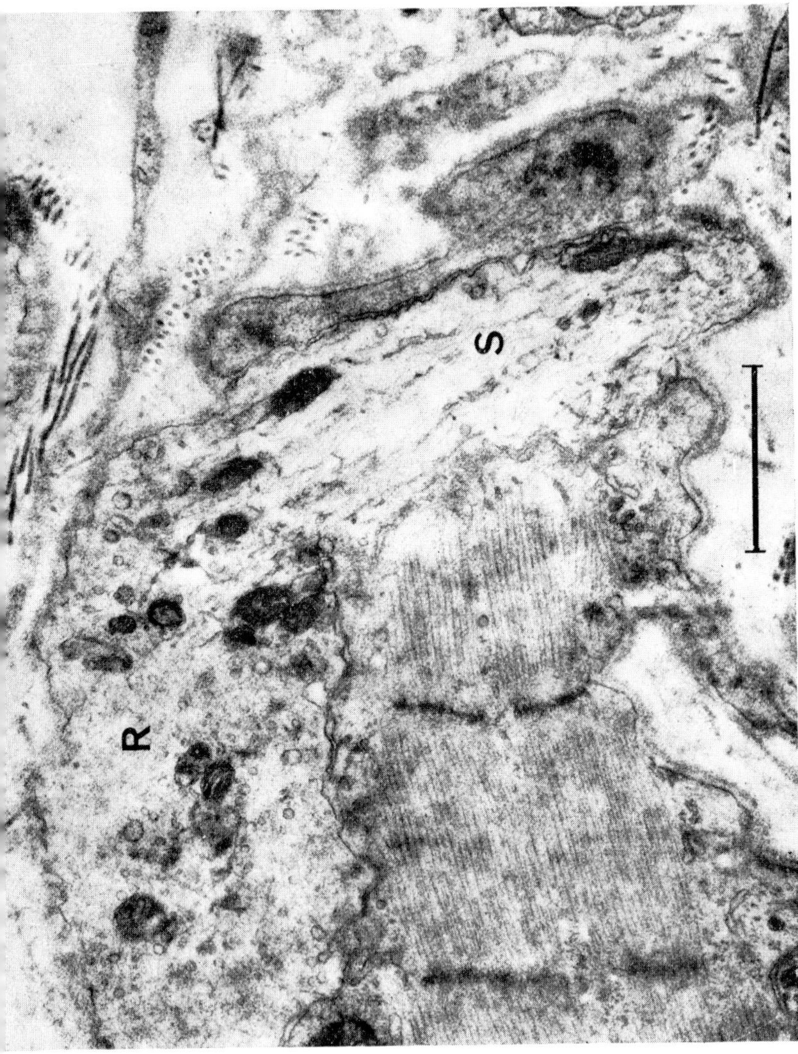

Fig. 12. A longitudinal section through the surface of a nuclear chain intrafusal fibre in the rat. Part of the secondary sensory ending is shown illustrating the abrupt transition between the "axonal" structure of the connecting stalk *S* and the amorphous dense cytoplasm of the receptor terminal proper *R* (scale = 1μ).

sarcolemma. Furthermore the physical nature of the cytoplasm of the terminals of both the primary and secondary afferents may itself play a role in the amplification of stresses induced by mechanical deformation of the ending. Whereas the parent afferent axon and the "connecting stalks" of the secondary terminals possess electron-lucent cytoplasm containing occasional mitochondria, vesicles and smooth endoplasmic reticular cisternae, together with neurofibrils and neurotubules, the terminals themselves generally lack both tubules and fibrils and, in common with other mechanoreceptors, possess a "fluffy" amorphous electron dense cytoplasm containing numerous small mitochondria and vesicles (Fig. 12). The striking difference in the appearances of the two cytoplasmic matrices presumably indicate some difference in the nature or concentrations of the materials present within them, and it is possible that such differences may be reflected in their inherent mechanical properties. To be more specific, the writer suggests that the cytoplasm of the terminal proper may possess greater stiffness than the axon, akin to a gel, such that it would initially resist deformation and then "crack" or "slip" to cause relatively large mechanical disturbances to localized areas of the axolemma, thus initiating a locally propagated potential in circumstances in which the more even distribution of forces to the axolemma through a fluid medium would fall below the threshold necessary to elicit a response.

VI. Role in Muscular Contraction

The early appreciation that the position of the muscle spindle in parallel with the extrafusal fibres enabled the receptor to detect not only change in tension but also change in length, stimulated much interest in its role in the control of skeletal muscular activity. The great importance of the afferent information from the muscle spindles (and the other stretch receptors in the limbs) in the control of muscle action by the central nervous system is shown by the severity of the impairment of both voluntary and reflexly induced movements which follows deafferentation by dorsal root section. On the other hand it is clear that the spindle plays very little part in conscious proprioception since joint position sense can be abolished by local anaesthesia or vascular occlusion which do not involve the spindles.

Apart from the presence of a direct monosynaptic reflex arc between the central ends of the primary afferent fibres and the large α motor neurones in the ventral columns of the cord that innervate the same muscle or its close synergists, little is known for certain of the central connections of the afferent neurones other than that the secondary afferents at least are projected to the cerebellum, and that the polysynaptic return circuit on the reflex pathway will permit both excitatory and inhibitory influences to be brought to bear upon the α motor neurones in the cord. This monosynaptic reflex arc forms the basis of the stretch reflex, i.e. the tendency for a muscle to contract reflexly to resist an applied stretch and maintain its resting length. It was suggested by Rossi (1927), and later by other workers, that the function of the intrafusal muscle fibres was to regulate the tone of a muscle, fusimotor activity exciting the spindle afferents which would then reflexly induce an increase in extrafusal muscle contraction. This fusimotor activity would have the effect of "tuning" the spindle so that it maintained optimal sensitivity and reflex function under all conditions of muscle length and tension.

In 1951 Merton proposed an alternative concept of spindle function. He suggested that the muscle spindle and its innervation formed part of a servo-system by means of which the length of a muscle might be reflexly adjusted to any required value. The basic control "loop" of this "follow-up length servo" would be the monosynaptic reflex by which tension applied to a muscle induces a primary afferent discharge, firing of the α motor neurones, and contraction of the muscle to resist the tendency to increase in length. Merton (1953) suggested that the fusimotor efferents, instead of acting so as to compensate the sensitivity of the spindle for changes in muscle length as proposed by Rossi, might provide a mechanism for initiating movements through their effect upon the primary afferent signal. An increase in fusimotor activity would cause an increased afferent discharge, and thus a reflex contraction of the extrafusal fibres which would be continued until the primary afferent discharge fell to its original value, while a reduction in fusimotor activity would result in an increase in muscle length to a new value appropriate to the reduced level of the primary discharge.

Although the basic concept of a servo-mechanism acting to maintain the length of a muscle against disturbing forces is a valuable one, and has considerable evidential support, subsequent studies have made it appear unlikely that muscle movements are initiated by fusimotor activity alone. Experiments on human volunteers have shown that although the fusimotor fibres are active throughout willed voluntary muscular contractions, their activity follows rather than precedes the onset of electrical activity and tension in the muscle. Matthews (1971) suggests that the role of the fusimotor system in muscular contraction may be one of "servo-assistance", rather than the "follow-up length servo" of Merton, the static fusimotor discharge being

centrally controlled to keep the spindles firing at much the same rate during the course of a movement provided it proceeds in the manner anticipated by the controlling higher centre. Any deviation of the movement from its planned course would then lead to an immediate change in spindle firing which would reflexly elicit the appropriate correction;

and he points out that the independence of the fusimotor pathway from the ordinary motor pathway provides the essential degree of freedom which will allow the fusimotor discharge to be held at a level appropriate to the trajectory of the planned movement, while the level of activity in the main motor pathway can be adjusted to meet the external load.

In conclusion, it must be emphasized that although we now know something of the detailed structure of the muscle spindle, and the functional properties of its component parts, the exact role or roles it plays in voluntary or reflex muscle movements are still a matter for further speculation and experiment, and further, that although the schemes of servo-mechanisms described have been of great value in providing conceptual first steps in our understanding of muscle movements, they are inevitably incomplete explanations since they largely neglect the role of the central nervous system in motor control, about which we know very little.

References

Banker, B. Q. and Girvin, J. P. (1971). The ultrastructural features of the mammalian muscle spindle. *J. Neuropathol. Exp. Neurol.* **30**, 155–195.

Barker, D. (1966). The motor innervation of the mammalian muscle spindle. In "Muscular Afferents and Motor Control. Nobel Symposium I" (R. Granit, ed.), pp. 51–58. Almquist and Wiksell, Stockholm.

Batten, F. E. (1897). The muscle spindle under pathological conditions. *Brain* **20**, 138–179.

Boyd, I. A. (1962). The structure and innervation of the nuclear bag muscle fibre system and the nuclear chain muscle fibre system in mammalian muscle spindles. *Phil. Trans. Royal Soc. Ser. B*, **245**, 81–136.

Brown, M. C., Crowe, A. and Matthews, P. B. C. (1965). Observations on the fusimotor fibres of the tibialis posterior muscle of the cat. *J. Physiol.* **177**, 140–159.

Brown, M. C. and Matthews, P. B. C. (1966). On the sub-division of the efferent fibres to muscle spindles into static and dynamic fusimotor fibres. In "Control and Innervation of Skeletal Muscle" (B. L. Andrews, ed.), pp. 18–34. Thompson and Co., Dundee.

Brzezinski, D. K. von (1961). Untersuchungen zur Histochemie der Muskelspindeln. II. Mitteilüng. Zur Topochemie und Funktion des Spindelraumes und der Spindelkapsel. *Acta Histochem.* **12**, 277–288.

Cooper, S. (1961). The responses of the primary and secondary endings of muscle spindles with intact motor innervation during applied stretch. *Quart J. exp. Physiol.* **46**, 389–398.

Cooper, S. (1966). Muscle spindles and motor units. In "Control and Innervation of Skeletal Muscle". (B. L. Andrew, ed.), pp. 9–16. Thomson and Co., Dundee.

Corvaja, N., Marinozzi, V. and Pompeiano, O. (1967). Close appositions and junctions of plasma membranes of intrafusal fibres in mammalian muscle spindles. *Pfluegers Archiv.* **296**, 337–345.

Crowe, A. (1968). A mechanical model of the mammalian muscle spindle. *J. Theor. Biol.* **21**, 21–41.

Crowe, A. and Matthews, P. B. C. (1964). The effects of stimulation of static and dynamic fusimotor fibres on the response to stretching of the primary endings of muscle spindles. *J. Physiol.* **174**, 109–131.

Granit, R. (1970). "The Basis of Motor Control". Academic Press, London and New York.

Jansen, J. K. S. and Matthews, P. B. C. (1962). The central control of the dynamic response of muscle spindle receptors. *J. Physiol.* **161**, 357–378.

Landon, D. N. (1966). Electron microscopy of muscle spindles. In "Control and Innervation of Skeletal Muscle", (B. L. Andrew, ed.). pp. 96–110. Thomson and Co., Dundee.

Matthews, P. B. C. (1962). The differentiation of two types of fusimotor fibre by their effects on the dynamic response of muscle primary endings. *Quart. J. exp. Physiol.* **47**, 324–333.

Matthews, P. B. C. (1964). Muscle spindles and their motor control. *Physiol. Rev.* **44**, 219–288.

Matthews, P. B. C. (1971). Recent advances in the understanding of the muscle spindle. In "Scientific Basis of Medicine Annual Review, 1971". (I. Gilliland and J. Francis, eds.). The Athlone Press, London.

Merton, P. A. (1951). The silent period in a muscle of the human hand *J. Physiol.*. **114**, 183–198.

Merton, P. A. (1953). Speculations on the servo-control of movement. In "The spinal cord. CIBA Symposium". (J. L. Malcolm and J. A. B. Gray, eds.), pp. 247–255. Churchill, London.

Rossi, G. (1927). Asimmetrie toniche posturali, ed asimmetrie motorie. *Arch. Fisiol.* **25**, 146–157.

Ruffini, A. (1898). On the minute anatomy of the neuromuscular spindles of the cat, and on their physiological significance. *J. Physiol.* **23**, 190–208.

Zelená, J. (1957). The Morphogenetic influence of innervation on the ontogenetic development of muscle spindles. *J. Embryol. exp. Morph.* **5**, 283–292.

Zelená, J. (1962). The effect of denervation on muscle development. *In* "The Denervated Muscle" (ed. E. Gutmann), pp. 103–126. Prague.

Further Reading

Barker, D., Stacey, M. J. and Adal, M. N. (1970). Fusimotor innervation in the cat. *Phil. Trans. Royal Soc. Ser. B.* **258**, 315–346.

Granit, R. (1970). "The Basis of Motor Control". Academic Press, London and New York.

Matthews, P. B. C. (1971). Recent advances in the understanding of the muscle spindle. In "Scientific Basis of Medicine Annual Review, 1971". (I. Gilliland and J. Francis, eds.). Athlone Press, London.

Matthews, P. B. C. (1972). "Mammalation Muscle Receptors and their Central Actions". Edward Arnold Ltd., London.

22. The Smooth Muscle Cell

W. R. KEATINGE and J. M. GRAHAM
London Hospital Medical College, England

I. Introduction

Fifteen years ago little could be said of the cellular biology of smooth muscle except that the tissue consisted of a network of small cells which often showed rhythmical contractions related to electrical discharges. It was believed, incorrectly, to be able to maintain an increase in tone without additional expenditure of energy and was known to react in complex ways and by largely unknown mechanisms to hormones and drugs.

The structure and physiology of striated muscle (see Chapter 20) was relatively well understood by that time. The excitation of striated muscle fibres by nerves had been shown to be brought about by acetylcholine which was released from motor nerve terminals and acted on the motor end plate of the muscle fibre to produce a non-specific increase in its permeability to ions (P. Fatt and B. Katz), thereby depolarizing the muscle membrane and triggering a conducted action potential, which spread over the entire membrane of the muscle cell but not to adjacent cells. The depolarization spreads into the transverse tubules and endoplasmic reticulum of the striated muscle cell (A. F. Huxley and R. E. Taylor) and releases Ca from reticulum into the cytoplasm. The consequent elevation of free Ca concentration activates myosin ATPase, and hydrolysis of ATP by this provides the energy to drive actin and myosin filaments over each other and so to shorten the cell. An important point in relation to smooth muscle was that the action potential in the striated muscle fibre was an essential step in the chain of events by which an action potential in the motor nerve led to contraction. Application of acetylcholine did not produce contraction if the action potential of the striated muscle cell was prevented by removal of external Na or by an increase in external K.

Like nerve cells, striated muscle cells were known to contain a high concentration of K and their resting potential, positive outside the cell, to be essentially a K equilibrium potential. The muscle cell's action potential, like that of nerve, was known to depend on the fact that moderate depolarization of the cell membrane induced a brief increase in the Na permeability of the membrane, allowing Na ions to enter the cell and reverse the membrane potential.

The relative ignorance about smooth muscle at this time resulted largely from technical problems, the most important being the small size of smooth muscle cells which made penetration of them by microelectrodes difficult. Some important electrical observations had nevertheless been made, notably in the 1930's and 1940's by Bozler (1948) using simple external electrodes applied to strips of intestinal smooth muscle suspended in air. Such recordings are now known to be possible because a good conducting pathway exists between the interiors of adjacent cells of this smooth muscle. Bozler was able to record spontaneous electrical discharges from intestinal smooth muscle strips, each discharge being followed by a contraction, and to show that acetylcholine increased the frequency of the discharges and the mean tension. Although the significance of these observations was not widely appreciated at the time they provided evidence that contractions of this tissue could be produced by electrical activity that was widely conducted through the tissue, that could occur spontaneously, and whose frequency was controlled by hormones.

Bozler recorded similar electrical activity from some of the other types of smooth muscle that he investigated, but could not do so from all types. He described those that did have widely conducted, co-ordinated activity as "single unit" smooth muscle. This group included intestinal, uterine and ureteric smooth muscle. It was assumed that local electrical activity controlled contraction in the remaining smooth muscles as well, but that in these tissues each cell fired action potentials asynchronously with its neighbours so that the discharges could not be recorded by extracellular electrodes. Such muscles were therefore described as "multiunit". They included the smooth muscle of blood vessels, nictitating membrane and iris as well as striated muscle. Subsequent work has shown that this division, into single- and multiunit muscles, is not valid in all respects. Some smooth muscles appear to have no electrical activity under physiological conditions; some have regular, conducted electrical activity but give smooth, non-rhythmical responses to it; and many occupy an intermediate position between the single and multiunit types. Perhaps most important of all, much of the mechanical response of all smooth muscles to hormones is produced by means that do not depend on electrical changes in the cell membrane at all. The terms are nevertheless convenient and are still commonly used to distinguish widely-conducting from poorly-conducting smooth muscles.

It was clear from such observations on isolated organs that some tissues containing smooth muscle had the capacity to initiate and co-ordinate their own electrical activity when free of external control. Whether such co-ordinated activity was initiated and conducted in the smooth muscle cells themselves, or in nerve networks in the tissue, was in many cases uncertain for a time. It was later shown that while nerves had an important role in some tissues such as the gut, initiation and conduction were purely myogenic in others such as the ureter. However achieved, co-ordinated contractions were of obvious value in hollow viscera that are required to propel their contents in response to local chemical and mechanical stimuli.

The different smooth muscles have many features in common and these will be considered next. A more detailed account will then be given of the properties of two particular smooth muscles, those of the intestine and of the arterial wall, followed by brief accounts of special features of other main varieties of smooth muscle.

II. Features Common to all Mammalian Smooth Muscles

A. Structure and Contractile Mechanism of the Smooth Muscle Cell

Mammalian smooth muscle cells are spindle-shaped, approximately 150μm long and 3μm wide. They are therefore much smaller than striated muscle fibres. Figure 1 shows a typical smooth muscle cell.

The cell membrane is often lined by small vesicles opening onto the surface, known as pinocytic vesicles. This name implies that the vesicles represent invaginations of the cell membrane with the function of incorporating extracellular material into the cell but there is no evidence that they have this function; since they remain in contact with the cell membrane and do not break away to float free in the cytoplasm it is not likely that they do behave thus, and their function is unknown. Endoplasmic reticulum is present and it's often plentiful.

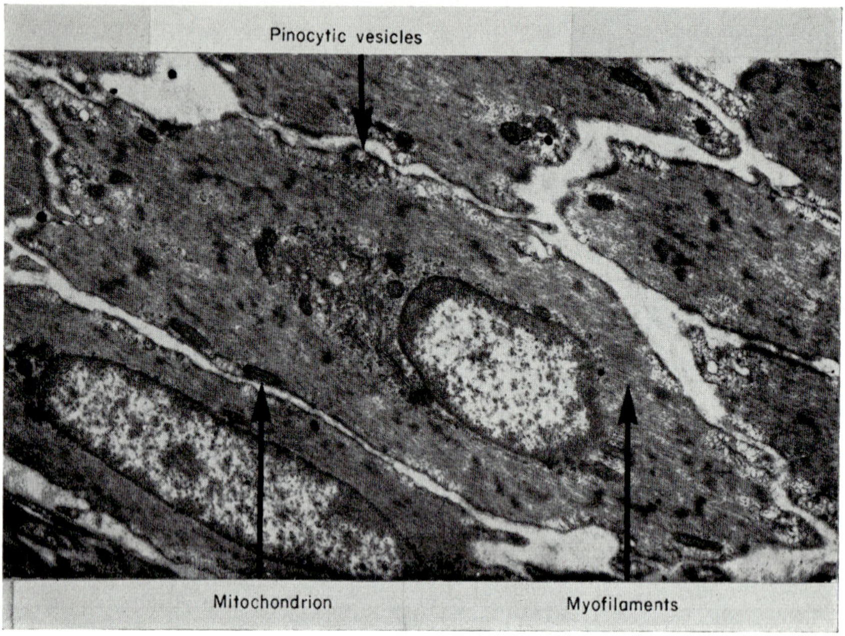

Fig. 1. Smooth muscle cell. Guinea-pig taenia coli. ×14 500. (Kindly provided by Professor F. R. Johnson.)

Most of it is of the rough variety. Smooth reticulum is often present near the cell membrane, and junctions with the cell surface that might conduct surface electrical changes into the reticulum of turtle oviduct have been reported by Somlyo *et al.*, (1971). A subcellular fraction has been obtained from uterine smooth muscle which can accumulate Ca in the presence of ATP in a similar way to the sarcoplasmic reticulum of striated muscle (Carsten, 1969). Although the amount of Ca taken up was very small, this suggests that the contraction of smooth muscle may be controlled to some extent by Ca released into and taken up from the cytoplasm by such reticulum.

The cytoplasm of smooth muscle cells contains large amounts of actin and myosin. A peculiarity of smooth muscle actomyosin is that while sections prepared for electron microscopy with conventional fixatives show the actin as filaments that resemble the actin filaments of striated muscle, they reveal no myosin filaments corresponding to those readily seen in striated muscle. This apparent absence of filaments of myosin in smooth muscle was puzzling for many years as myosin could readily be detected in the tissue by other means. Their absence made it difficult to see how the sliding filament theory of contraction, established for striated muscle (see Chapter 20), could apply to smooth muscle. It was later realized that the myosin of smooth muscle dissolves in the fixatives usually used for electron microscopy, and large myosin filaments are readily seen in suitably prepared sections.

Several lines of evidence suggest that smooth muscle actomyosin functions in a similar way to striated muscle actomyosin, in spite of the differences in the properties of its myosin component. A. Csapo first showed that actomyosin from uterine smooth muscle possessed ATPase activity. Actomyosins have since been extracted from many smooth muscles and have been shown both to possess ATPase activity and to be able to contract (e.g. Bohr *et al.*, 1962) in the presence of ATP and Ca. Myosin from the smooth muscle of chicken gizzard gives similar but shorter aggregates with actin to those obtained with striated muscle myosin (Kaminer, 1969). A protein behaving like the native tropomyosin that determines the Ca sensitivity of striated muscle myofibrils has been extracted from smooth muscle of the chicken gizzard (Esbashi *et al.*, 1966). Evidence that Ca plays an essential role in the contraction of intact smooth muscle is given by the fact that all smooth muscles lose the power to contract if they are immersed for long enough in Ca-free solutions containing a chelating agent. It is therefore probable that contraction of smooth muscle, like striated muscle, is brought about in life by an increase in the cytoplasmic concentration of free Ca which activates myosin ATPase: the resulting hydrolysis of ATP provides energy to drive actin over myosin filaments and shorten the cell. One of the differences from striated muscle actomyosin is that smooth muscle actomyosin has a lower ATPase activity. Like the difference in solubility of the myosin filaments, this may be related to the slower contraction of smooth muscle.

As might be predicted from the numerous mitochondria present in smooth muscle cells, most of the cells' metabolic energy is supplied by oxidative metabolism rather than glycolysis. Block of oxidative metabolism weakens the contraction and often alters the pattern of electrical activity in smooth muscle. However, sufficient energy can generally be obtained from glycolysis to enable the tissue to function to some extent without any O_2 at all. Arterial smooth muscle for example can contract in response to noradrenaline for 24 hr or more after deprivation of O_2 or exposure to cyanide provided that ample glucose is present in the perfusion medium (Keatinge, 1964).

When O_2 is present isometric contraction of smooth muscle results in increased O_2 consumption (Bülbring, 1953). Earlier findings that smooth muscle maintained increased tone without increased O_2 consumption (C. L. Evans) could be explained by shortening of the tissues in those experiments, since a decrease in the length of smooth muscle cells tends, other things being equal, to reduce their O_2 consumption.

B. RELATIONSHIP OF SMOOTH MUSCLE CELLS TO THEIR NERVE SUPPLY

The most striking feature of the innervation of smooth muscle is that the motor nerve fibres in it form a series of transmitter-releasing expansions that come into close relation with a succession of smooth muscle cells instead of ending blindly on a cell like the nerves supplying striated muscle (Fig. 2). Electron micrographs show that each expansion contains vesicles. It is believed that the vesicles contain transmitter and that some of the vesicles discharge their contents into the extracellular space whenever an action potential passes down the nerve fibre. The existence of successive transmitter-releasing sections enables a single branch of a nerve fibre to stimulate a large number of smooth muscle cells. The junctional regions formed by the non-terminal transmitter-releasing regions of smooth muscle motor nerves are often called "en passage" synapses (see Burnstock, 1970).

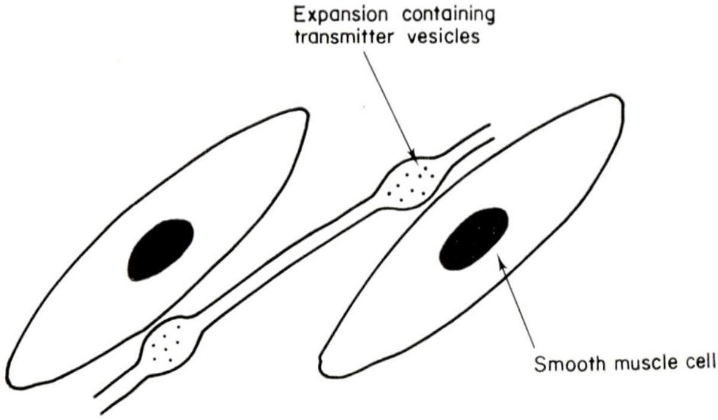

Fig. 2. "En passage" synapses. (Diagram not to scale.)

This arrangement has been demonstrated most clearly with adrenergic nerve fibres. These fibres can be seen with conventional ultraviolet fluorescence microscopy in tissues treated by the Falck method. In this the tissue is exposed to hot formaldehyde vapour to convert noradrenaline in the nerve terminals into a compound that fluoresces in ultraviolet light, enabling the vesicle-filled swellings to be seen as beads or varicosities along the adrenergic nerve fibres as they run through the smooth muscle.

The adrenergic nerve fibres can also be identified with reasonable confidence in electron micrographs by the existence of dense cores in their vesicles. The vesicle-filled expansions can be seen in these electron micrographs to come into close relationship with smooth muscle cells. In these regions the Schwann cells, that separate other portions of the nerve axon from surrounding tissues, stop to leave a cell-free space between the expansions and adjacent smooth muscle cells. The width of this space is generally wider than the neuro-muscular gap of striated muscle, often 80 nm or more, allowing diffusion of transmitter away from the junctional region. In unitary smooth muscle most of the junctional gaps are of this width, but multiunit smooth muscle generally contain a proportion of narrower neuromuscular gaps only 30-50 nm wide. Gaps as narrow as the 20 nm gap that is usual in striated muscle neuro-muscular junctions are occasionally seen in smooth muscle. Some, in the vas deferens for example, are of the terminal rather than "en passage" type and have a gap of this width.

The structure of "en passage" synapses provides strong indications that they behave as synapses, each vesicle-filled expansion releasing transmitter. More direct evidence of this has been obtained in one smooth muscle, the vas deferens, by Bennett and Merrillees (1966). Recording from a single cell they obtained evidence that activity in any one of several sympathetic nerve fibres could cause an electrical response in the cell, although histology showed that there was rarely more than one "terminal" synapse on each cell. There were several "en passage" synapses on each cell which must have been providing effective stimuli to the cell when the nerve fibres forming them were active.

Although the evidence regarding non-adrenergic motor innervation of smooth muscle is less complete, its structure and mode of operation are probably similar. Many smooth muscles have a cholinergic innervation and some appear to contain motor nerve fibres that release an unknown transmitter that is neither noradrenaline nor acetylcholine. While neither acetylcholine nor the unknown transmitter can be demonstrated histochemically, axons with only clear vesicles in an electron micrograph can at least be tentatively identified as non-adrenergic and the presence of cholinesterase in the nerve terminal, or adjacent smooth cell membranes, can provide evidence that a fibre is cholinergic. Such axons form "en passage" synapses similar, apart from the presence of clear rather than dense-cored vesicles in the axonal expansions, to those described above.

C. RELATIONSHIP OF THE SMOOTH MUSCLE CELLS TO EACH OTHER

In unitary smooth muscles an action potential set up in a group of smooth muscle cells will be propagated in neighbouring cells and, in some cases, in all the smooth muscle cells of the organ. Most smooth muscles show some degree of conduction of this kind. The best example is smooth muscle of the ureter which has a pacemaker at its cranial end, generating action potentials which pass down its entire length. Such propagation is brought about by electrically mediated excitation of successive areas of cell membrane very like propagation of the action potential along a nerve fibre. A difference is that in smooth muscle, composed of many separate cells, the current of excitation does not have a direct intracellular route, and propagation is made possible only by the existence of close junctions (nexuses) between the membranes of successive cells which allow current to flow between their interiors. These junctions are sometimes formed simply by appostion of small areas of adjacent cell membrane. In other cases, particularly in unitary smooth muscles, the junctions form a peg-and-socket arrangement with a projection of one cell pushed into another to place a large area of membrane of the two cells in contact (Fig. 3).

It will be obvious from Fig. 4 that when cell A is active current will flow from its interior to that of the adjacent inactive cell B. If the current is large enough and lasts a sufficient time the membrane potential of cell B will be reduced to the threshold level for an action potential to appear.

The exact nature of the close junction or "nexus" is not clear. Electron microscope studies on tissue prepared with certain fixatives show membrane fusion. However, Barr *et al.* (1968) found that in hypertonic solution the nexus is disrupted and propagation from cell-to-cell is abolished. Such disruption is completely reversible, suggesting that in life the cell membranes are not fused but closely opposed. In fact pemanganate fixed preparations show a gap of approximately 2 nm between cells at these junctions which are, accordingly, often described now as "gap" junctions.

The chief evidence that cell-to-cell transmission in smooth muscles is electrical, and not chemical as in mammalian nerve synapses, was

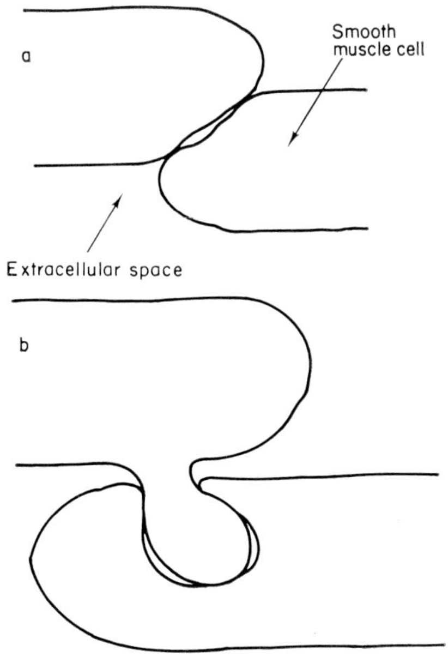

Fig. 3. Types of nexus (a) Simple. (b) Peg and socket.

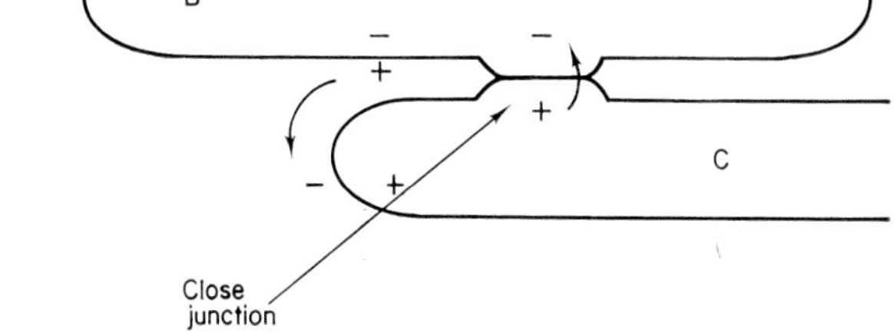

Fig. 4. Electrical transmission between smooth muscle cells. (Excitation of cell B by cells A and C.)

also provided by Barr *et al.* who found that if the external conducting pathway was broken by placing a section of the tissue in a non-ionic solution of high electrical resistance such as isotonic sucrose solution, conduction across this section ceased. When external conduction was restored by placing an artificial low resistance bridge across the part of the tissue immersed in sucrose, propagation across this section was restored.

Whether or not such propagation takes place in a given tissue will therefore depend on a number of factors. Figure 4 illustrates the situation. Transmission will occur if enough charge flows to the interior of cell B to depolarize the cell to the threshold level required for it to fire an action potential. If the threshold or membrane capacitance of cell B is high, transmission will require flow of a relatively large amount of electrical charge. The amount that flows will in turn depend on the size and shape of the action potential and on the number and electrical conductance of the nexuses between the cells. If the flow of charge is insufficient when only one cell fires, cell B might only fire an action potential if both A and C are active simultaneously. Because of this, in many smooth muscles the wave of propagation fades out after a short distance.

This type of nexal spread is responsible for the synchronization of contractions in many smooth muscles. The clearest example of this is in the chick amnion, which had been shown to be completely nerve-free (C. L. Prosser and N. S. Rafferty) so that cell-to-cell transmission of smooth muscle activity is its only means of synchronization. In other tissues such as the intestine nerves are responsible for some of the co-ordination.

The existence of a cell-to-cell conducting pathway via close junctions allows the sucrose-gap method to be used to record electrical activity of smooth muscles. This is often of great value as the presence of large amounts of connective tissue makes it difficult to use micro-electrodes in many smooth muscles, and the small size of smooth muscle cells always makes the use of micro-electrodes difficult. The sucrose-gap method was devised by R. Stampfli for use with nerves. A strip of tissue runs successively through three chambers which are each perfused with a different solution (Fig. 5). The central section contains isotonic sucrose solution. Since this is non-ionic, and therefore a very poor electrical conductor, flow of current through the extracellular space of the portion of tissue in this section is blocked.

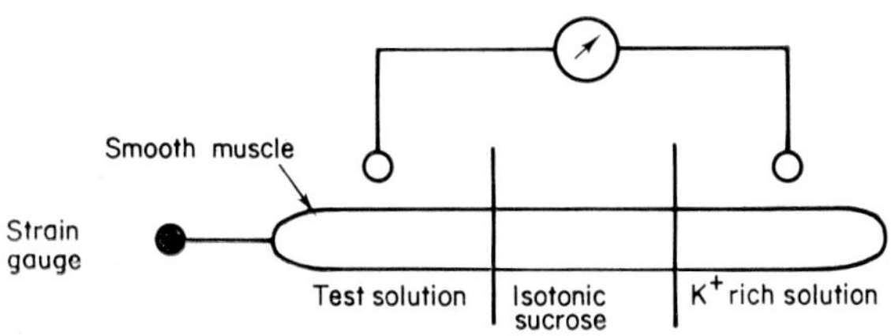

Fig. 5. Sucrose-gap recording of electrical activity.

One of the two end sections contains K-rich solution so that the normal cell membrane potential of the portion of tissue in this section is entirely eliminated. The other end section contains physiological saline and the portion of tissue in this section is the one available for investigation. The principle of the system is that an extracellular electrode in the K-rich solution will be at the same potential as the cell interiors in this section which in turn communicate electrically with the cell interiors of the test portion of tissue, in saline, via the cell interiors of the cells in the sucrose section. The extracellular electrode in the K-rich section therefore behaves as in intracellular electrode in the cells in the saline section.

As the cells of smooth muscle are too short to run through the sucrose section it will be clear that the method can be used most easily in smooth muscles whose tight junctions between cells provide good electrical conduction between adjacent cells. The technique as originally used on nerve has, however, been extensively modified for smooth muscles, mainly with the aim of preventing mixing of the different solutions and improving the stability of the tissue at the junctions of solutions. This reduces artefacts otherwise produced by flow potentials in the solutions, and it has increased the number of smooth muscles with relatively poor cell-to-cell conduction that can be studied by the sucrose-gap. The method does not give reliable absolute values for resting potentials because of interference by liquid junction potentials between the adjacent solutions of different ionic composition. It has however proved of great value for recording spontaneous electrical discharges in smooth muscles and the electrical responses of the muscles to hormones.

Desmosomes provide firm junctions between cells, usually at their ends, and are thought to be responsible for transmitting tension from cell to cell.

D. ELECTRICAL ACTIVITY

The resting potential of smooth muscle cells is essentially a K and Cl equilibrium potential maintained mainly by the K concentration gradient across the cell membrane, as in nerve and striated muscle. In all smooth muscles that have been tested elevation of the external K and decrease in the external Cl concentrations to those of the intracellular fluid virtually eliminate the normal resting potential.

Unitary smooth muscles often show spontaneous electrical activity. A type of electrical activity that is common in them, and has no close parallel in other mammalian tissues, is a sequence of slow waves of depolarization with a cycle often many seconds in length (Fig. 6). The waves are often surmounted by bursts of spikes during a period in each cycle in which the potential is below the threshold for spike activity. The mechanism of these waves has not been established. The slow waves in smooth muscle are normally freely conducted from cell to cell. Since they cause grouping of spike activity they enable the tissues to give slow and powerful rhythmical contractions. This is the usual pattern encountered for example in uterine smooth muscle, circular smooth muscle of the gut and some rhythmically contractile veins.

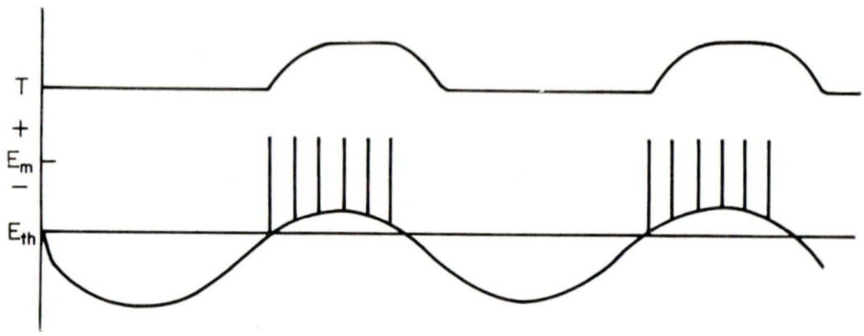

Fig. 6. Slow waves modulating spikes to produce rhythmical contractions.
Key:
T: Tension E_m: Membrane potential E_{th}: Threshold potential.

Other, less numerous, unitary smooth muscles achieve a similar mechanical result by single prolonged plateau-shaped action potentials sometimes as long as a minute in duration, which are again widely conducted through the cells of the tissue. Repeated discharge of such potentials at low frequency results again in slow, powerful waves of contraction. This pattern of electrical and mechanical activity closely resembles that of cardiac muscle. It occurs in the smooth muscle of the ureter, and in aortic smooth muscle of a number of reptiles and amphibians (Roddie, 1962).

Another pattern of activity that can be described as unitary from the electrical point of view consists of regular and relatively rapid discharge of brief action potentials. Although small increases in tone can sometimes be detected after each of such potentials, these are small and are often not recordable at all. The main effect of the activity is therefore to produce a sustained mechanical tone. This pattern often occurs in the longitudinal strips of smooth muscle on the wall of the colon known as taenia coli when they are highly stretched. The technical convenience of this is so great that it has been one of the most studied types of smooth muscle activity.

In multiunit smooth muscle, spontaneous electrical activity is unusual in the absence of stimulation, but nerves and hormones or artificial stimulation by electrical current will often induce action potentials. These may be either plateau shaped or take the form of spikes. They of course do not propagate far from cell to cell in multiunit muscle. They are seldom associated with slow waves.

Unlike nerve and striated muscle action potentials, action potentials of some smooth muscles including the intestine can be elicited in solutions containing little or no Na ions (Holman, 1957) and are difficult or impossible to elicit in solutions containing little or no Ca even if Na is present. Other smooth muscles readily produce action potentials in Ca-free solutions containing Na. The lines of evidence on this problem are considered in more detail in the sections on particular smooth muscles. The easiest interpretation is that both Ca or Na can carry the depolarizing current of the action potential through the cell membrane of all smooth muscles, but that some smooth muscles use one ion more readily and some the other.

Tetrodotoxin does not block smooth muscle action potentials whether these are predominantly Ca- or Na-dependent. This toxin is a specific blocker of Na-based but not of Ca-based action potentials in almost all other tissues, one of the few exceptions being the special case of nerve of the puffer-fish from which tetrodotoxin is extracted. These nerves have Na-dependent action potentials which are not readily blocked by the toxin.

Both Ca- and Na-dependent action potentials lead to contraction. This is most easily demonstrated in unitary smooth muscles where plateau action potentials or bursts of shorter action potentials are each regularly followed by contractions, but a similar sequence of electrical and mechanical activity is observed in multiunit muscles on the occasions when synchronized electrical discharges occur in them. In smooth muscles with Ca-dependent action potentials it is possible that enough Ca may enter the cell as a carrier of the depolarizing current of the action potential to lead directly to contraction. In smooth muscles with Na-based action potentials, the depolarization during action potentials is presumed to produce contraction by allowing Ca to enter the cytoplasm either from the extracellular space or from the endoplasmic reticulum.

It would be expected that slow waves of depolarization in smooth muscle, if large enough, might lead to contraction even when they do not induce action potentials. Such contractile responses of smooth muscle to slow waves without action potentials have been observed by Kuriyama. They are however small and the greater depolarization produced by action potentials is required for large mechanical responses. The main role of slow waves seems to be to modulate spike activity into intermittent bursts so as to produce rhythmical contractions.

E. MECHANISM OF RESPONSE TO NEUROTRANSMITTERS

At first sight the actions of neurotransmitters in different smooth muscles present a confusing pattern. Noradrenaline, for example, contracts arteries and relaxes most gut smooth muscle, while acetylcholine contracts most gut smooth muscle and many arteries but relaxes other arteries. There is a common pattern though, in that any excitatory substance that stimulates a given smooth muscle decreases the membrane potential, and tends to induce electrical activity if the membrane was initially quiescent or accelerate activity if the membrane was already spontaneously active. Stimulant hormones therefore probably all interact with the cell membrane to activate a common sequence of events. Transmitters acting to produce relaxation also produce a common pattern in so far as they tend to increase membrane potential and to slow or halt any electrical activity in progress.

Since electrical activity is produced by excitatory hormones and causes contraction, part of the mechanical response of smooth muscle is clearly secondary to their effect on electrical activity. Smooth muscle, however, shows the important difference from striated muscle that stimulant hormones will contract it and relaxant agents will relax it even when electrical responses are prevented by immersing the tissue in K-rich solution. This was shown for a large number of smooth muscles by Evans *et al.* (1958) and has proved to be true of all smooth muscles tested since. The organizational feature associated with this property is the wide neuromuscular gap often present in smooth muscle that allows wide diffusion of transmitter so that the entire area of a smooth muscle cell may be subjected to its action. The excitation sequence in striated muscle, which depends on electrical activity being conducted from the end-plate region via an action potential to activate contraction, has the advantage of permitting a rapid response. The smooth muscle cell is not required to respond so fast and so has the opportunity to use a process whereby successive portions of membrane are activated directly by transmitter diffusing from the nerve.

The main evidence about the way in which the hormones produce their action on the smooth muscle cell membrane has been obtained from experiments on smooth muscles that are already depolarized by immersion in K-rich solution. Since application of hormones then produces no change in membrane potential the effects of the hormones on ionic permeability can be investigated without the complication of changes produced secondarily to changes in membrane potential. R. P. Durbin and D. H. Jenkinson, and Jenkinson and Morton (1967) showed that carbachol, a stable analogue of acetylcholine, increased the transmembrane flux of ^{42}K and ^{24}Na in depolarized taenia coli. Since it was known that acetylcholine stimulated the motor end-plate of striated muscle by producing a non-specific increase in permeability to ions, the findings on smooth muscle made it likely that stimulant agents acted on smooth muscle by producing a similar non-specific increase in membrane permeability. By contrast noradrenaline, an inhibitory hormone in this tissue, increased the flux of ^{42}K but not that of ^{24}Na. Since it was known that inhibitory transmitters acted on the heart and on nerve cells by producing a specific increase in K permeability, this finding made it likely that such a change was the basis of the electrical action of relaxant agents on smooth muscle. It is presumed that prostaglandins and the many other agents that cause mechanical changes in different smooth muscles act by similar means.

These changes in permeability can provide a satisfactory explanation for the effects of the hormones on the membrane potential and spike frequency of smooth muscle. They can also go some way to explain the contractions produced by excitatory transmitter in the absence of electrical activity. Non-specific increase in the cell membrane's permeability produced by excitatory transmitter would be expected to allow Ca ions to enter the cell from the extracellular fluid and activate contraction. Unfortunately, transmembrane fluxes of Ca are generally obscured in smooth muscle by exchanges of dissolved extracellular Ca with Ca bound to extracellular connective tissue. Probably for this reason, most studies designed to produce direct evidence for or against an increase in transmembrane Ca flux after hormonal or electrical stimulation in K-rich solution have produced equivocal results. A possible additional means of excitation is the release of Ca from intracellular structures in smooth muscle by stimulant hormones. The main evidence that this may occur, first observed in uterine smooth muscle (Edman and Schild, 1962), is that smooth muscles often respond mechanically to such hormones after many minutes in Ca-free solution. They sometimes do so for a time even when a chelating agent such as EDTA is present.

In the case of inhibitory transmitters relaxation produced in the absence of electrical activity cannot be explained by the specific increase in K permeability that such agents cause. Jenkinson and Morton (1967) found that noradrenaline relaxed intestinal muscle in physiological saline solutions by acting on both α and β receptors. The α action was mediated through electrical stabilization of the cell membrane produced by an increase in its K permeability. The β action did not produce this increase in K permeability, but it was much more effective than the α action in relaxing depolarized intestine.

In this case the processes by which electrical and non-electrical inhibition were brought about were clearly different from the start, but the mechanism of non-electrical inhibition in smooth muscle is still being investigated.

III. Intestinal Smooth Muscle

A. GENERAL FUNCTION

The greater part of the alimentary canal contains an inner circular and an outer longitudinal layer of smooth muscle. Co-ordinated waves of contraction involving both of these layers propel the contents of the gut distally during the digestive process and ensure continuous mixing of them with the digestive juices. In the intestine distal to the duodenum this propulsive activity is dependent on nerves and only simple "segmenting activity" remains after nerve block. W. H. Bayliss and E. H. Starling observed that the local anaesthetic cocaine inhibited peristalsis. It was later shown that tetrodotoxin, known to block electrical activity in nerve fibres, also abolished the peristaltic reflex but left intestinal smooth muscle responsive to directly applied stimulant drugs. In the stomach and duodenum co-ordinated waves of contraction travel caudally even after nerve block. These waves are due to conducted electrical activity of the smooth muscle, but the ileum provides an example of complex activity co-ordinated partly by the properties of the smooth muscle itself and partly by an intrinsic nerve plexus.

Trendelenberg first described accurately the "peristaltic reflex", in which distension of an isolated section of small intestine causes relaxation of the longitudinal muscle and contraction of the circular muscle in the form of a wave travelling caudally. During this phase of the reflex there is active relaxation of the longitudinal layer at the point of the initial stimulus. The reflex is unaffected after section and degeneration of the extrinsic nerves to the gut but is dependent on the integrity of the intrinsic nerve plexus, situated between the two layers of smooth muscle.

B. STRUCTURE AND FUNCTIONAL ORGANIZATION

Intestinal muscle cells of both the inner circular and outer longitudinal layers are structurally very similar to many other varieties of smooth muscle cell. In the relaxed state they are fusiform, 100-150μm in length and 3-6μm in diameter at their central region where the single nucleus is situated. When contracted they become ellipsoid and their borders are indented at the points of attachment of myofilaments (Nagasawa and Suzuki, 1967). Adjacent cells are united by frequent nexal contacts.

Close (20 nm) neuromuscular junctions are not present in the intestine. Bennett and Rogers (1967) found that neuromuscular contacts were rarely less than 80 nm wide and were often more than twice this. Many seemingly contradictory observations on the detailed arrangements of the intrinsic plexus responsible for peristalsis have been reported. Some of these may be due to differences between the small intestine, colon, and stomach, and between corresponding regions of the alimentary canals of different species. Nevertheless some general principles of organization emerge.

The peristaltic reflex is initiated by sensory endings at the mucosal surface of the gut. Bülbring *et al.* (1958) described sensory endings presumed to be mechano- and chemo-receptors passing into the mucosa from ganglion cells in the submucous plexus. The relaxation that occurs during the peristaltic reflex is an active process. Schaumann *et al.* (1953) found that when a segment of intestine was contracted by acetylcholine, distention of the lumen would cause a further contraction of the circular and relaxation of the longitudinal coat. The process therefore appears to involve reciprocal excitatory and inhibitory innervation of the two muscle layers.

Several schemes have been postulated offering arrangements of nerve fibres within the intrinsic plexus capable of producing the peristaltic reflex. S. R. Kottegoda proposed parallel circuits, the sensory receptor of one having a low threshold, being activated by small distentions, and causing contraction of the longitudinal muscle. This further distorts the gut exciting the other receptors of the other circuit whose threshold is higher, and which activates the inhibitory phase of the reflex. Crema (1970) modified this simple scheme to explain the polarity of the peristaltic reflex. He suggested that inhibitory cells or their collaterals run longtitudinally in the gut wall in such a way as to relax the muscle caudal to an area of reflex activity, thus directing the movement of the bolus.

Five-hydroxytryptamine is released into the lumen of the gut during peristalsis, and Bülbring and Lin (1958) found that the amounts released are proportional to the intraluminal pressure. It has been proposed that 5HT acts as a neuromuscular transmitter agent in the peristaltic reflex, but its localization within the argentaffin cells of the epithelium suggests another role. 5HT is known to sensitize all sensory nerve endings, and it is possible that its action in the intestine is to lower the threshold of the mucosal receptors, so facilitating peristalsis.

The transmitter substances released by the motor nerves of the intrinsic plexus, and by the autonomic nerves which also supply the region, are not yet fully characterized. Paton (1955) showed that

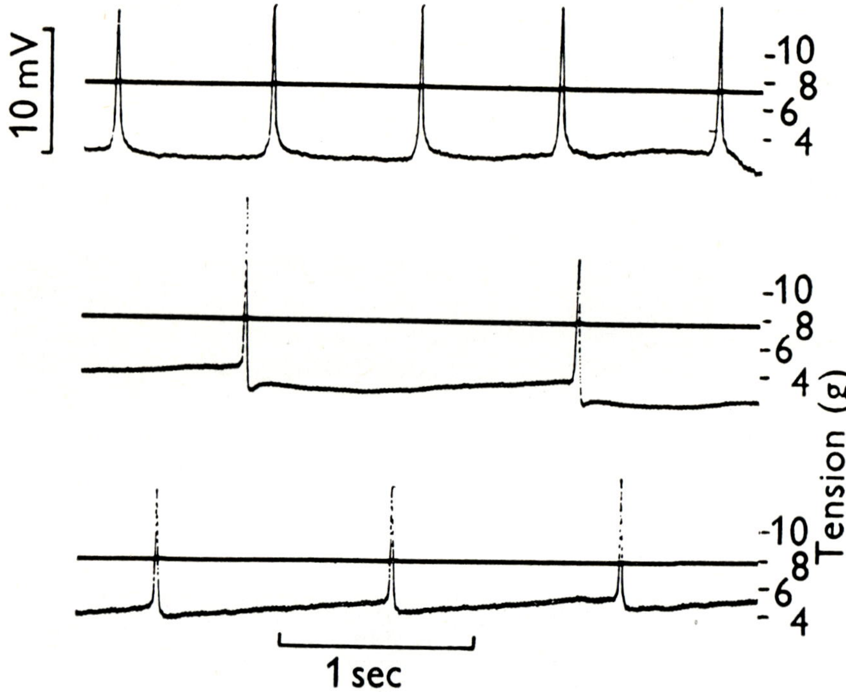

Fig. 8. Electrical activity of guinea-pig taenia coli 10, 45 and 100 minutes after perfusion with Na-free solution. From Holman, M. E. (1957).

Such results suggested that the depolarizing current of the action potential is carried by Ca rather than Na but they could be compatible with a Na-based action potential if small amounts of Na remained in the extracellular space even in Na-free solution and if intracellular Na concentration was very low. Precise measurements of intracellular Na concentration in taenia have been difficult because of the presence of bound Na in the connective tissue and difficulty in measuring the extracellular "Na space" (Goodford and Leach, 1966). Different markers such as inulin, sorbitol and EDTA give very different estimates of extracellular space. Some support for the possibility of a Na-based action potential was provided by the fact that the rate of rise of the action potential was increased in the presence of abnormally high concentrations of extracellular Na. There has also been reappraisal of the need for Ca, as action potentials have since been induced in taenia in simple Ca-free saline (Golenhofen and Petranyi, 1969) and Ca-free saline containing EDTA (E. Bülbring and T. Tomita) provided Mg as well as Na ions were present. While the evidence is still capable of several interpretations it can be explained most easily by supposing that either Na or Ca can carry the depolarizing current of action potentials in this tissue, Ca being the main carrier in physiological solutions.

The problem could be resolved by measuring influxes of Na and Ca during electrical activity but technical difficulties have hampered attempts to do this. Estimates even of the resting transmembrane Na fluxes in intestinal smooth muscle from ^{24}Na have given variable and often very large values. One reason lies in the fact that the phase of ^{24}Na washout representing intracellular Na has proved difficult to identify. Similar difficulties have attended the measurement of Ca flux. It has been reported that strong electrical shocks caused large increases of Ca entry in the muscle, but Von Lammel, *et al.*, (1969) have shown that influx does not increase measurably when stimuli just sufficient to cause contraction are used.

IV. Mammalian Arterial Smooth Muscle

A. General Function

Arterial smooth muscle resembles intestinal smooth muscle in producing contraction of a hollow organ, but in other respects its function is very different. Whereas intestinal muscle propels and mixes the intestinal contents, arterial muscle generally contracts to restrict or stop the flow of blood through the vessel lumen. In order to do this effectively it must contract smoothly rather than rhythmically and be able to sustain the contraction, steadily, at high tension for an indefinite period.

B. Structure and Functional Organization

The most obvious structural peculiarity of the arterial wall is the large amount of elastin and collagen intermingled with its smooth muscle. Not only is there a thick internal elastic lamina on the intimal side and a layer of elastin on the adventitial side of the smooth muscle coat of most arteries, but layers of elastin are interspersed through the muscle layer and collagen is distributed both within these layers and outside them, between individual smooth muscle cells. Penetration of arterial smooth muscle cells by micro-electrodes for electrical recording is therefore difficult. The smooth muscle cells of most mammalian arteries run circularly: longitudinally aligned smooth muscle cells are usually absent.

Although arteries differ in these respects from intestine, the microscopic and ultramicroscopic appearances of arterial smooth muscle cells are remarkably similar to those of intestinal smooth muscle cells. The dimensions of the cells though generally rather smaller are of the same order, the cell membrane is similarly lined by pinocytic vesicles, close junctions exist between adjacent cells and numerous mitochondria are present. Reliable counts of the close junctions are not easy to make but they appear to be less frequent in arterial than intestinal smooth muscle. This is probably the only important and demonstrable difference in structure between the cells of the two smooth muscles.

The excitatory motor innervation of most arteries is provided by adrenergic nerve fibres. This innervation resembles the innervation of intestinal muscle in that the nerves form "en passage" synapses, but differs in the important respect that the nerves do not penetrate the entire arterial wall. Falck (1962) used fluorescence histochemistry for noradrenaline to show that in small arteries the nerves do not enter the muscle coat but lie in the adventitia and form synapses on the adventitial surface of the smooth muscle layer. In some larger arteries such as common carotid arteries of sheep the nerves do penetrate the smooth muscle layer, but only the outer part, leaving the innermost quarter to half of the smooth muscle coat free of nerve fibres (Keatinge, 1966).

Adrenergic terminals can be identified with reasonable confidence in electronmicrographs of the artery wall by the presence of dense-cored vesicles within the terminals. These studies agree with the fluorescence studies in showing that the adrenergic nerves supplying arteries either remain in the adventitia or penetrate only a limited distance into the outer part of the media. Generally they also show a wide neuro-muscular gap of 80-400 nm separating nerve terminal areas from smooth muscle cells, although some probable neuromuscular junctions as narrow as 20 nm have been reported (Rhodin, 1962).

The cholinergic innervation of arteries is dilator in most cases. It seems, like the adrenergic innervation, to be confined to the adventitia and the outer part of the media, since electronmicrographs of the inner part of the arterial media have not shown nerve fibres of any kind. Electronmicrographs of nerve terminals containing clear vesicles, and presumed to be cholinergic because of the presence of acetylcholinesterase, show that the distribution of these fibres is similar to that described above for adrenergic fibres in the wall of the uterine artery (Bell, 1969).

C. ELECTRICAL ACTIVITY

Both microelectrode and sucrose-gap studies are technically difficult in mammalian arterial smooth muscle, because of the large amount of connective tissue and poor cell-to-cell conduction. The total number of electrical records is therefore still small, but a number have been made with both methods. Microelectrode studies (see Fig. 9) have shown that smooth muscle cells of the small mesenteric arteries of rats and guinea pigs, and small uterine arteries of guinea pigs, fire action potentials *in vivo* only when their adrenergic nerve supply is active (Speden, 1964; Steedman, 1966; Bell, 1969).

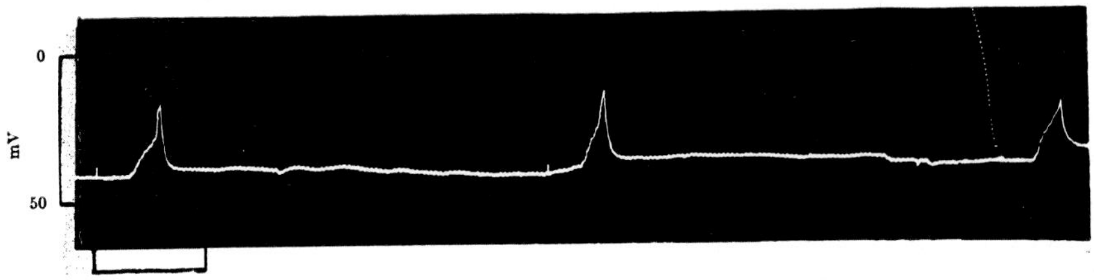

Fig. 9. Junction potentials and spikes induced by sympathetic nerve stimulation (small mesenteric arteries of guinea-pig). From Speden (1964).

Sucrose-gap studies on spirally cut strips show that isolated common carotid arteries of sheep behave similarly in that they are electrically quiescent in physiological saline, but respond by depolarization and action potentials to stimulation by noradrenaline (Keatinge, 1964). They have also provided evidence that these action potentials are largely responsible for the artery's immediate mechanical response to noradrenaline, but that the action potentials play little part in causing sustained contraction caused by the hormone. On the one hand electrical activity can be recorded for only a short time after noradrenaline is applied to the arteries, while contraction persists until the hormone is removed. On the other, when all electrical response is abolished by potassium-rich solution, noradrenaline produces a contraction that is slower in onset but ultimately similar in size to that which it produces in association with electrical activity in physiological saline.

Such stimulation by a sustained high concentration of catecholamines can occur in life as a result of release of adrenaline and noradrenaline into the circulation from the adrenal medulla, but it is not common. The more common pattern of stimulation is by small quantities of noradrenaline released asynchronously from nerve terminal regions at different points in the tissue as action potentials pass down different nerve fibres. Such activity can be induced in isolated arteries in the sucrose-gap apparatus by applying acetylcholine, which has no important direct effect on the smooth muscle of sheep carotid arteries but induces action potentials in the nerves.

The resultant action potentials in the groups of smooth muscle cells activated by different fibres are then recorded as a series of small and irregular spikes of depolarization (Keatinge, 1966). The electrical discharge is accompanied by contraction. The height of these spikes, combined with cell counts in the tissue indicates that a single adrenergic nerve fibre can fire more than 1000 smooth muscle cells. Some of these cells may be activated indirectly by electrical transmission from directly innervated cells, but the fact that they represented only a small proportion of the cells in the preparation shows that electrical activity was not conducted over long distances.

Poor conduction of electrical activity seems to be usual in small as well as large mammalian arteries, since stimulation of small arteries generally produces only localized contraction. However conduction does occur over short distances in some small arteries, since adrenaline produces localized rhythmical contractions in some of them. This was first observed by R. T. Grant in arterioles of the rabbit's ear. It is therefore reasonable to describe most arterial smooth muscle as functioning more as a multiunit than a single unit tissue under physiological conditions.

Arterioles of the bat's wing are exceptional in showing spontaneous and relatively well synchronized rhythmical behaviour (M. P. Weideman). In this they resemble a number of non-mammalian arteries, including turtle aorta (Roddie, 1962) and longitudinal smooth muscle from the adventia of fowl mesenteric arteries (Bolton, 1968). These vessels show spontaneous and well-conducted electrical activity which causes them to beat rhythmically like the heart. While mammalian arteries normally do not show well synchronized activity of this kind, they can be made to do so by special treatments, all of which probably act by lowering the resting potential until it is close to the spike threshold so that even a small flow of charge between an active and an inactive cell will fire an action potential in the latter. Anoxia or drug-induced inhibition of oxidative metabolism cause sheep carotid arteries to give rhythmical electrical and mechanical responses to noradrenaline (Keatinge, 1964). Solutions high in K and low in Ca produce rhythmical contractions in many mammalian arteries (Bohr, 1964). The rhythmical behaviour induced by anoxia may have a function in producing local vascular propulsion of blood in highly anoxic regions, although it cannot do so very efficiently as these vessels lack valves.

Repetitive electrical activity can readily be induced in sheep carotid arteries by exposure to Ca-free saline solution (Fig. 10), while complete replacement of Na in such solution by choline or Tris halts the activity and makes the tissue electrically inexcitable (Keatinge,

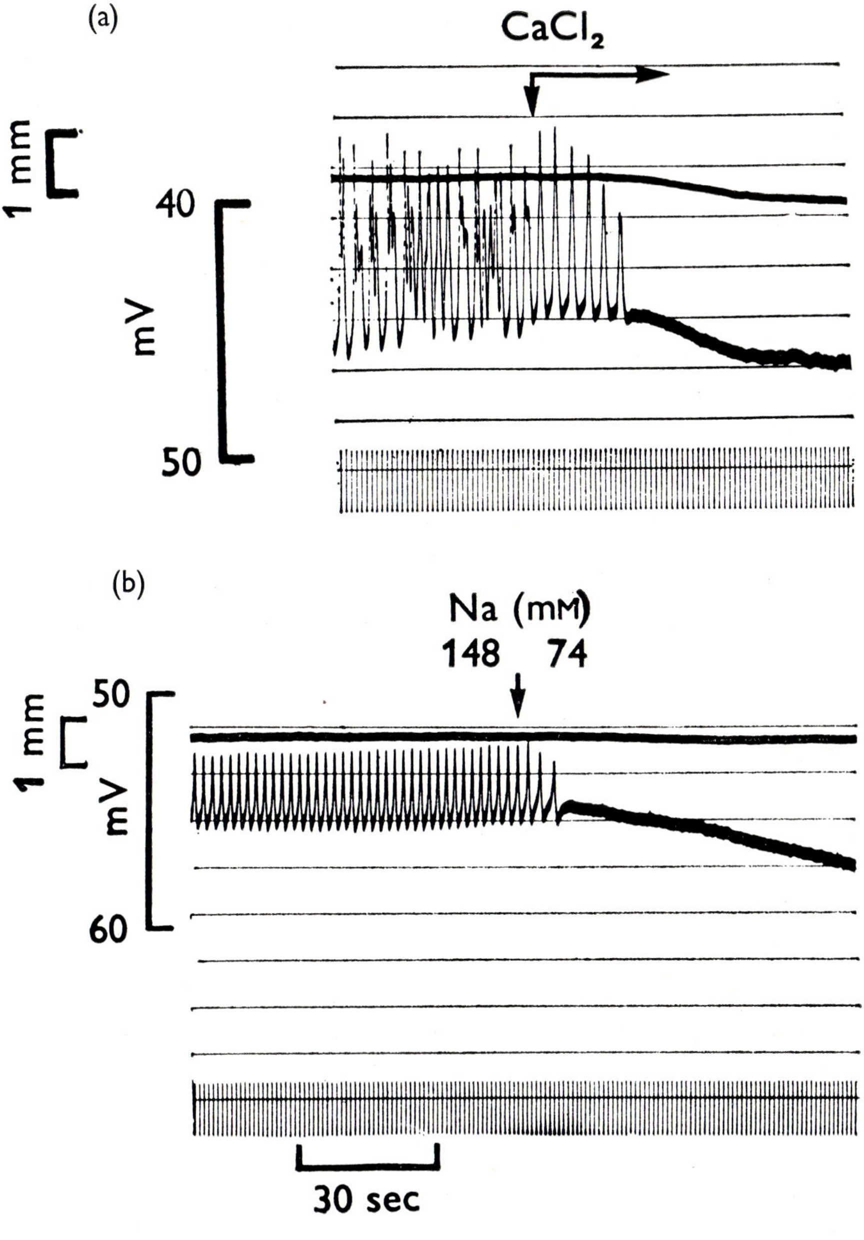

Fig. 10. Dependence of arterial action potential on Na and not on Ca (sheep carotoid artery). (a) Repetitive electrical activity in Ca-free saline, stopped by Ca 2·5 mM. (b) Cessation of activity in Ca-free saline on lowering Na. From Keatinge, W. R. (1968a).

1968a). This suggests that the action potentials of arterial muscle are Na-based, and not Ca-based as that of intestinal muscle appears to be. ^{24}Na flux studies support this, showing that electrical activity is associated with increased influx of Na ions into the arterial smooth muscle cells (Keatinge, 1968b). They were nevertheless not stopped by tetrodotoxin 10^{-6}-10^{-5}M.

Some arteries seem to have no action potentials even when contracting under physiological conditions in response to stimulant hormones. A few of the reports indicating this are attributable to technical recording difficulties, as later studies on the same arteries show both depolarization and action potentials under the same circumstances. However, Speden (1967), who recorded action potentials from some arterioles, recorded only smooth depolarization without action potentials in small arterioles from rabbit ear or mesentery when their sympathetic nerve supply was stimulated. It is difficult to exclude the possibility that such cells failed to produce action potentials because of damage to them by the electrode, but the findings make it probable that adrenergic nerves normally contract these vessels without the mediation of smooth muscle action potentials.

Spontaneous electrical activity has rarely been recorded in any mammalian artery either *in vivo* or *in vitro* in the absence of constrictor nerve activity or stimulant agents. Uchida and Bohr (1969) showed that even mechanical tone is absent in the majority of small and large arteries separated from their nerve supply and perfused by physiological saline. Small arteries from skeletal muscle did show some mechanical tone but it is not clear whether this had an electrical basis.

Vasodilator drugs and hormones act in part by non-electrical means, as relaxant agents do in other smooth muscles. Some vasodilator mechanisms, such as that responsible for dilating collateral vessels after arterial occlusion, are obscure. One of the most interesting of these dilator processes is a spreading dilatation that has been observed under many conditions. Hilton (1959), using cats, confirmed earlier reports that when vasodilatation was produced in leg muscle, by either exercise or injected acetylcholine, a wave of dilatation spread proximally from the muscle vessels up the femoral artery. The wave's slow rate of transmission, 10 cm/sec, suggested that it was transmitted by a muscular rather than a nerve conducting system. There is however no obvious way in which the known electrical properties of the smooth muscle cells of such arteries would provide for effective transmission of a wave of dilatation. Similar spreading dilatation has been observed in arterioles of the hamster cheek pouch (B. R. Duling and R. M. Berne). The process appears to be important in regulating peripheral vascular tone and the elucidation of its mechanism is likely to prove of interest.

V. Other Types of Smooth Muscle

A. VENOUS SMOOTH MUSCLE

Most veins, like arteries, contain a large amount of connective tissue and do not readily conduct electrical activity. They are consequently difficult to study by either microelectrodes or the sucrose-gap method. The main evidence of the electrical behaviour of such a vein was provided by a microelectrode study of frog cutaneous venules (Funaki, 1961). This showed stable resting membrane potentials, with direct electrical stimulation causing action potentials that resembled those of small arteries.

A number of veins, however, give rhythmical contraction. The longitudinal muscle of the anterior (superior) mesenteric and hepatic portal vein of many mammals is an extreme case in which there is spontaneous, widely conducted electrical and mechanical activity (Cuthbert *et al.*, 1965). This no doubt helps to propel blood in the portal circulation. The electrical activity usually takes the form of slow waves surmounted by spikes. The frequency of both slow waves and spikes is increased by stretch, noradrenaline and sympathetic nerve stimulation (Johansson *et al.*, 1967; Holman *et al.*, 1968). As with other vascular smooth muscles noradrenaline can produce large contractions by non-electrical means when the tissue is in K-rich solution.

The adrenergic innervation of veins is less often confined to the adventitia or outer part of the media than is that of arteries. It often penetrates the whole of the muscle coat of the larger veins. Although more widely distributed, this innervation is generally less dense than that of arteries (B. Ehinger, B. Falck and B. Sporrong).

B. UTERINE SMOOTH MUSCLE

Uterine muscle is unusual in that its properties vary greatly with its hormonal environment. As regards structure the nonpregnant uterus is small, while in gestation there is both hypertrophy and hyperplasia of its muscle coats. There are also changes in its sensitivity to adrenaline and other agents, which can be explained for the most part by changes in resting potential produced by circulating sex hormones during pregnancy.

Casteels and Kuriyama (1965) made microelectrode recordings that showed the membrane potential of the non-pregnant uterus to be low, approximately −40 mV, but to increase progressively during pregnancy until in late gestation values as high as −60 mV were observed. This increase depends upon the levels of the sex hormones, especially progesterone. J. M. Marshall treated ovariectomized cats with oestrogen and found that the average resting potential was increased from −32·5 to −57·6 mV. Addition of progesterone as well was needed to produce the high levels of potential associated with pregnancy. No changes can be measured in intracellular ionic concentrations during pregnancy and the increases in resting potential are probably a result of alterations of membrane permeability. Casteels and Kuriyama (1965) found that the membrane potential of the pregnant uterus was much more sensitive to changes of external potassium concentration than that of the non-pregnant uterus and suggested that during pregnancy there is a relative increase in potassium permeability.

Spontaneous electrical activity of the uterus increases greatly during pregnancy. In pregnancy, large spontaneous slow waves of depolarization pass over the entire mass of smooth muscle of the uterus. Each wave lasts between 10 and 60 sec, is surmounted by a burst of spikes, and is followed by a wave of contraction. The pregnant uterus therefore must have effective electrotonic coupling between cells and evidence of this was obtained by Abe (1970) who found that its space constant is longer than that of intestinal smooth muscle.

The action potential of uterine muscle is basically a sodium spike. Anderson (1969) applied the voltage clamp technique to the rat myometrium and found that the transient inward current associated with the spike was greatly reduced in sodium free solutions.

H. H. Dale first described the phenomenon of "adrenaline reversal". Stimulation of the adrenergic fibres of the hypogastric nerve caused relaxation of virgin cat uterus but contraction of the pregnant uterus. The relaxation is associated with hyperpolarization and the contraction with depolarization (Bülbring *et al.*, 1968). A possible explanation is that adrenaline or noradrenaline always changes membrane permeability towards the same equilibrium potential, but that this falls between the levels of membrane potential found in the pregnant and the non-pregnant uterus. The sensitivity of uterine muscle to oxytocin is greatly increased in pregnancy but falls again soon after parturition; this change may have a similar basis.

It has not yet been established how far the great growth of the uterus during pregnancy is a direct result of hormonal action and how far it is secondary to the increased electrical and mechanical activity.

C. BLADDER

Caesar *et al.* (1957) described the detailed structure and innervation of the mammalian bladder. The muscle cells are structurally similar to those of many other varieties of smooth muscle, and there are frequent nexal connections between adjacent cells. Large numbers of nerve fibres, many of them varicose, are observed in the bladder wall, and each cell receives at least one close (20 nm) neuro-muscular junction.

Ursillo (1961) made microelectrode recordings of single bladder smooth muscle cells. Their membrane potential fluctuated spontaneously, and small groups of spikes were generated when the threshold was exceeded. Spike activity could be accelerated by stretching the preparation or by stimulating its autonomic nerve supply.

The innervation of the detrusor muscle is thought to consist predominantly of parasympathetic motor fibres, but there are also sensory fibres which give rise to the spinal bladder reflexes. J. C. Thaemert suggested that these could be recognized in electron-micrographs as those axons containing an exceptional number of mitochondria.

The nature of the transmitter agent released by the parasympathetic nerves to the bladder has been the subject of great controversy. Carpenter and Rubin (1967) concluded that it was acetylcholine, because the normal acetylcholine content of the bladder wall disappeared after section and degeneration of its autonomic innervation. However, the contractions produced by parasympathetic nerve stimulation were shown by V. E. Henderson and M. H. Roepke to be largely atropine-resistant. Carpenter (1963) found that eserine greatly potentiated the response of the rat bladder to transmural (equivalent to post-ganglionic parasympathetic) stimulation and concluded that acetylcholine was indeed released. However, Ambache and Zar (1970) found that eserine, which blocks cholinesterase, had no effect on the response to nerve stimulation by the bladders of the guinea pig, cat and rabbit, indicating that transmission in these tissues is non-cholinergic. These results suggest that different transmitters are released in different species, but an alternative explanation is that the neuro-muscular junction region is isolated by a sheath of connective tissue cells in some animals, thus blocking the diffusion of eserine into the synaptic cleft.

D. URETER

Spontaneous waves of contraction conducted down the ureters aid the flow of urine between the renal pyramid and the bladder. A "pacemaker" region has been identified in the pelvis of the kidney. Autonomic innervation varies this activity but to different degrees in different species. In the cat and guinea pig, innervation of the ureter is sparse except in the pacemaker region but it is more abundant in the dog in which autonomic motor fibres are seen along its full length and there are numerous ganglia within the wall. Numerous close junctions between neighbouring smooth muscle cells provide the anatomical basis of the "unitary" behaviour of the tissue. Close apposition between nerve and muscle (less than 20 nm) is rare.

Microelectrode studies show that the action potential of the ureter consists of a spike followed by a prominent plateau. It was found by Kobayashi (1965) in cat ureter that increased external Ca caused an enhancement of spike amplitude but that in Na-free solutions the spike was unaffected while the plateau was lost. A possible explanation is that the spike is due to an influx of Ca and that depolarization is maintained by prolonged Na entry during the plateau.

E. VAS DEFERENS

The vas deferens is specialized for rapid simultaneous contraction. The hypogastric nerves provide a dense adrenergic innervation but detailed relationships between nerve and muscle vary between species. In rat vas deferens one or more nerve terminal or vesicle filled expansion is found within 20 nm of each cell, and there is little electrotonic coupling between neighbouring cells. In the vas of the guinea pig however only 20-50% of cells have close neuro-muscular junctions and full excitation is made possible by electrotonic coupling.

Microelectrode recordings from guinea pig vas deferens show spikes of 60 mV mean amplitude. These are thought to be predominantly Ca spikes because the action potentials are greatly depressed if the external Ca concentration is reduced, while they are not affected even if external Na falls below 30 mM (Bennett, 1967). Excitation of the muscle is seen when the hypogastric nerve is stimulated: graded excitatory junctional potentials are recorded which generate propagated spikes if the threshold is exceeded.

F. IRIS, CILIARY BODY AND NICTITATING MEMBRANE

The smooth muscles of the eye have resisted attempts to record electrical activity by microelectrodes but their structure suggests that they are predominantly multiunit tissues. Close (20 nm) appositions between nerve terminals and muscle cells are common and Taxi (1965) has estimated that there are an average of as many as four synapses per cell in the ciliary muscle of the cat. Close junctions are often present between adjacent cells in all of these types of smooth muscle in most species, and presumably aid the development of co-ordinated contractions, but in the cat nictitating membrane each cell is isolated by a dense 60 nm basement membrane and all excitation is believed to be brought about directly by nerves. None of these tissues show evidence of spontaneous activity.

References and Recommended Reading

Abe, Y. (1970). The hormonal control and the effects of drugs and ions on the electrical and mechanical activity of the uterus. In *"Smooth Muscle" (Bülbring, E., et al., eds.). Chap. 13, pp. 396–417. Edward Arnold Ltd., London.

Ambache, N. and Zar, M. A. (1970). Non-cholinergic transmission by post-ganglionic motor neurones in the mammalian bladder. *J. Physiol. London* **210**, 761–783.

Anderson, N. C. (1969). Voltage-clamp studies on uterine smooth muscle. *J. Gen. Physiol.* **54**, 145–165.

*Barr, L., Berger, W. and Dewey, M. M. (1968). Electrical transmission at the nexus between smooth muscle cells. *J. Gen. Physiol.* **51**, 347–369.

Bell, C. (1969). Fine structural localization of acetylcholinesterase at a cholinergic vasodilator nerve—arterial smooth muscle synapse. *Circ. Res.* **24**, 61–70.

Bennett, M. R. (1967). The effect of cations on the electrical properties of the smooth muscle of the guinea-pig vas deferens. *J. Physiol.* London **190**, 465–479.

Bennett, M. R., Burnstock, G. and Holman, M. E. (1966). Transmission from intramural inhibitory nerves to the smooth muscle of the guinea-pig taenia coli. *J. Physiol.* **182**, 541–558.

Bennett, M. R. and Merrillees, N. C. R. (1966). An analysis of the transmission of excitation from autonomic nerves to smooth muscle. *J. Physiol.* **185**, 520–535.

*Bennett, M. R. and Rogers, D. C. (1967). A study of the innervation of the taenia coli. *J. Cell. Biol.* **33**, 573–596.

Bohr, D. F. (1964). Electrolytes and smooth muscle contraction. *Pharmacol. Rev.* **16**, 85–111.

Bohr, D. F., Filo, R. S. and Guthe, K. F. (1962). Contractile protein in vascular smooth muscle. *Physiol. Rev.* 42 *Suppl.* **5**, 98–107.

Bolton, T. B. (1968). Electrical and mechanical activity of the longitudinal muscle of the anterior mesenteric artery of the domestic fowl. *J. Physiol.* **196** 283–292.

*Bozler, E. (1948). Conduction, automaticity and tonus of visceral muscles. *Experientia*, **4**, 213–218.

Brading, A. F., Bülbring, E. and Tomita, T. (1969). The effect of sodium and calcium on the action potential of the smooth muscle of the quinea-pig taenia coli. *J. Physiol.* **200**, 637–654.

Bülbring, E. (1953). Measurements of oxygen consumption in smooth muscle. *J. Physiol.* **122**, 111–134.

Bülbring, E. (1954). Membrane potentials of smooth muscle fibres of the taenia coli of the guinea-pig. *J. Physiol.* **125**, 302–315.

Bülbring, E., Casteels, R. and Kuriyama, H. (1968). Membrane potential and ion content in cat and guinea-pig myometrium and the response to adrenaline and nor-adrenaline. *Brit. J. Pharmacol.* **34**, 388–407.

Bülbring, E. and Kuriyama, H. (1963). Effects of changes in ionic environment on the action of acetylcholine and adrenaline on the smooth muscle cells of guinea-pig taenia. *J. Physiol.* **166**, 59–74.

Bülbring, E. and Lin, R. C. Y. (1958). The effect of intraluminal application of 5-hydroxytryptamine and 5-hydroxytryptophan on peristalsis; the local production of 5HT and its release in relation to intraluminal pressure and propulsative activity. *J. Physiol.* **140**, 381–407.

Bülbring, E., Lin, R. C. Y. and Schofield, G. (1958). An investigation of the peristaltic reflex in relation to anatomical observation. *Quart. J. Exp. Physiol.* **43**, 26–37.

*Burnstock, G. (1970). Structure of smooth muscle and its innervation. In "Smooth Muscle" (Bülbring, E. *et al.*, eds.) Chap. 1., pp. 1–69. Edward Arnold Ltd., London.

Caesar, R., Edwards, G. A. and Ruska, H. (1957). Architecture and nerve supply of mammalian smooth muscle tissue. *J. biophys. Biochem. Cytol.* **3**, 867–878.

Carpenter, F. G. (1963). Excitation of rat urinary bladder by coaxial electrodes and by chemical agents. *Amer. J. Physiol.* **204**, 727–731.

Carpenter, F. G. and Rubin, R. M. (1967). The motor innervation of the rat urinary bladder. *J. Physiol.* **192**, 609–618.

Carsten, M. E. (1969). Role of calcium binding by sarcoplasmic reticulum in the contraction and relaxation of uterine smooth muscle. *J. Gen. Physiol.* **53**, 414–426.

Cateels, R. and Kuriyama, H. (1965). Membrane potentials and ionic content in pregnant and non-pregnant rat myometrium. *J. Physiol.* **177**, 263–287.

Crema, A. (1970). On the polarity of the peristaltic reflex in the colon. In "Smooth Muscle" (Bülbring, E. *et al.*, eds) Chap. 18, pp. 542–548. Edward Arnold Ltd., London.

Cuthbert, A. W., Matthews, E. K. and Sutter, M. C. (1965). Spontaneous electrical activity in a mammalian vein. *J. Physiol.* **176**, 22–23P.

Ebashi, S., Iwakura, H., Nakajima, H. Nakamura, R. and Ooi, Y. (1966). New structural proteins from dog heart and chicken gizzard. *Biochem. Z.* **345**, 201–211.

Edman, K. A. P. and Schild, H. O. (1962). The need for calcium in the contractile responses induced by acetylcholine and potassium in the rat uterus. *J. Physiol.* **161**, 424–441.

Evans, D. H. L. and Evans, E. M. (1964). The membrane relationships of smooth muscles: an electronmicroscope study. *J. Anat.* **98**, 37–46.

Evans, D. H. L., Schild, H. O. and Thesleff, S. (1958). Effects of drugs on depolarised plain muscle. *J. Physiol.* **143**, 474–485.

Falck, B. (1962). Observations on the possibilities of cellular localization of monoamines by a fluorescence method. *Acta Physiol. Scand. Suppl.* **56**, 197, 1–25.

Funaki, S. (1961). Spontaneous spike-discharges of vascular smooth muscle *Nature, London.* **191**, 1102–1103.

Golenhofen, K. and Petranyi, P. (1969). Spikes of smooth muscle in calcium-free solution (isolated taenia coli of the guinea-pig). *Experientia*, **25**, 271–273.

Goodford, P. J. and Leach, E. M. (1966). The extracellular space of the smooth muscle of the guinea-pig Taenia coli. *J. Physiol.* **186**, 1–10.

Harry, J. (1962). The effect of cooling, local anaesthetic compounds and Botulinum toxin on the responses of and the acetylcholine output from the electrically transmurally stimulated isolated guinea-pig ileum. *Brit. J. Pharmacol. Chemother.* **19**, 42–55.

Harry, J. (1963). The action of drugs on the circular muscle strip from the guinea-pig isolated ileum. *Brit. J. Pharmacol. Chemother.* **20**, 399–417.

Hilton, S. M. (1959). A peripheral arterial conducting mechanism underlying dilatation of the femoral artery and concerned in functional vasodilatation of skeletal muscle. *J. Physiol.* **149**, 93–111.

Holman, M. E. (1957). The effect of changes in sodium chloride concentration on the smooth muscle of the guinea-pig's taenia coli. *J. Physiol.* **136**, 569–584.

Holman, M. E. (1958). Membrane potentials recorded with high resistance micro-electrodes and the effects of changes in ionic environment on the electrical and mechanical activity of smooth muscle of the taenia coli of the guinea-pig. *J. Physiol.* **141**, 464–488.

Holman, M. E., Kasby, C. B., Suthers, M. B. and Wilson, J. A. F. (1968). Some properties of the smooth muscle of rabbit portal vein. *J. Physiol.* **196**, 111–132.

Jenkinson, D. H. and Morton, I. K. M. (1967). The effect of noradrenaline on the permeability of depolarized intestinal smooth muscle to inorganic ions. *J. Physiol.* **188**, 373–386.

Johansson, B., Jonsson, O., Axelsson, J. and Wahlstrom, B. (1967). Electrical and mechanical characteristics of vascular smooth muscle response to norepinephrine and isoproterenol. *Circ. Res.* **21**, 619–633.

Kaminer, B. (1969). Synthetic myosin filaments from vertebrate smooth muscle. *J. Mol. Biol.* **39**, 257–264.

Keatinge, W. R. (1964). Mechanism of adrenergic stimulation of mammalian arteries and its failure at low temperatures. *J. Physiol.* **174**, 184–205.

*Keatinge, W. R. (1966). Electrical and mechanical responses of arteries to stimulation of sympathetic nerves. *J. Physiol.* **185**, 701–715.

Keatinge, W. R. (1968a). Ionic requirements for arterial action potential. *J. Physiol.* **194**, 169–182.

Keatinge, W. R. (1968b). Sodium flux and electrical activity of arterial smooth muscle. *J. Physiol.* **194**, 183–200.

Kobayashi, M. (1965). Effects of Na and Ca on the generation and conduction of excitation in the ureter. *Amer. J. Physiol.* **208**, 715–719.

Lüllmann, H. and Mohns, P. (1969). The Ca metabolism of intestinal smooth muscle during forced electrical stimulation. *Pflügers Arch.* **308**, 214–224.

Nagasawa, J. and Suzuki, T. (1967). Electron microscopic study of the cellular interrelationships in the smooth muscle. *Tohoku J. Exp. Med.* **91**, 299–313.

Ohashi, H. (1970). An estimate of the proportion of the resting membrane conductance of the smooth muscle of guinea-pig taenia coli attributable to chloride. *J. Physiol.* **210**, 405–419.

Paton, W. D. M. (1955). The response of the guinea-pig ileum to electrical stimulation by coaxial electrodes. *J. Physiol.* **127**, 40–41P.

*Rhodin, J. A. G. (1967). The ultrastructure of mammalian arterioles and precapillary sphincters. *J. Ultrastruct. Res.* **18**, 181–223.

Richardson, K. C. (1964). The fine structure of the albino rabbit iris with special reference to the identification of adrenergic and cholinergic nerves and nerve endings in its intrinsic muscles. *Amer. J. Anat.* **114**, 173–205.

Roddie, I. C. (1962). The transmembrane potentials associated with smooth muscle activity in turtle arteries and veins. *J. Physiol.* **163**, 138–150.

Schaumann, O., Jochum, K. and Schmidt, H. (1953). Analgetika und Darmmotorik. III Zum Mechanisms der Peristaltik. *Naunyn- Schmiedebergs. Arch. Exp. Path. Pharmakol.* **219**, 302–309.

Somlyo, A. P., Devine, C. E., North, S. and Somlyo, A. V. (1971). Sarcoplasmic reticulum in smooth muscle. Communication 1566, Proc. XXV Congress of the International Union of Physiological Sciences.

*Speden, R. N. (1964). Electrical activity of single smooth muscle cells of the mesenteric artery produced by splanchnic nerve stimulation of the guinea-pig. *Nature London*, **202**, 193–194.

Speden, R. N. (1967). Adrenergic transmission in small arteries. *Nature, London*, **216**, 289–290.

Steedman, W. M. (1966). Microelectrode studies on mammalian vascular muscle. *J. Physiol.* **186**, 382–400.

Taxi, J. (1965). Contribution a l'etude des connexion des neurones moteurs du systeme nerveux autonome. *Naturelles Zoologia* 12^{e} *serie*. **VII**, 413–674.

Uchida, E. and Bohr, D. F. (1969). Myogenic tone in isolated perfused vessels, (occurrence among vascular beds and along vascular trees). *Circ. Res.* **25**, 549–555.

Ursillo, R. C. (1961). Electrical activity of the isolated nerve-urinary bladder strip preparation of the rabbit. *Amer. J. Phvsiol.* **201**, 408–412.

Von Lammel, E., Golenhofen, R. and Schaumlöffel, E. (1969). Ca^{++} influx in the isolated taunia coli of the guinea-pig at 13°C, during rest and during electrical stimulation. *Pflügers Arch.* **307**, R85–86.

*Suitable for General Reading

Author Index

Subject Index